McGraw-Hill Networking and Telecommunications

Build Your Own

TRULOVE • *Build Your Own Wireless LAN* (with projects)

Crash Course

LOUIS • *Broadband Crash Course*
VACCA • *i-Mode Crash Course*
LOUIS • *M-Commerce Crash Course*
SHEPARD • *Telecom Convergence*, 2/e
SHEPARD • *Telecom Crash Course*
LOUIS • *Telecom Management Crash Course*
BEDELL • *Wireless Crash Course*
KIKTA/FISHER/COURTNEY • *Wireless Internet Crash Course*

Demystified

HARTE/LEVINE/KIKTA • *3G Wireless Demystified*
LaROCCA • *802.11 Demystified*
MULLER • *Bluetooth Demystified*
EVANS • *CEBus Demystified*
BAYER • *Computer Telephony Demystified*
HERSHEY • *Cryptography Demystified*
TAYLOR • *DVD Demystified*
HOFFMAN • *GPRS Demystified*
SYMES • *MPEG-4 Demystified*
CAMARILLO • *SIP Demystified*
SHEPARD • *SONET/SDH Demystified*
TOPIC • *Streaming Media Demystified*
SYMES • *Video Compression Demystified*
SHEPARD • *Videoconferencing Demystified*
BHOLA • *Wireless LANs Demystified*

Developer Guides

VACCA • *i-Mode Crash Course*
GUTHERY/CRONIN • *Mobile Application Development with SMS*
RICHARD • *Service and Device Discovery: Protocols and Programming*

Professional Telecom

SMITH/COLLINS • *3G Wireless Networks*
BATES • *Broadband Telecom Handbook*, 2/e
COLLINS • *Carrier Grade Voice over IP*
HARTE • *Delivering xDSL*
HELD • *Deploying Optical Networking Components*
MINOLI • *Ethernet-Based Metro Area Networks*
BENNER • *Fibre Channel for SANs*
BATES • *GPRS*

MINOLI • *Hotspot Networks: WiFi for Public Access Locations*
LEE • *Lee's Essentials of Wireless*
BATES • *Optical Switching and Networking Handbook*
WETTEROTH • *OSI Reference Model for Telecommunications*
SULKIN • *PBX Systems for IP Telephony*
RUSSELL • *Signaling System #7, 4/e*
SAPERIA • *SNMP on the Edge: Building Service Management*
MINOLI • *SONET-Based Metro Area Networks*
NAGAR • *Telecom Service Rollouts*
LOUIS • *Telecommunications Internetworking*
RUSSELL • *Telecommunications Protocols, 2/e*
MINOLI • *Voice over MPLS*
KARIM/SARRAF • *W-CDMA and cdma2000 for 3G Mobile Networks*
BATES • *Wireless Broadband Handbook*
FAIGEN • *Wireless Data for the Enterprise*

Reference

MULLER • *Desktop Encyclopedia of Telecommunications, 3/e*
BOTTO • *Encyclopedia of Wireless Telecommunications*
CLAYTON • *McGraw-Hill Illustrated Telecom Dictionary, 3/e*
RADCOM • *Telecom Protocol Finder*
PECAR • *Telecommunications Factbook, 2/e*
RUSSELL • *Telecommunications Pocket Reference*
KOBB • *Wireless Spectrum Finder*
SMITH • *Wireless Telecom FAQs*

Security

HERSHEY • *Cryptography Demystified*
NICHOLS • *Wireless Security*

Telecom Engineering

SMITH/GERVELIS • *Cellular System Design and Optimization*
ROHDE/WHITAKER • *Communications Receivers, 3/e*
SAYRE • *Complete Wireless Design*
OSA • *Fiber Optics Handbook*
LEE • *Mobile Cellular Telecommunications, 2/e*
BATES • *Optimizing Voice in ATM/IP Mobile Networks*
RODDY • *Satellite Communications, 3/e*
SIMON • *Spread Spectrum Communications Handbook*
SNYDER • *Wireless Telecommunications Networking with ANSI-41, 2/e*

BICSI

Network Design Basics for Cabling Professionals
Networking Technologies for Cabling Professionals
Residential Network Cabling
Telecommunications Cabling Installation

Wireless Network Performance Handbook

Clint Smith, P.E.

Curt Gervelis

McGraw-Hill

New York   Chicago   San Francisco   Lisbon   London   Madrid
Mexico City   Milan   New Delhi   San Juan   Seoul
Singapore   Sydney   Toronto

Library of Congress Cataloging-in-Publication Data

Smith, Clint.
  Wireless network performance handbook / Clint Smith, Curt Gervelis.
    p. cm.
  Includes bibliographical references and index.
  ISBN 0-07-140655-7 (alk. paper)
  1. Wireless communication systems—Management—Handbooks, manuals, etc. 2. Mobile communication systems—Management—Handbooks, manuals, etc. 3. Cellular telephone systems—Management—Handbooks, manuals, etc. 4. Wireless communication systems—Design and construction—Handbooks, manuals, etc. 5. Mobile communication systems—Design and construction—Handbooks, manuals, etc. 6. Cellular telephone systems—Design and construction—Handbooks, manuals, etc. I. Gervelis, Curt. II. Title.
  TK5103.2.S65257   2003
  621.382—dc21   2003044564

Copyright © 2003 by The McGraw-Hill Companies, Inc. All rights reserved. Printed in the United States of America. Except as permitted under the United States Copyright Act of 1976, no part of this publication may be reproduced or distributed in any form or by any means, or stored in a data base or retrieval system, without the prior written permission of the publisher.

1 2 3 4 5 6 7 8 9 0   DOC/DOC   0 9 8 7 6 5 4 3

ISBN 0-07-140655-7

The sponsoring editor for this book was Stephen S. Chapman, the editing supervisor was Stephen M. Smith, and the production supervisor was Sherri Souffrance. It was set in Century Schoolbook by Kim J. Sheran of McGraw-Hill Professional's Hightstown, N.J., composition unit.

Printed and bound by RR Donnelley.

McGraw-Hill books are available at special quantity discounts to use as premiums and sales promotions, or for use in corporate training programs. For more information, please write to the Director of Special Sales, McGraw-Hill Professional, Two Penn Plaza, New York, NY 10121-2298. Or contact your local bookstore.

 This book is printed on recycled, acid-free paper containing a minimum of 50% recycled de-inked fiber.

Information contained in this work has been obtained by The McGraw-Hill Companies, Inc. ("McGraw-Hill") from sources believed to be reliable. However, neither McGraw-Hill nor its authors guarantee the accuracy or completeness of any information published herein and neither McGraw-Hill nor its authors shall be responsible for any errors, omissions, or damages arising out of use of this information. This work is published with the understanding that McGraw-Hill and its authors are supplying information but are not attempting to render engineering or other professional services. If such services are required, the assistance of an appropriate professional should be sought.

This book is dedicated to my wife, Mary, and our two children, Sam and Rose, without whom this effort would not have been possible.

C.S.

I would like to dedicate this book to my brother, Aric, for his inspiration and encouragement; and to my favorite teacher, Dr. Markley, for his guidance and patience, which I have remembered throughout my life.

C.G.

Contents

Preface xv

Chapter 1. Introduction 1

    1.1 Communication History 2
    1.2 Cellular 3
    1.3 PCS 6
    1.4 WLL 8
    1.5 LMDS 8
    1.6 MMDS, MDS, ITFS 13
    1.7 Cable Systems 14
    1.8 WAP 14
    1.9 Bluetooth 17
    1.10 Wireless LAN (802.11) 18
    1.11 VoIP 20
    1.12 Typical Central Office 21

Chapter 2. Radio Engineering Topics 25

    2.1 Electromagnetic Waves 25
    2.2 Radio Systems 26
    2.3 Transmitters 26
    2.4 Transmitter System Building Blocks 27
    2.5 Information Bandwidth 29
    2.6 Modulation 29
    2.7 Antennas 32
    2.8 Filters 34
    2.9 Receivers 39
    2.10 Radio Performance Criteria 41
        2.10.1 Sensitivity 41
        2.10.2 Selectivity 41
        2.10.3 Dynamic Range 41
        2.10.4 Noise 43
        2.10.5 1-dB Compression 45
        2.10.6 Third-Order Intercept 45
        2.10.7 Desense 46

viii   Contents

2.11	Propagation Model	48
	2.11.1 Free Space	48
	2.11.2 Hata	49
	2.11.3 Cost231 Walfish/Ikegami	50
	2.11.4 Environmental Attenuation	55
2.12	Diffraction	55
2.13	Effective Radiated Power (ERP)	56
2.14	Link Budget	59
2.15	Path Clearance	60
2.16	Tower-Top Amplifiers	63
2.17	Intelligent Antennas	64

Chapter 3. Wireless Mobile Radio Technologies   67

3.1	AMPS	69
	3.1.1 AMPS Cell Site Configuration	70
	3.1.2 AMPS Call Setup Scenarios	71
	3.1.3 Handoff	71
	3.1.4 AMPS Spectrum Allocation	72
	3.1.5 AMPS Frequency Reuse	74
	3.1.6 AMPS Channel Band Plan	77
3.2	2G Digital Wireless Systems	80
3.3	IS-136 and TDMA (D-AMPS)	84
	3.3.1 Voice Channel Structure	86
	3.3.2 Offset between Transmit and Receive	88
	3.3.3 Speech Coding	89
	3.3.4 Time Alignment	90
	3.3.5 Control Channel	91
	3.3.6 MAHO	100
	3.3.7 Frequency Reuse	100
3.4	CDMA	101
	3.4.1 IS-95 System Description	103
	3.4.2 CDMA2000	104
	3.4.3 CDMA Radio Network	105
	3.4.4 PDSN	107
	3.4.5 CDMA Channel Allocation	110
	3.4.6 Forward Channel	111
	3.4.7 Reverse Channel	116
	3.4.8 SR and RC	119
	3.4.9 Power Control	121
	3.4.10 Walsh Codes	122
	3.4.11 Call and Data Processing	123
	3.4.12 CDMA Handoffs	132
	3.4.13 Pilot Channel PN Assignment	137
3.5	GSM	140
	3.5.1 GSM Air Interface	146
	3.5.2 Types of Air Interface Channels	147
	3.5.3 Air Interface Channel Structure	148
	3.5.4 Location Update	150
	3.5.5 Mobile-Originated Voice Call	153
	3.5.6 Mobile-Terminated Voice Call	155
	3.5.7 Handoff	157
	3.5.8 Traffic Calculation Methods	160
3.6	GPRS	161
	3.6.1 GPRS Services	161
	3.6.2 GPRS User Devices	163

	3.6.3 GPRS Air Interface	164
	3.6.4 GPRS Control Channels	164
	3.6.5 Packet Data Traffic Channels	165
	3.6.6 GPRS Network Architecture	166
	3.6.7 GPRS Network Nodes	166
	3.6.8 GPRS Traffic Scenarios	169
	3.6.9 GPRS Attach	170
	3.6.10 Combined GPRS and GSM Attach	173
	3.6.11 Establishing a PDP Context	173
	3.6.12 Inter-SGSN Routing Area Update	177
	3.6.13 Traffic Calculation and Network Dimensioning for GPRS	180
	3.6.14 Air Interface Dimensioning	180
	3.6.15 GPRS Network Node Dimensioning	181
3.7	iDEN	181
3.8	CDPD	190

Chapter 4. RF Design Guidelines   195

4.1	RF Design Process	197
4.2	Cell Site Design	201
4.3	Search Area Request (SAR)	201
4.4	Site Qualification Test (SQT)	205
4.5	Site Acceptance (SA)	208
4.6	Site Rejection (SR)	209
4.7	FAA Guidelines	211
4.8	EMF Compliance	212
4.9	Planning and Zoning Board	216
4.10	Design Guidelines	217
	4.10.1 Performance Criteria	217
	4.10.2 AMPS	218
	4.10.3 IS-136	219
	4.10.4 IS-95/CDMA2000 (1XRTT)	219
	4.10.5 iDEN	219
	4.10.6 GSM/GPRS	221
4.11	Link Budgets	221
4.12	Frequency Planning	224
	4.12.1 Frequency Plan and Alteration Test Plans	230
	4.12.2 System Radio Channel Expansion	233
4.13	Antenna Systems	235
	4.13.1 Base Station Antennas	235
	4.13.2 Diversity	236
	4.13.3 Installation Issues	237
	4.13.4 Wall Mounting	241
	4.13.5 Antenna Installation Tolerances	241
	4.13.6 Cross-Pole Antennas	242
	4.13.7 Antenna Change or Alteration	245
4.14	Site Types	246
4.15	Reradiators	246
4.16	Inbuilding and Tunnel Systems	247
	4.16.1 Antenna System	250
	4.16.2 Inbuilding Application	251
	4.16.3 Tunnel Applications	251
	4.16.4 Planning	253

4.17	Isolation		254
	4.17.1	Isolation Requirements	256
	4.17.2	Calculating Needed Isolation	257
	4.17.3	Isolation Requirements	259
	4.17.4	Free Space	259
	4.17.5	Antenna Patterns	260
	4.17.6	Vertical Separation	260
	4.17.7	Horizontal Separation	261
	4.17.8	Slant Separation	264
4.18	Base Station Site Checklist		265

Chapter 5. RF System Performance and Troubleshooting — 269

5.1	Key Factors		271
5.2	Performance Analysis Methodology		274
5.3	Lost Calls		281
5.4	Access Failures		292
5.5	Radio Blocking (Congestion)		298
5.6	Technology-Specific Troubleshooting Guides		304
5.7	IS-136		304
	5.7.1	Lost Calls	305
	5.7.2	Handoff Failures	307
	5.7.3	All Servers Busy (ASB)	308
	5.7.4	Insufficient Signal Strength (IS)	309
	5.7.5	Static	309
5.8	iDEN		310
	5.8.1	Lost Calls	311
	5.8.2	Access Problems	312
5.9	CDMA		315
	5.9.1	Lost Calls	316
	5.9.2	Handoff Failures (Problems)	319
	5.9.3	All Servers Busy	321
	5.9.4	Access Problems	323
	5.9.5	Packet Session Access	324
	5.9.6	Packet Session Throughput Problems	326
5.10	GSM		329
	5.10.1	Lost Calls	329
	5.10.2	Handoff Failures	336
	5.10.3	All Servers Busy	338
	5.10.4	Insufficient Signal Strength	339
	5.10.5	Packet Session Access	340
	5.10.6	Packet Session Throughput	343
5.11	Retunes		346
5.12	Drive Testing		358
5.13	Site Activation		364
5.14	Site Investigations		374
	5.14.1	New Sites	374
	5.14.2	Existing Cell Sites	379
5.15	Orientation		382
5.16	Downtilting		383
5.17	Intermodulation		385
5.18	System Performance Action Plan		388
	5.18.1	TIC Lists	388
	5.18.2	Weekly Reports	389
	5.18.3	Monthly Plan Format	389

Chapter 6. Circuit Switch Performance Guidelines — 393

- 6.1 Network Performance Measurement and Optimization — 393
  - 6.1.1 Switch CPU Loading — 393
  - 6.1.2 Switch Call-Processing Efficiency — 398
  - 6.1.3 Switch/Node Downtime (Service Outage) — 399
  - 6.1.4 Switch Service Circuit Loading — 400
  - 6.1.5 Switch/Node Total Erlangs and Calls Volume — 400
  - 6.1.6 Switch/Node Alarms — 401
  - 6.1.7 Switch Memory Settings and Utilization — 401
  - 6.1.8 Switch Timing Source Accuracy — 401
  - 6.1.9 Auxiliary Node Performances — 402
  - 6.1.10 Node Performance Summary — 402
- 6.2 Network Link Performance Measurement and Optimization — 403
  - 6.2.1 Network Link Performance — 403
  - 6.2.2 Link Traffic Loading — 403
  - 6.2.3 Link Retransmissions — 404
  - 6.2.4 Link Errors — 404
  - 6.2.5 Link Changeovers — 405
  - 6.2.6 Link Active Time — 405
  - 6.2.7 Link Performance Summary — 405
- 6.3 Network Routing Performance Monitoring and Management — 406
  - 6.3.1 Routing Efficiency (Voice and Data) — 406
  - 6.3.2 Network Routing Performance Summary — 409
- 6.4 Network Software Performance — 410
- 6.5 Network Performance (General Data) — 410
- 6.6 Network Call Delivery Troubleshooting — 411
  - 6.6.1 Network Call Delivery Troubleshooting Procedures (Initial Steps) — 411
  - 6.6.2 Troubleshooting Procedures (First Level) — 413
  - 6.6.3 Troubleshooting Procedures (Second Level) — 413
  - 6.6.4 Troubleshooting Call Testing Procedures — 415
  - 6.6.5 Network Call Delivery Troubleshooting Summary — 418
- 6.7 Network Call Delivery Troubleshooting Examples — 419

Chapter 7. Billing and Charging in a Wireless Network — 423

- 7.1 Basic Billing Process — 423
  - 7.1.1 Billing Process Description — 423
  - 7.1.2 Billing Cycles and Billing Data Filtering — 424
  - 7.1.3 Call Detail Record Description — 427
  - 7.1.4 Call Detail Record Generation and Collection — 428
  - 7.1.5 Call Detail Record Back-Office Processing — 429
- 7.2 Call Test Plan for Billing Verification — 431
- 7.3 Billing Verification Methods — 432
- 7.4 Nonbillable Billing Events — 433

Chapter 8. Revenue Assurance in a Wireless Network — 435

- 8.1 Revenue Assurance Basics — 435
- 8.2 Analysis and Reconciliation of Customer Databases within a Wireless System — 436
  - 8.2.1 Wireless Network Customer Database Types and Description — 436
  - 8.2.2 Customer Database Provisioning — 437
  - 8.2.3 Determining the Total Number of Subscribers in a System Customer Database — 438
  - 8.2.4 Postpaid and Prepaid Service Issues — 440

xii   Contents

8.3	Revenue Assurance Billing Verification	440
	8.3.1 CDR Volume Benchmarks	440
	8.3.2 Verification of Bill Contents	442
8.4	Comparison, Analysis, and Verification of System Usage and Performance Data	443
	8.4.1 CDR and Switch Statistical Call Data Comparison	443
	8.4.2 System Usage Data Applications	444
8.5	Summary	447

Chapter 9. System Documentation and Reports   449

9.1	Reports	451
9.2	Objectives	452
9.3	Bouncing Congestion Hour Traffic Report (Node and Service)	454
9.4	RF Network Performance Report	457
9.5	Packet Switch Performance Report	460
9.6	Circuit Switch/Node Performance Report	462
9.7	Telephone Number Inventory Report	464
9.8	IP Number Inventory Report	465
9.9	Facility Usage Report	465
9.10	Facilities Interconnect Report (Data)	467
9.11	System Circuit Switch Traffic Forecast Report	469
9.12	Network Configuration Report	470
9.13	System Growth Status Report	471
9.14	Exception Report	471
	9.14.1 Weekly and Regional Exception Report	475
9.15	Customer Care Report	476
9.16	System Status Bulletin Board	476
9.17	Project Status Report (Current and Pending)	479
9.18	System Software Report	480
9.19	Upper Management Report	480
9.20	Company Meetings	481
9.21	Network Briefings	485
9.22	Reporting Frequency	485

Chapter 10. Network and RF Planning   487

10.1	Planning Process Flow	487
10.2	Methodology	489
10.3	Traffic Tables	492
10.4	System Expansion	494
	10.4.1 New Wireless System Procedure	495
	10.4.2 2.5G or 3G Migration Design Procedure	496
10.5	Traffic Projections	497
10.6	Radio Voice Traffic Projections	498
	10.6.1 IS-136	500
	10.6.2 CDMA	501
	10.6.3 GSM/GPRS	505
10.7	Radio Data Traffic Projection	511
10.8	RF System Growth	515
	10.8.1 Coverage Requirements	517
	10.8.2 Capacity Cell Sites Required	518
	10.8.3 RF Traffic Off-Loading	521

10.9	Fixed Network	522
10.10	Circuit Switch Growth	524
	10.10.1 Switch Processor Capacity Study	524
	10.10.2 Switch Port Capacity Study	526
	10.10.3 Switch Subscriber Capacity Study	528
10.11	Network Interconnect Growth Study (Voice)	531
10.12	Network Interconnect Growth Study (Data)	532
10.13	PDSN	534
10.14	IP Addressing	536
10.15	Head Count Requirements	539
10.16	Budgeting	539
10.17	Final Report	541
10.18	Presentation	542

Chapter 11. Organization and Training 543

11.1	Technical Organization's Structure	543
11.2	Technical Organization's Departments	544
11.3	Engineering Organization	545
11.4	Operations	548
11.5	Real Estate and Implementation	548
11.6	New Technology and Budget Directorates	552
11.7	Head Count Drivers	552
11.8	Hiring	553
11.9	Smart Outsourcing	555
11.10	Training	556

Appendix A Erlang B Grade of Service   561
Appendix B Erlang C Grade of Service   565
Appendix C PCS 1900 GSM Channel Chart   567
Bibliography   569
Index   573

Preface

The wireless industry continues to be dynamic and invigorating, and we are glad to be part of it. For several years McGraw-Hill, via editor Steve Chapman, has been pursuing us to craft a wireless performance book. After much hemming and hawing the project was begun without our fully comprehending the level of complexity and number of variables that lay in front of us. In short order it became obvious that the book needed to focus on five wireless technology platforms (AMPS, IS-136, GSM/GPRS, CDMA, and iDEN), and so the amount of material it was to contain expanded greatly. Every attempt was made to make this book a very usable source of information by providing practical guidelines instead of the typical theory.

In this book we recorded real-life situations and solutions for wireless engineers, experienced and novice, to utilize. By conveying what has and has not been successful in engineering we hope to prevent many common mistakes. The information contained here is intended for the people who have to make a system work on a daily basis. As George Starace once said, "You must always concentrate on where you are getting your nickels from."

This book will cover both the RF and network aspects of the engineering efforts associated with a wireless mobile system. Design guidelines will be addressed and numerous troubleshooting examples will be given. The book will discuss how to monitor a system, put together a growth study, and organize engineering. The general format of this book is to introduce key topics first and then go over examples and/or case studies that can be used as general references. Since every system and situation is unique, we provide a process with which problems can be isolated, defined, and resolved. We have found that this seven-step process consistently brings about the technical excellence that achieves improvements in performance.

In short, system performance has always been the untold defense against churn in any market. With more consolidation taking place within the wireless industry, the valuation of a company is driven primarily by its revenue, value per pop, number of subscribers, and infrastructure. However, a fundamental assumption is that the technical performance of a wireless system will improve with the introduction of more technology, new features, system growth, and fewer technical personnel to do the tasks that seem to increase at an exponential rate.

Improving system performance through optimization is a continuous process of refinements to not only the existing network but also to the future network design. Improvement also requires a holistic view of the network. The benefits of optimization are not only the bettering of service quality for the subscriber, but also the minimizing of capital and expense dollars to provide the better service.

The inclusion of packet data services with the multiple wireless access platforms that many wireless operators are using has done nothing but complicate the design and, of course, the operation of the networks. Presently, packet services are emerging, although, as we have been told for over 10 years regarding wireless data, the possibilities that this can bring are limited by the imagination, spectrum, and, of course, capital.

Performance improvements should begin at the design phase for both the network and radio aspects of the system. The design phase incorporates all the inputs from marketing, sales, customer service, and the technical community and then coagulates the ideas into a meaningful plan. This is then followed closely by implementation efforts, where the design is transformed into reality. The implementation of the design can be either physical, software, or both. The next phase involves the integration of the hardware or services into the existing network in a manner that enhances the performance of the network. The integration phase is followed by the operations and maintenance phase, where corrective and preventative maintenance is performed. The next phase is measuring the system against performance goals and making the necessary adjustments to improve the quality of the system. Last, the performance activities are fed back into the design phase for additional corrections and the process begins again.

The consideration of all system parts is essential in improving performance since any and every component of the network can influence overall performance. System performance improvements, therefore, are brought to fruition by good design practices, defined procedures and processes, and a continuous regimen of training for the existing technical staff.

Chapter 1 of this book is an introduction to wireless mobile and other wireless access platforms. Included in this chapter are overviews of a general cellular network, MTSO, cell site configurations, LMDS, 802.11, DSL, VoIP, and cable.

Chapter 2 addresses topics basic to radio frequency engineering. It covers modulation, propagation, receivers, antennas, transmitters, filters, ERP, link budgets, and tower-top amplifiers.

Chapter 3 focuses on the different wireless mobile technologies, including AMPS, IS-136, GSM, GPRS, IS-95, CDMA2000, iDEN, and CDPD.

Chapter 4 discusses the RF engineering design guidelines for AMPS, IS-136, GSM/GPRS, CDMA, and iDEN. This chapter goes over the requirements for design reviews, required signatures, design change orders, tracking mechanisms, various design phases for a wireless mobile system, inbuilding and tunnel system guidelines, frequency planning, antenna selection and change, cell-site parameter setting and adjustment, and software and antenna system isolation calculations.

The performance aspects of an RF system and a network are covered in Chaps. 5 and 6. This is the area that tends to separate engineers into good, mediocre, and outright poor. These chapters focus on how you monitor and optimize the network on a continuous basis, and include recommended fixes that have worked, independent of vendor. Specific troubleshooting techniques are given in Chap. 5 related to AMPS, IS-136, CDMA, GSM, and iDEN systems.

Chapter 7 covers billing systems and their interaction with wireless mobile. Chapter 8 focuses on revenue assurance.

Chapter 9 addresses the important issue of which reports should be generated, how often, and who should be the recipient of each. A hierarchical approach to report generation is presented; for example, a report to an engineer should be different from the one a manager, director, or vice president uses. Reports that are discussed include network performance reports, RF performance reports, exception reports, reports for cross functional departments like operations engineering and construction, customer care reports, software configuration reports, and project reports (current and pending). Another focus of this chapter is meetings.

The methods and procedures needed to put together a network growth plan—specifically, what should be included, where you get the information, and how to actually do it, with a few examples—are covered in Chap. 10. Topics relating to circuit switch growth include defining the current growth of a switch's ports, CPU loading, subscriber database limits, and disaster recovery. Topics relating to cell-site growth include design criteria, frequency planning, physical equipment capacity, software loads, prioritization, and real estate acquisition.

Chapter 11 focuses on organization and training for an engineering department. It also suggests some basic resources and references for an engineering library.

The breadth and depth of the material contained in this book will greatly help the performance engineer, either used directly or augmenting existing knowledge, and we believe that we are doing our part to further the wireless engineering profession as it continues to mature and develop.

Acknowledgments

We would like to thank the many individuals who assisted us in putting together this important effort, especially Naz Abdool, Jim Davey, Larry Shutte, Don McElroy, and Keith Kuchenbecker.

Clint Smith, P.E.
Curt Gervelis

Chapter 1

Introduction

Since its inception, the mobile wireless industry has undergone tremendous changes and has seen a plethora of technologies introduced. While the technologies utilized may seem diverse, they all are based on similar concepts and have similar objectives: the delivery of voice and some data applications. By understanding the similarities and differences between available wireless systems, it is possible to determine how to improve their performance while offering unique services.

In the next few chapters we discuss numerous aspects pertaining to the design, deployment, and operation of mobile wireless systems. This chapter briefly covers some of the key concepts and terminology of mobile wireless communication commonly referred to as cellular communications. The term *cellular communications* is often interchanged with the terms *personal communication services (PCS),* and *third-generation (3G) services,* all of which are interrelated.

The wireless industry continues striving to augment or even replace the wired local loop. This effort has fostered the development of numerous radio technologies that operate over a vast range of spectrums from 400 megahertz (MHz) to 40 gigahertz (GHz). Initially wireless access involved delivering analog or digitized voice services utilizing a host of modulation techniques. The primary focus was on the deployment of radio base stations and then on the development of adjunct services to retain customers and enhance revenues. But as the Internet and other bandwidth-hungry services and products have become more prolific in society, the delivery of data taking precedence over the delivery of voice services is what is envisioned.

With the proliferation of the Internet protocol (IP) and its permeation into all aspects of life, both business and personal, the need to support additional services has become the driving force for all wired and wireless technologies. Society is being revolutionized through better access to information for the purposes of making business as well as purchasing decisions. The available information is so vast that traditional concepts of time and location no longer apply (that is, information is accessible anytime, anyplace). But the revolution is expected to take several years to fully unfold due to the current bandwidth bottleneck that exists at the "last mile" of any system, both wired and wireless. Numerous technology platforms that provide the needed bandwidth to facilitate the information revolution are currently being deployed or are being considered for deployment. The difficulty in the technology platform race is to determine which transport medium will meet both the current demand mode and any future demand models.

1.1 Communication History

Data communication began with Samuel Morse, who in 1844 invented and pioneered the telegraph, which used Morse code (consisting of interweaving dots and dashes) as its method for delivering communication over vast distances. This coding method was so good that it is still used extensively throughout the world today. Wireless data communication became possible thanks to the efforts of Guglielmo Marconi, who is credited with inventing radio.

A very condensed time line of major milestones for the telecommunication industry is provided here. There are numerous other milestones of equal importance for communications, but this list represents shifts in thinking.

1844	Samuel Morse invents the telegraph.
1876	Alexander Bell invents the telephone.
1901	Guglielmo Marconi sends Morse code using a radio.
1931	First U.S. television transmission takes place.
1946	AT&T offers mobile phone service.
1953	First microwave network installed.
1956	Transatlantic cable constructed.
1977	Bell Labs transmits TV signals on optical fibers.
1983	Cellular communication fosters another communications revolution.

Wireless systems have not been in existence long. The first systems were both two-way and broadcast. Radio communication at its onset focused primarily on voice communications, but with the advent of television it was used to deliver broadband video coupled with data for instructing the television on how to display the picture. Microwave communications fostered in the high-speed delivery of data where speeds of 155 megabits per second (Mbit/s) are now common.

The wired systems have also taken a major leap from just offering voice. Data applications initially involved use of a modem that operated at 300 baud, which then progressed to 1200 baud. Speeds now exceed 1 gigabit per second (Gbit/s) with higher speeds being reached every year as the need for more throughput increases.

Part of the need for more throughput was the invention of the fax machine and the proliferation of the computer and its numerous technological advances. The Internet has created the need for increased bandwidth so that a host of services, some still only being dreamed of, can be delivered to subscribers.

The transport platforms for voice and data are becoming similar and require not only bandwidth but design and management skills. The communications future at this time appears to be heading toward using IP as the primary edge or end-user protocol with other supporting transport protocols like asynchronous transfer mode (ATM) being used to deliver the information between similar and dissimilar networks.

1.2 Cellular

Cellular communication is one of the most prolific voice communication platforms that has been deployed within the last two decades. Cellular systems have always been able to transport data, and many advancements in different modulation formats allow for the delivery of narrowband data. However, cellular systems are unable to provide broadband data services because of bandwidth limitations. Typical data rates experienced by cellular applications are 9 kilobits per second (kbit/s).

Overall, cellular communication is the form of wireless communication that allows for

- Frequency reuse
- Mobility of the subscriber
- Handoffs

The cellular concept is employed in many different forms. Typically, when someone refers to cellular communication, the reference is to advanced mobile phone system (AMPS) or total-access communications system (TACS) technology. AMPS operates in the 800-MHz band: 821 to 849 MHz for the base station receive and 869 to 894 MHz for the base station transmit. For TACS the frequency range is 890 to 915 MHz for the base station receive and 935 to 960 MHz for the base station transmit.

Many other technologies also fall within the guise of cellular communication; these include both the domestic U.S. and the international PCS bands. In addition the same concept is applied to several technology platforms that are currently used in the specialized mobile radio (SMR) band. [IS-136 and integrated dispatch enhanced network (iDEN)]. However, cellular communication really refers to the AMPS and TACS bands but is sometimes interchanged with the PCS and SMR bands because of the similarities.

The concept of cellular radio was initially developed by AT&T at their Bell Laboratories to provide additional radio capacity for a geographic customer service area. The initial mobile systems from which cellular evolved were called mobile telephone systems (MTS). Later improvements to these systems occurred, and the systems were then referred to as improved mobile telephone systems (IMTS). One of the main problems with these systems was that a mobile call could not be transferred (handed off) from one radio station to another without loss of the signal. This problem was resolved by reusing the allocated frequencies of the system. With the handoff problem solved, the market was able to offer higher radio traffic capacity, which allowed for more users, in a geographic service area than with the MTS or IMTS. Cellular radio was thus a logical progression in the quest to provide additional radio capacity for a geographic area.

The cellular systems in the United States are broken into metropolitan statistical areas (MSAs) and rural statistical areas (RSAs). Each MSA and RSA has two different cellular operators that offer service. The two cellular operators are referred to as A-band and B-band systems. The *A band* is the nonwireline system, and the *B band* is the wireline system for the MSA or RSA. A brief configuration for a cellular system is shown in Fig. 1.1.

There are numerous types of cellular systems used in the United States and elsewhere. Here is a brief listing of some of the more common ones. All are similar in network layout in that they have base stations connected to a mobile switching center (MSC) that in turn connects to the public switched telephone network (PSTN) or postal, telegraph, and telephone (PTT) system.

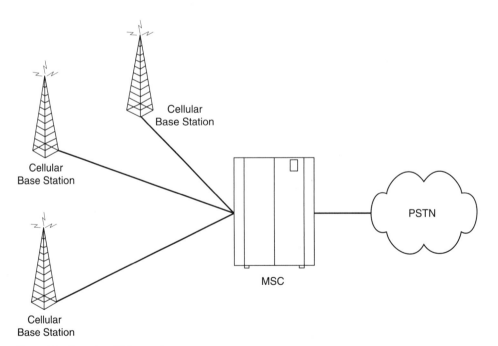

Figure 1.1 Generic cellular system.

1. *Advanced mobile phone system (AMPS).* The cellular standard developed for use in North America. This type of system operates in the 800-MHz frequency band and has also been deployed in South America, Asia, and Russia.

2. *Code-division multiple access (CDMA).* An alternative digital cellular standard developed in the United States. It utilizes the IS-95 standard and is implemented as the next generation for cellular systems. The CDMA system coexists with the current analog system.

3. *Digital AMPS (D-AMPS).* The digital standard for cellular systems developed for use in the United States, also called *North American digital cellular (NADC)*. When the existing analog systems needed to be expanded because they were growing at a rapid pace, instead of the creation of a new standard, the AMPS standard was developed into the D-AMPS digital standard. The D-AMPS is designed to coexist with current cellular systems and relies on both the IS-54 and the IS-136 standards.

4. *Global system for mobile communications (GSM).* The European standard for digital cellular systems operating in the 900-MHz band. This technology was developed out of the need for increased service capacity due to the analog system's limited growth. This technology offers international roaming, high speech quality, increased security, and the ability to develop advanced systems features. It was completed by a consortium of 80 pan-European countries working together to provide integrated cellular systems across different borders and cultures.

5. *Nordic mobile telephone (NMT).* The cellular standard developed by the Nordic countries of Sweden, Denmark, Finland, and Norway in 1981. This type of system was designed to operate in the 450- (NMT 450) and 900-MHz (NMT 900) frequency bands. NMT systems have also been deployed throughout Europe, Asia, and Australia.

6. *Total-access communications system (TACS).* A cellular standard derived from the AMPS technology. TACS operates in both the 800- and 900-MHz bands. The first system of this kind was implemented in England. Later these systems were installed in Europe, China, Hong Kong, Singapore, and the Middle East. A variation of this standard (*JTACS*) was implemented in Japan.

7. *Integrated dispatch enhanced network (iDEN).* The name for an alternative form of cellular communication which operates in the SMR band just adjacent to the cellular frequency band. iDEN is a blend of wireless interconnect and dispatch services which makes it very unique compared to existing cellular and PCS systems. iDEN utilizes a digital radio format called quadrature amplitude modulation (QAM) and is a derivative of GSM for the rest of the system with the exception of the radio link.

1.3 PCS

Personal communication services (PCS) is the next generation of wireless communications. It is a general name given to wireless systems that have recently been developed out of the need for more capacity and design flexibility than that provided by the initial cellular systems. The similarities between PCS and cellular lie in the mobility of the user of the service. The differences between PCS and cellular fall into the applications and spectrum available for PCS operators to provide to subscribers.

PCS, like its cellular cousin, is another narrowband service which offers many enhanced data services in conjunction with voice services. It was heralded in as providing many data services which would enable people to use one communication device for all their foreseeable needs. However, because of the bandwidth limitations associated with the PCS systems deployed, the data throughput remained at 9.6 kbit/s.

Wideband PCS has many promises for offering high-speed data but has not yet been deployed because there are particular problems that must be overcome. The obvious issue is coexistence with the current PCS system. Coupled with the coexistence problem is the need for more base stations due to reduced sensitivity caused by increased bandwidth. The third major problem that needs to be overcome is the offering of subscriber units that can act as dual band units in a vastly diverse PCS marketplace.

Figure 1.1, while labeled as a cellular system, has the same format and layout as a PCS system. The chief difference is that the frequency of operation is higher for PCS and therefore more base stations are required in order to cover the same geographic area.

The diverse PCS systems that an operator can possibly utilize are listed here. It is important to note that in several markets the same operator can and has deployed several types of PCS systems in order to capture market share.

1. *DCS1800.* A digital standard based upon the GSM technology with the exception that this type of system operates at a higher frequency range, 1800 MHz. This technology is intended for use in personal communication network (PCN) systems. Systems of this type have been installed in Germany and England. (DCS stands for digital cellular system.)

2. *PCS1900.* A GSM system which is the same as DCS1800 except that it operates in the PCS frequency band for the United States, 1900 MHz.

3. *Personal digital cellular (PDC).* A digital cellular standard developed in Japan. PDC-type systems were designed to operate in the 800-MHz and 1.5-GHz bands.

4. *IS-661.* The technology platform that is being promoted by Omnipoint. It is a spread-spectrum technology that relies on time-division duplexing (TDD).

5. *IS-136.* The PCS standard that relies on the NADC system except that it operates in the 1900-MHz band.

Introduction   7

6. *CDMA.* Another popular PCS platform utilizing the same standard as that for CDMA in cellular except that it too operates in the 1900-MHz band.

In the United States, PCS operators obtained their spectrum through an action process set up by the Federal Communications Commission (FCC). The PCS band was broken into A, B, C, D, E, and F blocks. The A, B, and C blocks involved a total of 30 MHz, while the D, E, and F blocks are allocated 10 MHz.

The spectrum allocations for both cellular and PCS in the United States are shown in Figs. 1.2 and 1.3. It should be noted that the geographic boundaries for PCS licenses are different than those imposed for cellular operators in the United States. Specifically PCS license boundaries are defined as metropolitan trading areas (MTAs) and basic trading areas (BTAs). The MTA has several BTAs within its geographic region. There are a total of 93 MTAs and 487 BTAs defined in the United States. Therefore, there are a total of 186 PCS MTA licenses, each with a total of 30 MHz of spectrum to utilize; this is in addition to the 1948 BTA licenses awarded in the United States. Regarding the PCS licenses, the A, B, and C bands will have 30 MHz of spectrum each while the D, E, and F blocks will each have only 10 MHz available. All the frequency allocations are duplexed.

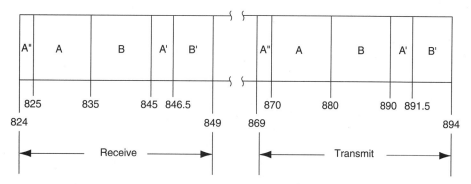

Figure 1.2 U.S. cellular spectrum chart (all frequencies in MHz).

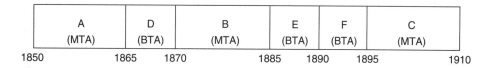

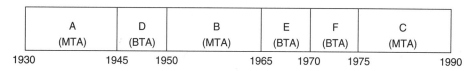

Figure 1.3 PCS spectrum allocation (all frequencies in MHz).

1.4 WLL

The wireless local loop (WLL) system utilizes many similar, if not the same, platforms as used in the cellular and PCS systems and is primarily focused on voice services. The WLL system, however, is different from the cellular or PCS systems in its application, which is fixed. Being a fixed service it is often referred to as a local multipoint distribution system (LMDS) or a fixed wireless point-to-multipoint (FWPMP) system. In fact, WLL in many cases is the same as LMDS or FWPMP in its deployment and application. WLL is most applicable in areas where local phone service is not available or cost effective. Primarily WLL is a system that connects a subscriber to the local telephone company (PTSN or PTT system) using a radio link as its transport medium instead of copper wires.

There is no specific band that WLL systems occupy or are deployed in. The systems can either operate in a dedicated, protected spectrum or in an unlicensed spectrum. Some of the services that fall within the definition of WLL include cordless phone systems and fixed cellular systems, as well as a variety of proprietary systems.

Given the wide choice of system types and the spectrum considerations, the choice of which combination of system types to use is directly dependent upon the application and services desired. Some of the additional considerations for choosing the right technology platform involve the determination of the geographic area needed to be covered, the subscriber density, the usage volume and patterns expected from the subscribers, and the desired speed of deployment.

Since no one radio protocol and service can do everything, the choice of which system to deploy will be driven by the desired market and applications required to solve a particular set of issues. Some of the more common types of WLL systems involve cellular, PCS, cordless telephone (CT-2), and digital European cordless telecommunication (DECT).

WLL has different implications when deployed in a developed country than when deployed in an emerging country. For a developed country WLL allows for use of a cordless phone as an extension of the house phone or private branch exchange (PBX), which is an added convenience. However, in an emerging country, which has areas without any access to a communication service, the use of WLL can create profound changes because it is quicker, easier, and less expensive to install than a regular landline system.

Figure 1.4 represents a typical WLL system. The WLL system has various nodes that are connected back to a main concentration point. The method for connecting the nodes to the concentration point can be by radio, wire, cable, or a combination of all three. Table 1.1 is a general representation of different technology platforms that can be used depending on the application involved. This generalization illustrates that there is no single platform or application to use when deploying WLL systems.

1.5 LMDS

The local multipoint distribution system (LMDS) is a unique wireless access system whose purpose is to provide broadband access to multiple subscribers

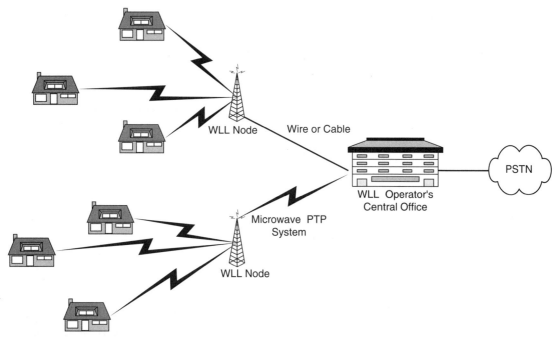

Figure 1.4 Wireless local loop (WLL).

TABLE 1.1 WLL Technology

Geographic area	WLL technology
Urban	Digital cellular
	DECT
	CT-2
	LMDS
	Proprietary radio system
Suburban	Digital cellular
	DECT
	LMDS
	Proprietary radio system
Rural	Analog cellular
	Digital cellular
	Proprietary radio system

in the same geographic area. Presently the majority of the systems that have deployed or started to deploy LMDSs are experiencing financial stress. However, there are numerous advantages to an LMDS that can help augment a mobile wireless system by providing a high bandwidth to selected areas or campus environments coupled with 802.11 (wireless LAN protocol). Additionally LMDSs can be used as an effective backhaul method for data traffic.

The LMDS utilizes microwave radio as the fundamental transport medium and is not really a new technology. It is an adaptation of existing technology for a new service implementation that allows multiple users to access the same radio spectrum. The LMDS is a wireless system that employs cellular-like design and reuse with the exception that there is no handoff. It can be argued that LMDS is in fact another variant to the WLL portfolio described previously and referenced as proprietary radio systems.

LMDS consists of two key elements, the physical transport layer and the service layer. The physical transport layer involves both radio and packet- and circuit-switching platforms. The radio platform consists of a series of base stations providing the radio communication link between the customers and the main concentration point, usually the central office of the LMDS operator. Figure 1.5 is a high-level system diagram of an LMDS. The system as shown has a similar layout as that for a cellular or PCS mobile system with the obvious difference being that the subscribers are fixed and, of course, operate at a different frequency.

The system diagram depicts multiple subscribers (customers) surrounding an LMDS hub or base station. The base station is normally configured as a sectorized site for frequency reuse purposes, and there are multiple subscribers

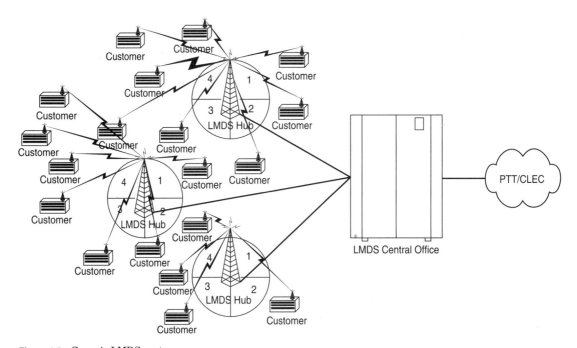

Figure 1.5 Generic LMDS system.

assigned to any sector. The amount of channels and the overall frequency plan for the system are driven by the spectrum available in any given market and the amount of capacity required in any geographic zone.

The LMDS is a point-to-multipoint system where multiple subscribers can access the same radio platform utilizing both a multiplexing method as well as queuing. Specifically a single radio channel may have 12-Mbit/s total throughput, but you might be able to offer 24 Mbit/s or greater for the same channel by allocating it to the entire sector and not to specific customers through overbooking. There, of course, are quality-of-service (QOS) issues and specific service delivery requirements with any commercial system. However, the concept is that an LMDS utilizing point-to-multipoint technology can provide vastly greater bandwidth and services to a larger population than a point-to-point system utilizing the same spectrum can.

Unlike mobile systems an LMDS has several key differences. The first is that ubiquitous coverage is not required; this is a key advantage. The LMDS if deployed properly can have the operator only provide service where the customers are actually located thereby maximizing the capital infrastructure effectiveness and minimizing operating expenses.

The other issue with LMDS relates to the fixed subscriber base that is potentially there. A primary concept of LMDS delivery is to provide the service not to one customer in a sector but to multiple customers. The concept is further carried to each building where the service is deployed. Specifically, LMDS is best positioned when there are multiple customers that utilize the same radio equipment, thereby maximizing the capital infrastructure installed at that location.

A brief example of a building having multiple customers is shown in Fig. 1.6. The simple concept of having multiple customers per geographic location will minimize the cost of acquisition for any customer and at the same time reduce operating and capital costs. It is important to note that initially the building where the equipment is to be deployed should be evaluated in order to properly establish its bandwidth potential. In this example, it is assumed that access to the wiring closet is achieved for distribution of the services offered. Also, Fig. 1.6 implies that there is LOS (line of sight, that is, no obstructions) with the hub site in order to ensure that the link is of sufficient quality for stable and reliable communication.

LMDS can be a very cost effective alternative for a competitive local exchange carrier (CLEC). With LMDS a CLEC can deploy a wireless system without having to experience the heavy capital requirements of laying down cable or copper to reach customers. The cost effectiveness is born out of the ability to focus the capital infrastructure where the customers are and at the same time being able to deploy the system in an extremely short period of time.

Some of the services that LMDS can offer customers are listed here. Note that the service offered cannot have a bandwidth requirement greater than what the radio transport layer can support.

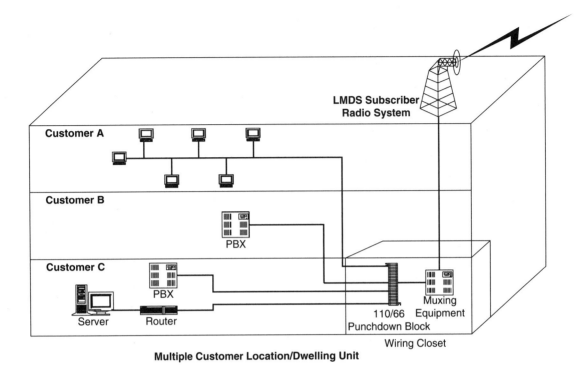

Figure 1.6 Host location.

Applications

- LAN/wide area network (WAN) [virtual private network (VPN)]
- Lease line (T1/E1) replacement (clear and channelized)
- Fraction T1/E1 (clear and channelized)
- Frame relay
- Voice telephony [plain old telephone service (POTS) and enhanced services]
- Videoconferencing
- Internet connectivity
- Web services [e-mail, hosting, virtual Internet service provider (ISP), etc.]
- E-commerce
- Voice over IP (VoIP)
- Fax over IP (FaxIP)
- Long-distance and international telephony
- Integrated systems digital network (ISDN) [basic rate interference (BRI) and primary rate interference (PRI)]

The host of services and perturbations to those just listed make an impressive portfolio. Of course, the necessary platforms and connectivity for the network need to be in place in order to ensure that these services can and will be offered and effectively delivered. It is interesting though that with an LMDS, as with any network, there are on-net and off-net traffic considerations. Ideally the traffic should be all on-net, but when the system initially goes on-line, most, if not all, the traffic goes off-net and the PTT system or another CLEC will need to be used almost exclusively to facilitate the delivery of the service.

As with all wireless systems, there are multiple LMDS from which an operator can choose to deploy. Some of the system architectures to pick from are frequency-division duplexing (FDD), TDD, time-division multiplexing (TDM)/ATM, ATM, FDD/TDM. Coupled with the transport method, the choice of modulation scheme as well as frequency planning options must all be weighed. Additionally another often overlooked aspect is the method for actually delivering service to a customer and the physical and electrical demarcation location and method.

1.6 MMDS, MDS, ITFS

Multichannel, multipoint distribution systems (MMDSs); instructional television fixed service (ITFS); and multipoint distribution service (MDS) are all sister bands to LMDS. The combination of MMDS, ITFS, and MDS bands make up what is referred to as *wireless cable*.

A total of 33 channels, each 6-MHz wide, make up the MMDS, MDS, and ITFS bands collectively. The bands, while currently being referenced together, were all developed for different reasons. However, the bands were originally broadcast related in that they were one-way oriented. The exception was the ITFS channels, which allocate a part of the band for upstream communication.

The MMDS, MDS, ITFS band has numerous subscribers utilizing its service. However, there has been increased activity in redefining the services the band can and will offer subscribers. The primary focus of the band is toward high-speed Internet traffic as compared to video services in conjunction with data. To make this happen the band has been allocated for two-way communication. But the channels are not paired as is done commonly in other bands. The two technology types now competing for use in this band are the FDD and TDD systems.

The technologies being deployed for the MMDS, MDS, and ITFS band are similar to that for the LMDS in that they involve a sectorized cell site which has multiple subscriber terminals associated with each channel in every sector. One of the key advantages the MMDS, MDS, and ITFS band has is the frequency this band operates within. The bands for operation are 2.15 to 2.162 GHz and 2.5 to 2.686 GHz, which do not require strict adherence to line of sight (LOS) for communication reliability as well as the elimination of rain fade considerations in the link budget.

The chief disadvantage with this band is the coordination an operator must achieve in order to utilize a particular frequency in a geographic area. The coordination is exceptionally tricky due to the present existence of MMDS, MDS, and ITFS operators that primarily utilize video as their service offering. The issue arises from both upstream and downstream frequency coordination since existing operators designed their systems based on a broadcast system basis.

1.7 Cable Systems

The proliferation of cable modems, primarily in the United States, has brought broadband service to many end users who were previously relying on dial-up IP. Cable operators have a unique advantage, as do PTT services, for delivering broadband services to the residential market because they already have a presence in many residential homes.

The common issue facing all broadband providers is the quality of their underlying transport layer. The quality of the cable plant itself dictates the delivery of services that can effectively be offered. The issues with cable plant quality are primarily driven by the number of drops (the wire or cable that connects to a house or building, as well as the wire or cable splits within the house or building) that are on any cable leg, which directly impacts the ingress noise problem that limits the ability for the cable plant to provide high-speed two-way communication. Since most of the information flow is from the head end to the subscriber, the system does not have to support symmetrical bandwidth requirements.

A hybrid fiber/coax (HFC) network is shown in Fig. 1.7, with the enhancement of providing two-way communication for both voice and data, besides the video service offering. The primary access method is physical media where the connection made to the subscriber at the end of the line is via coaxial cable. For increased distance and performance enhancements fiber-optic cables are often part of the cable network's topology.

Figure 1.8 is an example of a cable operator utilizing wireless access as the last leg in the access system. The wireless device listed can be a base station or a small remote antenna driver/remote antenna signal processor (RAD/RASP) unit installed on the coaxial cable itself. The figure depicts the potential for a cable operator and a wireless operator to utilize each other's infrastructure to deliver services. It should be noted that PCS is listed in Fig. 1.8 not only for mobility telephony but potentially also for better allocation for the PCS C band auctioned in the United States for delivering last-mile, high-bandwidth services.

1.8 WAP

The wireless application protocol (WAP) is one of the many broadband protocols being implemented into the wireless arena for the purpose of increasing mobility by enabling mobile users the ability to surf the Internet. WAP is being

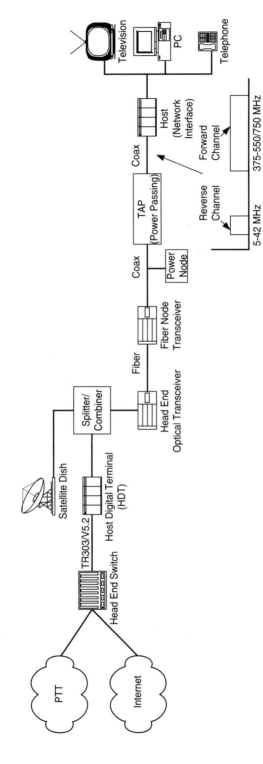

Figure 1.7 HFC.

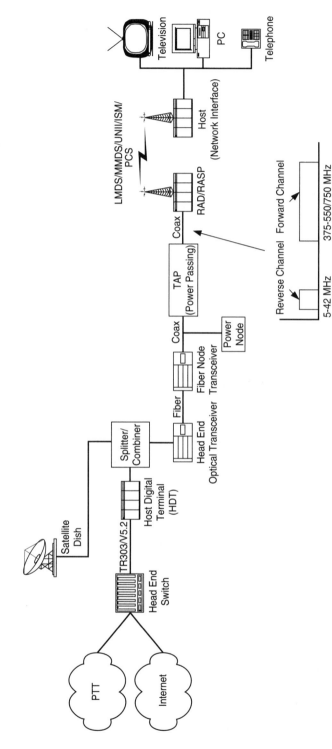

Figure 1.8 RAD/RASP.

implemented by numerous mobile equipment vendors since it is meant to provide a universal open standard for wireless phones (i.e., cellular/GSM and PCS) for the purpose of delivering Internet content and other value-added services. Besides various mobile phones, WAP is also designed to be utilized by personal digital assistants (PDAs).

WAP will enable mobile users to surf the Internet in a limited fashion; that is, they can send and receive e-mails and surf the net in a text-only format (without graphics). For WAP to be utilized by a mobile subscriber the cellular or PCS wireless operator needs to implement WAP in its system as well as ensuring that the subscriber units (i.e., the phones) are capable of utilizing the protocol. WAP is meant to be used by the following cellular/PCS system types:

- GSM-900, GASM-1800, GSM-1900
- CDMA IS-95
- TDMA IS-136
- 3G systems: IMT-2000, universal mobile telecommunications system (UMTS), wideband-CDMA (W-CDMA), wideband IS-95

WAP is fundamentally different than broadband technologies. While delivering wireless data, it does not have the bandwidth to deliver leased line replacement or to support broadband technologies. However, WAP has the potential to increase the mobility of many subscribers and enable a host of data applications to be delivered for enhanced services to subscribers.

1.9 Bluetooth

Bluetooth is a wireless protocol that operates in the 2.4-GHz industrial-scientific-medical (ISM) band allowing wireless connectivity between mobile phones, PDAs, and other similar devices for the purpose of information exchange. Bluetooth is meant to replace the infrared telemetry portion on mobile phones and PDAs enabling extended range and flexibility in addition to enhanced services.

Because Bluetooth systems utilize a radio link in the ISM band there are several key advantages to this transport protocol. Bluetooth can effectively operate as an extension of a local area network (LAN) or a peer-to-peer LAN and provide connectivity between a mobile device and the following other device types:

- Printers
- PDAs
- Mobile phones
- Liquid-crystal display (LCD) projectors
- Wireless LAN devices
- Notebooks and desktop personal computers

One of the key attributes that Bluetooth offers is the range that the system or connection can operate over. Since Bluetooth operates in the 2.4-GHz ISM band, it has an effective range going from 10 to close to 100 meters (m). The protocol does not require line of sight for establishing communication. Its pattern is omnidirectional, thereby eliminating orientation issues, and can support both isochronous and asynchronous services paving the way for effective use of TCP/IP communication.

Bluetooth is different than the wireless LAN protocol 802.11 and WAP but again looks at delivering data connectivity over radio. Bluetooth is also different because of the applications, use of the unlicensed band, and focus on end-user devices. Bluetooth is meant to be a LAN extension fostering communication connection ease for short distances.

1.10 Wireless LAN (802.11)

Wireless LAN (WLAN) is another wireless platform enabling various computers or separate LANs to be connected together into one LAN or WAN. A big advantage is that WLAN-enabled devices do not need to be physically connected to any wired outlet, which allows for location flexibility as shown in Fig. 1.9.

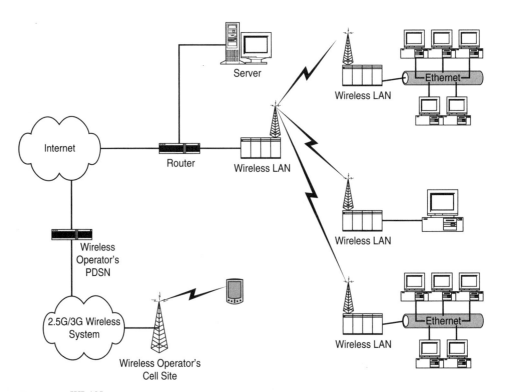

Figure 1.9 WLAN.

The convergence of WLAN 802.11 with wireless mobility has been described as the "real killer application." This means that it will truly allow the subscriber to take advantage of all the applications available on the World Wide Web while at the office, home office, or on the road at some unknown location, provided, of course, there is coverage. The issue of security and provisioning to make this a reality is not a trivial matter if true transparency (requiring no user intervention) is desired with the intranet of a company by its sales and support staff.

There are several protocols that fall into the WLAN arena. Not all of them are compatible, which leaves the possibility of local islands (when the protocols cannot communicate with each other unless a device provides translation) being established. The most prevalent WLAN protocol is IEEE 802.11, but Bluetooth is also referred to as a WLAN protocol. 802.11 is an Institute of Electrical and Electronics Engineers (IEEE) specification encompassing several standards; some of the more prevalent ones are 802.11a, 802.11b (WiFi), and 802.11g.

What is interesting is that 802.11a operates in the 5-GHz, unlicensed national information infrastructure (UNII) band, while 802.11b and 802.11g operate in the 2.4-GHz ISM band along with Bluetooth. 802.11g specifically is meant to increase the data rate to 54 Mbit/s while providing backward compatibility for 802.11b (WiFi) equipment. What this means is that 802.11g equipment operating in the 2.4-GHz band can operate at speeds previously enjoyed by 802.11a equipment in the 5-GHz band. To complicate matters there are a host of other 802.11 specifications, all which either exist or are in the process of being standardized.

The 802.11 specifications were designed initially as a wireless extension for a corporate LAN for enterprise applications, and numerous devices have been manufactured to this specification. For example, the 802.11b protocol is a shared medium and utilizes a listen-before-talk protocol called collision sense multiple access/collision avoidance (CSMA/CA).

Table 1.2 is a simple comparison between the key 802.11 protocols and Bluetooth. Both 802.11b and Bluetooth utilize the ISM band, but their formats and purposes are different. However, 802.11a operates in the UNII band and can operate at a much greater effective radiated power (ERP). Basically 802.11 devices are meant to cover a wider area than Bluetooth devices, and 802.11 devices have the potential for higher throughput. The data rate in the chart for 802.11a and b shows a range of speeds, which, of course, are dependent upon the modulation format used, available power, and interference experienced.

IEEE 802.11 is important for wireless mobility because it provides direct mobile data interoperability between the LAN of a corporation and the wireless operator's system. The inclusion of expending the corporate IP-PBX has great potential. Presently there have been many demonstrations and some operational systems regarding this integration of wireless mobility and wireless LANs which require application-specific programs to enable the interoperability.

TABLE 1.2 Comparison between Key 802.11 Protocols and Bluetooth

WLAN	802.11a	802.11b	Bluetooth
Transport	5-GHz UNII DSS	2.4-GHz ISM FHSS/DSS	2.4 GHz ISM FHSS
Data rate	6–54 Mbit/s	1–11 Mbit/s	1 Mbit/s
Range	*	50 m	1–10 m
Power	0.05/0.25/1 W	+20 dBm	0 dBm

*If used with an external antenna, the WLAN can be extended beyond the immediate office environment.

Note: DSS = direct sequence spread spectrum, FHSS = frequency-hopping spread spectrum, W = watts, dBm = decibels referenced to 1 milliwatt.

There is also another WLAN specification, HiperLan/2, which was developed under the European Telecommunications Standards Institute (ETSI). HiperLan/2 has similar physical layer properties as 802.11a in that it uses orthogonal frequency-division multiplexing (OFDM) and is deployed in the 5-GHz band. The media-specific access control protocol (MAC) layers are different; hence, the different technology specification in that HiperLan/2 uses a time-division multiple access (TDMA) format as compared to 802.11a which uses OFDM.

1.11 VoIP

Voice over IP has provided, and continues to provide, a viable alternative for call delivery of voice traffic. It is interesting that most of the initial VoIP implementations have not occurred over the Internet but rather over corporate LANs and private IP networks like long-distance providers. Private implementation has mitigated the QOS problems associated with VoIP on the Internet.

In many circles, the mention of VoIP invokes quality concerns due to delay and jitter problems when the access medium is over the public Internet. As mentioned previously though, the true application for VoIP is as a transport medium over private or dedicated pipes or networks where the QOS issue no longer is an issue.

The original standards activity for VoIP was defined in ITU H.323 which has the title "Packet-Based Multimedia Communication Systems." This standard's wide use was a direct result of its being offered as freeware by Microsoft. There is, however, an alternative standard in competition with H.323: the media gateway control protocol (MGCP), also called the single gateway control protocol (SGCP). SGCP assumes a control architecture similar to that of the current PTT voice system where the control elements are located outside the gateway itself. These external call control elements are referred to as call agents.

Wireless and CLEC operators that utilize the IP-based infrastructure only can also provide voice services as part of their offering if the proper QOS and delivery issues are addressed in the design and service offering. Wireless operators offering circuit emulation service (CES) voice services provide an attractive entry point for customers. However, the fact that VoIP is being used does not need to be conveyed to the customer if the proper delivery and QOS issues are addressed. A primary reason that VoIP is so attractive for a wireless operator is not solely related to the interconnect savings that may be achieved, but in saving the spectrum, since IP traffic is by itself dynamic in its bandwidth utilization.

Figure 1.10 depicts the major components involved with providing VoIP, either as a direct service or as an alternative transport medium that the wireless operator uses to be more cost competitive or better yet to improve the margin. As Fig. 1.10 suggests, VoIP can be delivered either directly to a public data network or via the Internet depending on the service-level agreement (SLA) used. In addition, the diagram depicts the issue of the operator using VoIP as a medium for handling voice traffic into the switching complex where it then converts the IP traffic into classical TDM traffic for interfacing to the PTT for call delivery.

Figure 1.11 depicts the connection between a wireless operator in one market and its operation in another market. The diagram can, of course, be meant for an ISP, CLEC, or large corporation.

Voice over IP is the most flexible choice for voice transport since it can run over any layer-one or layer-two infrastructure. This flexibility is particularly important in heterogeneous environments like LMDSs.

1.12 Typical Central Office

Figure 1.12 is a generic mobile telephone center configuration. The mobile switching center (MSC) is the portion of the network which interfaces the radio world to the public switched telephone network (PSTN). In mature systems there are often multiple MSC locations, and each MSC can have several cellular switches located within each building.

A mobile telephone switching office (MTSO) is commonly referred to as the MSC and is anything but typical. Although a MSC typically delivers voice services, the particular services that can be offered and delivered are extremely varied. For example, in AMPS only voice services could be delivered. GSM system voice along with short messaging service (SMS) services would be provided. However, with the advent of Internet and 2.5/3.0G (packet-based mobile) services, many residential MSCs which primarily delivered voice services are now transitioning from a circuit-switching to a packet-switching system.

A simplified example of a typical MSC layout is shown in Fig. 1.12. Naturally, when determining the dimensions and specific equipment required for the facility, one will need to factor in the type of services to be provided as well as the time frame the design is to encompass (i.e., the growth potential needed).

Typically a MSC consists of an equipment room, toll room, power room, and operations room. The functions of each room are unique. The equipment room has the switching and packet platforms for treating and servicing the

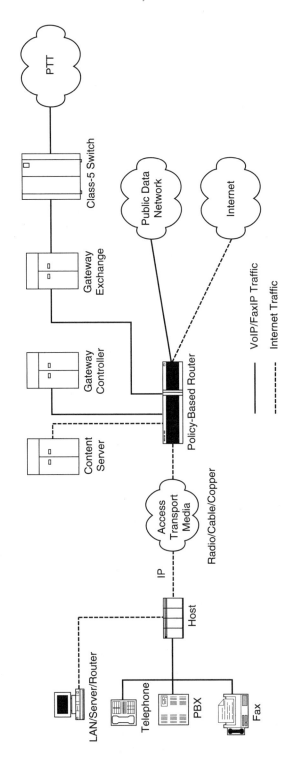

Figure 1.10 VoIP network.

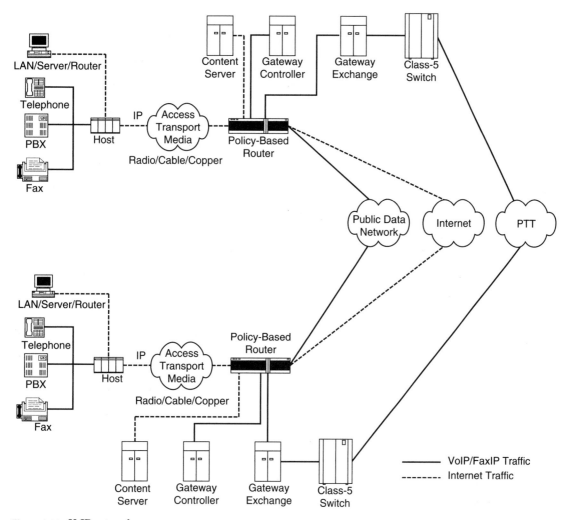

Figure 1.11 VoIP network.

subscriber's needs. The toll room, also referred to as the interconnect or telco room, is the area of the MSC where the system interfaces to the PSTN, CLECs, interexchange carriers (IXCs), and other outside carriers. The purpose of the toll room is to provide the portal for entry and exit of services for the central office. The power room usually houses the rectifiers, batteries, and generator for emergency backup purposes. The operations center is the area where the craft personnel perform the data entry and monitoring and maintenance of the network itself.

The following list of topics should prove helpful in establishing the resources and timing needed for a fixed network design to be successful.

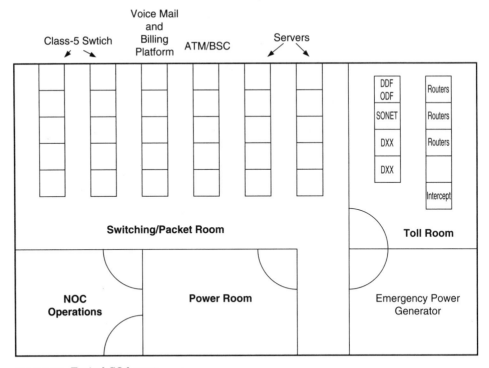

Figure 1.12 Typical CO layout.

Equipment room
- Class 5 switch
- ATM switches
- Voice mail system
- Servers
- Billing system

Toll room
- Signaling transfer point (STP)
- DXX (cross connect switch) equipment
- Routers
- Intercept equipment

This list does not address the issue of colocation with other service providers and the need to create a separate area for the operator to maintain and upgrade its equipment.

Chapter 2

Radio Engineering Topics

Because the fundamental transport mechanism for wireless communications is radio, the rapid expansion of wireless technology in the marketplace has resulted in a plethora of radio systems being deployed. This expansion of wireless technology has led to many engineers being thrust into the wireless arena with little or no formal training in radio engineering. A great deal of excellent reference material, including textbooks, is available for engineers to acquire radio engineering knowledge. However, these reference materials by and large focus on theory and not on the practical aspects of operating a wireless system.

This chapter reviews radio engineering principles, incorporating the basic building blocks of a radio communication system. Understanding these basic radio engineering concepts and topics, on more than a superficial basis, will allow the engineer to design or improve wireless communication systems while being aware of the technical and financial tradeoffs needed to be made.

2.1 Electromagnetic Waves

The ability for radio frequency (RF) energy to propagate through air enables wireless companies to deliver a service without physically connecting devices to the network. RF energy is comprised of electromagnetic energy which is best described by an illustration (see Fig. 2.1).

The frequency of oscillation for the electromagnetic wave is directly related to the physical wavelength as shown in Eq. (2.1).

$$C = f\lambda \tag{2.1}$$

where f = frequency, Hz
$C = 3 \times 10^8$ m/s
λ = wavelength, m

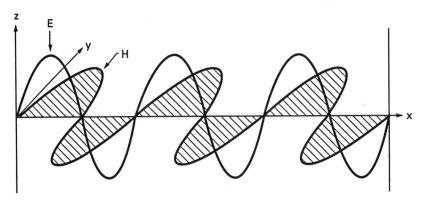

Figure 2.1 Electromagnetic wave.

We can establish the wavelength for a cellular base station transmitting site with a quick manipulation of Eq. (2.1). The frequency selected for this example is 880 MHz since it is right in the middle of both the A and B bands associated with U.S. cellular operators.

$$\lambda = \frac{C}{f} = \frac{3 \times 10^8 \text{ m/s}}{880 \times 10^6 \text{ Hz}} \tag{2.2}$$

where $\lambda = 0.34$ m [approximately 13 inches (in)]. However, the wavelength for PCS or enhanced specialized mobile radio (ESMR) can be easily determined through simple manipulation of Eq. (2.1).

2.2 Radio Systems

The fundamental building blocks of a communication system are shown in Fig. 2.2. This simplified drawing represents the major components in any communication system: antenna, filters, receivers, transmitter, modulation, demodulation, and propagation. Entire books have been written on each component of Fig. 2.2. In the following sections, we will briefly discuss each of these components because knowing the design characteristics of each component is essential to building a communication system that will provide the proper transport functions for the information content. The discussion will begin with the transmitter and then progress to general modulation types, to antennas, and finally to the receiver.

2.3 Transmitters

There are many types of transmitters and amplifiers in a communication system. Most of the amplifiers are located in the receiver for the communication cell site. The fundamental difference between the amplifiers in the transmit

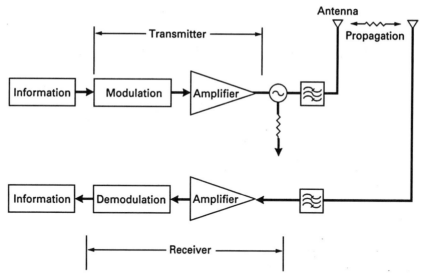

Figure 2.2 Basic radio system.

portion and in the receiver lies in the power type they are able to amplify and deliver to the desired load. The focus of this section will be on RF transmitters which are power oriented, but the same fundamental concepts apply to all amplifiers.

Knowledge of RF transmitter types and different combining techniques will enable an RF engineer to maximize the efficiency of a communication site through improving the amount of energy delivered to the antenna system or reducing the amount of physical antennas at a site. The amount of physical antennas available at a site may or may not be driven by economic reasons alone; local ordinances may have a more profound role in deciding the amount of antennas in the ultimate configuration for a communication site.

2.4 Transmitter System Building Blocks

The transmitter block diagrams for the three basic forms of radio communication are presented here. Figure 2.3 is a block diagram of an amplitude modulation (AM) transmitter. The AM transmitter changes the amplitude of the carrier as a function of the information content. Figure 2.4 is a brief block diagram of a frequency modulation (FM) transmitter. The FM transmitter modulates the information content by changing the frequency of the carrier as a function of the information content. Figure 2.5 is a block diagram of a phase modulation (PM) transmitter. The PM transmitter places the information onto the carrier much as is done with FM; however, the modulation is achieved through adjustment of the phase of the information that rides on the carrier.

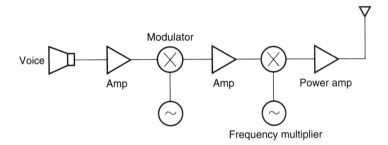

Figure 2.3 Amplitude modulation (AM).

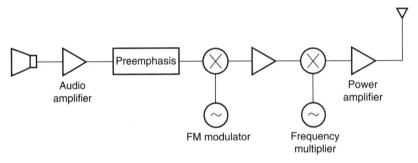

Figure 2.4 Frequency modulation (FM).

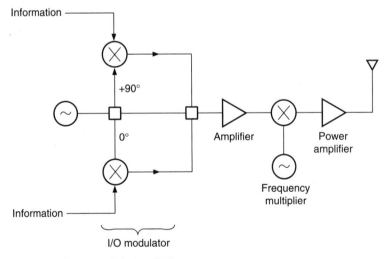

Figure 2.5 Phase modulation (PM).

2.5 Information Bandwidth

The information throughput and the channel bandwidth are critical parameters in determining which modulation scheme to utilize for the communication system. Often it is the channel bandwidth that is defined, and then the appropriate modulation technique must be applied. Conversely the manufacturer of a wireless system will state the information throughput of its system for a given channel bandwidth. The information bandwidth is even more important as the convergence of voice and data is taking place with wireless mobility and the proliferation of IP.

The channel's theoretical capacity is defined by the Shannon–Hartley equation:

$$C = B \log_2 (1 + \text{SNR})$$

where C = capacity, bit/s
 SNR = signal-to-noise ratio
 B = bandwidth, Hz

2.6 Modulation

To convey data and voice information from one location to another without a physical connection, as is done in wireless mobility, it is necessary to use another method to send the information. Many methods exist for conveying information to and from locations that are not physically connected. Some of these involve talking, flags, drums, and lights. Each of these methods has its advantages and disadvantages, but the problem common to them all is the relatively short physical distance the sender and receiver have to be from each other for the information to be successfully sent and received.

In order to increase this distance and increase the information transfer rate between the sender and receiver, the use of an electromagnetic wave is employed. However, to utilize an electromagnetic wave it is necessary to modulate the carrier wave at the transmitting source in order to put the information onto the carrier wave; then, at the receiving end, the receiver will need to demodulate the carrier wave in order to extract the information that was sent. The modulation and then demodulation of the carrier wave form the principle of a radio communication system. A generalized radio system is shown in Fig. 2.6.

The type of modulation and demodulation utilized for the radio communication system is directly dependent upon the information content to be sent, the spectrum available to convey the information, and the cost. The fundamental goal of modulating any signal is to obtain the maximum spectrum efficiency or rather information density per hertz.

There are many types of modulation and demodulation formats utilized for the conveyance of information. However, all the communication formats rely on one, two, or all three of the fundamental modulation types: amplitude modulation, frequency modulation, and phase modulation. Figure 2.7 highlights

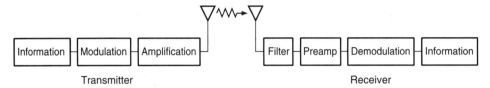

Figure 2.6 Basic radio system.

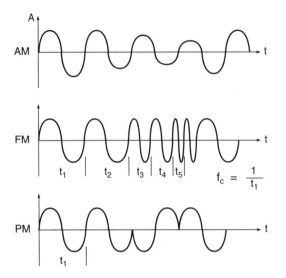

Figure 2.7 Modulation techniques.

the differences between the modulation techniques in terms of their impact on the electromagnetic wave itself.

The modulated electromagnetic wave for communications is represented by

$$E(t) = A \sin(2\pi f_c t + \phi) \quad (2.3)$$

where A = amplitude, μV
f_c = carrier frequency, Hz
ϕ = phase
t = time, s
E = instantaneous electric field strength

Equation (2.3) can be used for all modulation types: amplitude modulation amplifies A, frequency modulation modifies f_c, and phase modulation modifies ϕ.

Modulation techniques

Amplitude modulation has many unique qualities; however, this form of modulation is not utilized directly in mobile wireless communication systems primarily because it is more susceptible to noise. However, a variant of AM, quadrature amplitude modulation (QAM), is used predominantly.

Frequency modulation is utilized for many mobile communication systems that employ analog communication. One common use of FM is in analog cellular communication since it is more robust against interference.

Phase modulation is used for conveying digital information. There are many variations to phase modulation; specifically many digital modulation techniques rely on modifying the RF carrier's phase and amplitude as in the generation of quadrature phase-shift keying (QPSK) and QAM signaling formats.

Quadrature phase shift keying is one form of digital modulation which has a total of four unique phase states (hence the term *quadrature*) used to represent data. The four phase states are arrived at through different I (in phase) and Q (quadrature) values. Because four phase states are utilized, each phase state can represent two data bits. The two data bits are mapped on the IQ chart shown in Fig. 2.8.

The coordinate system for QPSK is best realized if you think in terms of an xy coordinate chart where x is now represented by I and y is represented by Q. The distinct IQ location (phase state) shown represents a symbol, and this symbol is made up of two distinct bits. The advantage of utilizing QPSK is the bandwidth efficiency. Since two data bits are now represented by a single symbol, less spectrum can be utilized to transport the information.

The symbol rate's ability to reduce the bandwidth requirement is best illustrated in the following equation:

$$\text{Symbol rate} = \frac{\text{bit rate}}{\text{number of bits per symbol}}$$

Differential quadrature phase-shift keying (DQPSK) is similar to QPSK. However, the primary difference between DQPSK and QPSK is that DQPSK does not require a reference from which to judge the transition. Instead DQPSK's data pattern is referenced to the previous DQPSK's phase state. DQPSK has four potential phase states with the data symbols defined relative to the previous phase state as shown in Table 2.1.

Pi/4 differential quadrature phase-shift keying (Pi/4DQPSK) modulation is very similar to that of DQPSK. However, the difference between Pi/4DQPSK and DQPSK is that the Pi/4DQPSK phase transitions are rotated 45° from that of DQPSK. Like DQPSK, Pi/4DQPSK has four transition states with the data symbols defined relative to the previous phase state as shown in Table 2.2.

For more in-depth information, see the multitude of textbooks and technical articles that abound in the industry that focus purely on each of the modulation schemes.

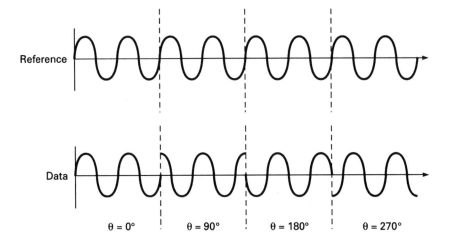
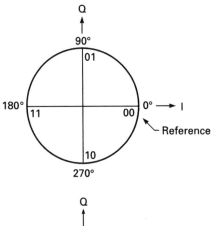
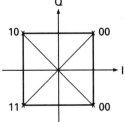

Figure 2.8 QPSK.

2.7 Antennas

The antenna system is one of the most critical and least understood parts of the radio communication system. The antenna system is the interface between the radio system and the external environment and, therefore, maintains the communication link. The antenna system for a wireless system can

TABLE 2.1 DQPSK

Symbol	DQPSK phase transition (°)
00	0
01	90
10	−90
11	180

TABLE 2.2 Pi/4DQPSK

Symbol	Pi/4DQPSK phase transition (°)
00	45
01	135
10	−45
11	−135

consist of a single antenna or multiple antennas at the base station and one at the host terminal station.

The two primary antenna classifications for a system are omni and directional. *Omni* antennas are used when the desire is to obtain a 360° radiation pattern. *Directional* antennas are used when a more refined pattern is desired. The directional pattern is usually needed to facilitate system growth through frequency reuse or to shape the system's contour.

An antenna can also be classified as active or passive. An *active* antenna usually has some level of electronics associated with it to enhance its performance. A *passive* antenna is more classical in that it does not have any electronics associated with its use; it consists entirely of passive elements.

The choice of which antenna to use will directly impact on the performance of the base station, subscriber, or the overall network. The RF engineer is primarily concerned in the design phase with the base stations since these are fixed locations and there is some degree of control over the performance criteria that the engineer can exert on a base station.

Coverage problems or other issues that the engineer is trying to prevent or resolve can be overcome by use of the correct antenna. The engineer must also take into account a multitude of design issues when choosing an antenna. Some of these issues include the antenna's gain, its pattern, the interface or matching to the transmitter, the receiver utilized for the site, the bandwidth and frequency range over which the signals desired to be sent will be applicable, the antenna's power handling capabilities, and the antenna's intermodulation distortion (IMD or intermod) performance. Ultimately the chosen antenna needs to match the system design objectives.

Many types of antennas are available on the commercial market, so there is no need to invent more. Two common types of antennas used in cellular communication systems are collinear and log periodic antennas.

The collinear antenna can be either omni or directional. This type of antenna has a series of dipole elements that operate in phase and is also referred to as a broadside radiator. The maximum radiation takes place along the dipole arrays axis, and the array consists of a number of parallel elements in one plane.

The log periodic antenna is also referred to as a log periodic dipole array (LPDA). The LPDA is a directional antenna whose gains, standing-wave ratio (SWR), and other key parameters remain constant over the operating band. The LPDA is used where a large bandwidth is needed, and the typical is 10 dBi (i = isotropic).

The performance of an antenna is not restricted to its gain characteristics and physical attributes (i.e., maintenance). There are many parameters that must be taken into account when looking at an antenna's performance. The parameters that define the performance of an antenna can be referred to as the antenna's *figures of merit* (FOMs). The following is a partial list of the FOMs that should be quantified by the antenna manufacturer. When choosing an antenna, the engineer may not be able to choose the desired values for all FOMs. Instead, tradeoffs will have to be made in order to arrive at the best possible antenna configuration.

1. Antenna pattern
2. Main lobe
3. Side lobe
4. Input impedance
5. Radiation efficiency
6. Beam width
7. Directivity
8. Gain
9. Antenna polarization
10. Bandwidth
11. Front-to-back ratio
12. Power dissipation
13. Intermodulation
14. Construction
15. Cost

2.8 Filters

Filters play an integral part in the design and operation of a radio and its associated communication system. More often than not the filter characteristics of an existing or new communication system are overlooked by the system design engineer. With the proliferation of wireless communication the need to pay particular attention to the filter characteristics of the base station and host terminals becomes even more paramount. With this proliferation comes the demand for a filter with a smaller physical size and an increased ability to attenuate unwanted signals but that will not distort the desired signal in any fashion.

Filters used in the network play a vital role in protecting the receiver from unwanted signals within a given cell site. There are many types of filters that can be deployed. The filter selection is based on the filter's mission statement and cost, either in terms of spectrum or actual monetary cost. Simply put, the purpose of a filter is to allow the desired energy or information to pass undistorted either in phase, amplitude, or time and at the same time completely suppress all other energy.

Each filter type has its own unique characteristics. The specific characteristics of a filter are driven by its physical construction. The physical construction of the filter is an important aspect of the selection process and the types of filters listed are passive filters only. Active filters, although having many good applications, are not covered in this section because they are not used presently in mobile systems.

There are four general classifications of filters used throughout radio communications:

1. Low-pass
2. High-pass
3. Bandpass
4. Bandreject

Figure 2.9 provides an example of each of these filter types. It should be noted that there are many perturbations regarding combinations of the general filter types listed. The specific configuration chosen is entirely dependent upon the application that it is meant to solve and the acceptable tradeoffs that come along with the filter choice made.

Ideally a filter will pass all the frequencies within a specified passband without attenuation and infinitely attenuate all those frequencies outside of the passband. Additionally the time response of the filter will ideally be such that the output is identical to the input with some time delay. In other words the transfer function for the filter would be equal to 1 for the frequencies of interest only.

As is often the case with a wireless system the filters used have been predefined by the manufacturer of the equipment, due to the form, fit, and function interrelationship. However, it is important to understand that the fundamental implications associated with filter techniques for both the base station and subscriber terminal could improve or degrade the system's performance. Since the ideal filter is not realizable at this particular time, the specific filter type

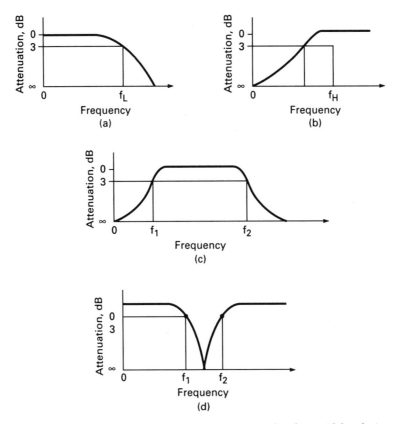

Figure 2.9 Ideal filters: (*a*) low-pass; (*b*) high-pass; (*c*) bandpass; (*d*) bandreject.

selected and used in a wireless system should be determined based on which tradeoffs or imperfections can be best tolerated.

Wireless systems utilize several types of filters in a normal communication path. As an example for cellular operations the A-band carrier usually employs two bandpass filters in series as the front-end filter for the receive path. Conversely the B-band carrier usually employs a bandpass filter with a notch to obtain its frequency selectivity. The actual filter performance needs to be paid attention to because the filter is meant to protect the receiver from unwanted signals. However, if the filter protects the receiver too much, then excess attenuation occurs over frequencies desired to be passed through. Figure 2.10 shows a typical bandpass filter response desired for the A-band carrier and Fig. 2.11 shows the response desired for the B-band carrier.

Table 2.3 displays which applications each type of filter is most appropriate for. Note that for certain applications several filters can be cascaded together to solve a particular design issue. We can summarize Table 2.3 as follows. The low-pass filter is meant to pass only those frequencies which are below its cutoff frequency. The high-pass filter rejects all frequencies below

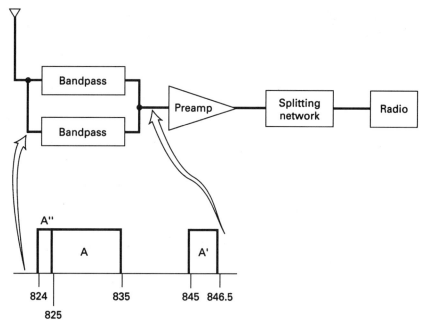

Figure 2.10 Cellular A-band receive.

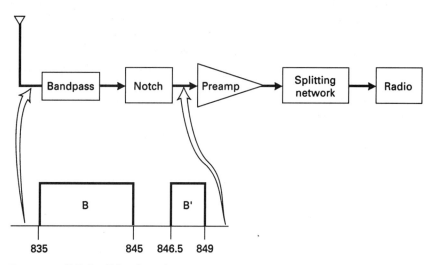

Figure 2.11 Cellular B-band receive.

its cutoff frequency and passes the rest. The bandpass filter passes frequencies within a specified band of frequencies and attenuates all the frequencies outside of the passband. The notch filter, or bandreject filter, has the ability to highly attenuate specific frequencies within a specified band and pass the rest.

TABLE 2.3 Filter Applications

Filter type	Appropriate applications
Low-pass	High-frequency interference rejection
	Band limiting
	Harmonic suppression
High-pass	Band limiting
	Nose reduction
	Interference elimination
	Broadcast signal conditioning
Bandpass	Banding limiting
	Comb filter
	Interference elimination
Notch	Selective frequency rejection
	Noise reduction
	Interference elimination

The filter used for the system's front end needs to meet several performance criteria. There are many aspects and criteria that define a filter's performance. The first and most important is the filter's design objective. Specifically, if the system design calls for a filter to pass frequencies between 1870 and 1885 MHz, a bandpass filter might be the best general type to utilize. However, the other needed filter attributes will be the determining factors in how well the filter performs in the application. For example, the ability to dissipate power might not be a criteria for a receive filter, but it will be very important for a transmit filter.

The following are some of the criteria that you should define during the design phase for the communication system.

1. Frequency response (passband, cutoff, transition band, stopband edge)
2. Insertion loss
3. Passband ripple
4. Attenuation floor
5. Shape factor
6. Phase error
7. Group delay
8. Selectivity (Q)
9. Temperature stability

2.9 Receivers

The receiver system utilized by a mobile or fixed wireless system is a crucial element of the network. Specifically, the receiver's job is to extract the desired signal from the plethora of other signals and noise that exist in its environment. A basic receiver block diagram is shown in Fig. 2.12.

The receiver system of the radio network includes everything in the receive path starting with the antenna system itself. In the more classical sense, a receiver involves only the portion of the network that is directly involved with down-converting and demodulating the signal to extract the initial information content. However, we include the rest of the components so that the receive system can be treated as a whole system since all these components directly influence the ultimate performance of the receiver.

There are many types of receivers utilized for wireless communication. The type of receiver utilized for the communication system should be selected so the information content desired to be received is done so in the most efficient method. "Most efficient" refers to both financial considerations and spectral efficiencies where the desired information content is delivered to the receiver.

Normally a communication system receiver must deal with a signal spectrum that contains more than just the desired signal. The multitude of signals that a receiver must simultaneously deal with puts a price on device linearity for each of the stages in a receiver's path. The receiver in its operation must select the desired carrier (signal) from a multitude of other signals, amplify the weak desired signal, and then demodulate it.

The top electrical performance and cost drivers for a receiver typically are

1. Frequency range
2. Dynamic range
3. Phase noise
4. Tuning resolution
5. Tuning speed

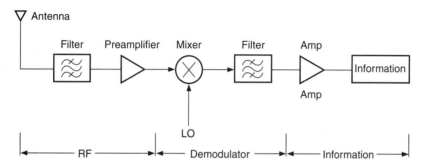

Figure 2.12 Basic radio receiver.

6. Sensitivity
7. Distortion (gain and phase)
8. Noise

The receiver design must incorporate the desired performance criteria and at the same time minimize the number of stages between the RF and intermediate-frequency (IF) portions of the receiver system. The number of stages between the RF and IF portions of the receiver system is dependent upon the modulation scheme selected.

The information content desired to be sent and received has a direct role in the selection of the type of modulation format that will be utilized. This modulation format in turn ultimately determines the type of receiver that will be utilized in the wireless network. There are many aspects to consider when determining the type of modulation format desired in a network. However, one key criterion is the information bandwidth needed as compared to the available spectrum. Specifically if you have an infinite amount of spectrum to utilize, then the type of modulation format chosen is done so purely on the cost constraints imposed on the system operator for operating a profitable business. However, as a matter of course you never have infinite spectrum (bandwidth) to utilize for the wireless system. Therefore, a tradeoff is needed to determine which modulation format will maximize the overall spectral and cost constraints of the system design.

There are three general types of receivers: AM, FM, and PM receivers. Many receivers utilize a combination of these three receiver types based on the wireless system's technology platform.

The *radio receiver*, whether located at the base station or the subscriber terminal itself, is the physical device that converts the RF energy into a usable form. The radio receiver can have one path or multiple receive paths connected to it. Usually there are two paths connected to the radio receiver in a cell site and only one path for a mobile or portable unit.

In Fig. 2.12 the radio receiver receives the RF energy, which is then passed through a filter for additional selectivity and is then amplified for additional gain. The RF energy is then put through a mixer which enables the signal to be down-converted to an intermediate frequency. The intermediate frequency is then filtered and then amplified again. The IF signal is now passed through another mixer and filter which now places the IF signal into the audio or information stage. The audio or information stage is where the initial information content is conveyed to the desired receiver, either a person or an electronic terminal device.

Obviously this example is a simplified version of the events that take place in the receive path. The specific sequence of events that take place as the signal goes through the receive path chain is dependent upon the type of modulation and information content that is trying to be interrupted by the end user or device.

2.10 Radio Performance Criteria

The performance criteria for a radio receiver are the key elements in determining if the radio selected will successfully extract the information content into a usable form depending on the environmental conditions it is placed within. The performance criteria that are covered in this section do not pertain to the physical environmental issues, such as power consumption, heat exchange requirements, or mean time between failure (MTBF). Instead the objective of this section is to provide a reference to the radio performance criteria that a design engineer should utilize in selecting radio for the network. These include sensitivity, selectivity, dynamic range, noise, 1-dB compression, third-order intercept, and desense.

2.10.1 Sensitivity

A receiver's ability to detect a weak signal is determined by its sensitivity. Thus, the sensitivity is a very important figure of merit for a receiver. The receiver sensitivity must be such that it can detect the minimal discernible signal from the background noise. The minimal discernible signal is a measure of sensitivity which incorporates the bandwidth of the system and will differ from one receiver to another based on the bandwidth of the signals received.

Specifically there is a relationship between thermal noise, the receiver's noise figure, and the bandwidth of the signal that the receiver is trying to detect.

$$\text{Sensitivity} = 10 \log_{10} kTB + 10 \log_{10} B + \text{NF}$$
$$= -174 \text{ dBm/Hz} + 10 \log_{10} B + \text{NF}$$

where k = Boltzmann constant = 1.38×10^{-23} joules/kelvin (J/K)
T = temperature, K
B = bandwidth, Hz
NF = noise figure of receiver, decibels (dB)

2.10.2 Selectivity

Receiver selectivity provides a measure of the protection that is afforded the radio from off-channel interference. The degree of selectivity is largely driven by the filtering system within the receiver. The IF portion of the receiver affords the most benefit for selectivity. The greater the selectivity, the better the receiver is able to reject unwanted signals from entering into it. However, if the receiver is too selective, it may not pass all the desired energy.

2.10.3 Dynamic range

The dynamic range is a very important figure of merit and defines the range of signals that the receiver can handle within its specified performance. There are several ways to specify the dynamic range of the system. One way is to define it as the range from the minimal discernible signal to the 1-dB compression point of the receiver; this is often called the blocking dynamic range. Another method of defining dynamic range is to define the range from the minimal discernible signal to where the third-order intermodulation distortion (IMD) equates the minimal discernible signal and is referred to as the spurious free dynamic range (SFDR), or it can be specified as the difference between the minimal discernible signal and a specified IMD level.

Figure 2.13 is a chart that can be used to determine the dynamic range of a radio system. Please note that the signal in the chart has a slope of 1:1, while the third-order intercept has a slope of 1:3. The actual intercept point referenced on the chart is a calculated value only.

The SFDR is a very important specification when the site is near other radio transmitters since it is a direct indication of how the signal interferes with adjacent channels. It provides a measurement of the performance of the radio as the desired signal approaches the noise floor of the receiver, providing an overall receiver SNR or bit error rate (BER). For example, if a radio can accurately digitize signals from -13 to -104 dBm in the presence of multiple signals, the dynamic range is -91 and it implies an SFDR of 95 to 100 dB. It is interesting to note that the SFDR can be improved when the signal level is reduced from the full scale, and this can improve the actual dynamic range even with the reduction in signal amplitude.

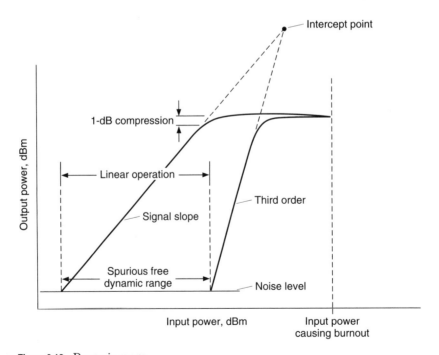

Figure 2.13 Dynamic range.

Dynamic range is usually defined as the range over which an accurate output will be produced. The lower level is called *sensitivity,* and the upper level is called the *degradation level.* The sensitivity is determined by the noise figure, IF bandwidth, and the method of processing. However, the degradation level is determined by whichever of the components in the receiver reaches its own degradation level first. Therefore, it is important to understand all the components in the receiver path since the first component to degrade will define the dynamic range of the system.

A large dynamic range for a receiver is a design priority. The following equation can be used when comparing the dynamic range of one receiver to that of another, assuming the receivers are operating in the same band and have the same bandwidth.

$$\text{Dynamic range} = \text{input IP3} - \text{NF}$$

where IP3 is the third-order intercept point. If the output IP3 is referenced for the receiver, then the input IP3 can be calculated once the gain of the device is known.

$$\text{Input IP3} = \text{output IP3} - \text{gain of device}$$

The dynamic range should always be referenced to the system input. Using this method a high dynamic range usually results in a positive FOM for the receiver. An example of how dynamic range is calculated is shown in Fig. 2.14.

2.10.4 Noise

Noise for a communication system directly affects its overall performance. All receivers need to have a certain carrier-to-interference ratio (C/I) or energy-per-bit to noise-density ratio (Eb/No) value to perform properly. If the overall noise that the receiver experiences or has to deal with increases, the desired signal needs to be increased but without an increase in the noise content, to ensure the proper ratio is maintained.

For the RF design engineer there are several components associated with the receive system that comprise noise. The three items are thermal, shot, and system noise. The latter can be reduced through use of proper frequency planning, power control, and appropriate use of selective filters along with isolation techniques. The thermal and short noise together comprise what is referred to as the *noise figure* for a receiver.

Noise figure. The noise figure is one of the fundamental measures of a receiver's performance and should be measured at a predetermined location for the receiver itself. The noise figure for a receiver degrades (i.e., increases) with each successive stage in the receive path. A common point at which to measure the noise figure is the receiver's audio output, but with digital radios there is no audio output and the measurement point is then the IF output.

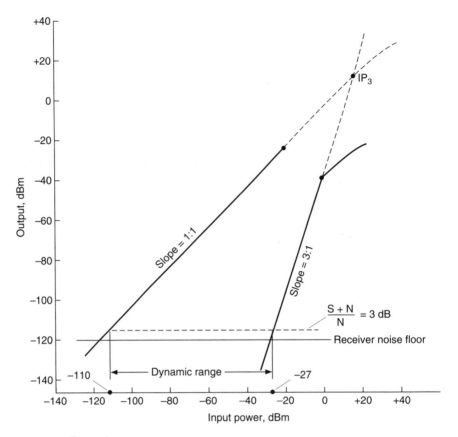

Figure 2.14 Dynamic range.

The noise figure of the receive system is directly related to the overall receiver sensitivity and is calculated as shown here:

$$\text{Noise figure} = \text{NF (dB)} = 10 \log_{10} F_n$$

where $F_n = \text{SNR}_{in}/\text{SNR}_{out}$. The noise figure for the system is normally set by the first amplifier in the receive path. The noise floor can be improved or attenuated by passive devices, but the SNR will not be improved unless the bandwidth is narrowed.

Thermal and shot noise. As previously stated, the two components of the noise figure are thermal and shot noise. Thermal noise is a direct result of kinetic energy and is directly related to the temperature. Shot noise on the other hand is caused by the quantized and random nature of current flowing in a device.

The noise temperature has a direct relationship to operating frequency and the bandwidth of the receiver or signal desired to be detected. The relationship is shown in the following equation:

Noise power per hertz = $10 \log_{10} kTB = -173.97$ dB/Hz   (25°C or 290 K)

What is of prime interest to the RF engineer is the effect each component in the receive system has on improving the overall sensitivity of the system. The first amplifier in the system sets what the noise figure for the system will be. The equation to determine the noise figure is shown here:

$$F = F_1 + \frac{F_2 - 1}{G_1} + \frac{F_1 - 1}{G_1 \times G_2}$$

where F_1 = noise figure (power) of first component in receive path
F_2 = noise figure (power) of second component in receive path
G_1 = gain/loss (power) of first component in receive path
G_2 = gain/loss (power) of second component in receive path

and inserting the value F into the next equation yields the noise figure for the receiver: NF (dB) = $10 \log F$.

However, using this example a quick relationship can be determined between the noise figure and receiver sensitivity.

Sensitivity (optimal, dB) = -174 dB + $10 \log_{10} B$ + NF

These calculations, however, do not factor in the effects of the antenna noise, feedline loss, and required SNR or Eb/No as part of the receiver sensitivity.

2.10.5 1-dB compression

The 1-dB compression point is a common reference term used to define the performance of a particular receiver, or rather the amplifier in the receiver itself. However, the 1-dB compression point can also reference an individual amplifier, like the preamp.

The 1-dB compression point is where the power gain for the receiver is down 1 dB from the ideal gain. That is, if the input signal goes up by 2 dB and the output goes up by only 1 dB, then this is the point where the 1-dB compression point occurs. Often the 1-dB compression point is referred to as the point of *blocking*. The blocking occurs in that the weaker signals are not amplified properly leading to them being potentially blocked from being detected. The 1-dB compression point is part of the component for determining the receiver's overall dynamic range.

The 1-dB compression point is shown in Fig. 2.15. The 1-dB compression can occur either as a result of trying to receive the desired signal, which is too hot, causing the overload condition, or through undesired signals overloading the receiver. In Fig. 2.15 the 1-dB compression point can be directly affected by the receiver's overall gain setting.

2.10.6 Third-order intercept

The third-order intercept point (IP3) is a figure of merit for a receiver. Specifically, the IP3 value directly determines the receiver's dynamic range. The

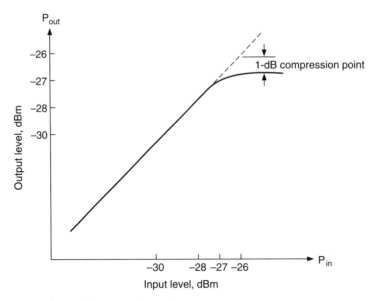

Figure 2.15 1-dB compression point.

third-order intercept value along with the second-intercept value, determine the receiver's linearity.

It should be noted that the IP3 value is a theoretical value which is achieved by extrapolation from the third-order curve. In addition, the IP3 value is frequently dependent, so based on the selection of the test frequencies a different IP3 value may be achieved. However, most IP3 tests are set up to produce a product that falls within the first IF of the receiver. Basically the IP3 value increases by 3 dB for every 1 dB increase in the desired signal. The slope of the IP3 line is therefore 3:1 and is shown in Fig. 2.13. The IP3 value is important for determining the receiver's performance since the presence of two larger signals can generate spurs caused by distortions that can override the weaker desired signals.

2.10.7 Desense

Any form of RF energy, whether it is artificially created or natural in origin, has the potential of adversely affecting the receiver. Any RF energy that adversely affects the receiver's ability is defined as radio frequency interference (RFI). *Desense* is technically a reduction in the receiver's overall sensitivity which is caused by artificial or natural RFI. Desense occurs when a very strong signal begins to overload the front end of the receiver and makes the detection of weaker signals more difficult.

All types of noise interference adversely affect the cell site's receiver performance, whether the desired signal is purely FM analog or TDMA. Noise interference can be attributed to out-of-band emissions from nearby transmitters whose sidebands contribute to the effective noise floor of the cell site, or it can come from in-band sources, cochannel, or IMD products.

The magnitude of the noise interference is determined by the amount of energy within the receiver passband. Depending on the amount of noise interference, the effective receiver sensitivity can be degraded during increases in noise interference level because of a reduced SNR or BER.

When colocated with many transmitters, the transmitter noise spectrum is noticeably close to the carrier at 60 dBc (c = carrier) and gradually falls off in power as a function of effective bandwidth. However, at some power level the noise level will degrade receive sensitivity. The noise interference power level caused by other transmitters can be reduced by increasing the physical separation between the offending transmitters and the offended receivers. Another alternative is to place a more narrow transmit filter on the transmitter or utilize a more selective receive filter to remove, or seriously reduce, the transmitter noise spectrum.

Whenever transmitter noise is present, there exists the potential for receiver desensitization. The cell site's receiver can be desensitized by undesired signals causing a reduction in the gain or an increase in noise level of certain amplifiers or mixers in the receive path of the cell site. It is possible, based on the receiver filter used, to pass out-of-band energy into the receive system and cause a reduction in receiver gain. The loss of gain in this manner is a form of desensitization.

Therefore, if a cell site is desensed due to out-of-band emissions, the installation of a more selective filter will make the site more sensitive. However, if the site is not being desensed due to out-of-band emissions, the filters introduction could show up as a reduction in effective noise floor for the cell site.

The basic function of any RF preamplifier is to increase the signal-to-noise ratio of the received signal. Since the RF preamp receives the desired signal at the lowest level of any receive stage in the cell site's receive path, any noise or other disturbances introduced in this stage have a proportionally greater effect.

The performance of the cell site's receiver with respect to weak signals depends on the performance of the preamp or rather the signal-to-noise ratio of its output. The key issue is that amplifiers do not discriminate between the signal and noise within their passband. In fact, the preamp will amplify the desired signal and any noise equally. Any amplifier has a power budget, regardless of its gain and robustness. The preamp for the cell site will amplify the out-of-band emissions that made it past the cell site's filters. Depending on the amount of out-of-band emissions allowed to pass into the receive path for the cell site, a decrease in overall receiver gain can be experienced.

The best method to overcome desense problems is to ensure that there is sufficient isolation between the potential offending transmitter and the receiver. Desense is and will continue to be more of a problem, especially in urban environments, as the use of wireless communications continues to expand.

2.11 Propagation Model

The radio wavelength for cellular and PCS is rather small in size, and as a result has some unique propagation characteristics which have been modeled over the years by numerous technical people. Some of the more popular propagation models used are Hata, Carey, Longley-Rice, Bullington, and Cost231. Each of these models has advantages and disadvantages. Specifically there are some baseline assumptions used with any propagation model which need to be understood prior to utilizing them. Most cellular operators use a version of the Hata model for conducting propagation characterization. The Carey model, however, is used for submitting information to the FCC with regards to cell-site filing information.

The model used for predicting coverage needs to factor in a large amount of variables which directly impact the actual RF coverage prediction of the site. The positive attributes affecting coverage are the receiver sensitivity, transmit power, antenna gain, and the antenna height above average terrain. The negative factors affecting coverage involve line loss, terrain loss, tree loss, building loss, electrical noise, natural noise, antenna pattern distortion, and antenna inefficiency, to mention a few.

With the proliferation of cell sites the need to theoretically predict the actual path loss experienced in the communication link is becoming more and more critical. To date, no overall theoretical model has been established that explains all the variations encountered in the real world. However, as cellular communication systems continue to grow, a growing reliance is placed on the propagation prediction tools. The reliance on the propagation tool is intertwined in the daily operation of the cellular communication system. The propagation model employed by the cellular operator is directly involved with the capital build program of the company for determining the budgetary requirements for the next few fiscal years.

Therefore, it is essential that the model utilized for the propagation prediction tool be understood in terms of what it can actually predict and what it cannot predict. Over the years there have been numerous articles written with respect to propagation modeling in the cellular communications environment. With the introduction of PCS there has been an increased focus on refining the propagation models to assist in planning out the networks.

Presently most of the propagation tools available utilize a variation of the Hata model or the Cost231 model. However, there are several variants to both of these models. In addition the individual propagation modeling software companies have crafted their own variants to the models. By understanding the advantages and disadvantages of each of the models, an engineer can design a better network. The specific models discussed here are free space, Hata, and Cost231.

2.11.1 Free space

Free-space path loss is usually the reference point for all the path loss models employed. Each propagation model points out that it more accurately predicts the attenuation experienced by the signal over that of the free-space model. The equation that is used for determining free-space path loss is based on $1/R^2$ or 20 dB per decade path loss:

$$L_f = 32.4 + 20 \log_{10} R + 20 \log_{10} f_c \qquad (2.4)$$

where R = distance from cell site, kilometers (km)
f_c = transmit frequency, MHz
L_f = free-space path loss, dB

The free-space path loss equation has a constant value that is used for the air interface loss (32.4). Using a frequency of 880 MHz and some basic distance values, different path loss values can be determined for comparison with the other models to be discussed (see Table 2.4). Looking at Table 2.4, it would be very nice if the frequency band utilized by cellular operators behaved with this path loss. However, radio frequency propagation whether it is cellular, PCS, or ESMR does not behave like free-space loss and, therefore, requires another equation to be utilized.

2.11.2 Hata

One of the most prolific path loss models employed in cellular, ESMR, and PCS is the empirical model developed by Hata or some variant of it. The Hata model is an empirical model derived from the technical report made by Okumura so the results could be used in a computational model. The Okumura report is a series of charts that are instrumental in radio communication modeling. The Hata model is given here.

$$L_H = 69.55 + 26.16 \log_{10} f_c - 13.82 \log_{10} h_b - a(h_m)$$
$$+ (44.9 - 6.55 \log_{10} h_b) \log_{10} R \qquad (2.5)$$

where L_H = path loss for Hata model, dB
h_b = base station antenna height, m
h_m = mobile or portable antenna height, m

TABLE 2.4 Free Space Path Loss

Distance, km	Path loss, dB
1.0	91.29
2.0	97.31
3.0	100.83
4.0	103.33
5.0	105.27

Utilizing the same values as for the free-space calculation, a similar table (2.6) is derived. It should be noted that there are some additional conditions that are applied when using the Hata model. The values utilized are dependent upon the range over which the equation is valid. If the equation is used with parameters outside the values, the equation defined for the results will be suspect to error. The ranges over which the Hata model is valid are listed here.

$$f_c = 150 \text{ to } 1500 \text{ MHz}$$
$$h_b = 30 \text{ to } 200 \text{ m}$$
$$h_m = 1 \text{ to } 10 \text{ m}$$
$$R = 1 \text{ to } 20 \text{ km}$$

Therefore the Hata model should not be employed when trying to predict path losses less than 1 km from the cell site or if the site is less than 30 m in height. This is an interesting point to note since at times cellular sites are being placed less than 1 km apart and are often below the 30-m height.

In the Hata model the value $a(h_m)$ is used to correct for the mobile antenna height. Interestingly, if you assume a height of 1.5 m for the mobile or portable, that value nulls out of the equation.

A critical point to mention here is that the Hata model employs three correction factors based on the environmental conditions that path loss prediction is evaluated over. The three environmental conditions are urban, suburban, and open.

The environmental correction values are easily calculated but vary for different values of mobile height. For the values shown in Table 2.5 the following conditions are assumed:

$$f_c = 880 \text{ MHz} \qquad h_b = 30 \text{ m} \qquad h_m = 1.5 \text{ m}$$

Table 2.6 shows a comparison of the free-space and Hata models. An open environment has been assumed to best compare the two equations.

The differences between the path loss estimates are profound.

2.11.3 Cost231 Walfish/Ikegami

The Cost231 Walfish/Ikegami propagation model is used for estimating the path loss in an urban environment for cellular, ESMR, and PCS communication. The

TABLE 2.5 Hata Environmental Correction Factors

Environment	Correction factor, dB
Urban	0
Suburban	−9.88
Open	−28.41

TABLE 2.6 Comparison of Free-Space and Hata Models

Distance, km	Path loss, dB		
	Free space	Hata—rural	Hata—urban
1	91.29	97.75	126.16
2	97.31	108.36	136.77
3	100.83	114.56	142.97
4	103.33	118.96	147.37
5	105.27	122.37	150.79

Cost231 model is a combination of empirical and deterministic modeling for estimating the path loss in an urban environment over the frequency range of 800 to 2000 MHz. The Cost231 model was initially used primarily in Europe for GSM/GPRS (general packet radio system) modeling but is now incorporated into numerous propagation models used for wireless mobility in the United States, South America, and Asia.

The Cost231 model is composed of three basis components:

1. Free-space loss (L_f), dB
2. Roof to street diffraction loss and scatter loss (L_{RTS}), dB
3. Multiscreen loss (L_{ms}), dB

$$L_c = L_f + L_{RTS} + L_{ms} \tag{2.6}$$

where $L_{RTS} + L_{ms} \leq 0$

$$L_f = 32.4 + 20 \log_{10} R + 20 \log_{10} f_c \tag{2.7}$$

$$L_{RTS} = -16.9 - 10 \log_{10} \omega + 10 \log_{10} f_c + 20 \log_{10} \Delta h_m + L_o \tag{2.8}$$

where ω = street width in meters and $\Delta h_m = h_r - h_m$. The street orientation correction factor L_o has the following values depending on the incident angle relative to the street, θ (see Fig. 2.16).

$$L_o = \begin{cases} -10 + 0.354 & 0 \leq \theta \leq 35 \\ 2.5 + 0.075 (\theta - 35) & 35 \leq \theta \leq 55 \\ 4.0 - 0.114 (\theta - 55) & 55 \leq \theta \leq 90 \end{cases}$$

$$L_{ms} = L_{bsh} + k_a + k_d \log_{10} R + k_f \log_{10} f_c - 9 \log_{10} b \tag{2.9}$$

where b = distance between buildings along radio path, m
      L_{bsh} = path loss correction for base station antenna height relative to roof, m
      k_a = path loss correction for increased path loss for lowered antenna heights

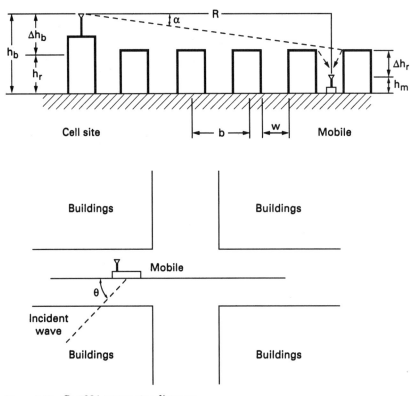

Figure 2.16 Cost231 parameter diagram.

k_d = path loss correction factor for multiscreen diffraction loss versus distance

k_f = path loss correction factor for multiscreen diffraction loss versus frequency

$$L_{\text{bsh}} = \begin{cases} -18 \log_{10}(1 + \Delta h_b) & h_b > h_r \\ 0 & h_b < h_r \end{cases} \quad (2.10)$$

$$k_a = \begin{cases} 54 & h_b > h_r \\ 54 - 0.8 h_b & d \geq 500 \text{ m}, h_b \leq h_r \\ 54 - 1.6 (\Delta h_b) R & d < 500 \text{ m}, h_b \leq h_r \end{cases} \quad (2.11)$$

Note that both L_{bsh} and k_a increase the path loss with lower base station antenna heights.

$$k_d = \begin{cases} 18 & h_b > h_r \\ 18 - 15 \dfrac{\Delta h_b}{\Delta h_r} & h_b \leq h_r \end{cases} \quad (2.12)$$

$$k_f = \begin{cases} 4 + 0.7\left(\dfrac{f_c}{925} - 1\right) & \text{midsized city and suburban area with moderate tree density} \\ 4 + 1.5\left(\dfrac{f_c}{925} - 1\right) & \text{metropolitan center} \end{cases} \quad (2.13)$$

In Eqs. (2.6) to (2.13), which comprise the Cost231 model, the following values bound the equation's useful range. It is important, as always, to know these valid ranges for the model.

$$f_c = 150 \text{ to } 1500 \text{ MHz}$$
$$h_b = 4 \text{ to } 50 \text{ m}$$
$$h_m = 1 \text{ to } 3 \text{ m}$$
$$R = 0.02 \text{ to } 5 \text{ km}$$

These parameter ranges show that when the range of the site is less than 1 km the Cost231 model would be a better choice than the Hata model.

There are some additional default values which apply to the Cost231 model when specific values are not known. The default values can and will significantly alter the path loss values arrived at. The recommended default values are listed here.

$$b = \text{distance between buildings} = 2 \text{ to } 50 \text{ m}$$
$$\omega = \text{width of street} = \dfrac{b}{2} \text{ m}$$
$$h_r = \text{height of roof} = 3 \text{ (number of floors)} + \text{roof, where}$$
$$\text{roof} = 3 \text{ m for pitched and } 0 \text{ m for flat}$$
$$\theta = \text{incident angle relative to the street} = 90°$$

Assuming the following values for the Cost231 and the same values used for the previous equations for the free-space and the Hata path loss models, another comparison table is presented (see Table 2.7) using these values:

$$f_c = 880 \text{ MHz} \quad h_b = 30 \text{ m} \quad h_m = 1.5 \text{ m}$$
$$\text{roof} = 0 \quad h_r = 30 \text{ m} \quad b = 30 \text{ m}$$
$$\omega = 15 \text{ m} \quad \theta = 90°$$

Obviously the assumptions made for entering values into the equations play a major role in defining the outcome of the path loss value.

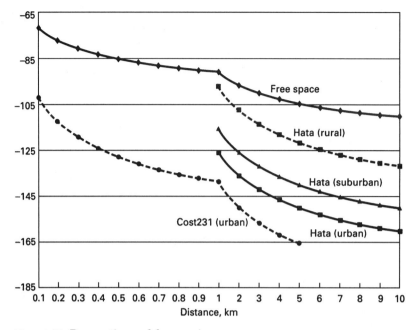

Figure 2.17 Propagation model comparison.

Figure 2.17 illustrates plots of the propagation models referenced here for several environments. The plots show the similarities and differences between models utilizing the same, or nearly the same, parameters. The assumptions made for the models employed in the graph utilize the baseline conditions listed in Table 2.7.

The propagation model or models employed by your organization must be chosen with extreme care, and a continuous vigil will be needed to ensure they are truly being a benefit to the company as a whole. The propagation model employed by the engineering department not only determines the capital build program but also plays a direct factor into the performance of the network.

The propagation model is normally used as part of the 1-, 2-, and 5-year growth studies performed by a cellular operator. The propagation model is used to determine how many sites are needed to provide a particular coverage requirement for the network. In addition, the coverage requirement is coupled into the traffic loading requirements. The traffic loading requirements rely on the propagation model chosen to determine the traffic distribution (off-loading) from exiting sites to new sites as part of the capacity relief program. The propagation model helps determine where the sites should be placed to achieve an optimal position in the network. If the propagation model used is not effective in helping place sites correctly, the probability of incorrectly justifying and deploying a site into the network is high.

The performance of the network is also affected by the propagation model chosen since it is used for interference prediction plots. If the propagation model is accurate by say 6 dB, then you could be designing for a 23 dB or 11 dB C/I. Based on your traffic loading conditions, designing for a very high C/I level could neg-

TABLE 2.7 Comparison of Free-Space, Hata, and Cost231 Models

Distance, km	Path loss, dB		
	Free space	Hata—urban	Cost231
1	91.29	126.16	139.45
2	97.31	136.77	150.89
3	100.83	142.97	157.58
4	103.33	147.37	162.33
5	105.27	150.79	166.01

atively impact your financials. On the other hand designing for a low C/I would have the obvious impact of degrading the quality of service to the very people who pay your salary. The propagation model is also used in a multitude of other system performance aspects that include handoff tailoring, power level adjustments, and antenna placements. To reiterate, no model can account for all the perturbations experienced in the real world; it is essential that you utilize one of several propagation models for determining the path loss of your network.

2.11.4 Environmental attenuation

We will now focus back on one of the more popular propagation models, the Hata model. The Hata model uses an alpha of about 3.5. There have been many debates regarding the particular propagation tool and the dB/decade slope used in determining path loss. However, it is the other variables associated with the propagation characteristics that have the largest impact on how the model truly predicts the propagation characteristics of the potential cell site.

The difference between an alpha of 3.5 and 3.8 is only 3 dB per decade. The 3 dB can easily be aborted in environmental attenuation issues. The environmental issues associated with propagation characteristics require the highest amount of focus. Table 2.8 is a listing of some generalized attenuation characteristics which are used in propagation models. Glancing at the table the difference between the urban and suburban environments is more than 3 dB. Obviously each area has its own unique propagation characteristic, and generalizing the characteristics of an area is a best-guess fit. Table 2.8 is meant to be used as a general guide for determining environmental effects on path loss. However, looking at the table it is obvious that the 3 dB per decade is not the leading indicator for modeling propagation.

2.12 Diffraction

Diffraction of the RF signal also has a very important role in predicting and attenuating the signal. How to calculate the actual attenuation from diffraction is shown in an excellent paper written by Bullington.

There are several types of diffraction methods modeled in RF: smooth and knife edge. Each diffraction method yields a different value. If you choose wisely for the terrain issues at hand, the calculation method presented will accurately

TABLE 2.8 Attenuation Table

Loss category and types	Attenuation, dB
Foliage	
Sparse	6
Light	10
Medium	15
Dense	20
Very dense	25
Building	
Water/open	0
Rural	5
Suburban	8
Urban	22
Dense urban	27
Vehicle	10–14

predict the attenuation experienced. The differences in attenuation involving antenna heights and a knife edge diffraction point are shown in Fig. 2.18.

One of the best uses for knowing how to calculate diffraction is to determine the signal loss expected in a major valley. Another use is to calculate the signal loss expected when a mountain is used for containing the signal through placement of the antenna. Still another valuable application for determining diffraction is when you reduce the tilt angle of a cell site and estimate the positive and negative impacts of improving the signal over a ridge or in a valley.

2.13 Effective Radiated Power (ERP)

The actual ERP for the radio site or transmitter used for the communication site will determine the transmit radius. The ERP setting should be balanced with the receive path to ensure that there is not a disparity between the talk-out, transmit and the talk-back, receive paths for a cell site.

The ERP for the site is set with reference to a dipole antenna. The method for calculating the ERP for a site is presented here with sample data.

$$\begin{aligned}
\text{Transmitter output} &= 44 \text{ dBM } [25 \text{ watts (W)}] \\
\text{Combining losses} &= -1 \text{ dB} \\
\text{Feedline loss} &= -3 \text{ dB} \\
\text{Antenna gain} &= \underline{10 \text{ dBd}} = \underline{12.14 \text{ dBi}} \\
& 50 \text{ dBm} = 52.14 \text{ dBm}
\end{aligned}$$

100 W ERP = 200 W EIRP (or, more exactly, 163.68 W EIRP)

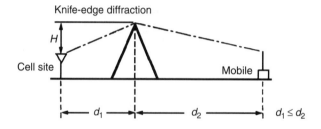

H	d_1 Feet	d_1 Meters	dB loss
50 ft (15 m)	50	15	23
	100	30	21
	200	60	17
	500	150	13
100 ft (30 m)	50	15	38
	100	30	37
	200	60	35
	500	150	31
200 ft (60 m)	50	15	47
	100	30	44
	200	60	40
	500	150	36

Figure 2.18 Diffraction loss (880 MHz).

where EIRP = effective isotropic radiated power. If you are using dBi, isotropic gain, or want to convert dBi to dBd, all that is involved is a simple conversion:

$$1 \text{ dBi} = 2.14 \text{ dB} + 1 \text{ dBd} \quad \text{or} \quad 1 \text{ dBd} = 1 \text{ dBi} - 2.14 \text{ dB}$$

The ERP for the site can have a dramatic impact on the cell-site radius. An increase in ERP could reduce the build program requirements. For example, altering the ERP by just 3 dB can alter the geographic area the cell site serves, assuming flat earth. In reality most major obstructions increasing the ERP have little if not real effect on the cell-site radius depending upon the actual terrain conditions. However, referring back to the propagation model which has a slope of 35 dB/decade, this value is then used to determine the effective cell radius (ECR). Example 2.1 evaluates a 3-dB increase and decrease in ERP and its effect on the ECR.

Example 2.1

$$\text{Area} = \pi R^2$$

3-db increase in ERP:

$$+3 \text{ dB} = 35 \log (R_{\text{new}})$$

$$\therefore R_{\text{new}} = 1.218 \quad R_{\text{old}} = 1.0$$

$$\frac{\text{Area (new)}}{\text{Area (old)}} = \frac{\pi R^2_{\text{new}}}{\pi R^2_{\text{old}}} = \frac{R^2_{\text{new}}}{R^2_{\text{old}}} = R^2_{\text{new}} = 1.484$$

$$\text{Area (new + 3 dB)} = 1.484\, \text{Area}_{\text{old}}$$

3-dB decrease in ERP:

$$-3 \text{ dB} = 35 \log(R_{\text{new}})$$

$$\therefore R_{\text{new}} = 0.820 \qquad R_{\text{old}} = 1.0$$

$$\frac{\text{Area (new)}}{\text{Area (old)}} = \frac{\pi R^2_{\text{new}}}{\pi R^2_{\text{old}}} = \frac{R^2_{\text{new}}}{R^2_{\text{old}}} = R^2_{\text{new}} = 0.674$$

$$\text{Area (new} - 3 \text{ dB)} = 0.674\, \text{Area}_{\text{old}}$$

The actual effect the ERP has on the build program is best shown in Example 2.2.

Example 2.2

$$\text{Total geographic area to cover} = 50 \text{ km}^2$$

$$\text{Area (old)} = \pi R^2_{\text{old}} = 3.14 \text{ km}^2 \qquad R_{\text{old}} = 1 \text{ km}$$

$$\text{Area } (+3 \text{ dB}) = \pi R^2_{\text{new}} = 4.66 \text{ km}^2 \qquad R_{\text{new} + 3 \text{ dB}} = 1.484 \text{ km}$$

$$\text{Area } (-3 \text{ dB}) = \pi R^2_{\text{new}} = 2.16 \text{ km}^2 \qquad R_{\text{new} - 3 \text{ dB}} = 0.674 \text{ km}$$

Approximate number of cells needed to cover 50 km²:

$$\text{Number of cell sites} = \frac{\text{geographical area}}{\text{cell site area}}$$

$$\text{Number of cell sites (old)} = \frac{50 \text{ km}^2}{3.14 \text{ km}^2} \approx 16$$

$$\text{Number of cell sites (new} + 3 \text{ dB)} = \frac{50 \text{ km}^2}{4.66 \text{ km}^2} \approx 11$$

$$\text{Number of cell sites (new} - 3 \text{ dB)} = \frac{50 \text{ km}^2}{2.16 \text{ km}^2} \approx 23$$

Example 2.2 points out that for a mere 3-dB increase in ERP across all the cells in the network there will be a 31 percent decrease in the total amount of cell sites required. Conversely a 3-dB reduction in ERP has the effect of increasing the cell-site build program by 43 percent. Therefore, it is clear that increasing or decreasing the ERP by itself should not be done without a careful analysis of the link budget for the communication system.

2.14 Link Budget

The link budget is a power budget which is used to determine from a modeling aspect what expected losses and gains will be experienced by the RF signal as it traverses from the transmitter of the cell to the mobile, or vice versa. The objective of defining the link budget is to arrive at the size of the cell sites needed for the network design. The link budget will determine if the system is limited in the uplink or downlink direction. Based on the link budget the distance to radius (D/R) ratio chosen for the system will determine the radius of the site and also the distance between the cells themselves. Also the link budget can be used to determine if the system can support in-car or in-building portables. The link budget has a direct impact on the number of cell sites required for a network and will determine the coverage requirements. For example, if the link budget indicates that a -85-dBm value is needed on the street for in-building coverage, this will require fewer cell sites to achieve the same objective as compared to if the link budget indicated that a -75-dBm value was needed instead.

There are many factors and parameters which go into establishing a link budget. The link budget needs to be calculated for both the cell site to subscriber (downlink) and subscriber to cell site (uplink) paths. The limiting path loss is then used to determine the range for the site using the propagation model for the network. The system setting for a wireless network will typically be established to facilitate a balanced path.

A balanced path, or system, occurs when the downlink and the uplink path losses for the system are the same. The balanced system is used as the design objective when determining the number of needed cell sites and the area they are expected to cover. However, depending on the technology used, a conscious decision can be made to make the downlink path stronger than the uplink path due to rescan issues associated with the phone.

$$\text{Downlink path loss (dB)} = \text{uplink path loss (dB)}$$

If the wireless system had a downlink path loss (dB) greater than the uplink path loss (dB), it is referred to as an uplink-limited system. Conversely if the wireless system has an uplink path loss (dB) greater than the downlink path loss, then it is referred to as a downlink-limited system.

Examples 2.3 and 2.4 are illustrations of very simplistic link budgets which account for the downlink and uplink paths. In these examples the receiver sensitivity value has the thermal noise, bandwidth, and noise figures factored into the final value presented. Also it should be noted that the ERP of the site and subscriber has antenna feedlines and filters and other insertion loss values included.

Example 2.3

Downlink Path		Uplink Path	
Cell-site ERP =	50 dBm	Subscriber maximum ERP =	28 dBm
Fade margin =	5 dB	Fade margin =	5 dB
Subscriber sensitivity =	−102 dBm	Cell-site receive antenna =	12 dBd
		Cell-site cable loss =	2 dB
		Cell-site receiver sensitivity =	−105 dBm
Maximum downlink path loss =	147 dB	Maximum uplink path loss =	138 dB

The maximum downlink path loss is greater than the maximum uplink path loss; therefore, the system is uplink-limited. One solution would be to operate the cell site at a lower power (−9 dB to 12.5 W). This, of course, is independent of other system performance issues.

Example 2.4 A downlink-limited system is one where the subscriber's coverage is greater than the cell site's.

Downlink Path		Uplink Path	
Cell-site ERP =	+10 dBm	Subscriber maximum ERP =	+28 dBm
Fade margin =	5 dB	Fade margin =	5 dB
Subscriber sensitivity =	−102 dBm	Cell-site receive antenna =	12 dBd
		Cell-site cable loss =	2 dB
		Cell-site receiver sensitivity =	−105 dBM
Maximum downlink path loss =	105 dB	Maximum uplink path loss =	138 dB

The maximum uplink path loss is greater than the maximum downlink path loss; therefore, the system is downlink-limited. One solution would be to operate the cell site at a higher power (increasing it by 33 dB to 20 W). This, of course, is independent of other system performance issues.

The maximum path loss, or limiting path, for any communication system used determines the effective range of the system.

2.15 Path Clearance

Radio path clearance is an essential criterion for any point-to-point communication system and in fixed wireless solutions. There are many different types of point-to-point communication systems that can be utilized in wireless mobility. Some of the technology platforms that require a path clearance analysis involve point-to-point microwave and reradiator systems.

The path clearance analysis needs to be performed for every RF point-to-point communication link in the network. The first step in determining if the point-to-point path is appropriate for the communication link is to determine the Fresnel zone. The Fresnel zone is shown in Fig. 2.19. There is effectively an infinite number of Fresnel zones for any communication link. The Fresnel zone is a function of the frequency of operation for the communication link. The primary energy of the propagation wave is contained within the first Fresnel zone. The Fresnel zone is important for the path clearance analysis since it determines the effect of the wave bending on the path above the earth and the reflections caused by the earth's surface itself. The odd-numbered Fresnel zones will reinforce the direct wave, while the even Fresnel zones will cancel it.

In a point-to-point communication system it is desirable to have at least a 0.6 first Fresnel zone clearance to achieve path attenuation approaching free-space loss between the two antennas. This clearance criterion applies to all sides of the radio beam, not just the top and bottom portions represented by the drawing in Fig. 2.20.

Environmental effects on the propagation path have a direct influence on the point-to-point communication system. Some environmental effects altering the communication system are foliage, atmospheric moisture, terrain, and antenna height of the transmitter and receiver.

The K factor utilized for point-to-point radio communication is 1.333 or four-thirds of the earth's radius. The K factor ties in the relationship between the earth's curvature and the atmospheric conditions that can bend electromagnetic waves.

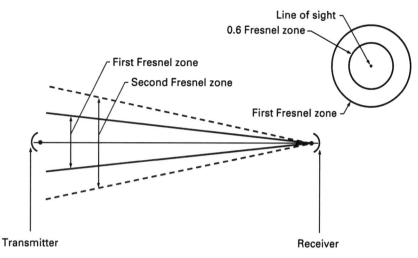

Figure 2.19 Fresnel zone.

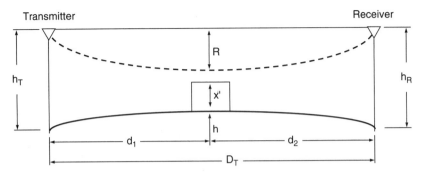

Figure 2.20 Path clearance.

A quick example of how the path clearance required for a point-to-point communication site is determined is shown in Example 2.5.

Example 2.5 To determine the path clearance required for a point-to-point communication site refer to Fig. 2.20.

First Fresnel zone:

$$R = 72 \sqrt{\frac{d_1 d_2}{D_T f}}$$

Earth curvature:

$$h = \frac{d_1 d_2}{1.5\, K}$$

with f = frequency in GHz; d_1, d_2, D_T in miles; x, h, h_T, h_R in feet; and where $d_1 = 1.6$ mi, $d_2 = 2.1$ mi, $D_T = 3.7$ mi, and $f = 0.88$ GHz (or 880 MHz).

$$R = 72 \sqrt{\frac{(1.6)(2.1)}{(8.7)(.88)}} = \sqrt{\frac{3.36}{3.256}} = 73.14 \text{ ft} \qquad R' = (0.6)\,R = 43.884$$

$$h = \frac{(1.6)(2.1)}{(1.5)(4/3)} = 1.68 \text{ ft} \qquad K = {}^4\!/_3$$

Assume that the transmitter, receiver, and obstruction have the same ASML.

Earth curvature	1.68
0.6 Fresnel zone	43.88
Obstruction height	100
	145.56 ft

The minimum T_x and R_x heights for the system are $h_T = h_R = 145.56$ ft.

2.16 Tower-Top Amplifiers

Tower-top amplifiers (TTA) have many applications in wireless communications. Until recently the TTA was a device utilized in two-way communication but not in cellular or PCS. However, based on the situation it is possible to exploit the advantages of a TTA for use in cellular and PCS.

The primary purpose of the TTA is to improve the receiver's sensitivity through elimination of the feedline loss component. The TTA effectively establishes the noise figure and ultimate receiver sensitivity for the communication site. Care, however, must be exercised to ensure that the installation of the TTA does not cause more problems than it was meant to solve. Specifically the TTA net gain must be such that it does not overdrive the front end of the cell site thereby reducing the real dynamic range of the site and also its sensitivity.

In addition the filtering used for the TTA needs to be studied since too wide a front end for the TTA can in fact cause it to go into compression when colocated with other transmitters, even the site's own transmitter. The location of the TTA is shown in Fig. 2.21.

The TTA net gain should be set so that it effectively removes the line loss component for the system and nothing more. There have been false reports issued that have indicated that an overall increase in gain from the TTA nets a similar gain in the cell's receive path. The latter is not true and usually comes about as a result of an incorrect analysis being performed.

Example 2.6 shows how to set the net gain for a TTA situation. This simple application is for when the site's coaxial cable run is excessive and it is desired to improve the site's performance by installation of a TTA.

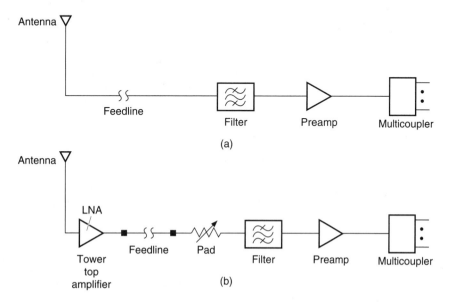

Figure 2.21 Tower-top amplifier (TTA): (a) no TTA; (b) with TTA.

64　Chapter Two

Example 2.6　The design parameters are

$$\text{Feedline for site} = 250 \text{ feet (ft)}$$

$$\text{Attenuation} = \text{⅞ in at 2000 MHz} = \frac{1.97 \text{ dB}}{100 \text{ ft}}$$

$$\text{Feedline loss} = \frac{250}{100} \times 1.97 = 4.925 \text{ dB} = 5 \text{ dB (reality)}$$

$$\text{TTA gain for linear operation} = 15 \text{ dB}$$

We subtract the gain from the loss.

$$\text{TTA gain} - \text{feedline loss} = 15 \text{ dB} - 5 \text{ dB} = 10 \text{ dB}$$

Therefore, a 10-dB pad is required.

Now assuming that you can only use ½-in coax cable which has a 3.45 dB/100 ft attenuation,

$$\text{Feedline loss} = \frac{250}{100} \times 3.45 = 8.625 \text{ (approximately 9 dB)}$$

We subtract the gain from the loss.

$$\text{TTA gain} - \text{feedline loss} = 15 \text{ dB} - 9 \text{ dB} = 6 \text{ dB}$$

Therefore, a 6-dB pad is required (could also use a 7-dB pad).

It is interesting to note that with the use of a TTA it is possible to relax the installation requirements for the site either in terms of overall cable length or changing the cable size itself. In Example 2.6 a smaller cable size was picked to reduce the cost and potential tower loading at the site. However, the relaxation of the installation requirements comes at the expense of increased operational costs and additional points for system failure.

2.17　Intelligent Antennas

Intelligent antenna systems are being introduced into commercial wireless communication systems. The concepts and implementation for intelligent antenna systems have been utilized in other industries for some time, primarily the military. This brief discussion regarding intelligent antennas is separate from the antenna section previously covered in this chapter, primarily to avoid commingling concepts.

Intelligent antenna systems can be configured for either receive-only or full-duplex operations. The configuration of the intelligent antenna systems affords them to be arranged in either an omni or a sector cell site depending on the application at hand.

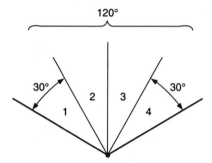

Figure 2.22 Improvement expected = $+10 \log_{10}(N)$.

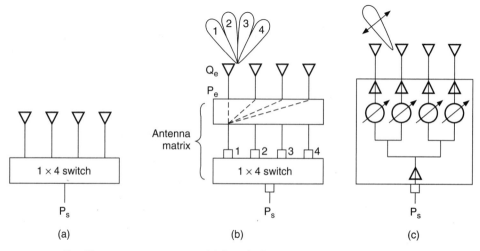

Figure 2.23 Intelligent antenna systems. (*a*) Switched antennas; $P_e = P_s$. (*b*) Multiple-beam array; $P_e = P_s/4$. (*c*) Steered-beam array, $P_e = P_s/4$.

The fundamental objective is to increase the SNR by reducing the amount of noise and interference and possibly increasing the serving signal in the same process. All the technologies referenced are based on the principle that a narrower radiation beam pattern will provide increased gain and can be directed toward the subscriber and at the same time will offer less gain to interfering signals that will arrive at an off-axis angle due to the reduced beam width.

A simple rule of thumb for determining the amount of improved SNR that can be achieved with an intelligent antenna system is shown in Fig. 2.22. The expected improvement = $10 \log_{10} 4 = 6$ dB, assuming the sector is divided into four even pieces from its original configuration. More specifically if the sector was 120°, a 6-dB improvement would be expected when the sectors were reduced to 30°.

Figure 2.23 illustrates three types of intelligent antenna systems, each with positive and negative attributes. All the illustrations shown can be either receive-only or full-duplex. The difference between the receive-only and the full-duplex systems involves the amount of antennas and potential number of transmitting elements in the cell site itself.

The beam-switching antenna arrangement shown is the simplest to implement. It normally involves four standard antennas of narrow-azimuth beam width, 30° for a 120° sector, and, based on the signal received, the appropriate antenna will be selected by the base station controller for use in the receive path.

The multiple-beam array shown involves utilizing an antenna matrix to accomplish the beam switching.

The beam-steering array, however, utilizes phase shifting to direct the beam toward the subscriber unit. The direction that is chosen by the system for directing the beam will affect the entire sector. Normally transmit and receive amplifiers are located in conjunction with the antenna itself. In addition, the phase shifters are located directly behind each antenna element. The objective of placing the electronics in the mast head is to maximize the receive sensitivity and exploit the maximum transmit power for the site.

Chapter 3

Wireless Mobile Radio Technologies

There are numerous wireless mobile technology platforms in existence today, and more are likely to emerge in the future. Some of the platforms are born from standards and others from proprietary systems. However, each particular wireless technology has its own unique advantages and disadvantages. Technology platforms are generally classified by generation, although some cross generation boundaries.

Presently the wireless industry is in transition between second generation and third generation. There is an interim set of technology platforms referred to as two and one-half generation. Although there is debate over the exact definition of each generation, we present a simple definition for each:

First generation (1G). Includes all the analog technologies, primarily advanced mobile phone system (AMPS) and total-access communications system (TACS).

Second generation (2G). Includes global system for mobile communications (GSM), IS-136, IS-95, and integrated dispatch enhanced network (iDEN).

Two and one-half generation (2.5G). Includes GSM/general packet radio services (GPRS)/enhanced data rates for GSM evolution (EDGE), code-division multiple access 2000 (CDMA2000) 1XRTT and 1xDO, iDEN.

Third generation (3G). Includes wireband CDMA (W-CDMA), CDMA2000 1XDV0.

For a service to claim to be 3G it must meet the IMT-2000 specification. This unifying specification allows for mobile and some fixed high-speed data services using one or several radio channels coupled with fixed network platforms for delivering the services envisioned.

The following are important characteristics of IMT-2000:

- Global standard
- Compatibility of service within IMT-2000 and other fixed networks
- High quality
- Worldwide common frequency band
- Small terminals for worldwide use
- Worldwide roaming capability
- Multimedia application services and terminals
- Improved spectrum efficiency
- Flexibility for evolution to the next generation of wireless systems
- High-speed packet data rates: 2 Mbit/s for fixed-environment traffic, 384 kbit/s for pedestrian traffic, 144 kbit/s for vehicular traffic

Some of the more salient issues for 3G involve the interoperability and data rates. True interoperability between technologies is as elusive as the wireless data killer application. Table 3.1 represents some of the different technology

TABLE 3.1 1G and 2G Systems

	Cellular and SMR bands					
	1G	2G				
	AMPS	IS-136	IS-136*	IS-95	GSM	iDEN
Base Tx, MHz	869–894	869–894	851–866	869–894	925–960	851–866
Base Rx, MHz	824–849	824–849	806–821	869–894	880–915	806–821
Multiple-access method	FDMA	TDMA/FDMA	TDMA	CDMA/FDMA	TDMA/FDMA	TDMA
Modulation	FM	Pi/4DPSK	Pi/4DPSK	QPSK	0.3 GMSK	16 QAM
Radio channel spacing	30 kHz	30 kHz	30 kHz	1.25 MHz	200 kHz	25 kHz
Users/channel	1	3	3	64	8	3/6
Number of channels	832	832	600	9 (A), 10 (B)	124	600
Codec	NA	ACELP/VSELP	ACELP	CELP	RELP-LTP	
Spectrum allocation, MHz	50	50	30	50	50	Varies by market

*SMR band.

Note: Tx = transmitter, Rx = receiver, FDMA = frequency-division multiple access, TDMA = time-division multiple access, CDMA = code-division multiple access, FM = frequency modulation, Pi/4DPSK = differential phase-shift keying offset by π/4, QPSK = quadrature phase-shift keying, GMSK = gaussian minimum shift keying, QAM = quadrature amplitude modulation, NA = not applicable, ACELP = algebraic code excited linear prediction, VSELP = vector sum excited linear prediction, RELP-LTP = residual excited linear prediction–long-term prediction.

platforms in the cellular, specialized mobile radio (SMR), and personal communication services (PCS) bands indicating both 1G and 2G systems.

Next we present discussions regarding individual major wireless technology platforms. Each technology platform is discussed separately since some of the standards cross generation boundaries.

3.1 AMPS

The advanced mobile phone system is one of the pioneering technologies which forms the foundation of today's wireless mobile systems. It is still in operation in the United States and many parts of the world offering voice services, which it does quite well. A generic AMPS cellular system configuration is shown in Fig. 3.1. The configuration shown involves all the high-level system blocks of a cellular network. There are many components comprising each of the blocks shown. Each of the individual system components of a cellular network are covered in later chapters of this book.

In Fig. 3.1 the mobile communicates to the cell site through use of radio transmissions. The radio transmissions utilize a full duplex configuration which has separate transmit and receive frequencies. The cell site transmits on the frequency the mobile unit is tuned to, while the mobile unit transmits on the radio frequency the cell site receiver is tuned to.

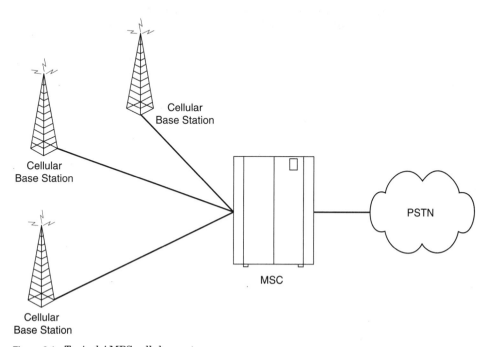

Figure 3.1 Typical AMPS cellular system.

The cell site acts as a conduit for the information transfer converting the radio energy into another medium. The cell site sends and receives information from the mobile and the mobile telephone switching office (MTSO), now commonly referred to as the mobile switching center (MSC). The MSC is connected to the cell site by either leased T1/E1 lines or through a microwave system. The cellular system is made up of many cell sites which all interconnect back to the MSC.

The MSC processes the call and connects the cell site radio link to the public switched telephone network (PSTN) or postal, telegraph, and telephone (PTT) service. The MSC performs a variety of functions involved with call processing and is effectively the brains of the network. It typically maintains the individual subscriber records, the current status of the subscribers, call routing, and billing information, to mention a few items. However, the subscriber records can be located remotely and accessed when the subscriber uses the system.

3.1.1 AMPS cell site configuration

The monopole cell site depicted in Fig. 3.2 is an example of a generic configuration typically associated with AMPS. However, the configuration can also apply to most wireless mobile and fixed systems. The equipment hut houses the radio transmission equipment. The monopole supports the antennas used for the cell site at the very top of the monopole. The cable tray between the equipment hut and the monopole supports the coaxial cables that connect the antennas to the radio transmission equipment.

A typical arrangement of the radio transmission equipment in the equipment room for a cellular base station is shown in Fig. 3.3. The cell site radio equipment consists of a base station controller (BSC), radio bay, and amplifier

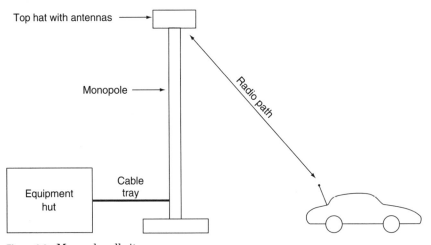

Figure 3.2 Monopole cell site.

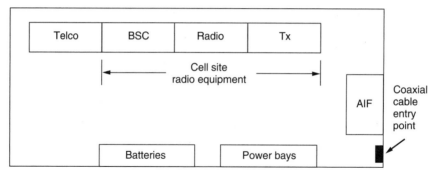

Figure 3.3 Generic cell site configuration. Key: BSC = base site controller; Radio = cellular radios; Tx = amplifier; AIF = antenna interface; Telco = T1/E1/microwave interconnect.

(transmitter) bay. The cell site radio equipment is connected to the antenna interface frame (AIF), which provides the receiver and transmitter filtering. The AIF is then connected to the antennas on the monopole through use of the coaxial cables which are located next to the AIF bay.

The cell site is also connected to the MSC through the telco bay. The telco bay either provides the T1/E1 leased line or the microwave radio link connection. The power for the cell site is secured through use of power bays (rectifiers), which convert alternating current (ac) electricity to direct current. In the event of a power disruption, batteries are used to ensure that the cell site continues to operate until power is restored or the batteries are exhausted. In many of the micro and pico cells the telco interconnection is enclosed in the cabinet itself leading to a lower physical footprint for the site, thereby easing land-use acquisition issues.

3.1.2 AMPS call setup scenarios

There are several general call scenarios which pertain to cellular AMPS (see Figs. 3.4 to 3.6). There are numerous algorithms utilized throughout the call setup and processing scenarios which are not included in the diagrams presented, including fraud prevention techniques employed by individual operators. However, the call scenarios presented here provide the fundamental building blocks for all call scenarios utilized in mobile cellular radio systems.

3.1.3 Handoff

The handoff concept is one of the fundamental principles of AMPS technology. Handoffs enable cellular to operate at lower power levels and provide high capacity. The handoff scenario presented in Fig. 3.7 is a simplified process. There are a multitude of algorithms that are invoked for the generation and processing of a handoff request and eventual handoff order. The individual algorithms are dependent upon the individual vendor for the network infrastructure and the software loads utilized.

72 Chapter Three

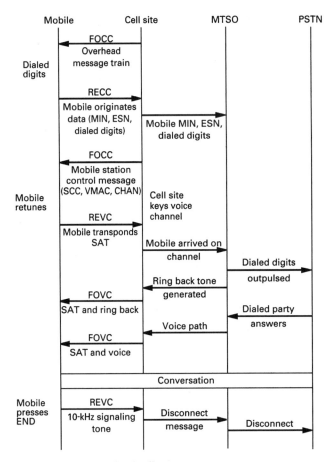

Figure 3.4 Mobile-to-land call setup.

Handoff from cell to cell is fundamentally the process of transferring a mobile unit that has a call in progress on a particular voice channel to another voice channel, all without interrupting the call. Handoffs can occur between adjacent cells or sectors of the same cell site. The actual need for a handoff is determined by the actual quality of the radio frequency (RF) signal received from the mobile into the cell site.

3.1.4 AMPS spectrum allocation

The cellular systems have been allocated a designated frequency spectrum to operate within. Both the A-band and B-band operators are allowed to utilize a total of 25 MHz of radio spectrum for their systems. The 25 MHz is divided into 12.5 MHz of transient frequencies and 12.5 MHz of receive frequencies for each operator. The cellular spectrum is shown in Fig. 3.8. The spectrum chart depicted has the location of the A-band and B-band cell site transmit and

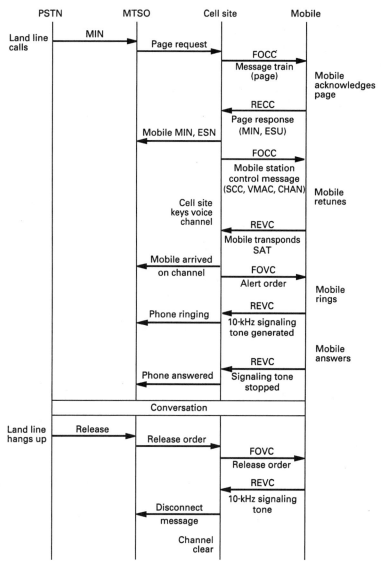

Figure 3.5 Land-to-mobile call setup.

receive frequencies indicated. There are currently a total of 832 individual Federal Communications Commission (FCC) channels available in the United States. Each of the radio channels utilized in cellular are spaced at 30-kHz intervals with the transmit frequency operating 45 MHz above the receive frequency. Both the A-band and B-band operators have a total of 416 radio channels (21 setup and 395 voice channels) available to them. Table 3.2 depicts the AMPS band channel assignment scheme.

Chapter Three

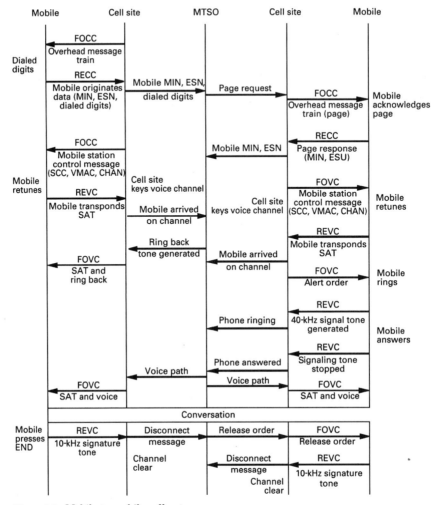

Figure 3.6 Mobile-to-mobile call setup.

3.1.5 AMPS frequency reuse

The concept and implementation of frequency reuse was an essential element in the quest for cellular systems to have a higher capacity per geographic area than mobile telephone systems (MTS) or improved MTS. Frequency reuse, reusing the same frequency in a system many times over, is the core concept defining a cellular system. The ability to reuse the same radio frequency many times is a result of managing the carrier-to-interference ratio (C/I) signal levels for an analog system. Typically the minimum C/I level designed for in a cellular analog system is 17 dB.

In order to improve the C/I, the reusing channel should be as far away from the serving site as possible so as to reduce the interference component of C/I.

Wireless Mobile Radio Technologies

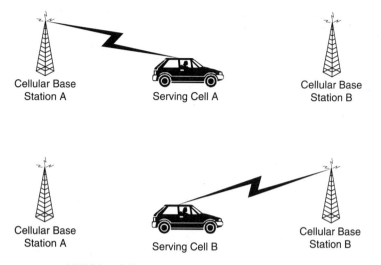

Figure 3.7 AMPS handoff.

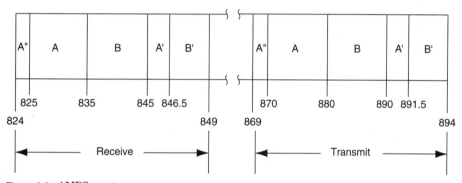

Figure 3.8 AMPS spectrum.

The distance needed between reusing base stations is found by the distance to radius (D/R) ratio, which defines the reuse factor for a wireless system. The D/R ratio (Fig. 3.9) is the relationship between the reusing cell site and the radius of the serving cell sites. Table 3.3 illustrates standard D/R ratios for different frequency reuse patterns N.

As Table 3.3 implies, there are several frequency reuse patterns currently in use throughout the cellular industry, each with advantages and disadvantages. The frequency reuse pattern ultimately defines the maximum amount of radios that can be assigned to an individual cell site. The most common frequency reuse pattern employed in cellular is the $N = 7$ pattern which is shown in Fig. 3.10. The $N = 7$ pattern can assign a maximum of 56 channels which are deployed using a three-sector design.

76 Chapter Three

TABLE 3.2 AMPS Channel Allocation

System	MHz	Number of channels	Boundary channel number	Transmitter center frequency, MHz	
				Mobile	Base station
(Not used)		1	(990)	(824.010)	(869.010)
A″	1	33	991	824.040	869.040
			1023	825.000	870.000
A	10	333	1	825.030	870.030
			333	834.990	879.990
B	10	333	334	835.020	880.020
			666	844.980	889.980
A′	1.5	50	667	845.010	890.010
			716	846.480	891.480
B′	2.5	83	717	846.510	891.510
			799	848.970	893.970

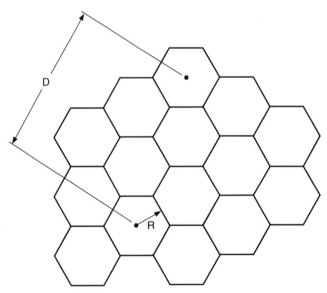

Figure 3.9 D/R ratio.

TABLE 3.3 D/R Ratios versus Reuse Patterns

D	N
3.46	4
4.6 R	7
6 R	12
7.55 R	19

Note: D/R = distance to radius; N = reuse pattern.

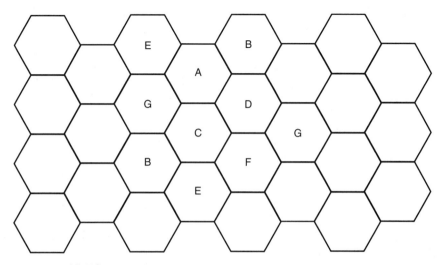

Figure 3.10 $N=7$ frequency reuse pattern.

3.1.6 AMPS channel band plan

The channel band plan is essential to any wireless system, especially one which reuses the spectrum at defined intervals. The channel band plan is a method of assigning channels, of fixed bandwidth, to a given amount of RF spectrum which is then grouped in a logical fashion.

An example of a channel band plan is given in Table 3.4. It's an AMPS (B-band) system utilizing an $N = 7$ frequency reuse pattern. The channels are all by definition 30 kHz in size. Therefore, if one were to count the individual channels listed in the chart, 12.5 MHz of spectrum would be accounted for. Since the cellular system is a duplexed system, 12.5 MHz is used for both transmitting and receiving and the total spectrum utilized is 25 MHz per operator.

TABLE 3.4 FCC Channel Chart for $N = 7$

				Wireline B-Band Channels						
Channel group:	A1	B1	C1	D1	E1	F1	G1	A2	B2	C2
Control channel:	334	335	336	337	338	339	340	341	324	343
	355	356	357	358	359	360	361	362	345	364
	376	377	378	379	380	381	382	383	366	385
	397	398	399	400	401	402	403	404	387	406
	418	419	420	421	422	423	424	425	408	427
	439	440	441	442	443	444	445	446	429	448
	460	461	462	463	464	465	466	467	450	469
	481	482	483	484	485	486	487	488	471	490
	502	503	504	505	506	507	508	509	492	511
	523	524	525	526	527	528	529	530	513	532
	544	545	546	547	548	549	550	551	534	553
	565	566	567	568	569	570	571	572	555	574
	586	587	588	589	590	591	592	593	576	595
	607	608	609	610	611	612	613	614	597	616
	628	629	630	631	632	633	634	635	618	637
	649	650	651	652	653	654	655	656	639	658
	717	718	719	720	721	722	723	724	725	726
	738	739	740	741	742	743	744	745	746	747
	759	760	761	762	763	764	765	766	767	768
	780	781	782	783	784	785	786	787	788	789
				Nonwireline A-Band Channels						
Channel group:	A1	B1	C1	D1	E1	F1	G1	A2	B2	C2
Control channel:	333	332	331	330	329	328	327	326	325	324
	312	311	310	309	308	307	306	305	304	303
	291	290	289	288	287	286	285	284	283	282
	270	269	268	267	266	265	264	263	262	261
	249	248	247	246	245	244	243	242	241	240
	228	227	226	225	224	223	222	221	220	219
	207	206	205	204	203	202	201	200	199	198
	186	185	184	183	182	181	180	179	178	177
	165	164	163	162	161	160	159	158	157	156
	144	143	142	141	140	139	138	137	136	135
	123	122	121	120	119	118	117	116	115	114
	102	101	100	99	98	97	96	95	94	93
	81	80	79	78	77	76	76	74	73	72
	60	59	58	57	56	55	54	53	52	51
	39	38	37	36	35	34	33	32	31	30
	18	17	16	15	14	13	12	11	10	9
	1020	1019	1018	1017	1016	1015	1014	1013	1012	1011
	999	998	997	996	995	994	993	992	991	716
	704	703	702	701	700	699	698	697	696	695
	683	682	681	680	679	678	677	676	675	674

				Wireline B-Band Channels						
D2	E2	F2	G2	A3	B3	C3	D3	E3	F3	G3
344	345	346	347	348	349	350	351	352	353	354
365	366	367	368	369	370	371	372	373	374	375
386	387	388	389	390	391	392	393	394	395	396
407	408	409	410	411	412	413	414	415	416	417
428	429	430	431	432	433	434	435	436	437	438
449	450	451	452	453	454	455	456	457	458	459
470	471	472	473	474	475	476	477	478	479	480
491	492	493	494	495	496	497	498	499	500	501
512	513	514	515	516	517	518	519	520	521	522
533	534	535	536	537	538	539	540	541	542	543
554	555	556	557	558	559	560	561	562	563	564
575	576	577	578	579	580	581	582	583	584	585
596	597	598	599	600	601	602	603	604	605	606
617	618	619	620	621	622	623	624	625	626	627
638	639	640	641	642	643	644	645	646	647	648
659	660	661	662	663	664	665	666			
727	728	729	730	731	732	733	734	735	736	737
748	749	750	751	752	753	754	755	756	757	758
769	770	771	772	773	774	775	776	777	778	779
790	791	792	793	794	795	796	797	798	799	
				Nonwireline A-Band Channels						
D2	E2	F2	G2	A3	B3	C3	D3	E3	F3	G3
323	322	321	320	319	318	317	316	315	314	313
302	301	300	299	298	297	296	295	294	293	292
281	280	279	278	277	276	275	274	273	272	271
260	259	258	257	256	255	254	253	252	251	250
239	238	237	236	235	234	233	232	231	230	229
218	217	216	215	214	213	212	211	210	209	208
197	196	195	194	193	192	191	190	189	188	187
176	175	174	173	172	171	170	169	168	167	166
155	154	153	152	151	150	149	148	147	146	145
134	133	132	131	130	129	128	127	126	125	124
113	112	111	110	109	108	107	106	105	104	103
92	91	90	89	88	87	86	85	84	83	82
71	70	69	68	67	66	65	64	63	62	61
50	49	48	47	46	45	44	43	42	41	40
29	28	27	26	25	24	23	22	21	20	19
8	7	6	5	4	3	2	1			
								1023	1022	1021
1010	1009	1008	1007	1006	1005	1004	1003	1002	1001	1000
715	714	713	712	711	710	709	708	707	706	705
694	693	692	691	690	689	688	687	686	685	684
673	672	671	670	669	668	667				

3.2 2G Digital Wireless Systems

Second-generation systems were deployed to improve the voice traffic throughput over existing analog cellular systems. However, there were numerous other enhancements for the mobile wireless operators and their subscribers. The major benefits associated with the introduction of 2G systems were

1. Increased capacity over analog
2. Reduced capital infrastructure costs
3. Reduced capital-per-subscriber cost
4. Reduced cellular fraud
5. Improved features
6. Encryption

The operators of the wireless systems gained the most from these benefits. Operating costs were reduced through either improved capital equipment and spectrum utilization or a decrease in cellular fraud. The improved features were centered around short message services (SMS) which the subscriber benefited from, but the primary benefit was that the overall cost to the subscriber was significantly reduced.

Second-generation mobile communication involved a variety of technology platforms as well as frequency bands. The issues regarding 2G deployment were

1. Capacity
2. Spectrum utilization
3. Infrastructure changes
4. Subscriber unit upgrades
5. Subscriber upgrade penetration rates

The fundamental binding issue with 2G was the utilization of digital radio technology for transporting the information content. It is important to note that while 2G systems utilized digital techniques to enhance their capacity over analog, their primary service was voice communication. At the time 2G systems were being deployed, 9.6 kbit/s was more than sufficient for existing data services, usually mobile fax. A separate mobile data system was deployed in the United States called cellular data packet data (CDPD), which was supposed to meet the mobile data requirements.

Digital radio technology was deployed in cellular systems using different modulation formats with the attempt to increase the quality and capacity of the existing cellular systems. As a point of reference, in an analog cellular system the voice communication is digitized within the cell site itself for transport over the fixed facilities to the MTSO. The voice representation and information transfer utilized in AMPS cellular was analog, and it is this part in the communication link that digital transition was focusing on.

The digital effort was meant to take advantage of many features and techniques that were not obtainable for analog cellular communication. There were several competing digital techniques that were being deployed in the cellular arena. The digital techniques for cellular communication fell into two primary categories, AMPS and TACS. For markets employing the TACS spectrum allocation, GSM was the digital modulation technique chosen. However, for AMPS markets the choice was between TDMA and CDMA radio access platforms. In addition to the AMPS and TACS spectrums, there was the iDEN radio access platform which operated in the specialized mobile radio band which is neither cellular nor PCS.

Currently with the introduction of the PCS licenses there are three fundamental competing technologies, each with its pros and cons: CDMA, GSM, and TDMA. Which technology platform is chosen depends on the application desired and whether there is a regulatory requirement to utilize a particular platform.

Initially, personal communication services (PCS) were considered the next generation of communication systems, and operators boasted about being able to provide new services which subscribers would want. However, PCS services took on the same look and feel as those originating from the cellular bands. By default, PCS has similarities and differences with its counterparts in the cellular band. The similarities between PCS and cellular lie in the mobility of the user of the service. The differences between PCS and cellular fall into the applications and spectrum available for PCS operators to provide to the subscribers. The license breakdown is shown in Fig. 3.11 and was discussed at the end of Sec. 1.3.

Currently there is no one standard for PCS operators to utilize for picking a technology platform for their networks. The choice of PCS standards is daunting, and each has its advantages and disadvantages. The current philosophy in the United States is to let the market decide which standards are best. This is significantly different than that used for cellular where every operator had a set interface for the analog system from which to operate from.

Table 3.5 represents various PCS systems which are used throughout the world and in particular the United States. The major standards utilized so far for PCS are DCS1900, IS-95, IS-661, and IS-136. DCS1900 utilizes a GSM format and is an upbanded DCS1800 system. IS-95 is the CDMA standard utilized by cellular operators, except that it is upbanded to the PCS spectrum. The IS-136 standard is an upbanded cellular TDMA system that is used by cellular operators. IS-661 is a time-division duplex system offered by Omnipoint Communications with the one notable exception that it was supposed to be deployed in the New York market as part of the pioneer preference license issued by the FCC, but is not a commercial system.

Digital communication is now prevalent throughout the entire wireless industry. *Digital communication* refers to any communication that utilizes a modulation format that relies on sending the information in any type of data format. More specifically, digital communication is where the sending location digitizes the voice communication and then modulates it. At the receiver the exact opposite is done.

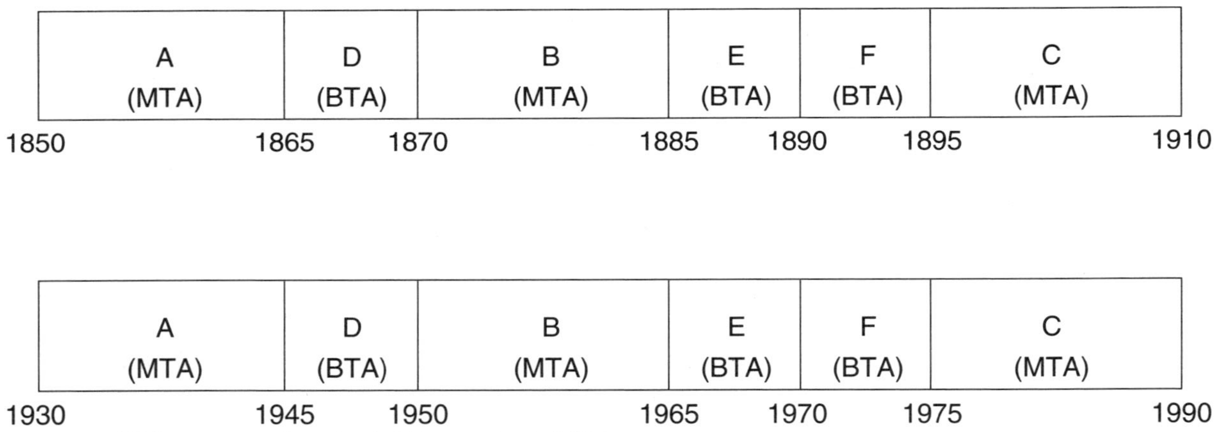

A (MTA)	D (BTA)	B (MTA)	E (BTA)	F (BTA)	C (MTA)
1850 – 1865	1865 – 1870	1870 – 1885	1885 – 1890	1890 – 1895	1895 – 1910

A (MTA)	D (BTA)	B (MTA)	E (BTA)	F (BTA)	C (MTA)
1930 – 1945	1945 – 1950	1950 – 1965	1965 – 1970	1970 – 1975	1975 – 1990

Figure 3.11 PCS spectrum allocation (all frequencies in MHz).

TABLE 3.5 PCS Systems

	IS-136	IS-95	DCS1800 (GSM)	DCS1900 (GSM)	IS-661
Base Tx, MHz	1930–1990	1930–1990	1805–1880	1930–1990	1930–1990
Base Rx, MHz	1850–1910	1850–1910	1710–1785	1850–1910	1850–1910
Multiple-access method	TDMA/FDMA	CDMA/FDMA	TDMA/FDMA	TDMA/FDMA	TDD
Modulation	Pi/4DPSK	QPSK	0.3 GMSK	0.3 GMSK	QPSK
Radio channel spacing	30 kHz	1.25 MHz	200 kHz	200 kHz	5 MHz
Users/channel	3	64	8	8	64
Number of channels	166/332/498	4–12	325	25/50/75	2–6
Codec	ACELP/VSELP	CELP	RELP-LTP	RELP-LTP	CELP
Spectrum allocation, MHz	10/20/30	10/20/30	150	10/20/30	10/20/30

Note: Tx = transmitter, Rx = receiver, TDMA = time-division multiple access, FDMA = frequency-division multiple access, CDMA = code-division multiple access, TDD = time-division duplexing, Pi/4DPSK = differential phase-shift keying offset by π/4, QPSK = quadrature phase-shift keying, GMSK = guassian minimum shift keying, ACELP = algebraic code excited linear/prediction, VSELP = vector sum excited linear prediction, RELP-LTP = residual excited linear prediction–long-term prediction.

The data are digital, but need to be converted into another medium in order to be transported from point *A* to point *B,* more specifically between the base station and the host terminal. The data between the base station and the host terminal are converted from a digital signal into RF energy whose modulation is a representation of the digital information which enables the receiving device (base station or host terminal) to properly replicate the data.

Digital radio technology is deployed in a cellular, PCS, or SMR system to increase the quality and capacity of the wireless system over its analog counterpart. The use of digital modulation techniques enables the wireless system to transport more bits per hertz than would be possible with analog signaling utilizing the same bandwidth. However, the service offering for 2G is for voice.

For the cellular operators there were several decisions to make for how to integrate the new system into the existing analog network. However, for PCS operators the integration with legacy systems did not present a problem since there was no legacy system. The PCS operators in the United States did have one obstacle to overcome, and that dealt with microwave clearance issues since the RF spectrum auctioned for use for the PCS operators was currently being used by 2-GHz point-to-point microwave systems.

The integration with the existing 1G (legacy) systems was therefore an issue that only affected the analog systems operating in the 800- and 900-MHz bands. The 2G technologies that were applicable involved GSM, TDMA, and CDMA radio access systems. There were several options available for the 1G operators to follow, and these are listed with the discussion of each access platform since the actual implementation is technology dependent.

3.3 IS-136 and TDMA (D-AMPS)

IS-136, an evolution of the IS-54 standard, is the digital cellular standard that was developed in the United States using TDMA technology and which enables a feature-rich technology platform to be utilized by the current cellular operators. Systems of this type operate in the same band as the AMPS and are used in the PCS spectrum also. IS-136, therefore, applies to both the cellular and PCS bands and in some unique situations to down-banded IS-136 which operates in the SMR band. IS-136 has enjoyed large-scale acceptance as a primary service offering by numerous wireless operators in North and South America.

TDMA technology allows multiple users to occupy the same channel through the use of time division. The TDMA format utilized in the United States follows the IS-54 and IS-136 standards and is referred to as North American digital-mode cellular (NADC). TDMA, utilizing the IS-136 standard, is currently deployed by several cellular operators in the United States. IS-136 utilizes the same channel bandwidth as analog cellular, 30 kHz per physical radio channel. However, IS-136 enables three and possibly six users to operate on the same physical radio channel at the same time. The IS-136 channel presents a total of six time slots in the forward and reverse directions. IS-136 at present utilizes two time slots per subscriber, with the potential to go to half-rate vocoders which require the use of only one time slot per subscriber.

The many advantages to the deployment of IS-136 in a cellular system are

Increased system capacity up to 3 times over analog

Improved protection for adjacent channel interference

Authentication

Voice privacy

Reduced infrastructure capital needed to deploy

Frequency plan integration over CDMA

Short message paging

Integrating IS-136 into an existing cellular system is easier than deploying CDMA. The use of IS-136 in a network requires the use of a guardband to protect the analog system from the IS-136 signal. However, the guardband consists of only a single channel on either side of the spectrum block allocated for IS-136 use. Depending on the actual location of the IS-136 channels in the operator's spectrum it is possible to require only one guardband channel or none.

The IS-136 has the unique advantage of affording the implementation of digital technology into a network without elaborate engineering requirements. The implementation advantages mentioned for IS-136 also facilitate the rapid deployment of this technology into an existing network.

The implementation of IS-136 is further augmented by requiring only one channel, per frequency group, as part of the initial system offering. The advantages of this are the minimization of capacity reduction for the existing analog network and elimination of the need to preload the subscriber base with dual-mode handsets.

Air interface signaling for an IS-136 system is shown in Fig. 3.12. The format of the signal involves a total of six potential conversations taking place over the same bandwidth that one 30-kHz voice conversation utilizes for analog cellular. Presently, full-rate vocoders are utilized by operators. The full-rate vocoder utilizes two time slots in both the forward and reverse links. The use

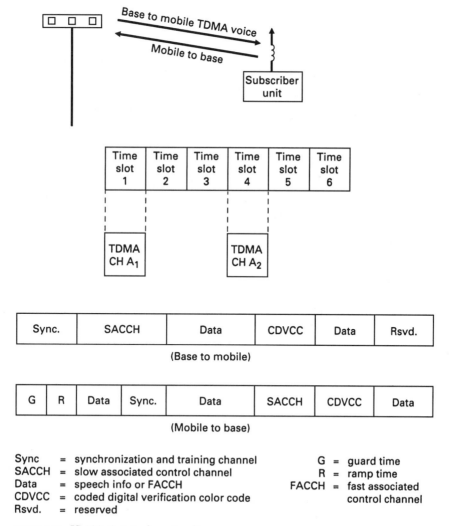

Figure 3.12 IS-136 air interface signaling.

of the full-rate vocoder limits to three the amount of TDMA users that can occupy one physical channel. In Fig. 3.12 a subscriber utilizing a full-rate vocoder would occupy two time slots, A1 and A2, while a half-rate vocoder user would only use A1 or A2. The time slots are paired for a full-rate vocoder system. Conversation A would utilize time slots 1 and 4, conversation B would use slots 2 and 5, and conversation C would utilize time slots 3 and 6.

Access to the system is either achieved through the primary control channel, utilized for analog communication, or the secondary dedicated control channel. During the initial acquisition phase the mobile reads the overhead control message from the primary control channel and determines if the system is digital-capable. If it is, a decision will be made whether to utilize the primary or secondary dedicated control channel. The secondary dedicated control channels are assigned as FCC channels 696 to 716 for the A-band system and channels 717 to 737 for the B-band system. The use of the secondary dedicated control channels enables a variety of enhanced features to be provided by the system operator to the subscribers. However, it is possible to create a system-defined control channel map which is unique to any system. Figure 3.13 is a call flowchart for IS-136 (800- or 1900-MHz band). Many of the details listed in the flowchart are covered in the following sections.

3.3.1 Voice channel structure

Associated with each digital traffic channel (DTC) are two other channels: the fast associated control channel (FACCH) and the slow associated control channel (SACCH). The FACCH is a signaling channel used for the transmission of control and supervisory information between the mobile and the network. For example, if a mobile is to send DTMF tones, then these are indicated on the FACCH. The SACCH is also used for transmission of control and supervisory information between the mobile and the network. Most notably, the SACCH is used by the mobile to transmit measurement information to the network describing the mobile's experience of the RF conditions. This information is used by the network to determine when and how handoff should occur.

Figure 3.14 shows the structure of the DTC. It is notable that the figure does not show the FACCH. This is because the DATA field, which is normally used to transmit voice, is also used to transmit FACCH information. In other words, if information is to be sent on the FACCH, then user data are briefly suspended while the FACCH information is being sent.

Figure 3.14 also shows six time slots within the frame structure. In fact, IS-136 allows for two types of mobiles: full rate and half rate. A full-rate mobile uses two of the time slots in the frame (1 and 4, 2 and 5, or 3 and 6), while a half-rate mobile uses just a single time slot. A full-rate mobile transmits 260 bits of speech per time slot (520 bits per frame). Since there are 25 frames per second, this means that the gross bit rate for speech is 13 kbit/s. In practice, only full-rate handsets are used.

In addition to the user data and SACCH within the DTC, we see a number of other fields:

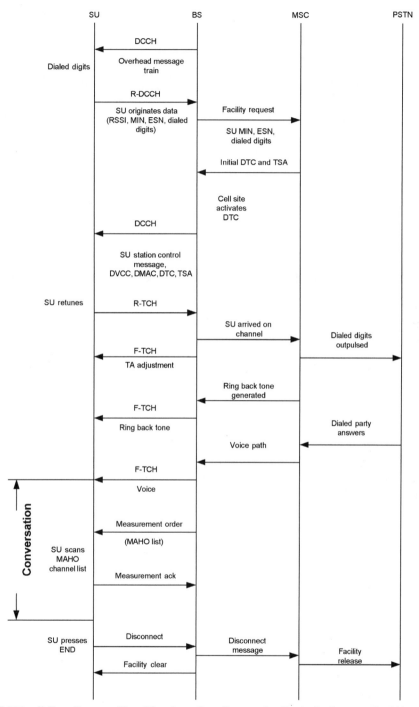

Figure 3.13 IS-136 call flow diagram. *Note:* The above flow diagram is effectively the same for 800 and 1900 MHz. For the 800-MHz systems it is assumed that there are DCCHs and the FCCH, analog, has the associated DCCH locator word in the overhead message. (a) 800-MHz subscriber units (SUs) keep the last-seen DCCHs in temporary memory and will scan these DCCHs in addition to the standard 21 analog control channel (ACC) and tune to the strongest signal. If an SU camps on an ACC it will look for the DCCH locator word and retune to the DCCH, if possible. The SU will camp on the DCCH, if available, and camp on the strongest DCCH signal. If the SU cannot find a DCCH it will tune to the strongest ACC. (b) 1900-MHz SUs will scan the initial set of DCCH frequencies it has programmed in which it can follow the standard list or be operator-specific. The SU will camp on the strongest signal. If it is not allowed to camp on that channel it will go to another channel on the list. If it exhausts the primary selection list it will then scan all channels looking for the DCCH.

88 Chapter Three

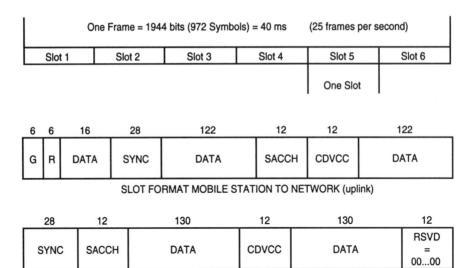

Figure 3.14 Digital traffic channel (DTC).

Guard time. This field is three symbols (6 bits) in duration. It is a buffer between adjacent time slots used by different mobiles and allows compensation for variations in distance between the mobile and the base station.

Ramp time. This is three symbols in duration allowing for ramp up to the RF power.

SYNC. This is a special synchronization pattern, which is unique for a given time slot. It is used for correct time alignment.

CDVCC. This is the coded digital voice color code, which is analogous to the supervisory audio tone used in analog AMPS. It is used to detect cochannel interference.

3.3.2 Offset between transmit and receive

IS-136 is a frequency duplex TDMA system. In other words, the mobile transmits on one frequency and receives on another frequency. In the uplink, the mobile transmits on a given pair of time slots, and on the downlink, it receives on the corresponding pair of time slots. If, for example, a given mobile transmits on time slots 1 and 4 on the uplink, then it receives on time slots 1 and 4 on the downlink. Time slots 1 and 4 on the downlink do not, however, correspond to the same instants in time as time slots 1 and 4 on the uplink. There is a time offset between the downlink and the uplink which corresponds to one time slot plus 45 symbol periods (207 symbol periods total, or 8.5185 ms), with the downlink lagging the uplink. Therefore, the mobile does not transmit and receive simultaneously. Rather, during a conversation, it receives a time slot on the downlink shortly after sending a time slot on the uplink. Figure 3.15 depicts this offset, showing the transmission and reception by a given mobile on time slots 1 and 4.

As can be seen from Fig. 3.15, there are times when the mobile is neither transmitting on a given time slot nor listening to the base station on the corresponding downlink time slot. So what does it do during these times? Rather than do nothing, the mobile tunes briefly to other base stations to measure the signals from them. As described later in this chapter, those measurements can be provided to the network to assist in determining when a handoff should take place.

3.3.3 Speech coding

Since the DTC is digital, it is necessary to convert the user speech from analog form to digital. In other words, the handset (and the network) must include a digital speech-coding scheme. In IS-136, the speech-coding technique uses vector sum excited linear prediction (VSELP). This is a linear predictive coding (LPC) technique that operates on 20-ms speech samples at a time. For each 20-ms sample, the coding scheme itself generates 159 bits. Thus, the coder provides an effective bit rate of 7.95 kbit/s.

The RF interface, however, is an error-prone medium. Therefore, to ensure high speech quality, it is necessary to include mechanisms that mitigate against errors caused in RF propagation. Consequently, the 159 bits are subject

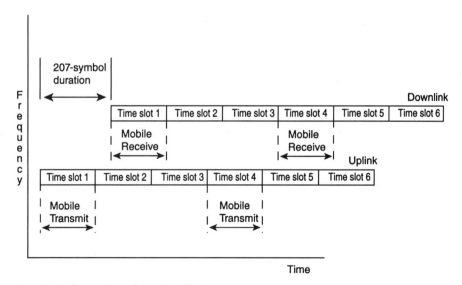

Figure 3.15 Transmit and receive offset.

to a channel coding scheme designed to minimize the effects of error. Of the 159 bits, 77 are considered class 1 bits (of greater significance to the speech perception) and 82 are considered class 2 bits. As shown in Fig. 3.16, the 77 class 1 bits are passed through a convolutional coder, which results in 178 bits. These 178 bits are combined with the 82 class 2 bits to give a total of 260 bits, and the 260 bits are allocated across the time slots used by the subscriber. Thus, each 20 ms of speech gives rise to a transmission of 260 bits, resulting in a gross rate of 13 kbit/s over the air interface.

However, Table 3.6 equates the voice quality to a particular bit error rate (BER). Both types of vocoders used for IS-136 are listed along with their relative voice quality perception levels. The lower the BER percent, the better the call regardless of the vocoder used.

3.3.4 Time alignment

Since three mobiles use a given RF channel on a time-sharing basis, it is necessary that they each time their transmissions exactly. Otherwise, their signals would overlap and cause interference at the base station receiver. Furthermore, a given cell may be many miles in diameter, and the time for transmission from one mobile to the base station may be different from the time taken by the transmission from another mobile. Therefore, if one mobile begins transmission immediately after another mobile stops transmission, it is possible that the two signals could collide at the base station. For example, consider a situation where mobile A is far away from the base station and mobile B is close to the base station. It takes longer for mobile B's transmission to reach the base station than that of mobile A. Therefore, if mobile A starts transmitting immediately after mobile B stops transmitting, the trans-

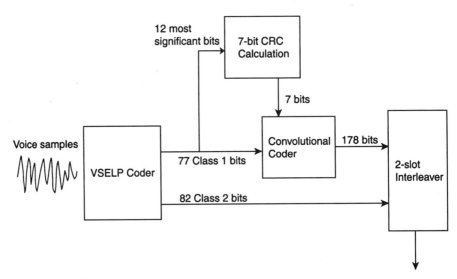

Figure 3.16 IS-136 speech coding.

TABLE 3.6 BER and Voice Quality Relationship

		Voice quality	
BER	BER%	ACELP	VSELP
0	<0.01	Good	Good
1	0.01–0.1	Good	Good
2	0.1–0.5	Good	Good
3	0.5–1.0	Good	Good
4	1.0–2.0	Good	Marginal
5	2.0–4.0	Marginal	Bad
6	4.0–8.0	Bad	Bad
7	>8.0	Bad	Bad

Note: BER = bit error rate, ACELP = algebraic code excited linear prediction, VSELP = vector sum excited linear prediction.

mission from mobile B could still be arriving at the base station when mobile A's transmission starts to arrive. Consequently, it is not just necessary to ensure that no two mobiles transmit at the same time, it is also necessary to time transmissions such that no two transmissions arrive at the base station at the same time. The methodology for this timing is called *time alignment,* which involves advancing or retarding the transmission from a given mobile so that the transmission arrives at the base station at the correct time relative to transmissions from other mobiles using the same RF channel.

When a mobile first accesses the system, the network assigns it a traffic channel, including a digital voice color code (DVCC). At this point, however, the network has not provided any time alignment information. Given that the mobile could be close to or far away from the base station, it needs the correct time alignment information before transmitting real user data, which means that the base station must determine roughly how far away the mobile happens to be and must send time alignment instructions. In order to help the base station determine what the time alignment instructions should be, the mobile sends a special sequence of 324-bit duration, called a *shortened burst* as shown in Fig. 3.17. The structure of the shortened burst is such that if the base station detects two or more sync words of the burst, it can determine the mobile's distance from the base station. The base station then sends a time alignment message instructing the mobile to adjust its transmission timing.

3.3.5 Control channel

Even though the control channel is analog, and IS-136 is designed to include a certain degree of compatibility with analog AMPS, the control channel contains a number of significant differences from the analog control channel. These changes were introduced to overcome known problems in AMPS and to

G1	R S D S D V S D W S D X S D Y S	G2

G1: 3-symbol-length guard time
R: 3-symbol-length ramp time
S: 14-symbol-length sync word; the mobile station uses its assigned sync word
D: 6-symbol-length CDVCC; the mobile station uses its assigned DVCC
G2: 22-symbol-length guard time

V = 0000
W = 00000000
X = 000000000000
Y = 0000000000000000

Figure 3.17 Shortened burst structure.

provide control channel support for digital voice channels. For example, when assigning a mobile to a given traffic channel, the downlink control channel must specify the time slots to be used by the mobile. Obviously, such capability does not exist in the standard AMPS control channel.

IS-136 brings to the table the digital control channel (DCCH), and it enables the delivery of adjunct features that in cellular were not really possible. The DCCH occupies two of the six time slots, and, therefore, if a physical radio also has a DCCH assigned to it, only two subscribers can use the physical radio for communication purposes.

The DCCH allocations for IS-136 in the cellular and PCS bands are shown in Tables 3.7 and 3.8, respectively. Both tables indicate that there is a preferred location for the assignment of DCCHs. The DCCHs can be located anywhere in the allocated frequency band; however, there are combinations of channels that are preferred to be used. The preference is based on the method that the subscriber unit uses to scan the available spectrum looking for the DCCH.

The preferred channel sets are broken down into 16 relative probability blocks for each frequency band of operation, both cellular and PCS. Relative probability block 1 is the first group of channels the subscriber unit uses to find the DCCH for the system and cell. The subscriber unit will then scan through the entire frequency band going through channel sets according to the relative probability blocks until it finds a DCCH. In the case of cellular if no DCCH is found, the subscriber unit reverts to the control channel for a dual-mode phone and then acquires the system through the control channel or is directed to a specific channel that has the DCCH.

Table 3.7 lists the preferred values (channels) on which the DCCH should reside for both A- and B-band cellular operators. The preferred location for the A- and B-band operators is in the extended spectrum portion of the frequency bands: A″ and upper B′ portions, respectively, of the band. Likewise the sug-

TABLE 3.7 Cellular DCCH Preferred Assignments

Block number	Channel number	Band	Number of channels	Relative probability
A block				
1	1–26	A	26	4
2	27–52	A	26	5
3	53–78	A	26	6
4	79–104	A	26	7
5	105–130	A	26	8
6	131–156	A	26	9
7	157–182	A	26	10
8	183–208	A	26	11
9	209–234	A	26	12
10	235–260	A	26	13
11	261–286	A	26	14
12	287–312	A	26	15
13	313–333	A	21	16 (lowest)
14	667–691	A'	25	3
15	692–716	A'	25	2
16	991–1023	A"	33	1 (highest)
B block				
1	334–354	B	21	16 (lowest)
2	355–380	B	26	15
3	381–406	B	26	14
4	407–432	B	26	13
5	433–458	B	26	12
6	459–484	B	26	11
7	485–510	B	26	10
8	511–536	B	26	9
9	537–562	B	26	8
10	563–588	B	26	7
11	589–614	B	26	6
12	615–640	B	26	5
13	641–666	B	26	4
14	717–741	B'	25	3
15	742–766	B'	25	2
16	767–799	B'	33	1 (highest)

gested locations for the DCCHs in the PCS bands are the upper portion of each of the PCS blocks shown in Table 3.8.

Table 3.9 represents the channel assignments for each PCS block that conforms to an $N = 7$ three-sector channel chart. It is obvious that the D, E, and F blocks do not have enough channels in the preferred locations for these groups. Therefore, in the figures listed for channel assignments the remaining channels that should make up the DCCH list include those in probability block 2. Please note that the channels on the top of all the charts represent the suggested channels that should be used for the initial control DCCHs for the PCS band.

TABLE 3.8 PCS DCCH Preferred Assignments

Block number	Channel number	Band	Number of channels	Relative probability
A block				
1	2–31	A	30	16 (lowest)
2	32–62	A	31	15
3	63–93	A	31	14
4	94–124	A	31	13
5	125–155	A	31	12
6	156–186	A	31	11
7	187–217	A	31	10
8	218–248	A	31	9
9	249–279	A	31	8
10	280–310	A	31	7
11	311–341	A	31	6
12	342–372	A	31	5
13	373–403	A	31	4
14	404–434	A	31	3
15	435–465	A	31	2
16	466–498	A	33	1 (highest)
B block				
1	668–698	B	31	16 (lowest)
2	699–729	B	31	15
3	730–760	B	31	14
4	761–791	B	31	13
5	792–822	B	31	12
6	823–853	B	31	11
7	854–884	B	31	10
8	885–915	B	31	9
9	916–946	B	31	8
10	947–977	B	31	7
11	978–1008	B	31	6
12	1009–1039	B	31	5
13	1040–1070	B	31	4
14	1071–1101	B	31	3
15	1102–1132	B	31	2
16	1133–1165	B	33	1 (highest)
C block				
1	1501–1531	C	31	16 (lowest)
2	1532–1562	C	31	15
3	1563–1593	C	31	14
4	1594–1624	C	31	13
5	1625–1655	C	31	12
6	1656–1686	C	31	11
7	1687–1717	C	31	10
8	1718–1748	C	31	9
9	1749–1779	C	31	8
10	1780–1810	C	31	7
11	1811–1841	C	31	6
12	1842–1872	C	31	5
13	1873–1903	C	31	4
14	1904–1934	C	31	3
15	1935–1965	C	31	2
16	1966–1998	C	33	1 (highest)

TABLE 3.8 PCS DCCH Preferred Assignments (*Continued*)

Block number	Channel number	Band	Number of channels	Relative probability
D block				
1	502–511	D	10	16 (lowest)
2	512–521	D	10	15
3	522–531	D	10	14
4	532–541	D	10	13
5	542–551	D	10	12
6	552–561	D	10	11
7	562–571	D	10	10
8	572–581	D	10	9
9	582–591	D	10	8
10	592–601	D	10	7
11	602–611	D	10	6
12	612–621	D	10	5
13	622–631	D	10	4
14	632–641	D	10	3
15	642–651	D	10	2
16	652–665	D	14	1 (highest)
E block				
1	1168–1177	E	10	16 (lowest)
2	1178–1187	E	10	15
3	1188–1197	E	10	14
4	1198–1207	E	10	13
5	1208–1217	E	10	12
6	1218–1227	E	10	11
7	1228–1237	E	10	10
8	1238–1247	E	10	9
9	1248–1257	E	10	8
10	1258–1267	E	10	7
11	1268–1277	E	10	6
12	1278–1287	E	10	5
13	1288–1297	E	10	4
14	1298–1307	E	10	3
15	1308–1317	E	10	2
16	1318–1332	E	15	1 (highest)
F block				
1	1335–1344	F	10	16 (lowest)
2	1345–1354	F	10	15
3	1355–1364	F	10	14
4	1365–1374	F	10	13
5	1375–1384	F	10	12
6	1385–1394	F	10	11
7	1395–1404	F	10	10
8	1405–1414	F	10	9
9	1415–1424	F	10	8
10	1425–1434	F	10	7
11	1435–1444	F	10	6
12	1445–1454	F	10	5
13	1455–1464	F	10	4
14	1465–1474	F	10	3
15	1475–1484	F	10	2
16	1485–1498	F	14	1 (highest)

TABLE 3.9 IS-136 PCS Channel Chart ($N = 7$)

A1	B1	C1	D1	E1	F1	G1	A2	B2	C2	D2	E2	F2	G2	A3	B3	C3	D3	E3	F3	G3
498	497	496	495	494	493	492	491	490	489	488	487	486	485	484	483	482	481	480	479	478
477	476	475	474	473	472	471	470	469	468	467	466	465	464	463	462	461	460	459	458	457
456	455	454	453	452	451	450	449	448	447	446	445	444	443	442	441	440	439	438	437	436
435	434	433	432	431	430	429	428	427	426	425	424	423	422	421	420	419	418	417	416	415
414	413	412	411	410	409	408	407	406	405	404	403	402	401	400	399	398	397	396	395	394
393	392	391	390	389	388	387	386	385	384	383	382	381	380	379	378	377	376	375	374	373
372	371	370	369	368	367	366	365	364	363	362	361	360	359	358	357	356	355	354	353	352
351	350	349	348	347	346	345	344	343	342	341	340	339	338	337	336	335	334	333	332	331
330	329	328	327	326	325	324	323	322	321	320	319	318	317	316	315	314	313	312	311	310
309	308	307	306	305	304	303	302	301	300	299	298	297	296	295	294	293	292	291	290	289
288	287	286	285	284	283	282	281	280	279	278	277	276	275	274	273	272	271	270	269	268
267	266	265	264	263	262	261	260	259	258	257	256	255	254	253	252	251	250	249	248	247
246	245	244	243	242	241	240	239	238	237	236	235	234	233	232	231	230	229	228	227	226
225	224	223	222	221	220	219	218	217	216	215	214	213	212	211	210	209	208	207	206	205
204	203	202	201	200	199	198	197	196	195	194	193	192	191	190	189	188	187	186	185	184
183	182	181	180	179	178	177	176	175	174	173	172	171	170	169	168	167	166	165	164	163
162	161	160	159	158	157	156	155	154	153	152	151	150	149	148	147	146	145	144	143	142
141	140	139	138	137	136	135	134	133	132	131	130	129	128	127	126	125	124	123	122	121
120	119	118	117	116	115	114	113	112	111	110	109	108	107	106	105	104	103	102	101	100
99	98	97	96	95	94	93	92	91	90	89	88	87	86	85	84	83	82	81	80	79
78	77	76	75	74	73	72	71	70	69	68	67	66	65	64	63	62	61	60	59	58
57	56	55	54	53	52	51	50	49	48	47	46	45	44	43	42	41	40	39	38	37
36	35	34	33	32	31	30	29	28	27	26	25	24	23	22	21	20	19	18	17	16
15	14	13	12	11	10	9	8	7	6	5	4	3	2							
A block																				
1165	1164	1163	1162	1161	1160	1159	1158	1157	1156	1155	1154	1153	1152	1151	1150	1149	1148	1147	1146	1145
1144	1143	1142	1141	1140	1139	1138	1137	1136	1135	1134	1133	1132	1131	1130	1129	1128	1127	1126	1125	1124
1123	1122	1121	1120	1119	1118	1117	1116	1115	1114	1113	1112	1111	1110	1109	1108	1107	1106	1105	1104	1103
1102	1101	1100	1099	1098	1097	1096	1095	1094	1093	1092	1091	1090	1089	1088	1087	1086	1085	1084	1083	1082
B block																				

1081	1080	1079	1078	1077	1076	1075	1074	1073	1072	1071	1070	1069	1068	1067	1066	1065	1064	1063	1062	1061
1060	1059	1058	1057	1056	1055	1054	1053	1052	1051	1050	1049	1048	1047	1046	1045	1044	1043	1042	1041	1040
1039	1038	1037	1036	1035	1034	1033	1032	1031	1030	1029	1028	1027	1026	1025	1024	1023	1022	1021	1020	1019
1018	1017	1016	1015	1014	1013	1012	1011	1010	1009	1008	1007	1006	1005	1004	1003	1002	1001	1000	999	998
997	996	995	994	993	992	991	990	989	988	987	986	985	984	983	982	981	980	979	978	977
976	975	974	973	972	971	970	969	968	967	966	965	964	963	962	961	960	959	958	957	956
955	954	953	952	951	950	949	948	947	946	945	944	943	942	941	940	939	938	937	936	935
934	933	932	931	930	929	928	927	926	925	924	923	922	921	920	919	918	917	916	915	914
913	912	911	910	909	908	907	906	905	904	903	902	901	900	899	898	897	896	895	894	893
892	891	890	889	888	887	886	885	884	883	882	881	880	879	878	877	876	875	874	873	872
871	870	869	868	867	866	865	864	863	862	861	860	859	858	857	856	855	854	853	852	851
850	849	848	847	846	845	844	843	842	841	840	839	838	837	836	835	834	833	832	831	830
829	828	827	826	825	824	823	822	821	820	819	818	817	816	815	814	813	812	811	810	809
808	807	806	805	804	803	802	801	800	799	798	797	796	795	794	793	792	791	790	789	788
787	786	785	784	783	782	781	780	779	778	777	776	775	774	773	772	771	770	769	768	767
766	765	764	763	762	761	760	759	758	757	756	755	754	753	752	751	750	749	748	747	746
745	744	743	742	741	740	739	738	737	736	735	734	733	732	731	730	729	728	727	726	725
724	723	722	721	720	719	718	717	716	715	714	713	712	711	710	709	708	707	706	705	704
703	702	701	700	699	698	697	696	695	694	693	692	691	690	689	688	687	686	685	684	683
682	681	680	679	678	677	676	675	674	673	672	671	670	669	668						

C block

1998	1997	1996	1995	1994	1993	1992	1991	1990	1989	1988	1987	1986	1985	1984	1983	1982	1981	1980	1979	1978
1977	1976	1975	1974	1973	1972	1971	1970	1969	1968	1967	1966	1965	1964	1963	1962	1961	1960	1959	1958	1957
1956	1955	1954	1953	1952	1951	1950	1949	1948	1947	1946	1945	1944	1943	1942	1941	1940	1939	1938	1937	1936
1935	1934	1933	1932	1931	1930	1929	1928	1927	1926	1925	1924	1923	1922	1921	1920	1919	1918	1917	1916	1915
1914	1913	1912	1911	1910	1909	1908	1907	1906	1905	1904	1903	1902	1901	1900	1899	1898	1897	1896	1895	1894
1893	1892	1891	1890	1889	1888	1887	1886	1885	1884	1883	1882	1881	1880	1879	1878	1877	1876	1875	1874	1873
1872	1871	1870	1869	1868	1867	1866	1865	1864	1863	1862	1861	1860	1859	1858	1857	1856	1855	1854	1853	1852
1851	1850	1849	1848	1847	1846	1845	1844	1843	1842	1841	1840	1839	1838	1837	1836	1835	1834	1833	1832	1831
1830	1829	1828	1827	1826	1825	1824	1823	1822	1821	1820	1819	1818	1817	1816	1815	1814	1813	1812	1811	1810
1809	1808	1807	1806	1805	1804	1803	1802	1801	1800	1799	1798	1797	1796	1795	1794	1793	1792	1791	1790	1789
1788	1787	1786	1785	1784	1783	1782	1781	1780	1779	1778	1777	1776	1775	1774	1773	1772	1771	1770	1769	1768
1767	1766	1765	1764	1763	1762	1761	1760	1759	1758	1757	1756	1755	1754	1753	1752	1751	1750	1749	1748	1747
1746	1745	1744	1743	1742	1741	1740	1739	1738	1737	1736	1735	1734	1733	1732	1731	1730	1729	1728	1727	1726
1725	1724	1723	1722	1721	1720	1719	1718	1717	1716	1715	1714	1713	1712	1711	1710	1709	1708	1707	1706	1705

(*Continued*)

TABLE 3.9 IS-136 PCS Channel Chart (N = 7) (Continued)

A1	B1	C1	D1	E1	F1	G1	A2	B2	C2	D2	E1	F2	G2	A3	B3	C3	D3	E3	F3	G3
1704	1703	1702	1701	1700	1699	1698	1697	1696	1695	1694	1693	1692	1691	1690	1689	1688	1687	1686	1685	1684
1683	1682	1681	1680	1679	1678	1677	1676	1675	1674	1673	1672	1671	1670	1669	1668	1667	1666	1665	1664	1663
1662	1661	1660	1659	1658	1657	1656	1655	1654	1653	1652	1651	1650	1649	1648	1647	1646	1645	1644	1643	1642
1641	1640	1639	1638	1637	1636	1635	1634	1633	1632	1631	1630	1629	1628	1627	1626	1625	1624	1623	1622	1621
1620	1619	1618	1617	1616	1615	1614	1613	1612	1611	1610	1609	1608	1607	1606	1605	1604	1603	1602	1601	1600
1599	1598	1597	1596	1595	1594	1593	1592	1591	1590	1589	1588	1587	1586	1585	1584	1583	1582	1581	1580	1579
1578	1577	1576	1575	1574	1573	1572	1571	1570	1569	1568	1567	1566	1565	1564	1563	1562	1561	1560	1559	1558
1557	1556	1555	1554	1553	1552	1551	1550	1549	1548	1547	1546	1545	1544	1543	1542	1541	1540	1539	1538	1537
1536	1535	1534	1533	1532	1531	1530	1529	1528	1527	1526	1525	1524	1523	1522	1521	1520	1519	1518	1517	1516
1515	1514	1513	1512	1511	1510	1509	1508	1507	1506	1505	1504	1503	1502	1501						
										C block										
665	664	663	662	661	660	659	658	657	656	655	654	653	652	651	650	649	648	647	646	645
644	643	642	641	640	639	638	637	636	635	634	633	632	631	630	629	628	627	626	625	624
623	622	621	620	619	618	617	616	615	614	613	612	611	610	609	608	607	606	605	604	603
602	601	600	599	598	597	596	595	594	593	592	591	590	589	588	587	586	585	584	583	582
581	580	579	578	577	576	575	574	573	572	571	570	569	568	567	566	565	564	563	562	561
560	559	558	557	556	555	554	553	552	551	550	549	548	547	546	545	544	543	542	541	540
539	538	537	536	535	534	533	532	531	530	529	528	527	526	525	524	523	522	521	520	519
518	517	516	515	514	513	512	511	510	509	508	507	506	505	504	503	502				
										D block										
1332	1331	1330	1329	1328	1327	1326	1325	1324	1323	1322	1321	1320	1319	1338	1337	1336	1335	1334	1333	1332
1331	1330	1329	1328	1327	1326	1325	1324	1323	1322	1321	1320	1319	1318	1317	1316	1315	1314	1313	1312	1311
1310	1309	1308	1307	1306	1305	1304	1303	1302	1301	1300	1299	1298	1297	1296	1295	1294	1293	1292	1291	1290
1289	1288	1287	1286	1285	1284	1283	1282	1281	1280	1279	1278	1277	1276	1275	1274	1273	1272	1271	1270	1269
1268	1267	1266	1265	1264	1263	1262	1261	1260	1259	1258	1257	1256	1255	1254	1253	1252	1251	1250	1249	1248
1247	1246	1245	1244	1243	1242	1241	1240	1239	1238	1237	1236	1235	1234	1233	1232	1231	1230	1229	1228	1227
										E block										

1226	1225	1224	1223	1222	1221	1220	1219	1218	1217	1216	1215	1214	1213	1212	1211	1210	1209	1208	1207	1206
1205	1204	1203	1202	1201	1200	1199	1198	1197	1196	1195	1194	1193	1192	1191	1190	1189	1188	1187	1186	1185
1184	1183	1182	1181	1180	1179	1178	1177	1176	1175	1174	1173	1172	1171	1170	1169	1168				

F block

1498	1497	1496	1495	1494	1493	1492	1491	1490	1489	1488	1487	1486	1485	1484	1483	1482	1481	1480	1479	1478
1477	1476	1475	1474	1473	1472	1471	1470	1469	1468	1467	1466	1465	1464	1463	1462	1461	1460	1459	1458	1457
1456	1455	1454	1453	1452	1451	1450	1449	1448	1447	1446	1445	1444	1443	1442	1441	1440	1439	1438	1437	1436
1435	1434	1433	1432	1431	1430	1429	1428	1427	1426	1425	1424	1423	1422	1421	1420	1419	1418	1417	1416	1415
1414	1413	1412	1411	1410	1409	1408	1407	1406	1405	1404	1403	1402	1401	1400	1399	1398	1397	1396	1395	1394
1393	1392	1391	1390	1389	1388	1387	1386	1385	1384	1383	1382	1381	1380	1379	1378	1377	1376	1375	1374	1373
1372	1371	1370	1369	1368	1367	1366	1365	1364	1363	1362	1361	1360	1359	1358	1357	1356	1355	1354	1353	1352
1351	1350	1349	1348	1347	1346	1345	1344	1343	1342	1341	1340	1339	1338	1337	1336	1335	1334			

3.3.6 MAHO

One of the unique features associated with IS-136 is the ability for a mobile assisted handoff (MAHO). The MAHO process enables the mobile to constantly report back to the cell site indicating its present condition in the network. The cell site also collects data on the mobile through reverse link measurements, but the forward link (base to mobile) is evaluated by the mobile itself, therefore, providing critical information about the status of the call.

For the MAHO process the mobile measures the received signal strength level (RSSI) from the cell site. The mobile also performs a BER test and a frame error rate (FER) test as other performance metrics. The mobile also measures the signals from a maximum of six potential digital handoff candidates utilizing either a dedicated control channel or a beacon channel. The channels utilized by the mobile for the MAHO process are provided by the serving cell site for the call. The dedicated control channel is either the primary or secondary control channel, and the measurements are performed on the forward link. The beacon channel is either a TDMA voice channel or an analog channel, both of which transmit continuously with no dynamic power control on the forward link. The beacon channel is utilized when the setup or control channel for the cell site has an omni configuration and not a dedicated setup channel per sector.

3.3.7 Frequency reuse

The modulation scheme utilized by the NADC TDMA system is a pi/4 DQPSK format. The C/I levels used for frequency management associated with IS-54 or IS-136 is the same for analog, 17 dB C/I. The C/I level desired is 17 dB and is the same for the DCCH and DTC. This is convenient since in all the cellular systems the majority of the channels are analog, and they too require a minimum of 17 dB C/I. The fundamental issue here is that the same D/R ratios can be used when implementing the radio channel assignments for digital.

The additional parameters associated with IS-136 and IS-54 involve digital color code (DCC), supplementary digital color code (SDCC), and digital verification color code (DVCC). DCC and SDCC must be assigned to each sector, cell, and control channel of the system that utilizes IS-136 or IS-54. The DCC is used by analog and dual-mode phones for accessing the system. The SDCC is used by dual-mode phones only and should be assigned to each control channel along with the DCC.

Parameter	Values
DCC	0, 1, 2, 3
SDCC	0 through 15

The reuse pattern for the DCC and SDCC is shown in Fig. 3.18.

The DVCC is assigned to each DTC. The DVCC values range from 1 to 255, which leaves much room for variations in the assignment methodology. However, the DVCC values should be assigned in a reuse pattern. The recom-

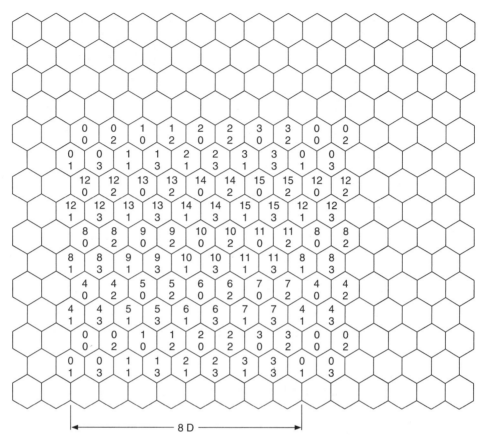

Figure 3.18 SDCC (top number) and DCC (bottom number) reuse pattern.

mended pattern is shown in Fig. 3.19. The same DVCC should be assigned to all the channels in the sector even if there are different frequency groups assigned to that sector.

3.4 CDMA

Code-division multiple access (CDMA) is now a widely accepted wireless mobile access platform. The term CDMA in this context is directed at what is referred to as narrowband CDMA using a 1.25-MHz channel spacing. CDMA is based on the principle of direct sequence and is a spread-spectrum technology which enables multiple users to occupy the same radio channel (frequency spectrum) at the same time. CDMA has been and is being utilized for microwave point-to-point communication, for satellite communication, and also by the military.

Subscribers of CDMA each utilize their own unique code to differentiate themselves from other subscribers. CDMA offers many unique features includ-

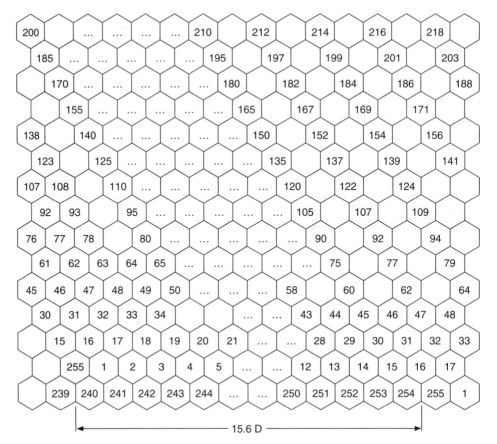

Figure 3.19 DVCC and traffic channel reuse.

ing the ability to thwart interference and improved immunity to multipath effects due to its bandwidth. The CDMA channel utilized is reused in every cell of the system and is differentiated by the pseudorandom number (PN) code that it utilizes. Depending on whether the system will be deployed in an existing AMPS or new PCS band system, the design concepts are fundamentally the same with the exception of frequency band particulars that are directly applicable to the channel assignments in an existing cellular band. Beyond the nuances, the design principles for CDMA are the same for cellular and PCS systems.

CDMA has seen many advancements since its introduction into commercial wireless mobile systems. It is now defined by several standards, IS-95, J-STD-008 (PCS band), and CDMA2000. CDMA is currently used primarily in the United States and Asia. Under CDMA2000 there are several variants which are covered later in this chapter.

The benefits associated with CDMA are

Increased system capacity over analog and TDMA systems

Improved interference protection

No frequency planning required between CDMA channels

Improved handoffs with MAHO and soft handoffs

Fraud protection due to encryption and authentication

Accommodates new wireless features, like data

What follows is a brief description of IS-95 which transitions into CDMA2000. The differences between IS-95 and CDMA2000 are numerous, but CDMA2000 is backward-compatible with IS-95. What CDMA2000 enables is a doubling of the voice-carrying capacity of a CDMA carrier, through the introduction of more Walsh codes, from 64 to 128 and then in the future to 256. In addition, CDMA2000 enables the introduction of high-speed data for mobility, *high speed* being defined by the IMT-2000 specification.

3.4.1 IS-95 system description

IS-95 has two distinct versions, IS-95A and IS-95B, besides the J-STD-008. The J-STD-008 is compatible with both IS-95A and B with the exception of the frequency band of operation. However, the primary difference between IS-95A and B is that IS-95B enables ISDN-like data rates to exist. The diagram shown in Fig. 3.20 is a simplified version of the IS-95A/B architecture.

Regardless of whether the system will be deployed in an existing AMPS or new PCS band system, the design concepts are fundamentally the same, with the exception that AMPS needs a guard zone to protect against analog interference.

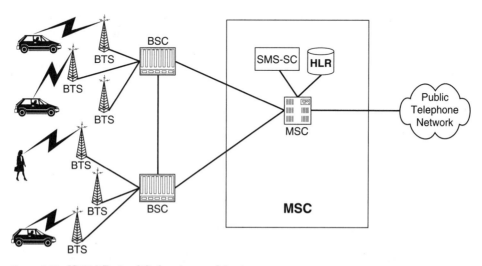

Figure 3.20 IS-95A/B simplified system architecture.

Utilizing the IS-95 standard the same functionalities, with the exception of frequency band particulars, apply to both the 800- and 1900-MHz bands. The primary difference between the 800- and 1900-MHz bands is that CDMA has a few nuances that are directly applicable to the channel assignments in an existing cellular band. Beyond that fundamental issue the discussion that follows, with the exception of the examples, will apply to both cellular and PCS systems.

As briefly mentioned, the introduction of CDMA into an existing cellular network is not plug and chug since there is the issue of immediate capacity reduction, but with a long-term upside. Also there is a requirement that PCS operators relocate existing microwave links to clear the spectrum for their use. The degree of ease or difficulty for implementing CDMA into the PCS market will be directly impacted by the operator's ability to clear microwave spectrum. This fact is complicated even further as more CDMA carriers are introduced into the wireless system.

3.4.2 CDMA2000

CDMA2000 comes under the specification of IS-2000, which is backward-compatible with the IS-95A and B and J-STD-008 specifications, which collectively are called CDMAOne. CDMA2000 is a 3G specification but is backward-compatible with CDMAOne systems, allowing operators to make strategic deployment decisions in a graceful fashion.

Since CDMA2000 is backward-compatible with existing CDMAOne networks, changes (upgrades) to the network from a fixed network aspect can be done in stages. More specifically the upgrades to the network involve the base transceiver station (BTS) with multimode channel element cards, the BSC with Internet protocol (IP) routing capability, and the introduction of the packet data serving network (PDSN). The radio channel bandwidth is the same for CDMA2000-1X as for existing CDMAOne channels, leading to a graceful upgrade. Of course, the subscriber units (mobiles) need to be capable of supporting the CDMA2000 specification, but this can be done in a more gradual fashion since the existing CDMAOne subscriber units can utilize the new network.

There are several terms used to describe CDMA2000 for the different radio carrier platforms, some of which exist and some of which are in the developmental phase, as follows:

CDMA2000-1X (1XRTT)
- 1XEV
- 1xDO
- 1XDV

CDMA2000-3X (3XRTT) (future)

The 1XRTT platform utilizes a single carrier requiring 1.25 MHz of radio spectrum, which is the same as for the existing CDMAOne system. However, the 1XRTT platforms utilize a different vocoder and introduce more Walsh codes,

256 and 128 versus 64, allowing for higher data rates and more voice conversions than are possible over existing CDMAOne systems. Under the 1XRTT platform, there are three primary methods, 1XEV, 1xDO, and 1XDV, which are not mutually exclusive. The term 1XEV is used to describe the first version of 1X. The other two are more specific; 1xDO means one carrier supporting data services only, and 1XDV means one carrier supporting both data and voice services. CDMA2000-3X uses 3.75 MHz of spectrum (3 times 1.25 MHz), with a change in the modulation scheme as well as vocoders, to mention a few of the salient issues that come about with the introduction of this platform.

Another important aspect of CDMA2000 is that it supports not only IS-41 system connectivity, as does IS-95, but also GSM mobile application part (GSM-MAP) connectivity requirements leading to the eventual harmonization of dual-system deployment in the same market by a wireless operator wishing to deploy W-CDMA and CDMA2000 concurrently.

The introduction of CDMA2000 into a network requires new or upgraded (from a CDMAOne system) major platforms. The platform upgrades involve the BTS and BSC, which can be facilitated by module additions or swaps, depending on the infrastructure vendor being used. Whether the system is new or upgraded from a CDMAOne system, the heart of the packet data services for a CDMA2000 network is the introduction of the PDSN, which is discussed in Sec. 3.4.4.

To understand the radio and network components required for the successful implementation of a CDMA2000 system, it is best to start with a simplified network layout for a CDMAOne system. Figure 3.20 is a stand-alone CDMAOne system employing several BTSs which are homed (connected) to two BSCs. Although the BSCs are not shown collocated with the MSC, they could be collocated depending on the specific interconnection requirements and commercial agreements arrived at. The home location register (HLR) is shown, but many of the supporting systems are left out of the picture for simplification purposes. The backhaul from the BTSs to the BSC and from the BSC to the MSC could be via microwave links or fixed facilities.

A general CDMA2000 network is shown in Fig. 3.21. To keep the diagram less cluttered the connectivities to other like networks are not shown. Figure 3.21 shows the new platforms required to support the CDMA2000 network over a CDMAOne system as depicted in Fig. 3.20.

3.4.3 CDMA radio network

The radio networks for IS-95 and CDMA2000 have many similarities, which ensures backward-compatibility. The following descriptions regarding the radio network of a CDMA system will utilize material directly associated with the CDMA2000 specification.

A CDMA2000 system has several enhancements over existing IS-95 and J-STD-008 wireless systems, including better power control, diversity transmit, modulation scheme changes, new vocoders, uplink pilot channel, expansion of the existing Walsh codes, and channel bandwidth changes. The CDMA2000

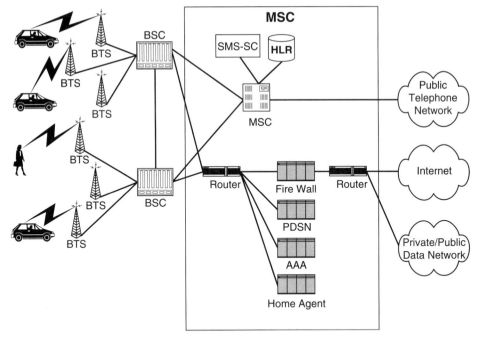

Figure 3.21 CDMA2000 system architecture.

radio systems following the IS-2000 specification are designed to allow an existing CDMAOne operator a phased entrance into the 3G arena.

The CDMA2000 radio network for phase 1 implementation, also called CDMA2000 1XRTT, is the same as that defined for the IS-95 and J-STD-008 systems where the channel bandwidth is 1.25 MHz. However, a channel bandwidth change takes place with the introduction of CDMA2000 phase 2, which is referred to as CDMA2000-3XRTT. A brief, simplified channel bandwidth diagram is shown in Fig. 3.22. The figure illustrates the radio carrier differences between the CDMA IS-95, 1XRTT, and 3XRTT systems.

CMDA2000 introduces several new channel types for the radio access scheme. The new channel types are implemented in both the 1XRTT and 3XRTT schemes and are introduced to support high-speed data as well as enhanced paging functions. To accomplish higher data rates, CDMA2000 uses a combination of expanded Walsh codes along with modulation and vocoder changes.

As in Fig. 3.22, a wireless operator can upgrade to CDMA2000 from either the IS-95A or IS-95B platforms using the same amount of existing spectrum. Here are two of the common progressive deployment paths for implementing CDMA2000 as an upgrade of the CDMAOne (IS-95A/B) platform:

1. CDMAOne (IS-95A) → CDMA2000 (phase 1) → CDMA2000 (phase 2)
2. CDMAOne (IS-95A) → CDMAOne (IS-95B) → CDMA2000 (phase 1) → CDMA2000 (phase 2)

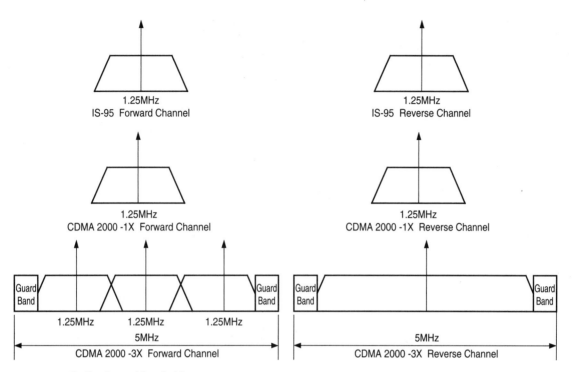

Figure 3.22 Radio channel bandwith.

The CDMA2000 radio access scheme has several enhancements over existing IS-95 systems:

Forward link
1. Fast power control
2. Quadrature phase-shift keying (QPSK) modulation rather than dual binary phase-shift keying (BPSK)

Reverse link
1. Pilot signal, to allow coherent demodulation for the reverse link
2. Hybrid phase-shift keying (HPSK) spreading in the reverse link

Table 3.10 shows the various relationships between the IS-95 and CDMA2000 radio channels.

3.4.4 PDSN

The packet data serving node (PDSN) is a new component associated with CDMA2000 systems as compared to CDMAOne systems. The PDSN is an essential element in the treatment of packet data services which will be offered, and its location in the CDMA2000 network is shown in Fig. 3.21. The purpose of the PDSN is to support packet data services. The PDSN performs the following major functions in the course of a packet data session.

TABLE 3.10 Relationships between IS-95 and CDMA2000

Platform	Description
IS-95A	Primarily voice with circuit switch speeds of 9600 bit/s or 14.4 kbit/s
	64 Walsh codes
	SR1 (1.2288 Mcps)
IS-95B	Primarily voice, data on forward link, improved handoff
	64 Walsh codes
	SR1 (1.2288 Mcps)
CDMA2000, phase 1 (1XRTT)	SR1 (1.2288 Mcps)
	Voice and data (packet data via separate channel)
	128 Walsh codes
	Closed-loop power control
CDMA2000, phase 2 (3XRTT)	SR3 (3.6864 Mcps)
	Primarily data
	Higher data rate
	256 Walsh codes

Note: SR = spreading rate, Mcps = million chips per second.

1. Establishes, maintains, and terminates point-to-point protocol (PPP) sessions with the subscriber
2. Supports both simple and mobile IP packet services
3. Establishes, maintains, and terminates the logical links to the radio network across the radio-packet (R-P) interface
4. Initiates authentication, authorization, and accounting (AAA) for the mobile station client to the AAA server
5. Receives service parameters for the mobile client from the AAA server
6. Routes packets to and from the external packet data networks
7. Collects usage data that are relayed to the AAA server

The overall capacity of the PDSN is determined by both the throughput and number of PPP sessions that are being served. The specific capacity of the PDSN is of course dependent upon the infrastructure vendor used as well as on the particular card population which is implemented. It is important to note that capacity is only one aspect of the dimensioning process and that the overall network reliability factor must be addressed in the dimensioning process.

AAA. The authentication, authorization, and accounting (AAA) server is a new component associated with CDMA2000 deployment. The AAA provides, as its name implies, authentication, authorization, and accounting functions for the packet data network associated with CDMA2000 and utilizes the RADIUS protocol.

The AAA server, as shown in Fig. 3.21, communicates with the PDSN via IP and performs the following major functions in its role in a CDMA2000 network.

- Authentication associated with PPP and mobile IP connections
- Authorization (service profile and security key distribution and management)
- Accounting

HA. The home agent (HA) is a component of the CDMA2000 packet data service network and should be compliant with IS-835 which is relevant to the home agent functionality within a wireless network. The HA performs many tasks, one of which is tracking the location of the mobile IP subscriber as it moves from one packet zone to another. In tracking the mobile, the HA will also ensure that the packets are forwarded to the mobile itself.

Router. The router shown in Fig. 3.21 has the function of routing packets to and from the various network elements within a CDMA2000 system. The router is also responsible for sending and receiving packets to and from the internal network to the off-net platforms. A firewall, not shown in the figure, is needed to ensure security is maintained when connecting to off-net data applications.

HLR. The home location register (HLR) used in existing IS-95 networks needs to store additional subscriber information associated with the introduction of packet data services. The HLR performs the same role for packet services as it does for voice services in that it stores the subscriber packet data service options and terminal capabilities along with the traditional voice platform needs. The service information from the HLR is downloaded in the VLR of the associated network (switch) during the successful registration process much the same as is done in existing IS-95 systems and other 1G and 2G voice-oriented systems.

BTS. The base transceiver site (BTS) is the official name of the cell site. The BTS is responsible for allocating resources (both power and Walsh codes) for consumption by subscribers. The BTS also has the physical radio equipment used for transmitting and receiving CDMA2000 signals.

The BTS controls the interface between the CDMA2000 network and the subscriber unit. It controls many aspects of the system which are directly related to the performance of the network. Some of the items the BTS controls are the multiple carriers which operate from the site; the forward power (allocated for traffic, overhead, and soft handoffs); and, of course, the assignment of the Walsh codes.

With CDMA2000 systems the use of multiple carriers per sector, as with IS-95 systems, is possible. Therefore, when a new voice or packet session is initiated, the BTS must decide how best to assign the subscriber unit to meet the services being delivered. The BTS in the decision process not only examines the service requested but also must consider the radio configuration and the subscriber type and, of course, whether the service requested is voice or packet. Therefore, the resources the BTS has to draw upon can be both physically and logically limited depending on the particular situation involved.

The BTS can perform a downgrade from a higher radio configuration (RC) or spreading rate, to a lower radio configuration if

1. The resource request is not a handoff.
2. The resource request is not available.
3. Alternative resources are available.

The following is a brief summary of some of the physical and logical resources the BTS must allocate when assigning resources to a subscriber.

Fundamental channel (FCH).   Number of physical resources available.

FCH power.   Power already allocated and available power.

Walsh codes.   Those required and those available.

Total FCHs.   Per sector.

The physical resources the BTS draws upon also involve the management of the channel elements which are required for both voice and packet data services.

Integral to the resource assignment scheme is the Walsh code management. For 1XRTT (whether 1XEV, 1xDO, or 1XDV) there are a total of 128 Walsh codes to draw upon. However, with the introduction of 3X the Walsh codes are expanded to a total of 256.

For CDMA2000-1X the voice and data distribution is handled by parameters that are set by the operator which involve

Data resources.   Percent of available resources, including the FCH and supplemental channel (SCH).

FCH resources.   Percent of data resources.

Voice resources.   Percent of total available resources.

A brief example to help facilitate the issue of resource allocation is given in Table 3.11. Obviously the allocation of data and FCH resources directly controls the amount of simultaneous data users on a particular sector or cell site.

BSC.   The base site controller (BSC) is responsible for controlling all the BTSs under its domain. The BSC routes packets to and from the BTSs to the PDSN. In addition, the BSC routes time-division multiplexing (TDM) traffic to the circuit-switched platforms and packet data to the PDSN.

3.4.5 CDMA channel allocation

The CDMA2000 channel allocations, just as with IS-95, have preferred locations and methods for deployment which are envisioned at this time to help

TABLE 3.11 Channel Resource Allocations Example

Topic	Percentage	Resources
Total resources	100	64
Voice resources	70	44
Data resources	30	20
FCH resources	40	8

facilitate the migration from 1X to 3X in the future. Tables 3.12 and 3.13 are for North America. Note that the channels defined in the tables are for 1.25-MHz channel spacing. The first carrier is used in 3X for access, and this then is used to help steer the subscriber to the correct carrier(s) to support the services being requested.

3.4.6 Forward channel

The forward link for a CDMA2000 channel (whether for IS-95, 1X, or 3X implementation) utilizes the structure shown in Fig. 3.23. Here the forward CDMA channel consists of the pilot channel, 1 sync channel, up to 7 paging channels, and 64 traffic channels for IS-95 and possibly 128 for CDMA2000. The cell site transmits the pilot and sync channels for the mobile to use when acquiring and synchronizing with the CDMA system. When this occurs, the mobile is in the mobile station initiation state. The paging channel, also transmitted by the cell site, is used by the subscriber unit to monitor and receive messages that might be sent to it during the mobile station idle state or system access state.

The pilot channel is continuously transmitted by the cell site. Each cell site utilizes a time offset for the pilot channel to uniquely identify the forward CDMA channel to the mobile unit. There are 512 possible different time-offset values for the cell site to utilize. If there are multiple CDMA channels assigned to a cell site, the cell will still utilize only one time-offset value. The time offset is utilized during the handoff process.

The sync channel is a forward channel that is used during the system acquisition phase. Once the mobile acquires the system, it will not normally reuse the sync channel until it powers on again. The sync channel provides the mobile with the timing and system configuration information. It utilizes the same spreading code time offset as the pilot channel for the same cell site. The sync channel frame is the same length as the pilot pseudorandom number (PN) sequence. The information sent on the sync channel is the paging channel rate and the time of the base station's pilot PN sequence with respect to the system time.

The cell site utilizes the paging channel to send overhead information and subscriber-specific information. The cell site will transmit, at minimum, one paging channel for each supported CDMA channel that has a sync channel.

Once the mobile has obtained the paging information from the sync channel, the mobile will adjust its timing and begin monitoring the paging channel. Each

TABLE 3.12 Cellular CDMA2000 1X and 3X Carrier Assignment Scheme

		Cellular system	
Carrier	Sequence	A	B
1	F1	283	384
2	F2	242*	425*
3	F3	201	466
4	F4	160	507
5	F5	119	548
6	F6	78	589
7	F7	37	630
8	F8 (not advised)	691	777

*Location where the first of three 1.25-MHz carriers is expected to be located for a 3X deployment.

TABLE 3.13 PCS CDMA2000 1X and 3X PCS Carrier Assignment Scheme

	PCS system					
Carrier	A	B	C	D	E	F
1	25	425	925	325	725	825
2	50	450	950	350*	750*	850*
3	75*	475*	975*	375	775	875
4	100	500	1000	NA	NA	NA
5	125	525	1025	NA	NA	NA
6	150*	550*	1050*	NA	NA	NA
7	175	575	1075	NA	NA	NA
8	200	600	1100	NA	NA	NA
9	225*	625*	1125*	NA	NA	NA
10	250	650	1150	NA	NA	NA
11	275	675	1175	NA	NA	NA

*Location where the first of three 1.25-MHz carriers is expected to be located for a 3X deployment.
Note: NA = not available.

mobile, however, only monitors a single paging channel. The paging channel conveys four basic types of information:

1. *Overhead information.* Conveys the system's configuration by sending the system and access parameter messages, the neighbor lists, and CDMA channel list messages.

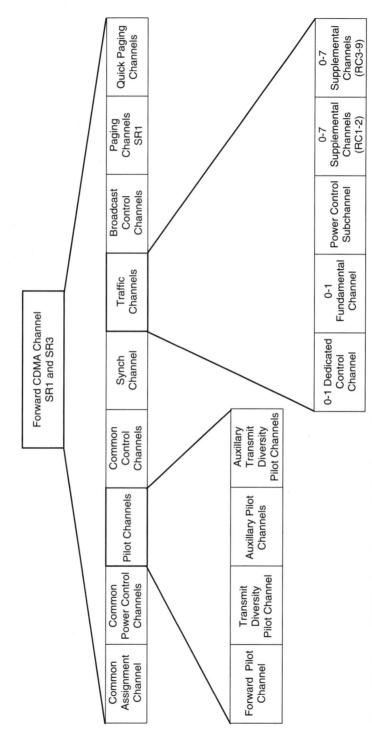

Figure 3.23 Forward CDMA channel transmitted by base station.

2. *Paging.* The cell site pages a mobile unit for a land-to-mobile or mobile-to-mobile call.

3. *Channel assignments.* The base station sends a message to assign a mobile to the traffic channel, alter the paging channel assignment, or redirect the mobile to utilize the analog FM system.

4. *Forward traffic channel.* Used for transmission of primary or signaling traffic to a specific subscriber unit during the duration of the call. The forward traffic channel also transmits the power control information on a subchannel continuously as part of the closed-loop system. The forward traffic channel will also support the transmission of information at 9600, 4800, or 1200 bit/s utilizing a variable rate which is selected on a frame-by-frame basis, but the modulation symbol rate remains constant.

To review, the base station transmits multiple common channels as well as several dedicated channels to the subscribers in their coverage area. Each CDMA2000 user is assigned a forward traffic channel which consists of the following combinations:

1	Forward fundamental channel (F-FCH)
0–7	Forward supplemental channels (F-SCHs) for both radio configuration 1 (RC1) and RC2
0–2	F-SCHs for RC3 to RC9

It is important to note that F-FCHs are used for voice, while F-SCHs are used for data.

When the channel is associated with a 3XRTT implementation, the data for the subscriber are mapped to each of the three different carriers enabling high throughput. However, the Walsh codes are the same for each carrier, meaning they share the same throughput distributing the traffic load evenly.

A CDMA2000 channel, unlike IS-95-only channels, utilizes different modulation schemes depending on the radio configuration employed. The modulation scheme used for RC1 and RC2 is BPSK, while that for RC3 to RC9 is QPSK. For RC3 to RC9 the data are converted into a 2-bit-wide parallel data stream which initially would seem counterintuitive since it reduces the data rate for each stream by a factor of 2. Each data stream, however, is then spread by a 128 Walsh code to get the spreading rate up to 1.2288 million chips per second (Mcps) which effectively doubles the processing gain allowing for greater throughput at the same effective power level. Descriptions of some of the forward channels follow:

Forward supplemental channel (F-SCH). Up to two F-SCHs can be assigned to a single mobile for high-speed data ranging from 9.6 to 153.6 kbit/s. It is important to note that each F-SCH assigned can be assigned at different rates. The F-SCH must be assigned with a reverse-SCH when only one F-SCH is assigned.

Forward quick paging channel (F-QPCH). Extends the mobile battery life by reducing the amount of time the mobile spends parsing pages that are not meant for it. The mobile monitors the F-QPCH, and when the flag is set, the mobile looks for the paging message. There are a total of 3 F-QPCHs per sector.

Forward dedicated control channel (F-DCCH). Replaces the dim and burst and the blank and burst. It is used for messaging and control for data calls.

Forward transmit diversity pilot channel (F-TDPICH). Used to increase RF capacity.

Forward common control channel (F-CCCH). Used to send paging, data, or signaling messages.

Table 3.14 lists many channel types and quantities for CDMA2000.

TABLE 3.14 Forward CDMA2000 Channel Descriptions

Channel type	Maximum number
SR1	
Forward pilot	1
Transmit diversity pilot	1
Sync	1
Paging	7
Broadcast control	8
Quick paging	3
Common power control	4
Common assignment	7
Forward common control	7
Forward dedicated control	1 per forward traffic channel
Forward fundamental	1 per forward traffic channel
Forward supplemental code (RC1 and RC2 only)	7 per forward traffic channel
Forward supplemental (RC3, RC4, and RC5 only)	2 per forward traffic channel
SR3	
Forward pilot	1
Sync	1
Broadcast control	8
Quick paging	3
Common power control	4
Common assignment	7
Forward common control	7
Forward dedicated control	1 per forward traffic channel
Forward fundamental	1 per forward traffic channel
Forward supplemental	2 per forward traffic channel

3.4.7 Reverse channel

The reverse link or channel for CDMA2000 has many properties similar to the forward link and, therefore, differs significantly from that used in IS-95. One of the major enhancements to CDMA2000 over IS-95 is the inclusion of a pilot on the reverse link. The structure of a reverse channel is shown in Fig. 3.24.

The cell site continuously monitors the reverse access channel to receive any message that the subscriber unit might send to the cell site during the system access state. The reverse CDMA channel consists of an access channel and the traffic channel. The access channel provides communication from the mobile to the cell site when the subscriber unit is not utilizing a traffic channel. One access channel is paired with a paging channel, and each access channel has its own pseudorandom number code. The mobile responds to the cell site's messages sent on the paging channel by utilizing the access channel.

The forward and reverse control channels utilize a similar control structure which can vary from 9600, 4800, 2400, or 1200 bit/s which enables the cell or mobile to alter the channel rate dynamically to adjust for the speaker. When there is a pause in the speech, the channel rate decreases to reduce the amount of energy received by the CDMA system increasing the overall system capacity.

There are four basic types of control messages on the traffic channel:

1. Messages that control the call itself
2. Handoff messages
3. Power control
4. Security and authentication

CDMA power control is fundamentally different than that utilized for AMPS or IS-54. The primary difference is that the proper control of total power coming into the cell site, if limited properly, will increase the traffic-handling capability of that cell site. As more energy is received by the cell site, its traffic-handling capabilities will be reduced unless it is able to reduce the power coming into it.

The forward traffic power control is composed of two distinct parts. The first part is that the cell site will estimate the forward links transmission loss utilizing the mobile subscriber's received power during the access process. Based on the estimated forward link path loss, the cell site will adjust the initial digital gain for each of the traffic channels. The second part of the power control involves the cell site making periodic adjustments to the digital gain which is done in concert with the subscriber unit.

The reverse traffic channel signals arriving at the cell site vary significantly and require a different algorithm to be used than that of the forward traffic power control. The reverse channel also has two distinct elements used for making power adjustments. The first element is the open-loop estimate of the transmit power which is performed solely by the subscriber unit without any feedback from the cell site itself. The second element is the closed-loop correction for these errors in the estimation of the transmit power. The power control subchannel is continuously transmitted on the forward traffic channel every 1.25

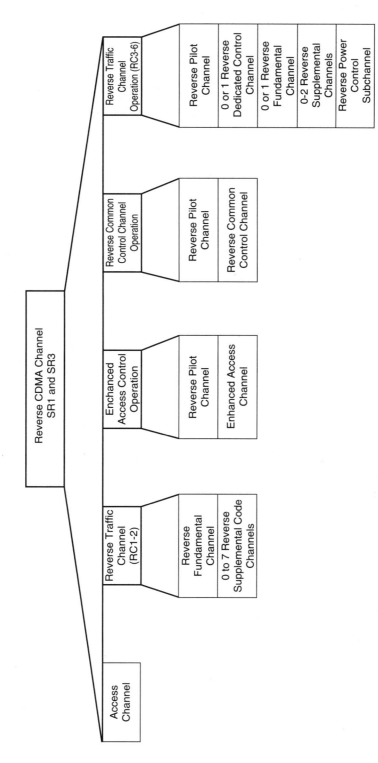

Figure 3.24 Reverse CDMA channel received at base station.

ms instructing the mobile to either power up or power down affecting the mean power output level. There are a total of 16 different power control positions. Table 3.15 illustrates the CDMA subscriber power levels available by station class.

Elaborating on the reverse channel, the subscriber (or mobile) is allowed to transmit more than one code channel to accommodate the high data rates. The minimum configuration consists of a reverse pilot channel to allow the base station to perform synchronous detection and a reverse fundamental channel (R-FCH) for voice. The inclusion of additional channels, such as the reverse supplemental channels (R-SCHs) and the reverse dedicated control channel (R-DCCH), can be used to send data or signaling information. The association between the radio configuration and the spreading rates is best shown in Table 3.18. It is important to note that the reverse channel for 3X is different than for 1X in that it is a direct spread but can be laid over a 1X implementation and then, depending on the subscribers operating in that sector, the appropriate SR and RC are selected. Descriptions of some of the reverse link channels follow:

Reverse supplemental channel (R-SCH). When data rates are greater than 9.6 kbit/s, a R-SCH is required and a R-FCH is also assigned for power control. A total of one or two R-SCHs can be assigned per mobile.

Reverse pilot channel (R-PICH). Provides pilot and power control information. The R-PICH allows the mobile to transmit at a lower power level and inform the base station of the forward power levels being received enabling the base station to reduce power.

Reverse dedicated control channel (R-DCCH). Replaces the dim and burst and the blank and burst. It is used for messaging and control for data calls.

Reverse enhanced access channel (R-EACH). Meant to minimize the collisions and, therefore, reduce the access channel's power.

Reverse common control channel (R-CCCH). Used by mobiles to send their data information after they have been granted access.

Table 3.16 lists many of the reverse channel types and quantities for CDMA2000.

TABLE 3.15 CDMA Subscriber Power Levels

Station class	EIRP (max), dBm
I	3
II	0
III	−3
IV	−6
V	−9

Note: EIRP = effective isotropic radiated power, dBm = decibels in milliwatts.

TABLE 3.16 CDMA2000 Channel Descriptions

Channel type	Maximum number
SR1	
Reverse pilot	1
Access	1
Enhanced access	1
Reverse common control	1
Reverse dedicated control	1
Reverse fundamental control	1
Reverse supplemental code (RC1 and RC2 only)	7
Reverse supplemental	2
SR3	
Reverse pilot	1
Enhanced access	1
Reverse common control	1
Reverse dedicated control	1
Reverse fundamental control	1
Reverse supplemental	2

3.4.8 SR and RC

CDMA2000 defines two spreading rates as compared to IS-95 which has only one. The spreading rates are referred to as spreading rate 1 (SR1) and spreading rate 3 (SR3). SR1 is utilized for IS-95A/B and CDMA2000 phase 1 (1XRTT) implementations, while SR3 is destined for CDMA2000 phase 2 (3XRTT).

For CDMA2000, SR1 has a chip rate of 1.2288 Mcps and occupies the same bandwidth as CDMAOne signals. The SR1 is a direct spread method and follows the same concept as that used for IS-95 systems. However, for 3XRTT an SR3 signal is introduced and has a rate of 3.6864 Mcps (3 × 1.2288 Mcps) and, therefore, occupies 3 times the bandwidth of a CDMAOne or 1XRTT channel. The SR3 system incorporates all the new coding implemented in an SR1 system while supporting even higher data rates. The 3XRTT channel scheme utilizes a multicarrier forward link and direct spread reverse link.

The IS-2000 specification also defines, for both 1XRTT and 3XRTT, a total of nine forward and six reverse link radio configurations as well as two different spreading rates. The radio configurations involve different modulation schemes, coding, vocoders, and spreading rates.

RC1 is backward-compatible with CMDAOne for 9.6 kbit/s voice traffic and supports circuit-switched data rates of 1.2 to 9.6 kbit/s. Although RC3 is based on the 9.6-kbit/s rate and supports variable voice rates from 1.2 to 9.6 kbit/s while also supporting packet data rates of 19.2, 38.4, 76.8, and 153.6 kbit/s, it operates using an SR1.

Tables 3.17 and 3.18 are meant to help illustrate the perturbations that exist with the different radio configurations and spreading rates. Table 3.17 is associated with the forward link, while Table 3.18 is associated with the reverse link.

TABLE 3.17 Forward Link RC and SR

RC	SR	Data rates, bit/s	Characteristics
1	1	1200, 2400, 4800, 9600	$R = 1/2$
2	1	1800, 3600, 7200, 14,400	$R = 1/2$
3	1	1500, 2700, 4800, 9600, 38,400, 76,800, 153,600	$R = 1/4$
4	1	1500, 2700, 4800, 9600, 38,400 76,800, 153,600, 307,200	$R = 1/2$
5	1	1800, 3600, 7200, 14,400, 28,800, 57,600, 115,200, 230,400	$R = 1/4$
6	3	1500, 2700, 4800, 9600, 38,400, 76,800, 153,600, 307,200	$R = 1/6$
7	3	1500, 2700, 4800, 9600, 38,400, 76,800, 153,600, 307,200, 614,400	$R = 1/3$
8	3	1800, 3600, 7200, 14,400, 28,800, 57,600, 115,200, 230,400, 460,800	$R = 1/4$ (20 ms) $R = 1/3$ (5 ms)
9	3	1800, 3600, 7200, 14,400, 28,800, 57,600, 115,200, 230,400, 460,800, 1,036,800	$R = 1/2$ (20 ms) $R = 1/3$ (5 ms)

Note: R = convolutional code rate.

TABLE 3.18 Reverse Link RC and SR

RC	SR	Data rates, bit/s	Characteristics
1	1	1200, 2400, 4800, 9600	$R = 1/3$
2	1	1800, 3600, 7200, 14,400	$R = 1/2$
3*	1	1200, 1350, 1500, 2400, 2700, 4800, 9600, 19,200, 38,400, 76,800, 153,600, 307,200	$R = 1/4$ $R = 1/2$ for 307,200
4*	1	1800, 3600, 7200, 14,400, 28,800, 57,600, 115,200, 230,400	$R = 1/4$
5*	3	1200, 1350, 1500, 2400, 2700, 4800, 9600, 19,200, 38,400, 76,800, 153,600, 307,200, 614,400	$R = 1/4$ $R = 1/2$ for 307,200 and 614,400
6*	3	1800, 3600, 7200, 14,400, 28,800, 57,600, 115,200, 230,400, 460,800, 1,036,800	$R = 1/4$ $R = 1/2$ for 1,036,800

*Reverse pilot.
Note: R = convolutional code rate.

3.4.9 Power control

Power control is a major enhancement of CDMA2000 over IS-95 because it enables higher data rates. The primary power control enhancement is with the fast forward link power control. As discovered through practical implementation issues, CDMA systems are interference-limited, which results in an improvement in system capacity.

Enabling better power control of both the forward and reverse links has several advantages:

- System capacity is enhanced or optimized
- Mobile battery life is extended
- Radio path impairments are properly or better compensated for
- Quality of service at various bit rates can be maintained

Obviously, as with any wireless system that is interference-limited, it is important to ensure that all transmitters, whether located at the mobile or base station, transmit at the lowest power level while maintaining a good communication link. To achieve this, CDMA2000 utilizes a fast response closed-loop power control on the reverse link. In summary, the BTS measures the reverse link from the mobile and sends power control commands to increase or decrease the mobile's power level, which is similar to IS-95. It is important to note that the mobile can also operate autonomously and make power corrections based on the frame error detection (FED) of the forward link, and from that it infers what it needs to do for the reverse link in terms of power control.

There is also a refinement to the closed-loop power control on the reverse link, and that is where the base station performs an outer-loop power control which is a refinement process for the inner power control process. Specifically, if the frame received from the mobile arrives without error, the base station instructs the mobile to power down while on the other side, and if the frame arrives in error, the mobile is instructed to power up.

With CDMA2000 the use of power control on the forward channel is possible with the introduction of the reverse pilot channel. The reverse pilot channel for power control was introduced to help reduce the interference caused by forward energy. Effectively, the mobile measures the received power and compares it against a threshold which the mobile then feeds back to the base station. Upon receipt of the power information, the mobile is then instructed to power up or power down.

In addition, as with the reverse power link, there is an outer-loop power control process which is used to dynamically adjust the target Eb/No, which is done by measuring the FER with a target FER, and if the FER is greater than the target, it is instructed to power up, and if it is below the target FER, it is instructed to power down.

3.4.10 Walsh codes

CDMA2000 introduces an increase in the number of Walsh codes from 64 with IS-95 to a total of 256 with 3XRTT. As with IS-95, CDMA2000 utilizes pseudo-random number long codes for both the forward and reverse directions. However, in CDMA2000 the introduction of variable-length Walsh codes is introduced to accommodate fast packet data rates.

The Walsh code chosen by the system is determined by the type of reverse channel. The R-SCH also uses a reserve Walsh code. If only one R-SCH is used, it utilizes a 2- or 4-chip Walsh code, but when the second R-SCH is used, it will utilize a 4- or 8-chip code. Therefore, in order to maintain or obtain the higher data rates on the F-SCH, the Walsh code must be shorter in order to maintain the same spreading rate.

Table 3.19 shows the relationship between Walsh codes, SR, RC, and, of course, data rates. One very important effect with utilizing variable-length Walsh codes is that if a shorter Walsh code is being used, then it precludes use of the longer Walsh codes which are derived from it. Table 3.19 helps in establishing the relationship between which Walsh code length is associated with a particular data rate. Table 3.20 is simplified to show the maximum number of simultaneous users for any data rate.

For an SR1 and RC1 there is a maximum number of users that have individual Walsh codes equating to 64, a familiar number from IS-95A. It is important to note that the shorter Walsh codes inhibit the use of longer Walsh codes because of the orthogonality required. Also all channel requests are allocated from the same Walsh code pool on a per sector basis. In addition, to achieve the

TABLE 3.19 Walsh Code Tree Table*

	Number of Walsh codes						
RC	256	128	64	32	16	8	4
				SR1			
1	NA	NA	9.6	NA	NA	NA	NA
2	NA	NA	14.4				
3	NA	—	9.6	19.2	38.4	76.8	153.6
4	NA	9.6	19.2	38.4	76.8	153.6	307.2
5	NA	NA	14.4	28.8	57.6	115.2	230.4
				SR3			
6	—	9.6	19.2	38.4	76.8	153.6	307.2
7	9.6	19.2	38.4	76.8	153.6	307.2	614.4
8	—	14.4	28.8	57.6	115.2	230.4	460.8
9	14.4	28.8	57.6	115.2	230.4	460.8	1036.8

*Values in this table are data throughput speeds in kbit/s.
Note: NA = not available.

TABLE 3.20 Simultaneous Users with SR1 and SR3*

RC	\multicolumn{7}{c}{Simultaneous users†}
	256	128	64	32	16	8	4
\multicolumn{8}{c}{SR1}
1	NA	NA	9.6	NA	NA	NA	NA
2	NA	NA	14.4				
3	NA	—	9.6	19.2	38.4	76.8	153.6
4	NA	9.6	19.2	38.4	76.8	153.6	307.2
5	NA	NA	14.4	28.8	57.6	115.2	230.4
\multicolumn{8}{c}{SR3}
6	—	9.6	19.2	38.4	76.8	153.6	307.2
7	9.6	19.2	38.4	76.8	153.6	307.2	614.4
8	—	14.4	28.8	57.6	115.2	230.4	460.8
9	14.4	28.8	57.6	115.2	230.4	460.8	1036.8

*Values in this table are data throughput speeds in kbit/s.
†Maximum number.
Note: NA = not available.

higher data rate, not only must the Walsh codes be implementation-modified but also the modulation scheme has to be changed.

3.4.11 Call and data processing

For IS-95 and CDMA2000 there are several types of call and data processing that take place. Depending on whether the service being requested or offered is circuit-switched or packet-switched, the call-processing methods are different. Regardless of whether the service being delivered is circuit- or packet-based, the system still will perform handoffs in addition to power control. There are a few nuances to be concerned about depending on if the service is packet versus voice, but the fundamental concepts of how neighbors are promoted and demoted is still the same.

We will now discuss various circuit-switched and packet-switched calls.

Call processing. The call flows for CDMA are shown in Figs. 3.25 and 3.26. It is important to note that IS-95 is primarily a voice-oriented and not a data-oriented system. Data are available to be sent via circuit-switched methods, but the call-processing flow is the same as that for voice since it still utilizes a traffic channel set up for voice transport. In addition, the call-processing flowcharts illustrated can easily be expanded to include the variations of subscriber types for Walsh code allocations along with different spreading rates. The first call-processing flowchart (Fig. 3.25) is for a mobile-to-land call (origination), while the second flowchart (Fig. 3.26) illustrates a land-to-mobile call (termination).

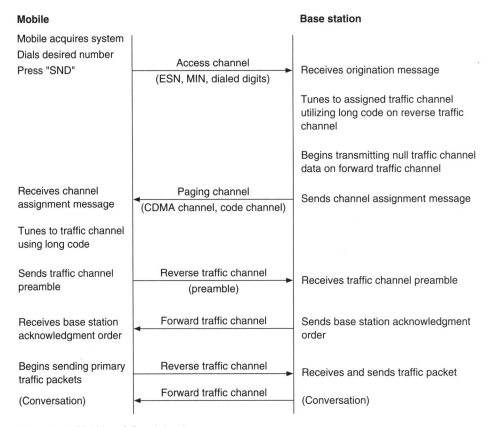

Figure 3.25 CDMA mobile origination.

Packet Data Transport Process Flow. CDMA2000 data services fall into two distinct categories, circuit-switched and packet-switched. Circuit-switched data are handled effectively the same as a voice call. But for all packet data calls a PDSN is used as the interface between the air interface data transport and the fixed network transport. The PDSN interfaces to the base station through a packet control function (PCF) which can be collocated with the base station.

The CDMA2000 has three packet data service states which need to be understood in the process:

Active/connective. A physical traffic channel exists between the subscriber unit and the base station with packet data being sent and received in a bidirectional fashion.

Dormant. No physical traffic channel exists, but a PPP link between the subscriber unit and the PDSN is maintained.

Null/inactive. Neither a traffic channel nor a PPP link is maintained or established.

Wireless Mobile Radio Technologies 125

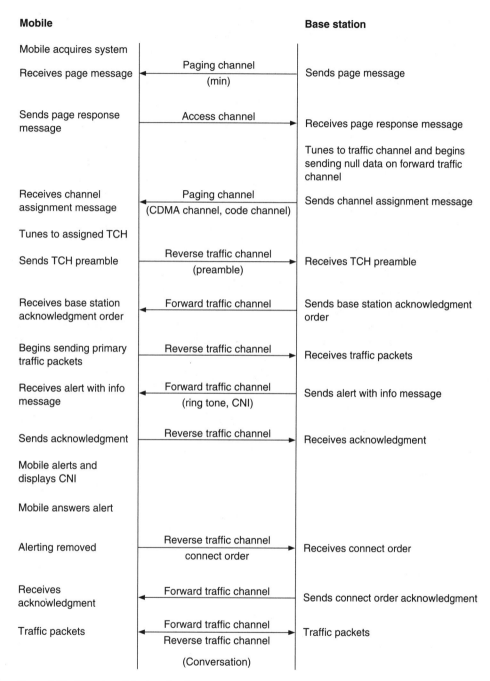

Figure 3.26 CDMA mobile termination.

The relationship between the three packet data states is best shown by the simplified state diagram in Fig. 3.27.

CDMA2000 introduces to the mobile environment real packet data transport and treatment at speeds that meet or exceed the ITM-2000 system requirements. The voice call processing that is implemented by CDMA2000 is functionally the same as that of existing CDMAOne networks with the exception that there is a vocoder change in the subscriber units. However, the key difference is that the network can now handle the packet data.

The mobile initiates the decision as to whether the session will be a packet data session, voice session, or concurrent (meaning voice and data). The network at this time cannot initiate a packet data session with the subscriber unit with the exception of a systems management server.

For call processing, the voice and data networks are segregated in general once the information (voice or data) leaves the radio environment at the BSC itself. Therefore, for packet data, the PDSN is central to all decisions made. Refer back to Fig. 3.21 which depicts a generalized network architecture.

The PDSN does not communicate directly with the voice network nodes like the HLR and visitor location register (VLR); instead it communicates via the AAA. As discussed previously, the voice and data networks normally are segregated once they leave the radio environment at the BSC. Additionally in a CDMA2000 network the system utilizes a PPP between the mobile and the PDSN for every type of packet data session that is transported and/or treated.

The notion of mobility is fundamental to the concept of CDMA2000. Therefore, the diagram shown in Fig. 3.28 illustrates some of the internetwork communication that needs to take place to establish a packet data session. The transport of the packets is not depicted in the figure—just the elements in the network which need to communicate in order to establish

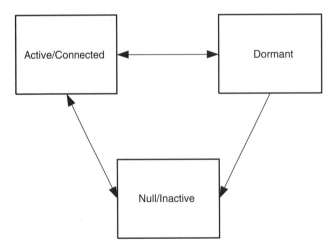

Figure 3.27 Packet data states.

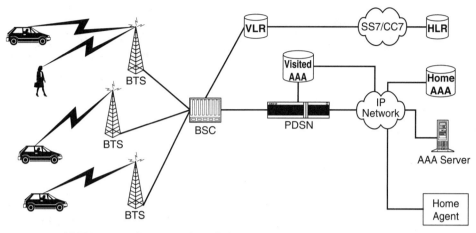

Figure 3.28 CDMA2000 packet network nonhome.

what services the subscriber is allowed to have and how the network is going to meet the service-level agreement that is expected for the packet session.

The VLR, which is normally colocated with the MSC, is shown in Fig. 3.28. When the subscriber initiates a packet data session, the BSC (via the MSC) checks the subscription information prior to the system granting the service request to the mobile subscriber. This will take place prior to the PDSN being involved with the packet session.

The two packet sessions available for use within a CDMA2000 network are simple and mobile IP. Each of these packet session types has two variants, where one uses a virtual private network (VPN) and the other does not. We now discuss each of these variants.

Simple IP. Simple IP is very similar to commonly used dial-up Internet connections over standard landline facilities. A PPP session is established between the mobile and the PDSN. The PDSN basically routes packets to and from the mobile in order to provide end-to-end connectivity between the mobile and the Internet. A diagram depicting simple IP is shown in Fig. 3.29.

When using simple IP, the mobile must be connected to the same PDSN for the duration of the packet session. If the mobile while in transit moves to a coverage area whose BSCs and BTSs are homed out of another PDSN, the simple IP connection is lost and needs to be reestablished. The loss of the existing packet session effectively is the same as when the Internet connection on the landline is terminated and you need to reestablish the connection.

Refer to Fig. 3.29, which is a simplified model of the simple IP implementation. Many of the details are left out, but the concept shows that the mobile is connected to the PDSN using a PPP connection in a best-effort data delivery method at an agreed-upon transfer rate. This transfer rate is determined by the subscriber's profile, radio resource availability, and, of course, the radio environment itself.

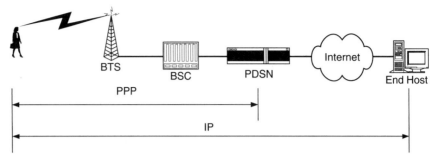

Figure 3.29 Simple IP.

The IP address of a mobile is linked to the PDSN which can be static or dynamic host configuration protocol (DHCP); for simple IP the choice is DHCP. A mobile with an active or dormant data call can traverse around the network going from cell to cell provided it stays within the PDSN's coverage area. Additionally, the PDSN should support both challenge handshake authentication protocol (CHAP) and password authentication protocol (PAP).

As indicated, simple IP does not allow the subscriber full mobility with packet data calls. After exiting the PDSN coverage area, the subscriber must negotiate for a new IP address from the new PDSN, which, of course, results in the termination of the existing packet session and requires a new session.

Regarding the radio environment, the CDMA2000 radio network provides the mobile with a traffic channel which consists of a fundamental channel and possibly a supplemental channel for higher traffic speeds. To help explain the use of simple IP, a packet session flowchart is shown in Fig. 3.30. The figure shows the situation where a subscriber is operating in the home PDSN network. Figure 3.31 shows a flowchart where the mobile is considered to be roaming.

Simple IP with VPN. An enhancement to simple IP is the ability to introduce a VPN to the path for security and also to provide connectivity to a corporate local area network (LAN) or other packet networks. With VPN the mobile user should have the appearance of being connected directly to the corporate LAN.

The PDSN establishes a tunnel using layer 2 tunneling protocol (L2TP) between the PDSN and the private data network. The mobile is effectively still using a PPP connection, but it is tunneled. The private network that the PDSN terminates to is responsible for assigning the IP address and, of course, authenticating the user beyond what the wireless system needs to perform for billing purposes.

Because the specific termination and authentication is performed by another network, the PDSN does not apply any IP services for the mobile and, therefore, except for the predetermined speed of the connection, that is all the system can provide.

Just as in the case of simple IP, the mobile must still be connected to the same PDSN for the packet session. If the mobile moves to another area of the network which is covered by a separate PDSN, the VPN is terminated and the

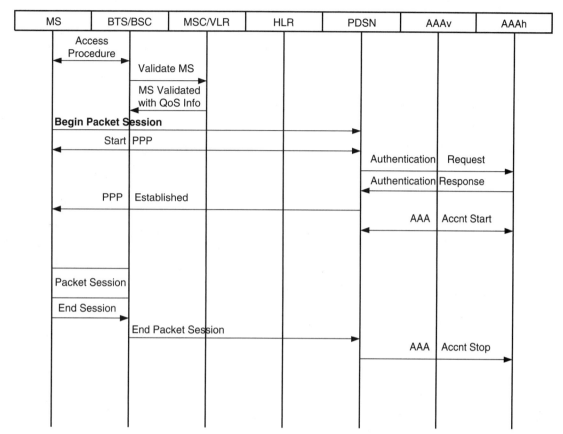

Figure 3.30 Simple IP flowchart.

mobile must reestablish the session. A simplified diagram is shown in Fig. 3.32. The packet session flowchart for simple IP with VPN is shown in Fig. 3.33 and assumes that the subscriber is *not* roaming.

Mobile IP (3G). Although mobile IP is a packet transport method, it is quite different than simple IP in that it actually transports the data. Mobile IP utilizes a static IP address that can be assigned by the PDSN. The establishment of a static IP address facilitates roaming during the packet session, provided the static IP address scheme is unique enough for the subscriber unit to be uniquely identified.

With mobile IP he PDSN is the foreign agent, and the home agent is set up as a virtual home agent. The mobile needs to register each time it begins a packet data session, whether it is originating or terminating. Also the PDSN on the visited network terminates the packet session using an IP-in-IP tunnel. The home agent delivers the IP traffic to the foreign agent through an IP-in-IP tunnel.

The mobile is responsible for notifying the system that it has moved to another service area. Once the mobile has moved to another service area, it

130 Chapter Three

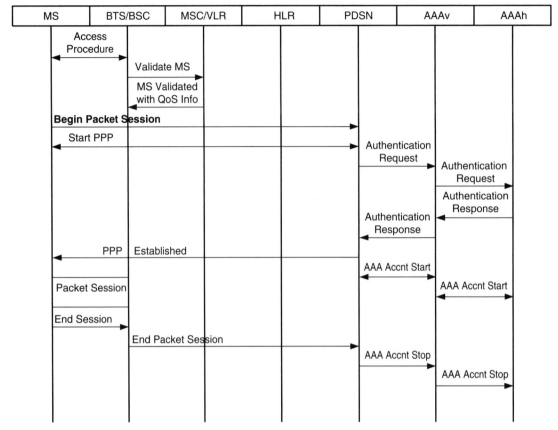

Figure 3.31 Simple IP roaming flowchart.

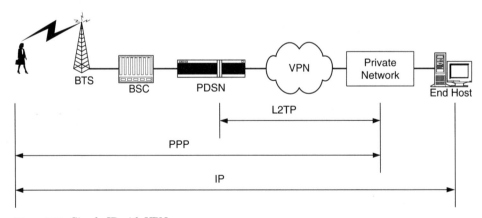

Figure 3.32 Simple IP with VPN.

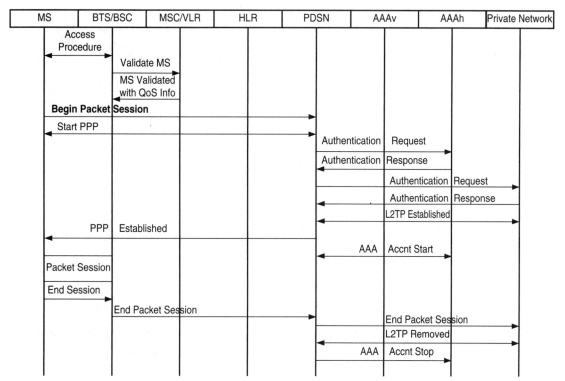

Figure 3.33 Simple IP with VPN session flowchart.

needs to register with another foreign agent. The foreign agent assigns the mobile a care-of address (COA). The home agent forwards the packets to the visited network for termination on the mobile and then encapsulates the original IP packet destined for the mobile using the COA. The foreign agent, using IP-in-IP tunneling, extracts the original packet and routes it to the mobile.

The IP address assignment is done via DHCP and is mapped to the home agent. However, PAP and CHAP are not used for mobile IP as in simple IP.

In the reverse direction, the routing of IP packets occurs the same as if on the home network and does not require an IP-in-IP tunnel unless the wireless operator decides to implement reverse IP tunneling. To summarize:

- The PDSN in the visited network always terminates the IP-in-IP tunnel.
- The home agent delivers the IP traffic through the mobile IP tunnel to the foreign agent.
- The foreign agent performs the routing to the mobile and assigns the IP address using DHCP.

The diagram shown in Fig. 3.34 is a simplified depiction of mobile IP. An example of a mobile IP packet session flowchart is shown in Fig. 3.35.

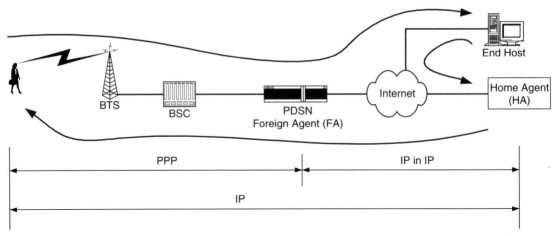

Figure 3.34 Mobile IP.

Mobile IP with VPN. Mobile IP with VPN affords greater mobility for the subscriber over simple IP with VPN since it can maintain a session when it moves from one PDSN area to another. As for mobile IP, the IP address assigned to the subscriber is static; however, the private network the mobile is connected to provide the IP address which needs to be drawn from a predefined IP scheme that is coordinated. The PDSN provides a COA when operating in a nonhome PDSN for routing purposes. The IP packets in both directions, however, flow between the home agent and the foreign agent using IP-in-IP encapsulation and no treatment with the exception of throughput speed allowed is performed by the wireless network. The brief diagram shown in Fig. 3.36 depicts the general packet flow for mobile IP with VPN.

3.4.12 CDMA handoffs

There are several types of handoffs available with CDMA: soft, softer, and hard. Each of the handoff scenarios is a result of the particular system configuration and where the subscriber unit is in the network.

There are several user-adjustable parameters that help the handoff process take place. The parameters that need to be determined involve the values to add or remove a pilot channel from the active list and the search window sizes. There are several values that determine when to add or remove a pilot from consideration. In addition, the size of the search window cannot be too small or too large.

The handoff process begins when a mobile detects a pilot signal that is significantly stronger than any of the forward traffic channels assigned to it. When the mobile detects the stronger pilot channel, the following sequence should take place. The subscriber unit sends a pilot strength measurement message to the base station instructing it to initiate the handoff process. The cell site then sends a handoff direction message to the mobile unit directing it

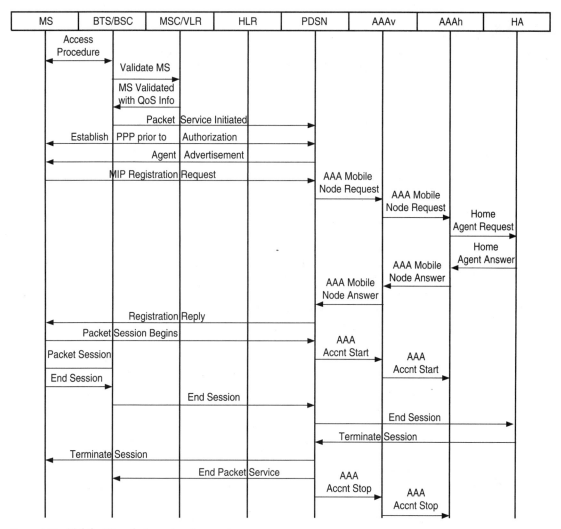

Figure 3.35 Mobile IP packet session flow.

to perform the handoff. Upon the execution of the handoff direction message, the mobile unit sends a handoff completion message on the new reverse traffic channel.

In CDMA a soft handoff involves an intercell handoff and is a make-before-break connection. The connection between the subscriber unit and the cell site is maintained by several cell sites during the process. Soft handoff can only occur when the old and new cell sites are operating on the same CDMA frequency channel. The advantages of the soft handoff is path diversity for the forward and reverse traffic channels. Diversity on the reverse traffic channel results in less power being required by the mobile unit, reducing the overall interference which increases the traffic-handling capacity.

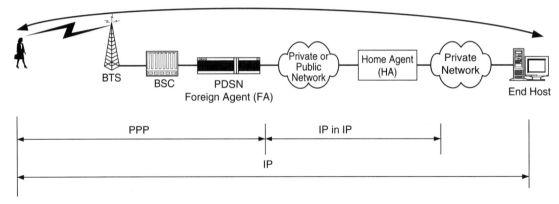

Figure 3.36 Mobile IP with VPN.

The hard handoff process is meant to enable a subscriber unit to hand a CDMA call to an analog call. It is functionally a break-before-make process and is implemented in areas where there is no longer CDMA services for the subscriber to utilize while on a current call. The continuity of the radio link is not maintained during the hard handoff. A hard handoff can also occur between two distinct CDMA channels operating on different frequencies.

Search window. There are several search windows in CDMA. Each of the search windows has its own role in the process, and it is not uncommon to have different search window sizes for each of the windows for a particular cell site. Additionally, the search window for each site needs to be set based on actual system conditions.

The search windows that need to be determined for CDMA involve the active, neighbor, and remaining windows. The search window is defined as an amount of time, in terms of chips, that the CDMA subscriber's receiver will hunt for a pilot channel. There is a slight difference in how the receiver hunts for pilots depending on its type.

If the pilot is an active set, the receiver center for the search window will track the pilot itself and adjust the center of the window to correspond to fading conditions. The other search windows are set as defined sizes (Table 3.21). The size of the search window is directly dependent upon the distance between the neighboring cell sites. You can extrapolate the correct search window for your situation from the example shown in Fig. 3.37.

The following simple procedure is used to determine the search window size:

1. Determine the distance in chips between sites A and B.
2. Determine the maximum delay spread in chips.
3. Search window = ±(cell spacing + maximum delay spread).

The search window for the neighbor and remaining sets consists of parameters SRCH_WIN_N and SRCH_WIN_R, which represent the search window

TABLE 3.21 Search Window Sizes

Search window	Window size, PN chips
0	2
1	4
2	6
3	8
4	10
5	14
6	20
7	28
8	40
9	56
10	80
11	114
12	160
13	226
14	320
15	452

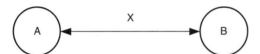

Figure 3.37 Search window. X = 10 chips; therefore, search window = ±10 chips = 6 (20 chips).

sizes associated, respectively, with the neighbor set and remaining set pilots. The subscriber unit centers its search window around the pilots' pseudorandom number offset and compensates for time variants with its own time reference.

The SRCH_WIN_N should be set so that it will encompass the whole area in which a neighbor pilot can be added to the set. The window should be set no larger than $1.75D + 3$ chips, where D is the distance in kilometers between the cells.

SRCH_WIN_A is the value that is used by the subscriber unit to determine the search window size for both the active and candidate sets. The difference between the search window for the active and candidate sets versus the neighbor and remaining sets is that the search window effectively floats with the active and candidate sets based on the first arriving pilot it demodulates.

Soft handoffs. Soft handoffs are an integral part of CDMA. The determination of which pilots will be used in the soft handoff process has a direct impact on the quality of the call and the capacity for the system. Therefore, setting the soft handoff parameters is a key element in the system design for CDMA.

The parameters associated with soft handoffs involve the determination of which pilots are in the active, candidate, neighbor, and remaining sets. The list of neighbor pilots is sent to the subscriber unit when it acquires the cell site or is assigned a traffic channel. A brief description of each type of pilot set follows:

Active set. The set of pilots associated with the forward traffic channels assigned to the subscriber unit. The active set can contain more than one pilot since a total of three carriers, each with its own pilot, could be involved in a soft handoff process.

Candidate set. The pilots that the subscriber unit has reported are of sufficient signal strength to be used. The subscriber unit also promotes the neighbor set and remaining set pilots that meet the criteria for the candidate set.

Neighbor set. A list of the pilots that are not currently on the active or candidate pilot list. The neighbor set is identified by the base station via the neighbor list and neighbor list update messages.

Remaining set. The set of all possible pilots in the system that can be possibly used by the subscriber unit. However, the remaining set pilots that the subscriber unit looks for must be a multiple of the Pilot_INC.

Figure 3.38 shows an example of a soft handoff region. The region shown is an area between cells A and B. Naturally, as the subscriber unit travels farther away from cell A, cell B increases in signal strength for the pilot. When the pilot from cell B reaches a certain threshold, it is added to the active pilot list. The process of how a pilot channel moves from a neighbor to candidate to active and then back to neighbor is best depicted in Fig. 3.39.

1. The pilot exceeds T_ADD, and the subscriber unit sends a PSMM and a transfer pilot to the candidate set.
2. The base station sends an extended handoff direction message.
3. The subscriber unit transfers the pilot to the active set and acknowledges this with a handoff completion message.
4. The pilot strength drops below T_DROP, and the subscriber unit begins handoff drop time.
5. The pilot strength goes above T_DROP prior to the handoff drop time expiring and T_DROP sequences stopping.
6. The pilot strength drops below T_DROP, and the subscriber unit begins the handoff drop timer.
7. The handoff drop timer expires, and the subscriber unit sends a PSMM.
8. The base station sends an extended handoff direction message.

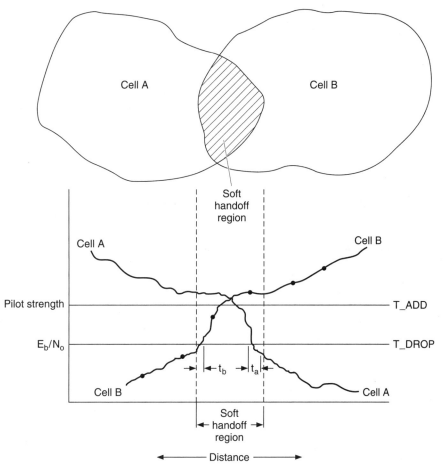

Figure 3.38 Soft handoff region.

9. The subscriber unit transfers the pilot from the active set to the neighbor set and acknowledges this with a handoff completion message.

To augment this description, Fig. 3.40 highlights how T_Comp is factored into the decision matrix for adding and removing pilots from the neighbor, candidate, and active sets.

3.4.13 Pilot channel PN assignment

The pilot channel carriers no data, but it is used by the subscriber unit to acquire the system and assist in the process of soft handoffs, synchronization, and channel estimation. A separate pilot channel is transmitted for each sector of the cell site. The pilot channel is uniquely identified by its pseudorandom number (PN) offset, or PN short code, that is used.

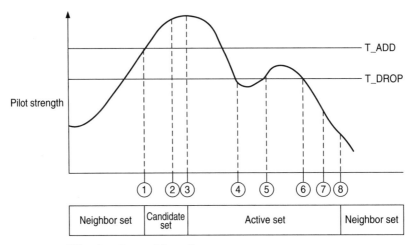

Figure 3.39 Pilot elevation and demotion process.

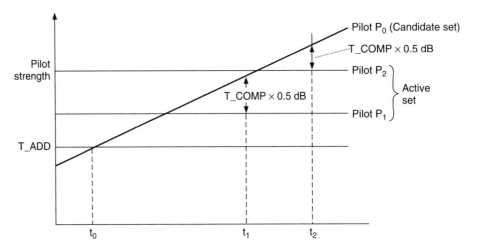

Figure 3.40 Active set.

The PN sequence has some 32,768 chips which when divided by 64 results in a total of 512 possible PN codes that are available for potential use. The fact that there are 512 potential PN short codes to pick from almost ensures that there will be no problems associated with the assignment of these PN codes. However, there are some simple rules that must be followed in order to ensure that there are no problems encountered with the selection of the PN codes for the cell and its surrounding cell sites.

$$\frac{32{,}768}{64} = 512 \text{ possible PN offsets}$$

TABLE 3.22 PN Reuse Scheme

Sector	PN code
Alpha	$3P(N - 2P)$
Beta	$3PN$
Gamma	$3P(N - P)$
Omni	$3PN$

Note: P = PN code increment, N = reusing PN cell.

$$f_{\text{chip}} = 1.228 \times 10^6 \text{ chips per second}$$

$$\text{Time} = \frac{1}{f_{\text{chip}}} = 0.8144 \text{ microsecond per chip}$$

$$\text{Distance} = 244 \text{ meters per chip}$$

There are numerous permutations of how to set the PN codes. However, it is suggested that a reuse pattern be established for allocating the PN codes. The rationale behind establishment of a reuse pattern lies in the fact that it will facilitate the operation of the network for maintenance and growth. In addition, when a second carrier is added, the same PN code should be used for that sector.

Table 3.22 can be used to establish the PN codes for any cell site in the network. The method that should be used is to determine whether you wish to have a 4, 7, 9, 19, etc., reuse pattern for the PN codes. The suggested PN reuse pattern is an $N = 19$ pattern for a new PCS system as shown in Fig. 3.41. If you are overlaying the CDMA system on a cellular system, an $N = 14$ pattern should be used when the analog system utilizes an $N = 7$ voice channel reuse pattern. Please note that not all the codes have been utilized in the $N = 19$ pattern. The remaining codes should be left in reserve for use when a PN code problem arises. In addition, a PN_INC value of 6 is also recommended for use.

The PN short code used by the pilot is an increment of 64 from the other PN codes, which require an offset to be defined. The Pilot_INC is the value that is used to determine the amount of chips (the phase shift one pilot has versus another pilot). To calculate the PN offset, use the equations in Example 3.1.

Example 3.1

$$C/I = 10 \log_{10} \left(\frac{D(P, P_0)}{D(P, P_1)} \right)^{-3} \geq a$$

$$M \geq (R + S) \cdot (10^{a/(\alpha)10} - 1)$$

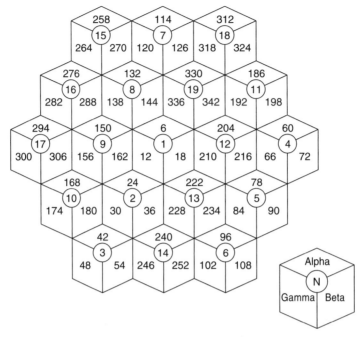

Figure 3.41 PN reuse pattern.

where M = offset
R = radius in chips
S = $\frac{1}{2}$ search window_A
a = C/I
α = attenuation factor, propagation exponent

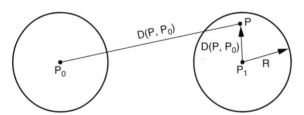

Pilot_INC is valid in the range 0 to 15. Pilot_INC is the PN sequence offset index and is a multiple of 64 chips. The subscriber unit uses the Pilot_INC to determine which are the valid pilots to be scanned. Table 3.23 can be used to determine the Pilot_INC as a function of the distance between reusing sites.

3.5 GSM

Global system for mobile communications (GSM) is the European standard for digital cellular systems operating in the 900-MHz band. This technology was

TABLE 3.23 Pilot Offsets

R, km	R (chips)	S	C/I	m (chips)	Pilot_INC	No. of offsets
25	103	14	24	622	10	50
20	82	12	24	499	8	64
15	61	12	24	390	6	85
12.5	51	10	24	325	5	102
10	41	10	24	271	4	128
7	29	10	24	207	4	128
5	21	10	24	165	3	170
3	12	10	24	117	2	256
2.5	10	10	24	106	2	256
2	8	10	24	96	2	256

developed out of the need for increased service capacity due to the limited growth of analog systems. This technology offers international roaming, high speech quality, increased security, and the ability to develop advanced systems features. The development of this technology was completed by a consortium of 80 pan-European countries working together to provide integrated cellular systems across different borders and cultures.

GSM has achieved worldwide success. It has many unique features and attributes that make it an excellent digital radio standard to utilize, which has resulted in its being the most widely accepted radio communication standard at this time. GSM was developed in Europe as a communication standard that would be utilized throughout all of Europe in response to the problem of multiple and incompatible standards that still exist there today.

GSM consists of the three following major building blocks: the switching system, the base station system (BSS), and the operations and support system (OSS). The BSS comprises both the base station controller (BSC) and the base transceiver stations (BTS). In an ordinary configuration several BTSs are connected to a BSC and then several BSCs are connected to the MSC.

The GSM radio channel is 200-kHz wide. GSM has been deployed in several frequency bands, namely, the 900-, 1800-, and 1900-MHz bands. Both the 1800- and 1900-MHz bands required some level of spectrum clearing before the GSM channel could be utilized. However, the 900-MHz spectrum was used by an analog system, ETACS, which occupied 25-kHz channels. The introduction of GSM into this band required the reallocation of channels to accommodate GSM.

Figure 3.42 shows the basic architecture of a GSM network. Working our way from the left, we see that the handset, known in GSM as the mobile station, communicates over the air interface with a BTS. Strictly speaking, the mobile station is composed of two parts—the handset itself, known as the mobile equipment, and the subscriber identity module (SIM), a small card containing an integrated circuit. The SIM contains user-specific information, including the

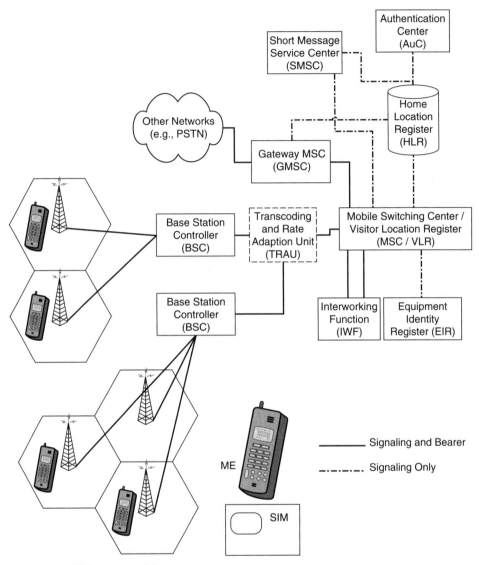

Figure 3.42 GSM system architecture.

identity of the subscriber, subscriber authentication information, and some subscriber service information. It is only when a given subscriber's SIM is inserted into a handset that the handset acts in accordance with the services subscribed to. In other words, my handset only acts as my handset when my SIM is inserted.

The BTS contains the radio transceivers that provide the radio interface with mobile stations. One or more BTSs are connected to a BSC. The BSC provides a number of functions related to radio resource management, some functions related to mobility management for subscribers in the coverage area of the BTSs, and a number of operation and maintenance functions for the overall radio network. Together, BTSs and BSCs are known as the base station subsystem.

The interface between the BTS and the BSC is known as the Abis interface. Many aspects of that interface are standardized. One aspect, however, is proprietary to the BTS and BSC vendor: the part of the interface that deals with configuration, operation, and maintenance of the BTSs. This is known as the operation and maintenance link (OML). Since the internal design of a BTS is proprietary to the BTS vendor, and since the OML needs to have functions that are specific to that internal design, the OML is also proprietary to the BTS vendor. The result is that a given BTS must be connected to a BSC of the same vendor.

One or more BSCs are connected to a mobile switching center (MSC). The MSC is the *switch*—that node that controls call setup, call routing, and many of the functions provided by a standard telecommunications switch. The MSC is no ordinary PSTN switch, however. Because of the fact that the subscribers are mobile, the MSC needs to provide a number of mobility management functions. It also needs to provide a number of interfaces that are unique to the GSM architecture. When we speak of an MSC, a visitor location register (VLR) is also usually implied. The VLR is a database that contains subscriber-related information for the duration that a subscriber is in the coverage area of an MSC. There is a logical split between an MSC and a VLR, and the interface between them has been defined in standards. No equipment vendor, however, has ever developed a stand-alone MSC or VLR. The MSC and VLR are always contained on the same platform and the interface between them is proprietary to the equipment vendor. Although early versions of GSM standards defined the MSC-VLR interface (known as the B interface) in great detail, later versions of the standards recognized that no vendor complies with the standardized interface. Therefore, any "standardized" specification for the B interface should be considered informational.

The interface between the BSC and the MSC is known as the A interface. This is an SS7-based interface using the signaling connection control part (SCCP) as depicted in Fig. 3.43. In layer 3 in the signaling stack, we find the BSS application part (BSSAP), which is the protocol used for communication between the MSC and the BSC and also between the MSC and the mobile station. Since the MSC communicates separately with both the BSC and the mobile station, the BSSAP is divided into two parts—the BSS management application part (BSSMAP) and the direct transfer application part (DTAP). BSSMAP contains those messages that are either originated by the BSS or need to be acted upon by the BSS. The DTAP contains those messages that are passed transparently through the BSS from the MSC to the mobile station, or vice versa. Note that there is also a BSS operation and maintenance application part (BSSOMAP). Although this is defined in standards, it is normal for the BSC to be managed through a vendor proprietary management protocol.

Referring back to Fig. 3.42, we find (in dashed outline) the transcoding and rate adaptation unit (TRAU). In GSM, the speech from the subscriber is usually

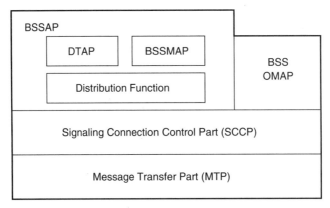

Figure 3.43 BSSAP protocol layers.

coded at either 13 (full rate) or 12.2 kbit/s [enhanced full rate (EFR)]. In some cases we also find half-rate coding at a rate of 5.6 kbit/s, but that is rare in commercial networks. In any case, it is clear that the speech to and from the mobile station is very different from the standard 64-kbits pulse-code modulation (PCM) used in switching networks. Since the MSC interfaces with the PSTN network, it needs to send and receive speech at 64 kbit/s. The function of the TRAU is to convert the coded speech to and from the standard 64 kbit/s. Strictly speaking, the TRAU is a part of the BSS. As far as the MSC is concerned, voice to and from the BSS is passed at 64 kbit/s and the BSS takes care of the transcoding. In practice, however, it is common for the TRAU to be physically separate from the BSC and placed near the MSC. This reduces the bandwidth required between the MSC location and the BSC location and can mean significant savings in transport cost, particularly if the BSC and MSC are separated by a significant distance. In case where the BSC and TRAU are separated, the interface between them is known as the Ater interface. This interface is proprietary to the BSS equipment vendor. Hence, the BSC and TRAU must be from the same vendor.

In Fig. 3.42 we also find a home location register (HLR)—a node found in most mobile networks. The HLR contains subscriber data, such as the details of the services to which a user has subscribed. Associated with the HLR, we find the authentication center (AuC). This is a network element that contains subscriber-specific authentication data, such as a secret authentication key called the Ki. The AuC also contains one or more sophisticated authentication algorithms. For a given subscriber, the algorithm in the AuC and the Ki is also found on the SIM card. Using a random number assigned by the AuC and passed down to the SIM via the HLR, MSC, and mobile equipment, the SIM performs a calculation using the Ki and authentication algorithm. If the result of the calculation on the SIM matches that in the AuC, then the subscriber has been authenticated. The interface between the HLR and AuC is not standardized. Although there are implementations where the HLR and AuC are separate, it is more common to find the HLR and AuC integrated on the same platform.

Calls from another network, such as the PSTN, first arrive at a type of MSC known as a gateway MSC (GMSC). The main purpose of the GMSC is to query the HLR to determine the location of the subscriber. The response from the HLR indicates the MSC where the subscriber may be found. The call is then forwarded from the GMSC to the MSC serving the subscriber. A GMSC may be a full MSC/VLR such that it may have some BSCs connected to it. Alternatively, it may be a dedicated GMSC whose only function is to interface with the PSTN. The choice is dependent upon the amount and types of traffic in the network and the relative cost of a full MSC-VLR versus a pure GMSC.

Figure 3.42 also includes the short message service center (SMSC). Strictly speaking, the correct term is short message service–service center (SMS-SC), but that is a bit of a mouthful and is usually shortened to SMSC. The SMSC is a node that supports the storing and forwarding of short messages to and from mobile stations. Typically, these are short text messages up to 160 characters in length.

Logically, there are three components to an SMSC. First, there is the service center itself, which stores messages and interfaces with other systems such as e-mail or voice mail equipment. Second, there is the SMS–gateway MSC (SMS-GMSC), which is used for delivery of short messages to a mobile subscriber. Much like a GMSC, the SMS-GMSC queries the HLR for the subscriber's location, and then forwards the short message to the appropriate visited MSC where it is relayed to the subscriber. Third, there is the SMS–interworking MSC (SMS-IWMSC), which receives a short message from the MSC serving the subscriber. It forwards such messages to the SC, which then passes them on to the final destination. It is very common for the service center, SMS-GMSC, and SMS-IWMSC to be included within the same platform, though there are implementations where there is a stand-alone service center. In such implementations, the SMS-GMSC function may be included within a GMSC, and the SMS-IWMSC function may be included with an MSC-VLR.

In a GSM network, we may also find a node known as the equipment identity register (EIR). As mentioned, it is not the handset that identifies a subscriber, but the information on the SIM. Therefore, to some degree, the handset used by a particular subscriber is not relevant. On the other hand, it may be important for the network to verify that a particular handset or model of mobile equipment is acceptable. For example, a network operator might want to restrict access from a handset that has not been fully type-approved. Also, a network operator might want to restrict access from a handset that is known to be stolen. An international mobile identity (IMEI, 15 digits) number or an international mobile equipment identity and software version (IMEISV, 16 digits) number is stored in each handset. Both the IMEI and IMEISV have a structure that includes the type approval code (TAC) and the final assembly code (FAC). The TAC and FAC combine to indicate the make and model of the handset and the place of manufacture. The IMEI and IMEISV also include a specific serial number for the mobile equipment in question. The only difference between IMEI and IMEISV is the software version number. Within the EIR, there are three lists—black, gray, and white. These lists contain values of TAC, TAC and FAC, or complete IMEI or IMEISV. If a given TAC, TAC and FAC combination, or complete IMEI appears in the black list, then calls from the mobile equipment are barred. If it appears in the gray list, then calls may or may not be barred at the discretion of the network operator. If it appears in the white list, then calls are allowed. Typically, a given TAC is included in the white list if the handset model is one that has been approved by the handset manufacturer. The EIR is an optional network element, and some network operators have chosen not to deploy an EIR.

Finally, we find the interworking function (IWF). This is used for circuit-switched data and fax services and is basically a modem bank. Typical dial-up modems and fax machines are analog. For example, when one uses a computer with a 28.8-kbit/s modem on a regular telephone line, the modem modulates the digital data from the computer to an analog format that appears like analog speech. The same cannot be done directly for a digital system such as GSM because all transmissions are digital and it is not possible to transmit data over the air in a manner that emulates analog voice. Furthermore, a remote dial-up modem, such as at an Internet service provider, expects to be called by another modem. Therefore, a circuit-switched data call from a mobile station is looped through the IWF before being routed onwards by the IWF. Within the IWF, a modem is placed in the call path. The same applies for fax services, where a fax modem would be used farther than a data modem. GSM supports data and fax services up to 9.6 kbit/s.

3.5.1 GSM air interface

GSM is a time-division multiple-access (TDMA) system, with frequency-division duplexing (FDD). It uses gaussian minimum shift keying (GMSK) as the modulation scheme. TDMA means that multiple users share a given RF channel on a time-sharing basis. FDD means that different frequencies are used in the downlink (from network to mobile station) and uplink (from mobile station to network) directions.

GSM has been deployed in numerous frequency bands, including the 900-, 1800-, and 1900-MHz bands (in North America). Table 3.24 shows the frequency allocations for these three bands. In GSM, a given band is divided into 200-kHz carriers or RF channels in both the uplink and downlink. In addition, at the end of each frequency band, there is a guard band of 200 kHz. For example, in standard GSM900, the first uplink RF channel is at 890.2 MHz and the last uplink RF channel is at 914.8 MHz, allowing for a total of 124 carriers. Similarly, DCS1800 has a maximum of 374 carriers, and PCS1900 has a maximum of 299 carriers.

As mentioned, in GSM, a given band is divided into a number of RF channels or carriers, each 200 kHz in both the uplink and downlink. Thus, if a handset is transmitting on a given 200-kHz carrier in the uplink, then it is receiving on a corresponding 200-kHz carrier in the downlink. Since the uplink and downlink are rigidly associated, when one talks about a carrier or RF channel, both the uplink and downlink are usually implied. A given cell can have multiple RF

TABLE 3.24 GSM Frequency Bands

	Frequency bands, MHz			
	GSM900	Extended GSM (E-GSM)	DCS1800	PCS1900
Uplink (MS to network)	890–915	880–915	1710–1785	1850–1910
Downlink (network to MS)	935–960	925–960	1805–1880	1930–1990

Note: MS = mobile station.

carriers—typically one to three in a normally loaded system, though as many as six carriers might exist in a heavily loaded cell in an area of very high traffic demand. Note that, when we talk about a cell in GSM terms, we mean a sector. Thus, a three-sector BTS implies three cells. This is a somewhat confusing distinction between GSM and some other technologies.

Each RF carrier is divided into eight time slots, numbered 0 to 7, and these are transmitted in a frame structure. Each frame lasts approximately 4.62 ms, such that each time slot lasts approximately 576.9 microseconds (μs). Depending on the number of RF carriers in a given cell, all eight time slots on a given carrier might be used to carry user traffic. In other words, the RF carrier might be allocated to eight traffic channels. There must be, however, at least one time slot in a cell allocated for control channel purposes. Thus, if there is only one carrier in a cell, then there is a maximum of seven traffic channels, such that a maximum of seven simultaneous users can be accommodated for circuit-switched applications, i.e., voice.

3.5.2 Types of air interface channels

The foregoing description of the RF interface suggests that there are just traffic channels and control channels. This is only partly correct. In fact, there are traffic channels, numerous types of control channels, and a number of other channels. To begin with, there are a number of broadcast channels:

Frequency correction channel (FCCH). Broadcast by the BTS and used for frequency correction of the mobile station.

Synchronization channel (SCH). Broadcast by the BTS and used by a mobile station for frame synchronization. In addition to frame synchronization information, it also contains the base station identity code.

Broadcast control channel (BCCH). Used to broadcast general information regarding the BTS and the network in general. It is also used to indicate the configuration of the common control channels (CCCHs) described next.

The CCCH is a bidirectional control channel used primarily for functions related to initial access by a mobile station. It has a number of components.

Paging channel (PCH). Used for paging of mobile stations.

Random-access channel (RACH). Used only in the uplink direction. It is used by a mobile station to request allocation of a stand-alone dedicated control channel (SDCCH).

Access grant channel (AGCH). Used in the downlink in response to an access request received on the RACH. Used to allocate a mobile station to an SDCCH or directly to a traffic channel.

Notification channel (NCH). Used with voice group call and voice broadcast services to notify mobile stations regarding such calls.

There are a number of dedicated control channels. These are channels that are used by one mobile station at a time, typically either during call establishment or while a call is in progress. The dedicated control channels are as follows:

Stand-alone dedicated control channel (SDCCH). A bidirectional channel used for communication with a mobile station when the mobile station is not using a TCH. The SDCCH is used, for example, for short message service when the mobile station is not in a call. It is also used for call establishment signaling prior to allocation of a traffic channel (TCH) for a call.

Slow associated control channel (SACCH). A unidirectional or bidirectional channel, used when the mobile station is using a TCH or SDCCH. For example, when a mobile station is engaged in a call on a TCH, power control messages from a BTS to a mobile station are sent on the SACCH. In the uplink, the mobile station sends measurement reports to the BTS on the SACCH. These reports indicate how well the mobile station can receive transmissions from other BTSs, and the information is used to determine if and when handoff should occur. The SACCH is also used for short message transfer when the mobile station is in a TCH.

Fast associated control channel (FACCH). Associated with a given TCH and, thus, used when the mobile is involved in a call. It is typically used to transmit nonvoice information to and from the mobile station. Such information would include, for example, handoff instructions from the network, commands from the mobile station for generation of dual-tone multifrequency (DTMF) tones, and supplementary service invocations.

3.5.3 Air interface channel structure

Clearly, it does not make sense for these different types of channels to each be allocated one of the eight time slots. First, there would simply not be enough time slots. Moreover, different data rates apply to the various types of channels. Instead, a sophisticated framing structure is used on the air interface to allocate the various channel types to the available time slots. There are frames, multiframes, superframes, and hyperframes.

As mentioned above, a single frame lasts approximately 4.62 ms and contains eight time slots. In standard GSM (as opposed to GPRS), there are two types of multiframe: a 26-multiframe (containing 26 frames and having a duration of 120 ms) and a 51-multiframe (containing 51 frames and having a duration of 235.4 ms). The 26-multiframe is used to carry TCHs and the associated SACCH and FACCH. The 51-multiframe is used to carry broadcast, common (including PCH, RACH, and AGCH), and stand-alone dedicated (and its associated SACCH) control channels. A superframe lasts 6.12 s, corresponding to 51×26-multiframes or 26×51-multiframes. A hyperframe corresponds to 2048 superframes [a total of 2,715,648 frames, lasting just under 3 hours (h), 28 minutes (min), 54 s]. When numbering frames over the air interface, each frame is numbered modulo its hyperframe. In other words, a frame can have a frame number from 0 to 2,715,467. The reason for the large hyperframe is to allow for a large frame number value, which is used as part of the encryption over the air interface.

Certain time slots on a given RF carrier may be allocated to control channels, while the remaining time slots are allocated for traffic channels. For example, time slot 0 on the first carrier in a cell is used to carry the broadcast and common control channels. It may also carry four SDCCH channels. It is also common to find that time slot 1 on the first RF carrier in a cell is used to carry eight SDCCH channels (with the associated SACCHs), with the remaining time slots allocated as TCHs. Exactly how much SDCCH capacity is allocated is dependent upon the number of carriers and amount of traffic in the cell. Figure 3.44 shows two typical arrangements.

BCCH/ CCCH/ SDCCH/4	TCH	TCH	TCH	TCH	TCH	TCH	TCH

SDCCH sharing time slot zero with BCCH and CCCH; common when only one carrier per cell.

BCCH/ CCCH	SDCCH/8	TCH	TCH	TCH	TCH	TCH	TCH
TCH	TCH	TCH	TCH	TCH	TCH	TCH	TCH

SDCCH using time slot one on first carrier; common when more than one carrier per cell. Second carrier dedicated to traffic channels.

Figure 3.44 Example GSM air interface time slot allocations.

As mentioned, the 26-multiframe is used for the TCH. The structure is depicted in Fig. 3.45 where only one time slot per frame is shown (only a full-rate traffic channel is considered in the figure). A given time slot carries user traffic (e.g., voice) for 24 out of 26 frames. One of the 26 frames is idle and one carries the SACCH. The FACCH is transmitted by preempting half or all of the user traffic in a TCH.

This overall structure allows a TCH to have a gross bit rate of 22.8 kbit/s. Of course, this rate is not allocated completely to user data (e.g., speech). Rather, a sophisticated coding and interleaving scheme is applied. This scheme adds a significant number of bits for error detection and correction, which reduces the bandwidth available for raw user data. In fact, for standard GSM full-rate voice coding, the speech is carried at 13 kbit/s, and for enhanced full-rate voice coding, the speech is carried at 12.2 kbit/s. While it may seem that a great deal of the gross 22.8 kbit/s is consumed by coding overhead, it is worth remembering that an RF interface is unreliable at best and error correction overhead is necessary to overcome the limitations of the medium.

Since the control channels (with the exception of FACCH and SACCH) are carried on different time slots from the TCH, it is possible to have a different framing structure. In fact, a 51-multiframe structure is used for transmitting the control channels, and this structure applies to any time slot that is allocated to control channels.

3.5.4 Location update

When a mobile station is first turned on, it must first "camp on" (find, decode, and wait for further instructions) a suitable cell. This largely involves scanning the air interface to select a cell with a suitably strong received signal strength and decoding the information broadcast by the BTS on the BCCH. Generally, the mobile station will camp on the cell with the strongest signal strength provided that cell belongs to the home public land mobile network (HPLMN) and provided that the cell is not barred. The mobile station then registers with the network, which involves a process known as *location updating* as shown in Fig. 3.46.

The sequence begins with a channel request issued by the mobile station on the random-access channel (RACH). This includes an establishment cause, such as location updating, voice call establishment, or emergency call establishment. In the example in Fig. 3.46, the cause is location updating.

The BSS allocates an SDCCH for the mobile station to use. It instructs the mobile station to move to the SDCCH by sending an immediate assignment

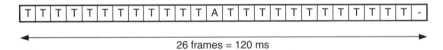

26 frames = 120 ms

Figure 3.45 TCH/SACCH framing structure. T = TCH; A = SACCH; - = idle frame.

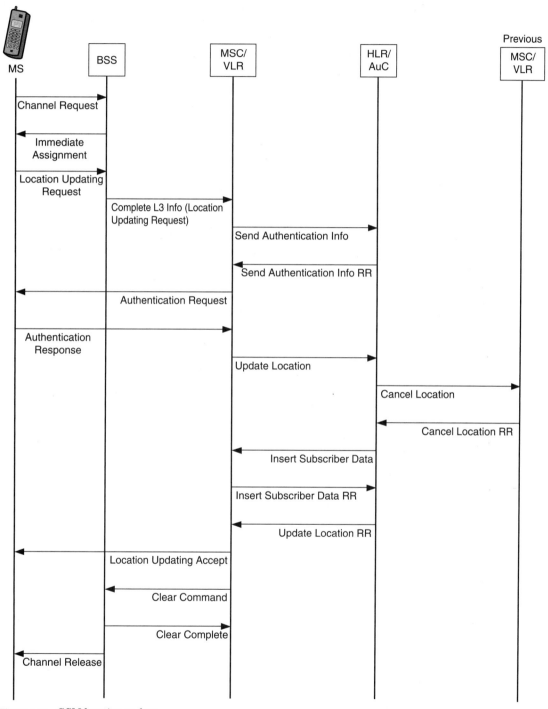

Figure 3.46 GSM location update.

message on the AGCH. The mobile station moves to the SDCCH and sends the location updating request. This contains a set of information including the location area identity (as received by the mobile station on the BCCH) and the mobile identity. The mobile identity is usually either the international mobile subscriber identity (IMSI) or temporary mobile subscriber identity (TMSI). This is sent through the BSS to the MSC, using a generic message known as Complete Layer 3 Info. This message is included as part of an SCCP connection request. Hence, it uses connection-oriented SCCP.

If the subscriber attempts to register with a TMSI and the TMSI is unknown in the MSC-VLR, then the MSC-VLR may request the mobile station to send the IMSI (not shown in the example). Equally, the MSC-VLR may request the mobile station to send the IMEI so that it can be checked (not shown in the example).

Upon receipt of the location updating request, the MSC-VLR may attempt to authenticate the subscriber. If the MSC-VLR does not already have authentication information for the subscriber, then it requests that information from the HLR, using the mobile application part (MAP) operation Send Authentication Info. The HLR-AuC sends a MAP return result (RR) with up to five authentication vectors, known as triplets. Each triplet contains a random number (RAND) and a signed response (SRES).

The MSC sends an authentication request to the mobile station. This contains only the RAND. The mobile station performs the same calculations as were performed in the HLR-AuC and sends an authentication response containing a SRES parameter. The MSC-VLR checks to make sure that the SRES received from the mobile station matches that received from the HLR-AuC. If there is a match, then the mobile station is considered authenticated.

At this point, the MSC-VLR uses the MAP operation Update Location to inform the HLR of the subscriber's location. The message to the HLR includes the subscriber's IMSI and the SS7 global time address of the MSC and VLR. The HLR immediately sends a MAP cancel location message to the VLR (if any) where the subscriber had previously been registered. That VLR deletes any stored data related to the subscriber and issues a return result to the HLR.

The HLR uses the MAP operation Insert Subscriber Data to the VLR to inform the VLR about a range of data regarding the subscriber in question, including information regarding supplementary service. The VLR acknowledges receipt of the information, and then the HLR issues a return result to the MAP update location.

Upon receipt of that return result, the MSC-VLR sends the DTAP message Location Updating Accept to the mobile station. It then clears the SCCP connection to the BSS. This causes the BSS to release the mobile station from the SDCCH by sending a channel release message to the mobile station.

A number of optional messages have been excluded in Fig. 3.46. For a complete understanding of all options, the reader is referred to GSM specification 04.08 [digital cellular telecommunications system (phase 2+); mobile radio layer 3 specification]. A number of messages shown in this figure (e.g., channel request, immediate assignment, channel release) are common to many traffic scenarios. For the sake of brevity, they are not necessarily shown in the following call examples.

3.5.5 Mobile-originated voice call

Figure 3.47 shows a basic mobile-originated call to the PSTN. After the mobile station has been placed on an SDCCH by the BSS (not shown), the mobile station issues a connection management service request to the MSC. This includes information about the type of service that the mobile station wishes to invoke (a mobile-originated call in this case, but it could also be another service such as SMS). Upon receipt of the connection management service request, the MSC may optionally invoke authentication of the mobile. Typically, an MSC is configured to authenticate a mobile whenever it performs an initial location update and every N transactions thereafter (e.g., every N calls). Next the MSC initiates ciphering so that the voice and data sent over the air are encrypted. Since it is the BSS that performs the encryption and decryption, the MSC needs to pass the authentication key (Kc) to the BSS. The BSS then instructs the mobile station to start ciphering. The mobile station, of course, generates the Kc independently, so that it is not passed over the air. Once the mobile station has started ciphering, it informs the BSS, which, in turn, informs the MSC.

Next, the mobile station sends a setup message to the MSC. This includes further data about the call, including information such as the dialed number and the required bearer capability. Once the MSC has determined that it has received sufficient information to connect the call, it lets the mobile station know by sending a call proceeding message.

Next, using the assignment request message, the MSC requests seizure of a circuit between the MSC and BSS. That circuit will be used to carry the voice to and from the mobile station. At this point, the BSS sends an assignment command message to the mobile station, instructing it to move from the SDCCH to a TCH. Further signaling between the mobile station and the network will now occur on the FACCH associated with the assigned TCH. The mobile station responds with an assignment complete message, indicating that it has moved to the assigned TCH. Upon receipt of that message, the BSS sends an assignment complete message to the MSC, which indicates that there is now a voice path available from the mobile station through to the MSC.

Upon receipt of the assignment complete message from the BSS, the MSC initiates call setup toward the PSTN. This starts with the issuance of an initial address message. Subsequent receipt of an address complete message from the destination end indicates that the destination phone is now ringing. The MSC informs the mobile station of that fact by sending an alerting message. In addition, the address complete message triggers a one-way path to be opened from the destination PSTN switch through to the mobile station, and the ring-back tone heard at the mobile station is actually being generated at the destination PSTN switch.

When the called phone is answered, an answer message is returned. This leads the MSC to open a two-way path to the mobile station and also causes the MSC to send a connect message to the mobile station. Upon receipt of the connect message, the mobile station responds with a connect acknowledge message. The two parties are now in conversation and, from a billing perspective, the clock is now ticking.

154 Chapter Three

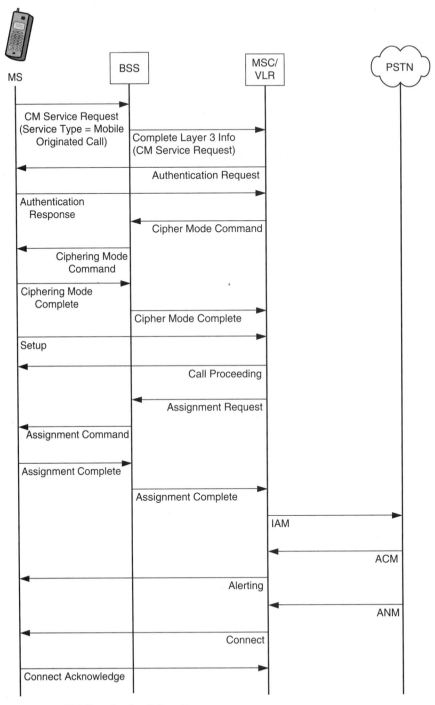

Figure 3.47 Mobile-to-land call flow diagram.

3.5.6 Mobile-terminated voice call

Figure 3.48 shows a basic mobile-terminated call from the PSTN. It begins with the arrival of an initial address message at the GMSC. The initial address message contains the director number of the called subscriber, known as the mobile station ISDN (MSISDN) number. The GMSC uses this information to determine the applicable HLR for the subscriber and invokes the MAP operation Send Routing Information (which contains the subscriber's MSISDN number) toward the HLR.

The HLR uses the MSISDN to retrieve the subscriber's IMSI from its database. Through a previous location update, the HLR knows the MSC-VLR that serves the subscriber, and it queries that MSC-VLR using the MAP operation Provide Roaming Number, which contains the subscriber's IMSI. From a pool, the MSC-VLR allocates a temporary number, known as a mobile station roaming number (MSRN) for the call and returns that number to the HLR. The HLR returns the MSRN to the GMSC.

The MSRN is a number that appears to the PSTN as a dialable number. Thus, it can be used to route a call through any intervening network between the GMSC and the visited MSC-VLR. In fact, that is exactly what the GMSC does. It routes the call to the MSC-VLR by sending an initial address message, with the MSRN as the called party number. Upon receipt of the initial address message, the MSC-VLR recognizes the MSRN and knows the IMSI for which the MSRN was allocated. At this point the MSRN can be returned to the pool for use with another call.

Next, the MSC requests the BSS to page the subscriber using the paging request message, which indicates the location area in which the subscriber should be paged. The BSS uses the PCH to page the mobile station. Upon receipt of the page, the mobile station attempts to access the network using a channel request message on the RACH. The BSS responds with an immediate assignment message, instructing the mobile station to move to an SDCCH. The mobile station moves to the SDCCH and, once there, indicates to the network that it is responding to the page. The BSS passes the response to the MSC.

At this point, the MSC may optionally authenticate the MS (not shown). It will then proceed to initiate ciphering, which is done in the same manner as was described previously for a mobile-originated call. Once ciphering is started, the MSC sends a setup message to the mobile station. This is similar to the setup message that is sent from a mobile station for a mobile-originated call, including information such as the calling party number and the required bearer capability.

Upon receipt of the setup message, the mobile station sends a call confirmed message to the MSC, indicating that it has the information it needs to establish the call. The call confirmed message acts as an instruction to the MSC to establish a path through to the mobile station. Therefore, the MSC begins the assignment procedure, which establishes a circuit between the MSC and the BSS and a TCH between the BSS and the mobile station (rather than an SDCCH). Further signaling between the mobile station and the network will now use the FACCH associated with the TCH to which the mobile station has been assigned.

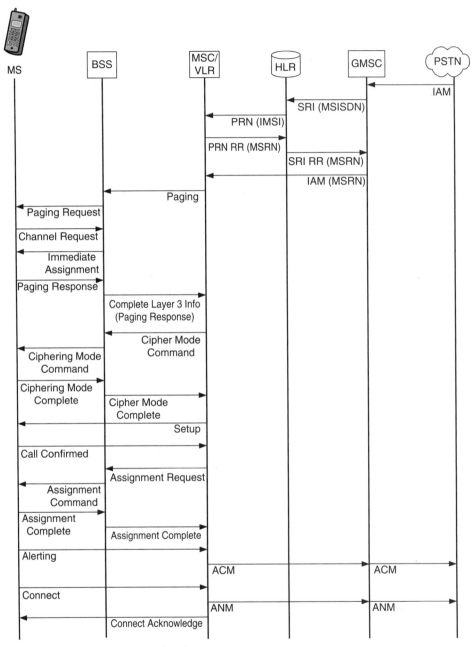

Figure 3.48 Land-to-mobile call flow diagram.

Once established on the TCH, the mobile station starts ringing to alert the user and informs the network by sending the alerting message to the MSC. This triggers the MSC to open a one-way path back to the original caller, generate a ring-back tone, and send an ACM message back to the originating PSTN switch via the GMSC.

Once the called user answers, the mobile station sends a connect message to the MSC. This triggers the MSC to send an ANM message back to the originating switch and to open a two-way path. Finally, it sends a connect acknowledge message to the mobile station and conversation begins.

3.5.7 Handoff

Handoff (also known as handover) is the process by which a call in progress is transferred from a radio channel in one cell to another radio channel, either in the same cell or in a different cell. Handoff can occur within a cell, between cells of the same BTS, between cells of different BTSs connected to the same BSC, between cells of different BSCs, or between cells of different MSCs. Not only can handoff occur between TCHs, but handoff is also possible from a SDCCH on one cell to a SDCCH on another cell, and is possible from a SDCCH on one cell to a TCH on another cell. The most common, however, is handoff from one traffic channel to another.

Depending on the source (i.e., original cell) and target (i.e., destination cell) involved in the handoff, the handoff may be handled completely within a BSS or may require the involvement of an MSC. In the case where a handoff occurs between cells of the same BSC, the BSC may execute the handoff and simply inform the MSC after the handoff has taken place. If, however, the handoff occurs between BSCs, then the MSC must become involved, since there is no direct interface between BSCs.

Handoff in GSM is known as mobile-assisted handover (MAHO). This means that it is the network that decides if, when, and how a handoff should take place. The mobile station, however, provides information to the network to enable the network to make the decision.

Recall that GSM is a TDMA system, with eight time slots per frame in the case of full-rate speech. This means that the mobile station is transmitting for one-eighth of the time and receiving for one-eighth of the time. In fact, at the BTS, a given time slot on the uplink is three time slots duration later than the corresponding downlink time slot, which means that the mobile station is not required to receive and transmit simultaneously. We note that this offset is specified at the BTS rather than at the mobile station since the distance of the mobile station from the BTS influences the exact instant at which the mobile station should transmit. For example, when a mobile station is close to the BTS, it should transmit slightly later than if it were further from the BTS. This variation is known as *time alignment* and is controlled by the BSS. In other words, the BSS periodically instructs the mobile station to change its time alignment as necessary.

Nonetheless, it is clear, that for most of the time, the mobile station is neither transmitting nor receiving. During this time, the mobile station has the opportunity to tune to other carrier frequencies and determine how well it can receive those signals. It can then relay that information to the network to allow the network to make a determination as to whether the mobile station would be better served by a different cell. Because of frequency reuse, it is possible that a number of nearby cells might be using the same BCCH frequency. Therefore, it is not sufficient for the mobile station to simply report signal strength for specific frequencies. Rather, the mobile station must be able to synchronize to the BCCH of neighboring cells and decode the information being transmitted. Exactly which frequencies the mobile station should check for are specified in system information messages transmitted by the BTS on the BCCH and the SACCH. The mobile station sends measurement reports to the BSS on the SACCH as often as possible. These reports include information on how well the mobile station can "hear" the serving cell as well as information about signal strength measurements on up to six neighboring. Specifically, for the serving cell, the mobile station reports RXLEV (received-signal level; an indication of received signal strength) and RXQAL (received-signal quality; an indication of bit error rate on the received signal). For neighboring cells, the mobile station reports the base station identification code (BSIC), the BCCH frequency, and RXLEV.

In addition to the measurements reported by the mobile station, the BTS itself makes measurements regarding the RXLEV and RXQUAL received from the mobile station. These measurements and those from the mobile station are reported to the BSC. Based on its internal algorithms, the BSC makes the decision as to whether a handoff should occur and, if so, to what cell.

Figure 3.49 shows an inter-BSC handoff. In this case, it is not sufficient for the BSC to handle the handoff autonomously—it must involve the MSC. Therefore, once the serving BSC determines that a handoff should take place, it immediately sends a handoff required message to the MSC. This message contains information abut the desired target cell (or cells in preferred order) plus information about the current channel that the mobile station is using. The MSC analyzes the information and identifies the target BSC associated with at least one of the target cells identified by the source BSC. It then sends a handoff request message to the target BSC. This contains, among other items, information about the target cell; the type of channel required; and, in the case of a speech or data call, the circuit to be used between the MSC and the target BSC.

If the target BSC can accommodate the handoff (e.g., if resources are available), then it allocates the necessary resources and responds to the MSC with a handoff request acknowledge message. This message contains a great deal of information regarding the cell and channel to which the mobile station is to be transferred, such as the cell identity, the exact channel to be used (including the type of channel), synchronization information, power level to be used by the mobile station when accessing the new channel, and a handoff reference. The MSC then sends the handoff command message to the

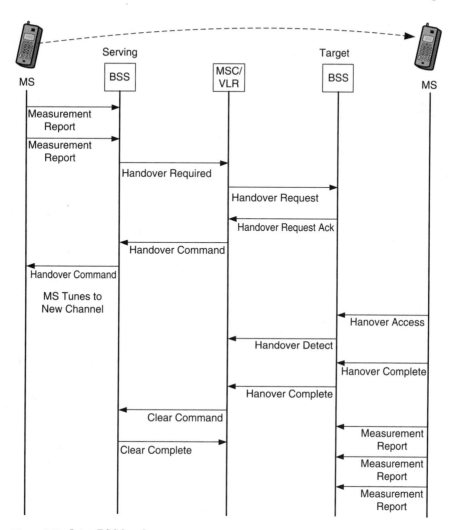

Figure 3.49 Inter-BSC handover.

serving BSC. This message is used to relay the information received from the target BSC. On receipt of the handover command message from the MSC, the serving BSC passes the information to the mobile station in a handoff command message over the air interface.

Upon receipt of the handoff command message, the mobile station releases existing RF connections, tunes to the target channel, and attempts to access that channel. Upon access, it may send a handoff access message to the target BSS. It will do so if it was commanded to do so in the handoff command message. If the handoff access message is received by the target BSS, then it sends a handoff detect message to the MSC. When the mobile station has established all lower-layer connections on the target channel, it sends a handoff complete message to the target BSC, which, in turn, sends a handoff complete message to the MSC. At this point, the mobile station again starts taking measurements of neighboring cells. Meanwhile the MSC instructs the original BSC to release all radio and terrestrial resources related to the mobile station.

3.5.8 Traffic calculation methods

Like any mobile communications technology, traffic calculation and system dimensioning for GSM begins with the estimation of how much traffic demand there will be and from where it will come. In other words, one must estimate the traffic demand in the coverage of each cell. This is rather an inexact science. One can certainly acquire demographic data such as population density and average household income. One can also acquire data related to vehicular traffic in order to estimate traffic demand for cells that cover roads. Based on these factors and others (such as how many competing operators exist), one makes an estimate of the peak traffic demand per cell. This estimate may well be incorrect. Fortunately, however, time is an ally. In a new network, traffic demand grows gradually, which provides the operator with sufficient time to monitor usage and more accurately predict traffic demand over time.

Because all GSM traffic is circuit-switched, network dimensioning is a relatively straightforward process, once the traffic demand per cell is specified. The process largely involves determining the amount of traffic to be carried in the busy hour and dimensioning the network according to erlang tables. (An erlang is a unit of measure used to express traffic volume and is dimensionless. In wireless communications, 1 erlang is 1 hour or 60 minutes of use for a circuit or facility.)

The air interface, which represents the scarcest resource in the network, is dimensioned with the highest blocking probability. Typically, network designers dimension the air interface according to a 2 percent blocking probability (erlang B). For a 1 TRX cell, with 7 traffic channels (BCCH, CCCH, and SDCCH/4 sharing time slot 0), the cell can accommodate approximately 2.9 erlangs. For a 2 TRX cell, with 14 traffic channels (time slot 0 on one carrier used for BCCH and CCCH and time slot 1 used for SDCCH/8), the cell can accommodate approximately 8.2 erlangs. For a 3 TRX cell with 22 traffic channels (one time slot allocated for SDCCH/8), the cell can accommodate approximately 14.9 erlangs. It is important to note that the traffic-carrying capacity of each cell must be calculated independently.

Other interfaces in the network are usually dimensioned at much lower blocking probabilities. For example, the A interface would typically be designed for a 0.1 percent blocking probability. Similar blocking would apply to other network-internal interfaces such as the interface between the MSC and IWF. Typically, interfaces to other networks, such as the PSTN, are dimensioned at slightly higher blocking probabilities—such as 0.5 percent. Of course, the choice of blocking probability for any interface is a balance between cost and quality. The lower the blocking probability, the lower the quality and the lower the cost. The higher the blocking probability, the higher the quality and the higher the cost.

3.6 GPRS

General packet radio service (GPRS) is an enhancement to GSM-type systems which will allow a GSM system to resemble a subscriber's LAN or wide area network (WAN) and, of course, the Internet. The key benefit of GPRS is that it integrates higher-throughput packet data to mobile networks, fully enabling mobile Internet applications and a range of other advanced data services.

GPRS enables GSM operators to offer customers better wireless access to the Internet as well as a wide range of other IP packet-based services including e-mail, Web browsing, and better e-commerce features. GSM operators wanting to deploy GPRS will need to implement various system upgrades and enhancements.

GSM provides voice and data services that are circuit-switched. For data services, the GSM network effectively emulates a modem between the user device and the destination data network. Unfortunately, however, this is not necessarily an efficient mechanism for support of data traffic. Moreover, standard GSM supports user data rates of up to 9.6 kbit/s. In these days of the Internet, such a speed is considered very slow. Consequently, there is an obvious need for a solution that will provide more efficient packet-based data services at higher data rates. One solution is GPRS. Although GPRS does not offer the high-bandwidth services envisioned for 3G, it is an important step in that direction.

3.6.1 GPRS services

GPRS is designed to provide packet data services at higher speeds than those available with standard GSM circuit-switched data services. In theory, GPRS could provide speeds of up to 171 kbit/s over the air interface, although such speeds are never achieved in real networks (since, among other considerations, there would be no room for error correction on the RF interface). In fact, the practical maximum is actually a little over 100 kbit/s, with speeds of about 40 or 53 kbit/s being more realistic. Nonetheless, one can see that such speeds are far greater than the 9.6 kbit/s maximum provided by standard GSM.

The greater speeds provided by GPRS are achieved over the same basic air interface (i.e., the same 200-kHz channel, divided into eight time slots). With GPRS, however, the mobile station can have access to more than one time slot. Moreover, the channel coding for GPRS is somewhat different than that of GSM. In fact, GPRS defines a number of different channel-coding schemes. The most commonly used coding scheme for packet data transfer is coding scheme 2 (CS-2), which allows for a given time slot to carry data at a rate of 13.4 kbit/s. If a single user has access to multiple time slots, then speeds, such as 40.2 or 53.6 kbit/s, become available to that user.

Table 3.25 lists the various coding schemes available and the associated data rates for a single time slot. The air interface rates in this table give the user rates over the RF interface. As we shall see, however, the transmission of data in GPRS involves a number of layers above the air interface, with each layer adding a certain amount of overhead. Moreover, the amount of overhead generated by each layer depends on a number of factors, most notably the size of the application packets to be transmitted. For a given amount of data to be transmitted, smaller application packet sizes cause a greater net overhead than larger packet sizes. The result is that the rate for usable data is approximately 20 to 25 percent less than the air interface rate.

As mentioned, the most commonly used coding scheme for user data is CS-2. This scheme provides reasonably robust error correction over the air interface. Although CS-3 and CS-4 provide higher throughput, they are more susceptible to errors on the air interface. In fact, CS-4 provides no error correction at all on the air interface. Consequently, CS-3 and particularly CS-4 will generate a great deal more retransmission over the air interface. With such retransmission, the net throughput may well be no better than that of CS-2.

Of course, the biggest advantage of GPRS is not simply the fact that it allows higher speeds but that it is a packet-switching technology. This means that a given user consumes RF resources only when sending or receiving data. If a user is not sending data at a given instant, then the time slots on the air interface can be used by another user. Consider, for example, a user that is browsing the Web. Data are transferred only when a new page is being requested or sent. Nothing is being transferred while the subscriber contemplates the contents of a page. During this time, some other user can have access to the air interface resources, with no adverse impact to our Web-browsing friend. Clearly, this is a very efficient use of scarce RF resources.

The fact that GPRS allows multiple users to share air interface resources is a big advantage. This means, however, that whenever a user wishes to transfer data, the mobile station must request access to those resources and the network must allocate the resources before the transfer can take place. While this appears to be the antithesis of an "always-connected" service, the functionality of GPRS is such that this request-allocation procedure is well hidden from the user and the service appears to be always on. Imagine, for example, a user that downloads a Web page and then waits for some time before downloading another page. In

TABLE 3.25 GPRS Coding Schemes and Data Rates

Coding scheme	Air interface data rate, kbit/s	Approximate usable data rate, kbit/s
CS-1	9.05	6.8
CS-2	13.4	10.0
CS-3	15.6	11.7
CS-4	21.4	16.0

order to download the new page, the mobile station requests resources, is granted resources by the network, and then sends the Web page request to the network, which forwards the request to the external data network (i.e., the Internet). This happens quite quickly, however, so that the delay is not great. Quite soon, the new page appears on the user's device and at no point did the user have to dial-up to the Internet service provider.

3.6.2 GPRS user devices

GPRS is effectively a packet-switching data service overlaid on the GSM infrastructure, which is primarily designed for voice. Furthermore, while there is certainly a demand for data services, voice is still the big revenue generator—at least for now. Therefore, it is reasonable to assume that users will require both voice and data services and that operators will want to offer such services either separately or in combination. Consequently, there are three classes of GPRS.

Class A. Supports simultaneous use of voice and data services. Thus, a class A user can hold a voice conversation and transfer GPRS data at the same time.

Class B. Supports simultaneous GPRS and GSM attach, but not simultaneous use of both services. A class B user can be "registered" on GSM and GPRS at the same time, but cannot hold a voice conversation and transfer data simultaneously. If a class B user has an active GPRS data session and wishes to establish a voice call, then the data session is not cleared down. Rather it is placed on hold until such time as the voice call is finished.

Class C. Can attach to either GSM or GPRS, but cannot attach to both simultaneously. Thus, at a given instant, a class C device is either a GSM device or a GPRS device. If the device is attached to one service, the device is considered detached from the other.

In addition to the three classes described above, there are other aspects of the mobile system that are important. Most notable is the multislot capability of the device, which directly affects the data rate supported. For example, one device might support three time slots, while another might only support two. Note also that GPRS is asymmetric—it is possible for a single mobile station to have different numbers of time slots in the downlink and uplink. Normal usage patterns (e.g., Web browsing) generally require more data transfer in the downlink direction. Consequently, it is common for a user device to have different multislot capabilities between the uplink and downlink. For example, many of today's handsets support just a single time slot in the uplink direction, while supporting three or four time slots in the downlink direction.

3.6.3 GPRS air interface

The GPRS air interface is built upon the same foundations as the GSM air interface—the same 200-kHz RF carrier and the same eight time slots per carrier. This allows GSM and GPRS to share the same RF resources. In fact, if one considers a given RF carrier, then at a given instant, some of the time slots may be carrying GSM traffic while some are carrying GPRS data. Moreover, GPRS allows for dynamic allocation of resources, such that a given time slot may be used for standard voice traffic and subsequently for GPRS data traffic, depending on the relative traffic demands. Therefore, no special RF design or frequency planning is required by GPRS above that required for GSM. Of course, GPRS demand may require additional carriers in a cell. In such a situation, an additional frequency planning effort may be required, but this is no different to the frequency planning that is required with the addition of an RF carrier to support additional GSM voice traffic.

While GPRS does use the same basic structure as GSM, the introduction of GPRS means the introduction of a number of new logical channel types and new channel coding schemes to be applied to those logical channels. When a given time slot is used to carry GPRS-related data traffic or control signaling, then it is known as a packet data channel (PDCH). As shown in Fig. 3.50, such channels use a 52-multiframe structure as opposed to the 26-multiframe structure used for GSM channels. In other words, for a given time slot (i.e., PDCH), the information that is being carried at a given instant is dependent upon the position of the frame within an overall 52-frame structure. Of the 52 frames in a multiframe, there are 12 radio blocks, which carry user data and signaling; there are two idle frames; and there are two packet timing control channels (PTCCHs). Each radio block occupies four TDMA frames, such that there are 12 radio blocks in a multiframe. In other words, a radio block is equivalent to four consecutive instances of a given time slot. The idle frames in the multiframe can be used by the mobile station for signal measurements.

3.6.4 GPRS control channels

Similar to GSM, GPRS requires a number of control channels. To begin, there is the packet common control channel (PCCCH), which like the CCCH in

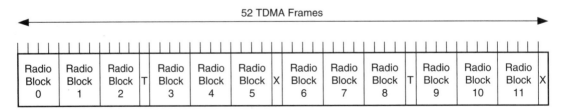

Figure 3.50 GPRS air interface frame structure. X = idle frame; T = frame used for PTCCH.

GSM, comprises a number of logical channels. The logical channels of the PCCCH include

Packet random-access channel (PRACH). Applicable only in the uplink and used by a mobile station to initiate transfer of packet signaling or data.

Packet paging channel (PPCH). Applicable only in the downlink and used by the network to page a mobile station prior to a downlink packet transfer.

Packet access grant channel (PAGCH). Applicable only in the downlink and used by the network to assign resources to the mobile station prior to packet transfer.

Packet notification channel (PNCH). Used for point-to-multipoint–multicast (PTM-M) notifications to a group of mobile stations.

The PCCCH must be allocated to different RF resources (i.e., a different time slot) from the CCCH. The PCCCH, however, is optional. If it is omitted, then the necessary GPRS-related functions are supported on the CCCH.

Similar to the BCCH in GSM, GPRS includes a packet broadcast control channel (PBCCH). This is used to broadcast GPRS-specific system information. Note, however, that the BBCCH is optional. If the PBCCH is omitted, then the BCCH can be used to carry the necessary GPRS-related system information. If the PBCCH is provisioned in a cell, then it is carried on the same time slot as the PCCCH in the same way that a CCCH and BCCH can be carried on the same time slot in GSM.

In the case where a given time slot is used to carry control channels (PBCCH, PCCCH), radio block 0 is used to carry the PBCCH, with up to three additional radio blocks allocated for PBCCH. The remaining radio blocks are allocated to the various PCCCH logical channels such as PPCH and PAGCH.

Similar to GSM, GPRS supports some dedicated control channels (DCCHs). In GPRS, these DCCHs are the packet associated control channel (PACCH) and the packet timing control channel (PTCCH). The PTCCH is used for control of timing advance for mobile stations. The PACCH is a bidirectional channel used to pass signaling and other information between the mobile station and the network during packet transfer. It is associated with a given packet data traffic channel (PDTCH) described next. The PACCH is not permanently assigned to any given resource. Rather, when information needs to be sent on the PACCH, part of the user packet data is preempted, in much the same manner as is done for the FACCH in GSM.

3.6.5 Packet data traffic channels

The PDTCH is the channel that is used for transfer of actual user data over the air interface. All PDTCHs are unidirectional—either uplink or downlink. This corresponds to the asymmetric capabilities of GPRS. One PDTCH occupies a time slot, and a given mobile station with multislot capabilities may use multiple PDTCHs at a given instant. Furthermore, a given mobile station may use a different number of PDTCHs in the downlink versus the uplink. In fact, a mobile station could be assigned a number of PDTCHs in one direction and zero PDTCHs in the other.

If a mobile station is assigned a PDTCH in the uplink, it must still listen to the corresponding time slot in the downlink, even if that time slot has not been assigned to the mobile station as a downlink PDTCH. Specifically, it must listen for any PACCH transmissions in the downlink. The reason is the bidirectional nature of the PACCH, which in the downlink is used to carry signaling from the network to the mobile station, such as acknowledgments.

3.6.6 GPRS network architecture

GPRS provides packet data channels on the air interface and a packet data-switching and transport network that is largely separate from the standard GSM switching and transport network.

3.6.7 GPRS network nodes

Figure 3.51 shows the GPRS network architecture. One can see that there are a number of new network elements and interfaces. In particular, we find the packet control unit (PCU), the serving GPRS support node (SGSN), the gateway GPRS support node (GGSN), and the charging gateway function (CGF).

The PCU is a logical network element that is responsible for a number of GPRS-related functions such as air interface access control, packet scheduling on the air interface, and packet assembly and reassembly. Strictly speaking, the PCU can be placed at the BTS, BSC, or SGSN. Logically, the PCU is considered a part of the BSC, and in real implementations, one finds the PCU physically integrated with the BSC.

The SGSN is analogous to the MSC-VLR in the circuit-switched domain. Just as the MSC-VLR performs a range of functions in the circuit-switched domain, the SGSN performs the equivalent functions in the packet-switched domain. These include mobility management, security, and access control functions.

The service area of an SGSN is divided into routing areas (RAs), which are analogous to location areas in the circuit-switched domain. When a GPRS mobile station moves from one RA to another, it performs a routing area update, which is similar to a location update in the circuit-switched domain. One difference, however, is that a mobile station may perform a routing area update during an ongoing data session, which in GPRS terms is known as a packet data protocol (PDP) context. In contrast, for a mobile station involved in a circuit-switched call, a change of location area does not cause a location update until after the call is finished.

A given SGSN may serve multiple BSCs, whereas a given BSC interfaces with only one SGSN. The interface between the SGSN and the BSC (in fact the PCU within the BSC) is the Gb interface. This is a frame relay–based interface, which uses the BSS GPRS protocol (BSSGP). The Gb interface is used to pass signaling and control information to and from the SGSN as well as user data traffic.

Wireless Mobile Radio Technologies 167

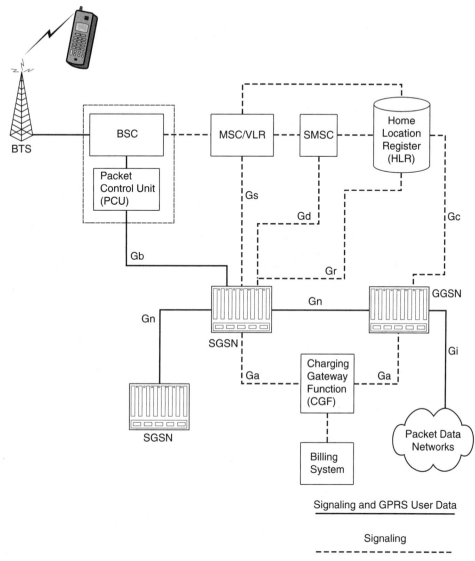

Figure 3.51 GPR network architecture.

The SGSN also interfaces to an HLR via the Gr interface. This is an SS7-based interface and it uses the mobile application part (MAP), which has been enhanced for support of GPRS. The Gr interface is the GPRS equivalent of the D interface between a VLR and HLR. The Gr interface is used by the SGSN to provide location updates to the HLR for GPRS subscribers and to retrieve GPRS-related subscription information for any GPRS subscriber that is located in the service area of the SGSN.

An SGSN may optionally interface with an MSC, via the Gs interface. This is an SS7-based interface that uses the signaling connection control part (SCCP). Above the SCCP, there is a protocol known as BSSAP+, which is a modified version of the base station subsystem application part (BSSAP) as used between an MSC and BSC in standard GSM. The purpose of the Gs interface is to allow coordination between an MSC-VLR and SGSN for those subscribers that support both circuit-switched services controlled by the MSC-VLR (i.e., voice) and packet data services controlled by the SGSN. For example, if a given subscriber supports both voice and data services, and is attached to an SGSN, then it is possible for an MSC to page the subscriber for a voice call via the SGSN by using the Gs interface.

The SGSN interfaces with a short message service center (SMSC) via the Gd interface. This allows GPRS subscribers to send and receive short messages over the GPRS network (including the GPRS air interface). The Gd interface is an SS7-based interface using MAP.

A GGSN is the point of interface with external packet data networks (e.g., the Internet). Thus, user data enters and leaves the PLMN via a GGSN. A given SGSN may interface with one or more GGSNs, and the interface between an SGSN and GGSN is known as the Gn interface. This is an IP-based interface used to carry signaling and user data. The Gn interface uses the GPRS tunneling protocol (GTP), which tunnels user data through the IP backbone network between the SGSN and GGSN.

A GGSN may optionally use the Gc interface to an HLR. This interface uses MAP over SS7. This interface would be used in the case where a GGSN needs to determine the SGSN currently serving a subscriber, similar to the manner in which a GMSC queries an HLR for routing information for a mobile-terminated voice call. One difference between the scenarios, however, is the fact that a given data session is usually established by the mobile station rather than by an external network. If the mobile station establishes the session, then the GGSN knows what SGSN is serving the mobile station because the path from the mobile station to the GGSN passes via the serving SGSN. Therefore, in such situations, there is no need for the GGSN to query the HLR. A GGSN will query the HLR in the case where the session is initiated by an external data network. This is an optional capability, and a given network operator may choose not to support that capability. In many networks the capability is not implemented as it requires that the mobile station have a fixed packet protocol address (e.g., IP address). Given that address space is often limited (at least for IP version 4), a fixed address for each mobile station is not often possible.

An SGSN may interface with other SGSNs in the network. This inter-SGSN interface is also termed the Gn interface and also uses GTP. The primary function of this interface is to allow tunneling of packets from an old SGSN to a new SGSN when a routing area update takes place during an ongoing PDP context. Note that such forwarding of packets from one SGSN to another occurs only briefly—just as long as it takes for the new SGSN and the GGSN to establish the PDP context directly between them, at which point the old

SGSN is removed from the path. This is different from, for example, inter-MSC handoff for a circuit-switched call, where the first MSC remains as an anchor until the call is finished.

3.6.8 GPRS traffic scenarios

Sections 3.6.9 and 3.6.10 provide some straightforward examples of GPRS traffic. Here, we briefly describe some unique terms associated with GRPS traffic flow:

Temporary block flow (TBF). The physical connection between the mobile station and the network for the duration of data transmission. A number of radio blocks are used over the air interface.

Temporary flow identity (TFI). An identifier assigned to a given TBF and used for distinguishing one TBF from another. A TFI is used in control messages (e.g., acknowledgments) related to a given TBF, so that the entity receiving the control message can correlate the message with the appropriate TBF.

Temporary logical link identity (TLLI). An identifier that uniquely identifies a mobile station within a routing area. The TLLI is sent in all packet transfers over the air interface. The TLLI is derived from the packet temporary mobile station identity (P-TMSI) assigned by an SGSN, provided that the mobile station has been assigned a P-TMSI. When the mobile station has never been assigned a P-TMSI, the mobile station may generate a random TLLI.

Uplink state flag (USF). An indicator used by the network to specify when a given mobile station is entitled to use a given uplink resource. In GPRS, resources are shared in both the downlink and the uplink. The downlink is under the control of the network, which can schedule transmissions for a given user on a given downlink PDTCH as appropriate. On the uplink, however, a mechanism is necessary to ensure that only a given mobile station transmits on a given uplink resource at a given time. There are two ways to do this—fixed allocation and dynamic allocation. With fixed allocation, the network allocates some number of uplink time slots to a user, allocates some number of radio blocks that the mobile station may transmit, and specifies the TDMA frame when the user may begin transmission. Thus, the mobile station is provided with exclusive access to the time slot for a particular period of time. With dynamic allocation, the network does not allocate a specific time upfront for the user to transmit. Rather, it allocates the user a particular value of USF for each time slot that the user may access. Then on the downlink, the network transmits a USF value on each radio block. This value indicates which mobile station has access to the next radio block on the corresponding time slot in the uplink. Thus, by examining the value of USF received on the downlink, the mobile station can schedule its uplink transmissions. The USF is a 3-bit field and thus has eight possible values. Thus, with dynamic allocation, up to eight mobile stations can share a given uplink time slot.

3.6.9 GPRS attach

GPRS functionality in a mobile station may be activated either when the mobile station itself is powered on or perhaps when the browser is activated. Whatever the reason for the initiation of GPRS functionality within the mobile station, the mobile station must attach to the GPRS network so that the GPRS network (and specifically the serving SGSN) knows that the mobile station is available for packet traffic. In the terms used in GPRS specifications, the mobile station moves from the idle state (not attached to the GPRS network) to the ready state (attached to the GPRS network and in a position to initiate a PDP context). When in the ready state, the mobile station can send and receive packets. There is also a standby state, which the mobile station enters after a time-out in the ready state. If, for example, the mobile station attaches to the GPRS network but does not initiate a session, then it will remain attached to the network, but move to standby state after a time-out.

Figure 3.52 shows the simple case of a class C mobile station performing a GPRS attach. In this diagram we have included a great deal of the air interface signaling. Many of the air interface messages shown in the figure are applicable to any access to or from the mobile station, even if that access is just for signaling the transfer of user packets. For the sake of brevity, the messages will not be repeated in every subsequent scenario we describe.

A GPRS attach is somewhat similar in functionality to a location update in GSM. The process begins with a packet channel request from the mobile station. In the request, the mobile station indicates the purpose of the request—e.g., page response, mobility management procedure, or a two-phase access, which would be used in the case of transferring user data. In the scenario of Fig. 3.52, a mobility management procedure is indicated. The network responds with a packet uplink assignment, which allocates a specific time slot or time slots to the mobile station for the message that the mobile station wishes to send. The network includes a TFI to be used by the mobile, a USF value for the mobile on the time slots assigned (in the case of dynamic allocation), and an indication of the number of radio link control (RLC) blocks granted to the mobile station for the TBF in question.

The mobile station proceeds to send the attach request in one or more radio blocks to the network on the assigned resources. The mobile station can send no more than the number of blocks that have been allocated by the network. In the case of mobility management messages, the assigned resources will typically be sufficient for the mobile station to send the necessary data. If not, as might be the case where the mobile station wishes to send user packet data, the mobile station can request additional resources through a packet resource request message.

Upon receipt of the attach request at the BSS, the BSS uses the PACCH to acknowledge the receipt. In the case where the mobile station has sent all the information it wishes to send, which would be the case in our example, then this is indicated in the transmission from mobile station to network. In that case, the acknowledgment from the network is a final acknowledgment, which

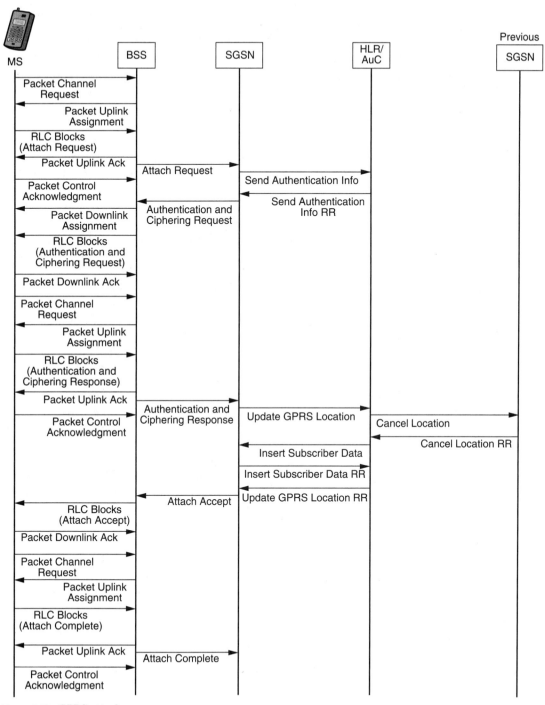

Figure 3.52 GPRS attach.

is indicated in the acknowledgment message itself. This causes the mobile station to send a packet control acknowledgment message back to the network and release the assigned resources.

Meanwhile, the BSS forwards the attach request to the SGSN. The SGSN may choose to invoke security procedures, in which case it fetches triplets from the HLR. Note, however, that there is a slight difference in GPRS regarding authentication and ciphering. Specifically, ciphering in GPRS takes place between the mobile station and the SGSN such that the whole link from mobile station to SGSN is encrypted. In standard GSM, only the air interface is encrypted. The authentication and ciphering is initiated by the issuance from the SGSN of the authentication and ciphering request to the mobile station via the BSS.

The BSS first sends a packet downlink assignment message to the mobile station. This message can be sent either on the PCCCH or PACCH. Which is chosen depends upon whether the mobile station currently has an uplink PDTCH. If it does, then the PACCH is used. The packet downlink assignment instructs the mobile station to use a given resource in the downlink—including time slots to be used and a downlink TFI value. The BSS subsequently forwards the authentication and ciphering request as received from the SGSN.

Upon receipt of the request, the mobile station acknowledges the downlink message and then requests uplink resources so that it can respond. Thus, it sends another packet channel request, much like the one it sent initially. Once again, the network assigns resources to the mobile station, which the mobile station uses to send its authentication and ciphering response to the network. That response is forwarded from the BSS to the SGSN. The BSS also sends an acknowledgment to the mobile station, and the mobile station confirms receipt of the acknowledgment, just as it did for the acknowledgment associated with the initial attach request.

Once the mobile station is authenticated by the SGSN, the SGSN performs a GPRS update location towards the HLR. This is similar to a GSM location update, including the download of subscriber information from the HLR to the SGSN. Once the update location message is accepted by the HLR, the SGSN sends an attach accept message to the mobile station. As for other messages, the BSS first assigns resources so that the mobile station can receive the message. Similarly, once the mobile station receives the message, it requests resources in the uplink so that it can respond with an attach complete message. The BSS acknowledges receipt of the RLC data containing the attach complete message, and forwards the message to the SGSN. The mobile station confirms receipt of the acknowledgment.

Note that, throughout the procedure just described, the mobile station requests access to resources for each message that it sends toward the network. This is typical of the manner in which GPRS manages resources and is one of the main reasons why GPRS enables multiple users to share limited resources. Of course, in our example, only signaling is occurring, which consumes very little RF capacity (i.e., very few radio blocks). In the case of packet data transfer, many more data blocks would be transmitted for a given TBF. Not every block needs to be acknowledged, however. In fact, GPRS allows for both acknowledged and unacknowledged operation. In the case of acknowledged operation, acknowledgments are sent only periodically, with each acknowledgment indicating all correctly received RLC blocks up to an indicated block sequence number.

3.6.10 Combined GPRS and GSM attach

Figure 3.53 depicts a very simple GPRS attach scenario that would apply to a class C mobile station. In the case of a class A or class B mobile station, the mobile station may wish to simultaneously attach to the GSM network and the GPRS network. In this case, the mobile station can attach to the MSC-VLR during the GPRS attach procedure. This assumes, of course, that the network supports a combined attach (which it broadcasts in system information messages) and that the network includes the Gs interface.

In the case, for example, that a class B mobile station is powered up and needs to attach to both the GSM and GPRS services, the sequence would be as depicted in Fig. 3.53. For the sake of brevity, we have omitted air interface signaling, which would be the same in the example of Fig. 3.53 as already shown in Fig. 3.52.

In this case, the mobile station instigates an attach to the SGSN, but also indicates that it wishes to perform a GSM attach. In this case, the new SGSN, in addition to performing the procedures required of a GPRS attach, also interacts with the VLR to initiate a GSM attach. Specifically, we note the use of the BSSAP+ messages Location Update Request and Location Update Accept between the SGSN and VLR. The location update request message from the SGSN is similar to the equivalent message that would be received from a mobile station that performs a normal GSM location update. Therefore, the MSC-VLR performs similar mobility management functions (see Fig. 3.46)—such as performing a MAP update location to the HLR. One difference in this scenario, however, is the fact that the MSC-VLR does not attempt to authenticate the mobile station itself, as the authentication has already been performed by the SGSN.

Note that, in Figs. 3.52 and 3.53, certain optional functions have not been shown. These functions include an IMEI check and the allocation of a new packet temporary mobile subscriber identity.

3.6.11 Establishing a PDP context

The transfer of packet data is through the establishment of a packet data protocol (PDP) context, which is effectively a data session. Normally, such a context is initiated by the mobile station, as would happen, for example, when a browser on the mobile station is activated and the subscriber's home page is retrieved from the Internet. When a mobile station or the network initiates a PDP context, the mobile station moves from the standby state to the ready state. The initiation of a PDP context is illustrated in Fig. 3.54.

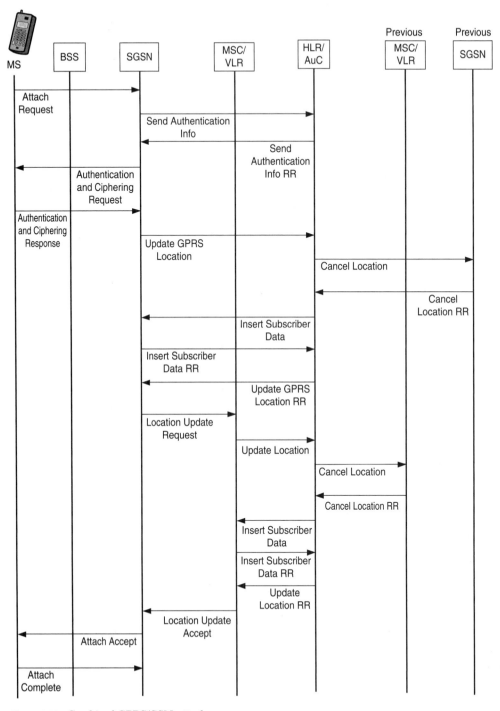

Figure 3.53 Combined GPRS/GSM attach.

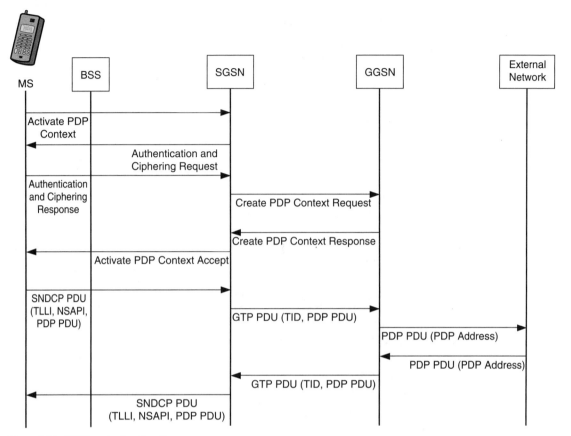

Figure 3.54 PDP context activation.

A mobile station–initiated PDP context begins with a request from the mobile station to activate a PDP context. This request includes a number of important information elements, including a requested network service access point identifier (NSAPI), a requested logical link control (LLC) service access point identifier (SAPI), a requested quality of service (QOS), a requested PDP address, and a requested access point name (APN).

The NSAPI indicates the specific service within the mobile station that wishes to use GPRS services. For example, one service might be based on IP, while another might be based on X.25.

The LLC SAPI indicates the requested service at the LLC layer. LLC is used both during data transfer and during signaling. Consequently, at the LLC layer, it is necessary to identify the type of service being requested—e.g., GPRS mobility management signaling, user data transfer, or short message service (which can be supported over GPRS as well as over GSM).

The requested QOS indicates the desires of the mobile station regarding how the session should be handled. Components of QOS include reliability (including maximum acceptable probabilities of packet loss, packet corruption, and out-of-sequence delivery), delay, mean throughput, peak throughput, and precedence (which is used to determine the priority of the mobile station's packets in cases of network congestion where packets may need to be discarded).

The requested PDP address will typically be either an IP address or empty. The network will interpret an empty address as a request that the network should assign an address. In such a case, the dynamic host configuration protocol (DHCP) should be supported in the network. The address is assigned by the GGSN, which must either support DHCP capabilities itself or interface with a DHCP server.

The access point name indicates the GGSN to be used and may indicate the external network to which the mobile station should be connected. The APN contains two parts: the network identifier and the operator identifier. The APN network identifier appears like a typical Internet uniform record locator (URL) according to domain name service (DNS) conventions, i.e., a number of strings separated by dots, such as host.company.com. The APN operator identifier is optional. When present, it has the format operator.operator-group.gprs. For each operator, there is a default APN operator identifier, which has the form MCC.MNC.GRPS. The mobile country code (MCC) and the mobile network code (MNC) are part of the IMSI that identifies the subscriber and is available at the SGSN. The default APN operator identifier is used to route packets from a roaming subscriber to a GGSN in the home network in the case where the APN from the subscriber does not include an APN operator identifier.

Based on the APN received from the subscriber, the SGSN determines the GGSN that should be used. The SGSN normally does this by sending a query to a DNS server (not shown in Fig. 3.54). The query contains the APN, and the DNS server responds with an IP address for the appropriate GGSN.

Next the SGSN creates a tunnel identifier for the requested PDP context. The tunnel identifier combines the subscriber IMSI with the NSAPI received from the mobile station and uniquely identifies a given PDP context between the SGSN and GGSN. The SGSN sends a create PDP context request message to the GGSN. This contains a number of information elements, including the tunnel identifier, the PDP address, the SGSN address, and the QOS profile. Note that the QOS profile sent from the SGSN to the GGSN may not match that received from the mobile station. The SGSN may choose to override the QOS parameters received from the mobile station based upon the subscribed QOS (as received from the home location register) or based upon the resources available at the SGSN. If the PDP address is empty, then the GGSN is required to assign a dynamic address.

The GGSN returns the create PDP context response message to the SGSN. Provided that the GGSN can assign a dynamic address and provided that it can support connection to the external network as specified by the APN, the response is a positive one. In that case, the response includes, among other items, GGSN addresses for user traffic and for signaling, and end-user address (as received from DHCP), the tunnel identifier, the QOS profile, a charging identifier, and a charging gateway address.

Upon receipt of the create PDP context response message from the GGSN, the SGSN sends an activate PDP context accept message to the mobile station. This contains the PDP address for the mobile station (in the case that a dynamic address has been assigned by the network), the negotiated QOS, and the radio priority (which indicates the priority the mobile station shall indicate to lower layers, and which is associated with the negotiated QOS). Note that the network shall attempt to provide the mobile station with the requested QOS, or at least come close. If the QOS returned by the SGSN is not acceptable to the mobile station, then the mobile station can deactivate the PDP context.

Once the mobile station has received the PDP context accept message from the SGSN, then everything necessary is in place to route packets from the mobile station through the SGSN to the GGSN and on to the destination network. The mobile station sends the user packets as subnetwork-dependent convergence protocol protocol data units (SNDCP PDUs). Each such PDU contains the temporary logical link identity for the subscriber and the NSAPI indicating the service being used by the subscriber, plus the user data. The temporary logical link identity and NSAPI allow the SGSN to identify the appropriate GTP tunnel toward the correct GGSN. The SGSN encapsulates the user data within a GTP PDU, including a tunnel identifier, and forwards the user data to the GGSN. At the GGSN, the GTP tunnel "wrapper" is removed and the user data are passed to the remote data network (e.g., the Internet).

Packets from the external network back to the mobile station first arrive at the GGSN. These packets include a PDP address for the mobile station (e.g., an IP address), which allows the GGSN to identify the appropriate GTP tunnel to the SGSN. The GGSN encapsulates the received PDU in a GTP PDU, which it forwards to the SGSN. The SGSN uses the tunnel identifier to identify the subscriber and service in question (i.e., the temporary logical link identity and NSAPI). It then forwards an SNDCP PDU to the mobile station via the base station system.

Note again, that each access to and from the mobile station over the air interface requires the request and allocation of resources for use by the mobile station. In other words, the PDTCH(s) that the mobile station may be using are not dedicated solely to the mobile station either during the PDP Context establishment or during packet transfer to or from the external packet network.

3.6.12 Inter-SGSN routing area update

In GPRS, each PDU to or from the mobile station is passed individually, and there is no permanent resource established between the SGSN and mobile station. Thus, if a subscriber moves from the service area of one SGSN to that of another, it is not necessary for the first SGSN to act as an anchor or relay of packets for the duration of the PDP context. This is fortunate as the PDP context could last for a long time. Thus, there is no direct equivalent of handoff as it is known in circuit-switching technology, where the first MSC acts as an anchor until a call is finished. Nonetheless, as a mobile station moves from one SGSN to another during an active PDP context, there are special functions that need to be invoked so that packets are not lost as a result of the transition.

The process is illustrated in Fig. 3.55, where a mobile station moves from the service area of one SGSN to that of another during an active PDP context. The mobile station notices, from the PBCCH (or BCCH) that it is in a new routing area. Consequently, it sends a routing area update message to the new SGSN. Among the information elements in the message are the temporary logical link identity, existing P-TMSI, and old routing area identity (RAI). Based on the old RAI, the new SGSN derives the address of the old SGSN and sends an SGSN context request message to the old SGSN. This is a GTP message, passed over an IP network between the two SGSNs.

The old SGSN validates the P-TMSI and responds with an SGSN context response message, with information regarding any PDP context and mobility management context currently active for the subscriber, plus the subscriber's IMSI. PDP context information includes GTP sequence numbers for the next PDUs to be sent to the mobile station or tunneled to the GGSN, the APN, the GGSN address for control plane signaling, and QOS information. The old SGSN stops transmission of PDUs to the mobile station, stores the address of the new SGSN, and starts a timer.

The mobility management context sent from the old SGSN to the new SGSN may include unused triplets, which the new SGSN will use to authenticate the subscriber. If the old SGSN has not sent such triplets, then the new SGSN can fetch triplets from the home location register in order to perform authentication and ciphering.

The new SGSN responds to the old SGSN with the GTP message SGSN context acknowledge. This indicates to the old SGSN that the new SGSN is ready to take over the PDP context. Consequently, the SGSN forwards any packets that may have been buffered at the old SGSN so that the new SGSN can forward them. The old SGSN continues to forward any additional PDUs that are received from the GGSN.

The new SGSN sends an update PDP context request to the GGSN, to inform the GGSN of the new serving SGSN for the PDP context. The GGSN responds with the update PDP context response message. Any subsequent PDUs from the GGSN to the mobile station are now sent via the new SGSN.

The new SGSN then invokes (requests) an update GRPS location operation message to the home location register. This causes the home location register to send a MAP cancel location message to the old SGSN. Upon receipt of the cancel location message, the old SGSN stops the timer and deletes any information regarding the subscriber and the PDP context.

Once the MAP update location procedure is complete, the new SGSN accepts the routing area update from the mobile station, which the mobile station

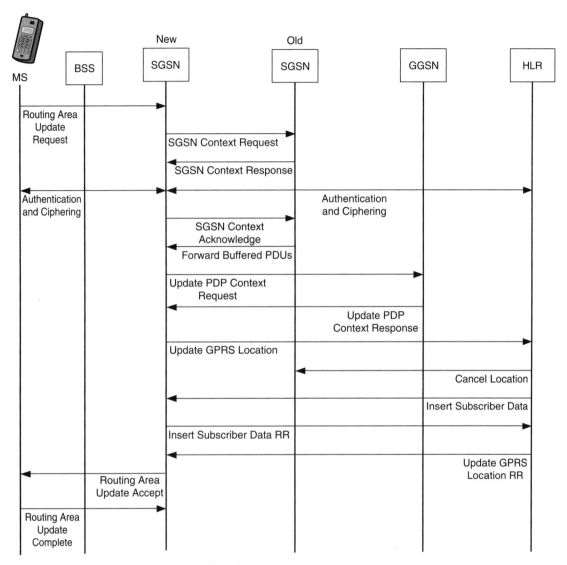

Figure 3.55 Inter-SGSN routing areas update during active context.

acknowledges with a routing area complete message. The new SGSN proceeds to send and receive PDUs to and from the mobile station.

In a combined GSM and GPRS network, it is common for location area boundaries and routing area boundaries to coincide. In such a case, an inter-SGSN routing area update might also coincide with the need to perform a location update toward a new MSC-VLR. In that case, the SGSN can communicate with the MSC over the Gs interface and can trigger a location update at the MSC, in much the same manner as shown in Fig. 3.55 for a combined GSM and GPRS attach.

3.6.13 Traffic calculation and network dimensioning for GPRS

Dimensioning a GPRS network involves the dimensioning of a number of network elements (e.g., SGSN, GGSNs) and a number of network interfaces (e.g., air interface, Gb, Gn). Of each of the resources (nodes and interfaces) available in a GPRS network, the most limited is the air interface. Moreover, the air interface resources must be shared with the GSM. One does not want to inhibit GSM voice traffic for the sake of GPRS data traffic, or vice versa, so some planning is required.

3.6.14 Air interface dimensioning

The most straightforward way to determine the required GPRS air interface capacity is to estimate the amount of data traffic (in terms of bits per second) that a given cell will be required to handle in the busy hour. This can be done by estimating the number of GPRS users in the cell and the usage requirements of those users (which will be linked to handset capabilities and the commercial agreements between users and the network operator). From this demand estimate, we can estimate an average GPRS throughput requirement in the busy hour. In order to allow for usage spikes within the busy hour, it is appropriate to add an overhead of 20 to 25 percent. From this, we can then determine the number of channels that are needed to support that load. For example, using CS-2, a single time slot can carry about 10 kbit/s of user data.

This approach, however, does not account for the fact that a given cell will most likely be used to support both GPRS data traffic and GSM voice traffic. When a cell's resources are shared between GPRS and GSM, it is quite inefficient to independently determine GPRS resource requirements and GSM resource requirements (based on some blocking criteria) and simply add the two together. To do so would result in overdimensioning of the cell. The reason is that voice traffic follows an erlang distribution, which requires that there be more channels in a cell than are used, on average, by the voice traffic. If, for example, we have a cell with three RF carriers and a total of 22 traffic channels (one traffic channel for BCCH and one for SDCCH/8), then at 2 percent blocking the 22 traffic channels can carry approximately 15 erlangs. In other words, at any given instant, we can expect 15 of the 22 traffic channels to be occupied with voice traffic, leaving 7 channels available. This is not to say that voice traffic will never use more than 15 traffic channels during the busy hour, but there will be an average of 7 traffic channels available during the busy hour. These 7 traffic channels can be used for GPRS traffic. At CS-2, this corresponds to a gross data rate on the air interface of over 90 kbit/s for GPRS traffic and a usable rate of about 70 kbit/s. Thus, we can accommodate an average of 70 kbit/s of GPRS traffic in the cell during the busy hour without increasing the number of RF channels. Whether this will be sufficient to accommodate the needs of the GPRS users (including any buffer for usage spikes) is dependent upon what those needs happen to be. If it is insufficient, then more RF capacity will need to be added. This RF capacity can be dedicated for GPRS or can be shared between GSM and GPRS. For that matter, any cell that supports both GSM and GPRS can be configured so that all resources are shared or so that certain resources are reserved for one service or the other, with any remaining resources shared.

The approach just described, whereby inefficiently used GSM capacity is used by GPRS, does not necessarily tell the whole story of RF dimensioning. First, the approach assumes that the GSM network is correctly dimensioned for voice to begin with, which may not be the case in heavily loaded cells. Second, what we have described implies an assumption that may not be true in reality—that the GPRS and GSM busy hours coincide. If they do not coincide, then the approach just described will err on the conservative side.

As of this writing, relatively few GPRS networks have been deployed (when compared with the number of GSM networks) and there are relatively few GPRS subscribers. Therefore, there is not a great deal of real-world experience to draw upon. This is unfortunate as it means that real-world rules of thumb have not yet been developed. On the other hand, it is fortunate that we have not had to deal with a very sudden explosion in the number of GPRS subscribers. As the number of subscribers grows, we will be able to monitor traffic patterns to see the types of transactions subscribers require, typical file sizes, burstiness, etc. Such monitoring will allow for trending so that RF dimensioning decisions can be made in advance of subscriber demand.

3.6.15 GPRS network node dimensioning

Among the nodes that need to be dimensioned for GPRS traffic are the BSC, SGSN, and GGSN. Generally, the capacity of a BSC is limited by the number of cells, BTS sites (or interfaces to BTS sites), transceivers (regardless of whether those transceivers are used for voice or data), and simultaneous PDCHs. In addition, one needs to dimension the Gb interface, which is related to the number of Gb ports (T1 or E1) supported by the BSC. In most implementations one finds that the number of supported PDCHs is sufficiently large that other limitations, such as the maximum number of transceivers, will be reached first.

For the SGSN, there are a number of capacity limitations—the number of attached subscribers, cells, routing areas, and Gb ports, and the total throughput capacity. Typically, one finds that the key capacity limitations are the number of attached subscribers and the total throughput as these limits are likely to be met before any of the others. For the GGSN, the key limitations are the number of simultaneous PDP contexts and the total throughput. These key dimensioning factors are listed in Table 3.26.

3.7 iDEN

The integrated dispatch enhanced network (iDEN) system is a unique wireless access platform because it involves the integration of several mobile phone

TABLE 3.26 GPRS Node Dimensioning Factors

Network node	GPRS-specific dimensioning factors
BSC	Number of PDCHs
	Number of Gb ports
SGSN	Number of attached subscribers
	Total throughput
	Number of Gb ports
	Number of cells
	Number of routing areas
GGSN	Number of simultaneous PDP contexts
	Total throughput

technologies based on a modified GSM platform. The services that are integrated into the iDEN involve dispatch, full-duplex telephone interconnect, data transport, and short messaging services.

The dispatch system involves a feature called *group call* where multiple people can engage in a conference. The user list is preprogrammed, and the conference call can be set up just like it is done in a two-way or specialized mobile radio system with the exception that the connection can take place utilizing any of the frequencies that are available from the pool of channels where the subscriber is physically located.

The telephone interconnect and data transport are meant to offer conventional mobile communications. The short messaging service enables the iDEN phones to receive up to 140 characters for an alphanumeric message. An example of a typical iDEN system is shown in Fig. 3.56. The elements which comprise the iDEN system shown in the figure are briefly listed here:

Dispatch application processor (DAP)

Enhanced base transceiver (EBTS)

Home location register (HLR)

Metro packet switch (MPS)

Mobile switching center (MSC)

Operations and maintenance center (OMC)

Short message service–service center (SMS-SC)

Transcoder (XCDR)

Figure 3.56 shows several differences between an iDEN system and a typical mobile wireless system. Although it integrates several platforms together, iDEN is effectively two distinct systems overlaid on each other. The two distinct systems are the interconnection and dispatch systems which share some common elements, like EBTS radios.

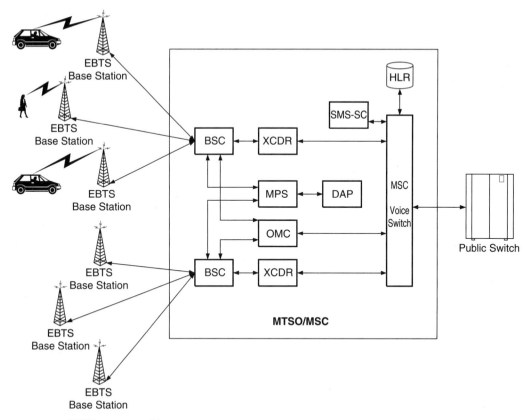

Figure 3.56 iDEN system architecture.

The BSC is responsible for traffic and control channel allocations in addition to handoff data collection and controlling handoffs between other BSCs. The MPS provides the connectivity for the dispatch calls; it distributes the dispatch packets as well as the ISMI assignment. The DAP is the processing entity responsible for overall coordination and control of the dispatch services. It enables the following types of calls to take place:

Talk group

Private call

Call alert

The radio access system used by an iDEN system is TDMA. The channel bandwidth is 25 kHz which consists of four independent sidebands, each being a 16-QAM baseband signal. The center frequencies of these sidebands are 4.5 kHz apart from each other and are spaced symmetrically about a suppressed RF carrier frequency resulting in a 16-point data symbol constellation that carries 4 data bits per symbol. The location of where an iDEN system is utilized in the spectrum is shown in Fig. 3.57. The RF channel structure is shown in Fig. 3.58.

The iDEN system was introduced using a 6:1 interleave for both dispatch and interconnect services. Later the system was upgraded enabling a 3:1 interleave for interconnect-only service. The wireless operator has the choice of offering 6:1 or 3:1 voice service in addition to dispatch. There are numerous impacts to capacity with the selection of which interconnection method is used and the amount of dispatch traffic that is carried on a system. Looking at a simplistic example, the 3:1 voice call requires two traffic channels, while a 6:1 (dispatch call) requires only a single traffic channel. There are, of course, other issues related to signaling and call quality which are factored into this.

iDEN systems utilize several control channels, similar in nature to GSM systems. These control channels are listed next for reference. In addition to the control channels, there are two other channels used in iDEN: traffic and paging channels.

1. *Primary control channel (PCCH).* A multiple-access channel used for the transmission of general system parameters. The outbound PCCH contains the broadcast control channel (BCCH) and the common control channel (CCCH), while the in-bound PCCH is referred to as the random-access channel (RACH).
    a. Inbound—service requests
    b. Outbound—service grants
    c. BCCH
        (1) Neighbor cells
        (2) Control channel

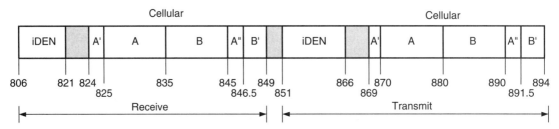

Figure 3.57 iDEN spectrum location.

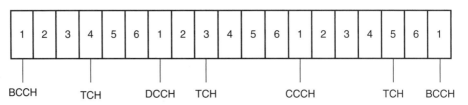

Figure 3.58 iDEN RF channel structure.

(3) Packet channels
(4) Location areas
(5) Common control channel
(6) Paging subchannel
(7) Service grants
2. *Temporary control channel (TCCH)*
   a. Inbound—dispatch reassignment requests
   b. Outbound—handoff target
3. *Dedicated control channel (DCCH)*
   a. Inbound—location updating
      (1) Authentication
      (2) SMS
      (3) Registration
   b. Outbound
      (1) Location updating
      (2) Authentication
      (3) SMS
      (4) Registration
4. *Associated control channel (ACCH)*
5. *Traffic channel (TCH).* Provides circuit mode transmission for voice and data
   a. Inbound—dispatch reassignment requests
   b. Outbound—handoff target
6. *Packet channel (PCH).* Provides for multiple-access packet mode transmission.

There are many interesting issues associated with the iDEN call processing for dispatch and interconnection calls. From the time an iDEN mobile subscriber is initially powered up, until it is powered down, a series of procedures are executed between the EBTS and the mobile to control the radio communications link. Before a call description flowchart is shown, a few terms or processes used in iDEN systems associated with the mobile will be briefly covered.

Cell selection. At power-up, the mobile scans a preprogrammed list of system frequencies, called a *bandmap,* looking for a PCCH. When the mobile hears a PCCH, outbound power and signal quality estimate (SQE) measurements are taken and the frequency is added to a list. The mobile continues scanning channels until either 32 PCCHs are found or until the bandmap list is exhausted, which is market specific. The PCCH list is sorted based on SQE and the receive signal strength indicator (RSSI), and the subscriber then attempts to camp on the first cell on the list. If the subscriber fails, an attempt will be made to camp on the next cell, etc., until the subscriber either succeeds in the camping or exhausts the list, requiring a new cell selection process to begin.

Cell reselection. Each serving cell will transmit its neighbor cell list to all the subscribers it serves, and the mobile will take SQE measurements of the received power of the serving cell and of each neighbor cell and will sort the neighbor cell list according to received signal strength. When the mobile determines that the best neighbor cell is a better candidate than the current serving cell, a reselection occurs making the formerly best neighbor cell the new serving cell.

Fast reconnect. During a dispatch call, the mobile continues to monitor the SQE and signal strength of the serving and neighbor cells. Under certain conditions, the mobile may decide to change its serving cell. When the mobile is on the traffic channel (during the talk phase of a call), the mobile initiates a reconnect if the serving cell's outbound SQE is less than desired or upon the failure or disconnect of the serving cell.

Power control. The mobile periodically adjusts its transmit power based on the power received at the fixed network equipment (FNE). It also periodically receives a power control constant and periodically measures the serving cell output power. It then calculates the desired mobile transmit power by subtracting the serving cell output power from the power control constant and adjusts its transmit power accordingly.

Handoff. The iDEN utilizes MAHO to assist in the handoff process. The handoff can be initiated by either the mobile or base station depending on the parameter settings. Handoffs are only possible with interconnection calls. However, for dispatch the location information supplied in the response also includes the neighbor list from cells that are on the beacon channel list, and, therefore, if the dispatch location area (DLA) is set up incorrectly, it is possible that subscribers will need to reacquire the system if they move outside of the coverage area of the sites in the list.

The mobile-assisted handoff (MAHO) process is as follows:

1. The mobile monitors the BCCH to determine which cells to include in the MAHO list.
2. The mobile continues to monitor SQE, the RSSI for the primary serving channel, as well as the channels in the MAHO list.
3. If the subscriber detects trouble on a primary service or a better neighbor cell, the mobile sends a sample of its measurements.
4. The subscriber signals in the ACCH with SQE measurements.
5. The MSC/BSC/EBTS finds a new server to hand off to and allocates a traffic channel for this process.
6. The MSC/BSC/EBTS senses a hand off command on the ACCH with initial power setting and channel and traffic channel to tune to.
7. The mobile station changes to the assigned channel.
8. The mobile station uses a random-access procedure, to get its timing information from the target EBTS.
9. The channel changes to a traffic channel, and conversation continues.

Last, SQE is used extensively in various cell site selection decisions, primarily for outbound RSSI measurements of the serving cell as well as for neighboring cells which are potential handoff candidates. SQE is very similar to $C/(I + N)$ in the range of 15 to 23 dB.

The dispatch system (part of the iDEN system shown in Fig. 3.59) basically has three primary service offering or functions:

1. Private
2. Talk group
3. Call alert (twiddle)

Private dispatch is where the originating call uses PTT between one subscriber unit and another (classic two way). Call alert is used to notify a subscriber that a voice communication is desired. Talk groups involve a more extensive look.

Talk groups are defined by service areas, shown in Fig. 3.60. The service area (SA) is used for dispatch group calls. When a dispatch call takes place, a single voice channel slot is used in any coverage area for a cell when there are one or more members of the call group in that coverage area. Fleets are assigned to the same group, and a mobile can be grouped into several talk groups in order to communicate between specific groups which comprise the entire fleet.

Briefly stated, a mobile can be grouped into several talk groups used to communicate with a group of mobiles in the fleet at the same time. For example, say there is a fleet for all of New York City, but the subscriber only wants to talk with the Queens fleet. The mobile for the Queens fleet is assigned its own talk group which is part of the overall fleet group. A mobile can, therefore, be part of numerous talk groups.

To help clarify, a flowchart for dispatch calls is shown in Fig. 3.60. The following text better explains some of the sequences in the flowchart.

1. A dispatch call request is made by push to talk.
2. The call request packet is routed to the dispatch application processor (DAP).
3. The DAP recognizes the subscriber unit's group affiliation and tracks the group members' current location area.
4. The DAP sends a location request to each group member location area to obtain the various subscribers' cell and sector location information.
5. The subscriber units in the group respond with their current cell and sector location information.

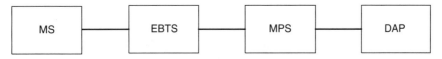

Figure 3.59 Dispatch only.

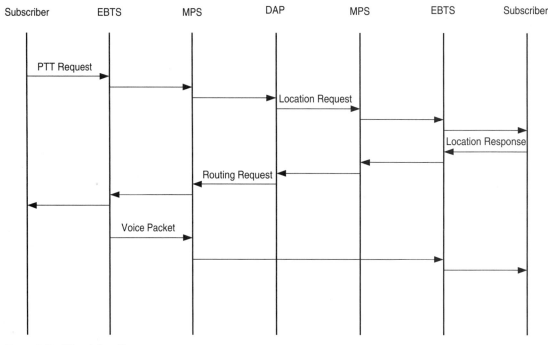

Figure 3.60 Dispatch call sequence.

6. The DAT instructs the originating EBTS with packet routing information for all group members.

7. Call voice packets are received by the packet duplicator (PD) which then are replicated and distributed to the group's end node.

Another portion of the iDEN system is utilized for interconnection after the radio access. The general sequence of events for an interconnection call is the same for 3:1 and 6:1 calls, with the exception of the amount of TCHs assigned. The interconnection sequence for a mobile-to-land call is briefly listed here:

- Call initiation
- Random-access protocol/procedure (RAP) on PCCH
- DCCH assignment
- Authentication
- Call setup transaction
- Traffic channel assignment
- Conversation
- Call termination request via ACCH
- Call release

Figure 3.61 is a flowchart for a mobile land interconnection call sequence which should help bring the components together. It is interesting to note the differences between the interconnection call diagram and that for the dispatch sequence.

The interaction of sharing resources for the radio access for both dispatch and interconnection involves establishment of dispatch location areas (DLAs) and

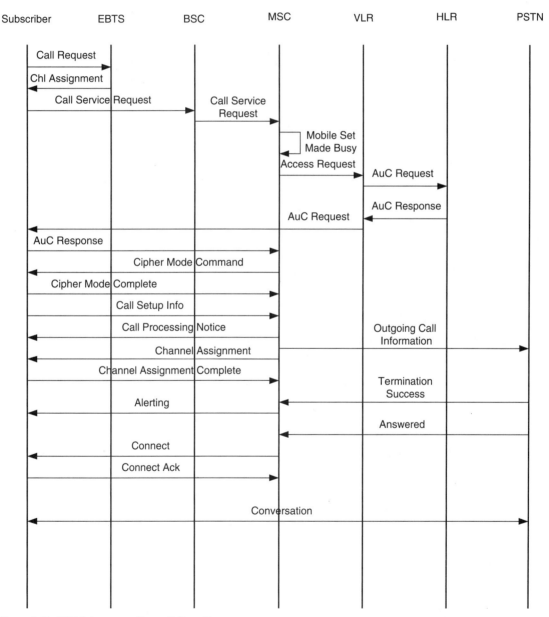

Figure 3.61 M-L interconnection call flow diagram.

interconnection location areas (ILAs). The DLAs and ILAs are usually designed independently but have interactions which require that joint considerations be made for the selection of both the DLA and ILA boundaries. The DLA and ILA boundaries are design considerations in addition to BSC boundaries; however, the ILA and DLA boundaries need to be inclusive of the EBTSs, which are connected to a BSC.

An example of a DLA boundary is shown in Fig. 3.62; it shows a total of four location areas associated with dispatch. Each location area is then folded into a service areas. Care must be taken by the design engineer not only in the selection of location areas but in determining what constitutes the service area. The location area is where the dispatch call is broadcast, while the service area defines which location areas are possible for inclusion in the dispatch call.

Figure 3.63 is the corollary to the DLA boundaries and shows ILAs for the same sample system. The ILA is used for call delivery to and paging of the subscriber unit. The ILA boundaries should not be set up such that the subscriber units regularly transition from one ILA to another; this increases the amount of overhead signaling required to keep track of the mobile.

Looking at Figs. 3.62 and 3.63, the differences between the ILA and DLA boundaries become evident. Figure 3.64 shows the composite view of both the ILA and DLA boundaries.

3.8 CDPD

Cellular digital packet data (CDPD) is a packetized data service with its own air interface standard. The CDPD system utilized by the cellular operators is

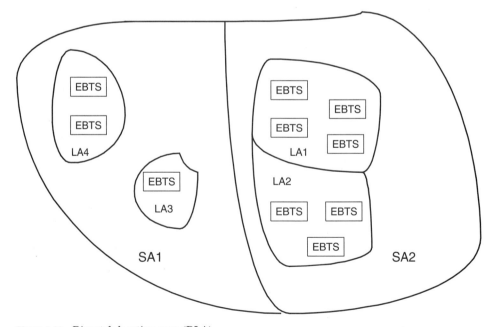

Figure 3.62 Dispatch location area (DLA).

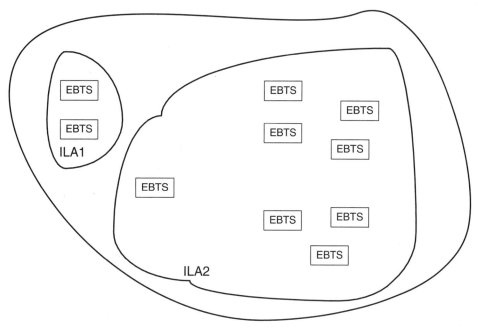

Figure 3.63 Interconnection location area (ILA).

functionally a separate data communication service which physically shares the cell site and cellular spectrum.

CDPD services have many applications but are most applicable for short, bursty type data applications and not large file transfers. Some CDPD applications of the short messages are e-mail, telemetry applications, credit card validation, and global positioning.

CDPD does not establish a direct connection between the host and server locations. Instead it relies on the open standards interface model for packet-switching data communications which routes the packet data throughout the network. The CDPD network has various layers which comprise the system, layer 1 is the physical layer, layer 2 is the data link itself, and layer three is the network portion of the architecture. CDPD utilizes an open architecture and has incorporated authentication and encryption technology into its air-link standard.

The CDPD system consists of several major components, and a block diagram of a CDPD system is shown in Fig. 3.65. The mobile end system (MES) is a portable wireless computing device that moves around the CDPD network communicating to the mobile data base station. The MES is typically a laptop computer or other personal data device which has a cellular modem.

The mobile data base station (MDBS) resides in the cell site itself and can utilize some of the same infrastructure that the cellular system does for transmitting and receiving packet data. The MDBS acts as the interface between the MES and the mobile data intermediate system. One MDBS can control several physical radio channels depending on the site's configuration and loading requirements. The MDBS communicates to the mobile data intermediate

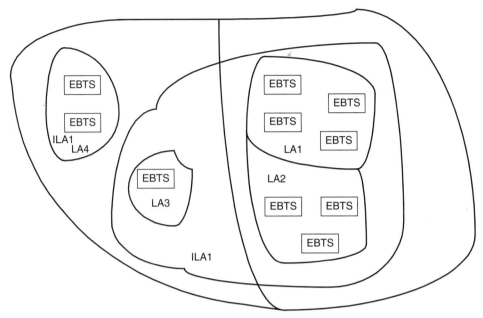

Figure 3.64 ILA and DLA composite view.

system via a 56-kbit/s data link. Often the data link between the MDBS and mobile data intermediate system utilizes the same facilities as that for the cellular system, but occupies a dedicated time slot.

The mobile data intermediate system (MDIS) performs all the routing functions for CDPD. The MDIS performs the routing tasks utilizing the knowledge of where the MES is physically located within the network itself. Several MDISs can be networked together to expand a CDPD network. The MDIS also is connected to a router or gateway which connects the MDIS to a fixed end system (FES). The FES is a communication system that handles layer 4 transport functions and other higher layers.

The CDPD system utilizes a gaussian minimum-shift keying (GMSK) method of modulation and is able to transfer packetized data at a rate of 19.2 kbit/s over the 30-kHz-wide cellular channel. The frequency assignments for CDPD can take on two distinct forms. The first is a method of dedicating specific cellular radio channels to be utilized by the CDPD network for delivering the data service. The other method is to utilize channel hopping where the CDPDs mobile data base station utilizes unused channels for delivering its packets of data. Both methods of frequency assignment have advantages and disadvantages.

When a dedicated channel assignment is utilized for CDPD, the CDPD system does not interfere with the cellular system it is sharing the spectrum with. By enabling the CDPD system to operate on its own set of dedicated channels, there is no real interaction between the packet data network and the cellular voice network. However, the dedicated channel method reduces the overall

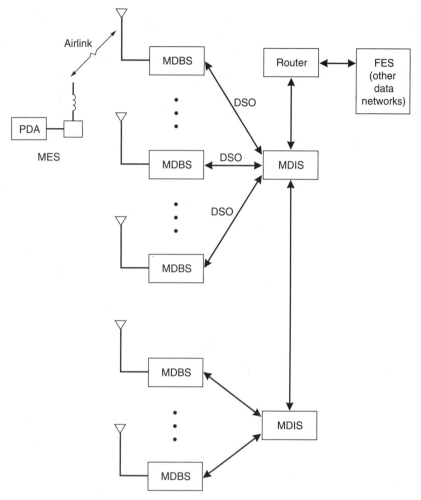

Figure 3.65 CDPD.

capacity of the network, and thus, depending on the system loading conditions, this might not be a viable alternative.

If the method of channel hopping is utilized for CDPD, and this is part of the CDPD specification, the MDBS for that cell or sector will utilize idle channels for the transmissions and reception of data packets. In the event the channel being used for packet data is assigned by the cellular system for a voice communication call, the CDPD mobile data base station detects the channel's assignment and instructs the MES to retune to another channel before it interferes with the cellular channel. The mobile data base station utilizes a scanning receiver, or sniffer, which scans all the channels it is programmed to scan to determine which channels are idle and which are in use.

The disadvantage of the channel hopping method is that when the channels coexist with those of the cellular system, mobile to base station interference can be created. This interference occurs because of the different handoff boundaries for CDPD and cellular for the same physical channel. The difference in handoff boundaries occurs largely due to the fact that CDPD utilizes a BER for handoff determination and the cellular system utilizes RSSI at either the cell site, analog, or MAHO for digital.

Chapter 4

RF Design Guidelines

No true radio frequency (RF) engineering, either design or performance based, can take place without some form of RF design guidelines. The RF design guidelines can be either formal or informal. However, with the level of complexity rising every day in the wireless communication systems the lack of a clear, definitive set of design guidelines is fraught with potential disaster. While this concept seems straightforward, many wireless engineering departments have a difficult time defining exactly what their design guidelines are.

The RF design guideline is a set of rules followed by the RF engineering department not only when designing the network and the new components which are added, e.g., cell sites, but also to improve the performance of the network. The design criteria should be structured so that the system will be configured to offer the best service within monetary and technological constraints.

Therefore, the design criteria for the radio access part of a wireless mobile system are extremely important to establish at the onset of the design, whether it is for a new system, migration to a new platform, or expansion of an existing system. There are many aspects associated with an RF design, and surprisingly they are similar, in concept, to those of any radio access platform that is being utilized by a wireless operator.

This chapter consolidates many of the more important issues with the design criteria associated with the radio access portion of a system.

- RF system design procedures
- Methodology
- Propagation model
- Link budget
- Cell site design
- RF design report

The chapter will conclude with a recommended format for presenting the design criteria in a formalized report that will list the design criteria, assumptions, and other key issues.

The RF design process for a wireless network is an ongoing process of refinements and adjustments based on a multitude of variables, most of which are not under the control of the engineering department. The RF system design process involves both RF and network engineering efforts with implementation, operations, customer care, marketing, and, of course, operations. However, it is important to note that while many issues are outside the control of the technical services group of a wireless company, it is essential to stipulate a design and its associated linkages if there is any desire to obtain an operating system which meets customers' requirements.

Therefore, the RF system design process that should be followed is listed here in summary form. This process can be used for an existing system or a new system since the material needs to be revisited for each of the topics when any system design takes place.

- Determine marketing requirements
- Determine methodology to be used
- Decide on technology to be used
- Define the types of cell sites
- Establish a link budget
- Define coverage requirements
- Define capacity requirements
- Complete RF system design
- Issue search area
- Perform site qualification test (SQT)
- Follow procedure for site acceptance or site rejection as appropriate
- Complete land-use entitlement process
- Integrate into network
- Hand over responsibility to operations

It is important to note that the design process involves not only the establishment of the criteria but also the realization of the design itself.

The information needed for a system design varies from market to market, and, of course, there are nuances between different technology platforms. However, there is commonality between markets and also technology platforms. The following is a brief listing of the most important pieces of information needed for a system design.

1. Time frames for the report to be based on
2. Subscriber growth projections (current and future by quarter)

3. Subscriber voice usage projection (current and forecasted by quarter)
4. Subscriber packet usage projection (current and forecasted by quarter)
5. Subscriber types (mobile, portable, packet capable, blend)
6. New features and services offered
7. Design criteria (technology-specific issues)
8. Baseline system numbers for building on the growth study
9. Cell site construction expectations (ideal and with land-use entitlement issues factored in)
10. FNE ordering intervals
11. New technology deployment and time frames
12. Budget constraints
13. Due date for design
14. Maximum and minimum off-loading for cell sites when a new cell is added to the design

There are many sources and types of information required for an RF design. The basic inputs pertaining to subscriber growth, usage, and the associated features and services offered are normally obtained from the marketing and sales organization.

The output from the RF design process determines the requirements and fundamental structure of the radio access portion of a wireless system. A simplified radio access structure is shown in Fig. 4.1 but can apply to any situation by adding individual components as needed for different technology platforms.

In order to design a new system or establish the migration path for an existing system, the specific access method that the subscriber will have with the wireless system must be determined. The subscriber and base stations both have a transmitter and receiver incorporated into their fundamental architecture, regardless of the technology access method used. Figure 4.2 is an illustration of the various components which need to be factored into the design of a system and can be used for any technology platform.

All wireless transport systems, whether fixed or mobile, have similar design issues. Obviously there are technology-specific issues, which will be identified in this chapter. To avoid duplication due to commonality issues a general process will be described first. However, the commonalities for a design cannot be overlooked since they, as is commonly known by experienced engineers, play a major role in the design and function of any wireless system.

4.1 RF Design Process

The RF design process plays a critical role in the success of any wireless company. The design process is an interactive and iterative process, which can and will change based on the local marketing and regulatory conditions. There are

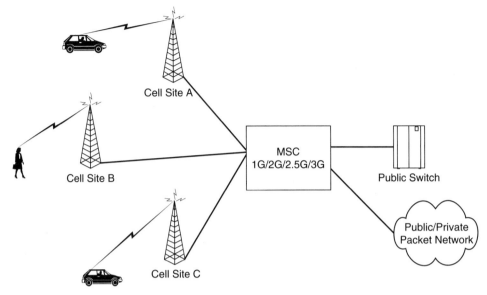

Figure 4.1 Generic radio access system.

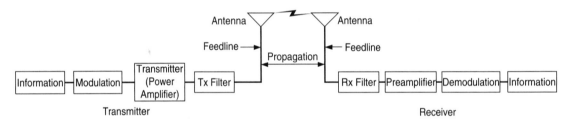

Figure 4.2 Generic radio system.

some fundamental issues that are relevant regardless of the situation and location involved. The purpose of the design review is to ensure that there is an integrated design that spans across all the functional departments.

The overall RF design process is made up of three fundamental areas or groupings. Within each design phase there are numerous subtasks or areas which can either be elaborated upon or eliminated depending on if the design is related to a new system deployment, new service delivery within an existing system, or simply an expansion of an existing network.

A key concept is that the RF design process should be the fundamental structure used in the performance improvement process. Using a design process, whether it is for hardware, software, or any potential alteration to a wireless system, will only result in improvements, and, more importantly, the avoidance of major errors.

The three major components of the RF design process are the

1. High-level design review (HDR)
2. Preliminary design review (PDR)
3. Critical design review (CDR)

The first step in the design process is the high-level design which is completed with the high-level design review (HDR). The HDR involves multiple departments within the company. Specifically the high-level design review is meant to ensure that the company's marketing, financial, sales, customer care, and technical groups have an integrated approach to deploying a system or implementing a new service.

The RF engineering objective of the HDR is to create a formal review of the system solution for the radio environment aspect. Issues that will be addressed involve the interface, capacity, quality, and availability as compared to the company's marketing, financial, and sales objectives, taking into account legal and regulatory requirements.

Some of the suggested RF design issues that need to be included in the HDR are

- Topology and architecture selected
- Available capacity and quality levels supported
- Technology choices
- Integration into existing network architecture
- Vendor, equipment, and system selections
- Network management architecture (if applicable)
- Growth concepts
- System reliability concepts (i.e., redundancy and disaster recovery)
- Future services and platforms concept

The preliminary design and preliminary design review (PDR) process falls primarily under the technical service responsibility. This is a review of the various platforms involved with supporting the concepts and decisions made in the HDR. The purpose of the PDR is to identify and open issues or areas where critical decisions need to be made and discuss the alternatives. The PDR is meant to define the direction of the technical design.

The components of an RF design process for an existing wireless system are as follows:

1. Obtain the marketing plan.
2. Identify coverage problem areas.
3. Establish technology platform decisions [e.g., cellular data packet data (CDPD)].

4. Determine the maximum radius per cell (link budget).
5. Establish environmental corrections.
6. Determine the desired signal level.
7. Establish the maximum number of cells to cover area(s).
8. Generate a coverage propagation plot for the system and areas showing before and after coverage.
9. Determine the subscriber usage.
10. Allocate a percent of system usage to each cell.
11. Determine the maximum number of cells for the capacity (technology dependent).
12. Establish which cells need capacity relief.
13. Determine how many new cells are needed for capacity relief.
14. Establish the total number of cells required for coverage and capacity.
15. Generate a coverage plot incorporating coverage and capacity cell sites (if different).
16. Reevaluate the results and make assumption corrections.
17. Determine the revised (if applicable) number of cells required for coverage and capacity.
18. Check the number of sites against the budget objective; if it exceeds the number of sites, reevaluate the design.
19. Using the known database of potential site candidates overlay them on system design and check for matches or close matches to desired design locations.
20. Adjust the system design using site-specific parameters from known database matches.
21. Generate propagation and usage plots for the system design.
22. Evaluate the design objective with time frame and budgetary constraints and readjust if necessary.
23. Issue search rings.

The conclusion of the PDR allows the RF design team to continue with the design effort. If during the PDR the design or methodology required alteration due to design assumption flaws or changes to the business plan, the PDR, or even the HDR, may need to be redone.

Assuming that the PDR was successful, the next step in the design process is the critical design phase, where the decisions made in the PDR are used to complete and refine the final design. The final design then undergoes a final design review called the critical design review (CDR). The CDR is the part of the design phase where no more changes are made to the design, unless there is a flaw in the design process which requires the revisiting of critical design material resulting in a rescheduling of the CDR. The CDR should have the same format as the PDR in terms of desired content with the exception that design options are no longer relevant to the discussion since those decisions were made in the PDR.

With the completion of the CDR the system design is finished. The design review process and subsequent revisions to the design itself will then be included in the quarterly or 6-month design review process that should be required by any wireless service company. The overall design process flow is depicted in Fig. 4.3.

4.2 Cell Site Design

While cell site design is not necessarily the first step in a design process after the completion of the CDR, it is one of the most important for the RF engineering department. The reason is that it is where the bulk of the capital is spent. The cell site design process takes on many facets, and each company's internal processes are different. However, no matter what internal process the company has, the following items are needed as a minimum. The cell site design guidelines listed here can be utilized directly or modified to meet your own particular requirements.

1. Search area
2. Site qualification test (SQT)
3. Site acceptance form (SAF)
4. Site rejection (SR)
5. FCC guidelines
6. FAA guidelines
7. EMF compliance
8. Site activation
9. Post turn-on (PTO)

The use of this defined set of criteria will help facilitate the cell site building program by improving interdepartmental coordination and providing the proper documentation for any new engineer to review and understand the entire process with ease. Often when a new engineer comes onto a project, all the previous work done by the last engineer is reinvented primarily due to a lack of documentation and/or design guidelines from which to operate from.

4.3 Search Area Request (SAR)

A critical first step in the cell site design process is the definition of a search area. The search area request includes information about the search area and is a key source document used by the real estate acquisition department of the

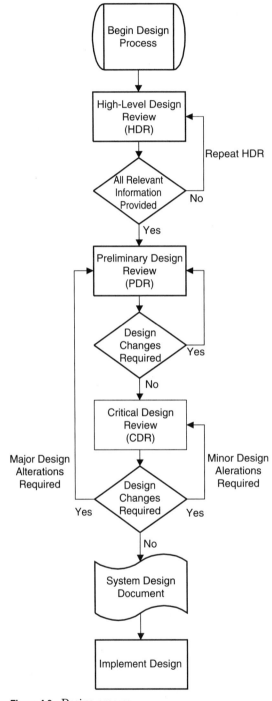

Figure 4.3 Design process.

company. The selection and form of the material presented should not be taken lightly since more times than not the RF engineers rely heavily upon the real estate group to find a suitable location for the communication facility. If the search area definition is not accurate and thorough, it should not be a surprise when the selection of candidate properties is poor.

The search area request issued needs to follow the RF system design objectives. The search area request should be put together by the RF engineer responsible for the site's design. The final paper needs to be reviewed and signed by the appropriate reviewing process, usually the department manager, to ensure that there is a check and balance in the process. The specifications for the search area document need to not only meet the RF engineering department's requirements but also the real estate and construction group's needs. In summary, the proposed form is issued by the RF engineering department, but it is imperative that the search area request issued undergoes a design review by the various groups prior to its issuance.

Figure 4.4 is an example of a proposed search area request (SAR) form. Naturally each company has its unique set of requirements driven by management and accounting practices. However, the SAR form presented here and all the forms in the site design process should be compared to the forms used internally.

```
Search area code: _____            Capital funding code: _____
Issuance date:
Target on-air date:


                         Search Area Map


# sectors:
AGL:         ASML:
Size of equipment room (ft² or m²):
# antenna:
Maximum cable length (ft or m):
Comments:




                                  Search Area Request
                                  Document #: _____ Date: _____
                                  Design engineer: _____
                                  Reviewed by: _____
                                  Revision: _____
```

Figure 4.4 Search area request (SAR) form.

The map included with the SAR in Fig. 4.4 needs to include as much information as practical for the real estate acquisition group to help locate the proper sites. This map will minimize the amount of inappropriate sites that are presented to the RF engineering department for consideration in the system design. Therefore, the SAR map needs to include area-specific information. This area-specific information varies from location to location. The variations which you can use for the map format are largely dependent upon your design criteria for the site. If the search area ring is very defined, as is the case with mature systems, then it is imperative that the adjacent existing sites and search areas are identified on the map itself. The rationale behind including adjacent sites and the current search areas is that this will better define options available to the real estate acquisition group.

A propagation plot and/or desired coverage area also needs to be generated and included with the search area folder maintained by the RF engineering department. Having a defined coverage area helps define the search area and coverage rings put forth in the map provided. The propagation plots or coverage objectives will be used as part of the site acceptance procedure, which will be discussed later. The coverage area definition is one of the steps taken to ensure the proposed site meets its desired objective.

On the search area request form the search area code should be identified along with its capital funding number. The capital funding number and the search area code can and should be the same.

The on-air, or system-ready, target date identifies when the site needs to be placed into commercial service and helps prioritize the internal resources of the company by helping define the importance of the site. The on-air date should match the date put forth in the RF system design plan.

It is also important to define the total number of antennas and their type in advance of the site search. The approach the real estate acquisition group takes with the potential landlord will be different if a site requires 3 antennas, than if it requires 16 antennas.

The rationale behind listing the above ground level (AGL) value, and above mean sea level (AMSL) value is that this will help define the site's location options for the real estate acquisition group. The AGL will help structure the search for suitable properties that fit the design parameters. The rationale here is that sometimes sites are available that are 10 m tall and have a willing landlord and permit accessibility. However, the design specification calls for a 40 m antenna height, disqualifying the site for consideration and resource expenditure. If a site needs to be 30 m tall, about 100 ft, it is important to define its AMSL. For example, if a 30-m height is desired but the location of the property is in a gully 20 m down, a 50-m-tall antenna will need to be installed.

The comment section of the SAR form is meant to provide an area for the designer to specify any particulars desired for the site.

On the SAR form itself is a section for documentation control. This part tracks who issued the search area request, what revision it is, and the dates associated with it. Also on the document control portion of the form is a section for the design reviewer to sign off on the request.

It cannot be overstressed that the information provided on the SAR form will largely determine the success or failure of the property search. All search area requests need to undergo a design review.

4.4 Site Qualification Test (SQT)

The site qualification test (SQT) is an integral part of any RF system design. Even in the age of massive computer modeling it is still essential that every site have some form of transmitter or site qualification test conducted at it. The fundamental reason behind requiring a test is to assure that the site is a viable candidate before a large amount of company capital is spent on building the site. This test is also required to make sure the site will operate well within the network.

Based on the volume of sites required, within a specified time frame, it may not be possible to physically test every cell site candidate. Therefore, it is essential that a defined goal of how many sites should be physically tested, e.g., 75 percent, be established. The establishment of a goal for physically testing or using a propagation model evaluation will help determine the risk factors associated with building out of the network.

Regardless of whether a site is to be physically tested or evaluated through use of a computer simulation there are several stages to this process. It is strongly recommended that the RF engineer responsible for the final site design visit the location prior to the SQT taking place. The engineer will then have a better idea of the potential usefulness of the site and be able to provide more accurate instructions to the testing team.

It is also strongly recommended that the RF engineer not design the test on the fly by instructing the testing team on the day of testing where to place the transmitter and which routes to drive. Instead the RF engineer should put together his or her test plan, identifying the location of the transmitter antenna, the effective radiated power (ERP), drive routes, and any particular desired variations. The test should be submitted to the manager of the department for approval and then be passed on to the SQT team. The SQT is a very critical step in the process and needs to be well thought out in advance. Improperly defining the test routes or transmitter locations can lead to a poorly performing site being accepted or to a potentially well performing site not being accepted. The driving issue here as always is that a well-planned test will save time and money many times over.

The proposed form to use for the SQT process is shown in Fig. 4.5. The format of the SQT form needs to directly match the input requirements of the RF engineers, the SQT measurement team, and the group responsible for post-processing the data.

It is imperative that the testing plan be reviewed and signed off on as part of the design review process. The actual test conducted, antenna placement,

```
                                                    Date: __/__/__
                     Site Qualification Test

    Search area code: _____    Capital funding no.: _____

    Address of test site: _____
                         _____
                         _____
                         _____

    Site contact    Name: _____
                    Phone no.: (____) _____

    Requested test date: __/__/__

            7.5-minute map: _____
            Test antenna: _____
            Test antenna height: _____
            Test ERP: _____ watts

    Test frequency/channel: _____

    Test mounting information: roof, tower, crane, water tank, misc.

            Rigger required: _____ (Y/N)
            Test location sketch attached: _____
            Test routes plan attached: _____
            SQT team leader: _____
            Postprocessing information:
                Scale: _____
                Color code: _____
                Data reduction method: _____

    SQT test calibration date: _____
    SQT transmitter calibration date: _____
```

Site Qualification Test		
Document No.:		Date:
Test No.	Design	Reviewed

Figure 4.5 Site qualification test (SQT).

physical routes traversed, and postprocessing criteria directly determine the viability of the location. A poorly designed test plan can cause the failure of a site's acceptance even though the site may meet the needs set forth in the growth plan. Of course, the opposite situation can also occur, where a site meets the design criteria but could negatively impact the network once it goes commercial.

The SQT needs to include a sketch of the test location defining where to place the actual test transmitter antenna. For example, when a crane is used to place the transmitter antenna, it is important to define where the crane should be parked and the test antenna height actually used. If the test location is not properly defined in advance, an error could occur in the test transmitter placement significant enough to incorrectly pass or fail the SQT requirements. Figure 4.6 has several example diagrams of test transmitter antenna locations.

Figure 4.7 is an example drive test route for an SQT. The RF engineer working on the SQT must make certain that the defined drive test route matches the design criteria for the specified site. The individual line items of the SQT form shown in Fig. 4.7 are self-explanatory. The 7.5-minute map portion is meant to help specify the actual grid location of the test which can be used as part of the archiving process for SQTs. The test antenna type, height, and ERP obviously should be defined in advance to ensure the testing is conducted in

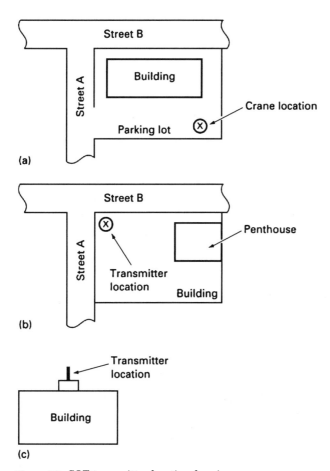

Figure 4.6 SQT transmitter location drawings.

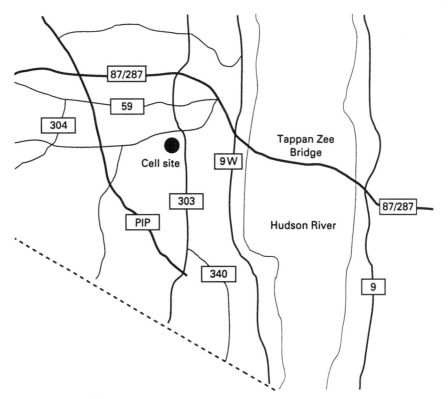

Figure 4.7 SQT drive route.

accordance with the design specifications. The calibration requests on the form are meant to ensure that the equipment used is within calibration. It is important to always check that the calibration for any test equipment is within specification to ensure that the equipment will provide reliable measurements.

4.5 Site Acceptance (SA)

The testing of a site for its potential use in the network determines whether the site is acceptable or not. For this section the assumption will be that the site is acceptable for use by the RF engineering department as a communication facility. It is imperative that the desires of the RF engineering department be properly communicated to all the departments within the company in a timely fashion. The method of communication can be verbal at first, based on time constraints, but written documentation must follow that will ensure that the design objectives are properly communicated.

It is strongly recommended that the RF engineer responsible for the final site design visit the site prior to acceptance. This site visit will facilitate several items; the engineer will have a better idea of the potential usefulness of the site and its ability to be built, and will be able to provide more accurate instructions to the implementation team.

Before the SAF is released it is imperative that it go through the design review process to ensure that nothing is overlooked. The SAF will be used to communicate the RF engineering's intention for the site and will be a key source document used by real estate, construction, network engineering, operations, and the various subgroups within engineering itself. The SAF will also need to be sourced with a document control number to ensure that changes in personnel during the project's life are as transparent as possible.

The site acceptance form (SAF) shown in Fig. 4.8 is meant to be used as a general guide. This sample SAF can be easily expanded upon or modified to ensure that all the relevant information required within your organization is provided. However, whatever form or method you ultimately utilize, it is important to at the minimum include the information listed in the sample form. Most of the information included in the SAF is self-explanatory.

As done for all the other steps in the design process, a design review and signoff must take place establishing a formal paper flow. A copy of the predicted propagation used to generate the search area request must be included with the SAF. In addition, a copy of the expected coverage area versus the design objective needs to be included with the SAF to approve the site for RF engineering.

A copy of the proposed antenna installation configuration must be included in the SAF. The proposed antenna configuration is used by equipment engineering and construction to evaluate the feasibility of the proposed installation. The antenna configuration drawing is also used as part of the lease exhibit information.

Copies of the SAF and of the supporting documents need to be stored in a central filing location, which is secure, preferably on a server. The use of a central filing location will enable all the information pertaining to this location and search area to be stored in one location and not distributed among many people's cubes.

4.6 Site Rejection (SR)

In the unfortunate event that a potential site has been tested and is determined not to be suitable for potential use in the network, a site rejection form needs to be filled out. The issuance of a site rejection form may seem trivial until there is a change of personnel and the site is tested again at a later date. The site rejection form, Fig. 4.9, serves several purposes. The first purpose is that it formally lets the real estate acquisition team know that the site is not acceptable for engineering to use and it needs to pursue an alternative location. The second purpose is that this process identifies why the site did not qualify as a potential communication site. The third purpose ties into future use. The SQT data are stored and can be used for future system designs when the site might be more favorable for the network.

```
Search area code:                              Capital authorization code:
SAF document #:
Site address:




Lat:                                    AGL:
Long:                                   ASML:
PtP backhaul (Y/N):
Regulatory issues:
  PtP license needed (Y/N):
  PtP frequency secured (Y/N):
  FAA approval attached (Y/N):
  FAA marking and lighting required (Y/N):

                        Site-Specific Information
(a) Antenna configuration attached
(b) Radio equipment location defined
(c) Network equipment location defined
(d) Equipment room and location sketch attached
      Radio equipment
      Network equipment (if required)
(e) Antenna structure (roof, tower, monopole, water tank)
(f) Equipment room [prefab, interior fitup (TI), exterior]
(g) Approx cable length (ft, m)

              Type and Quantity of Antennas (include PtP)
Sector      Antenna type      # antennas      Orientation        ERP




Existing transmitters on structure (Y/N):
  If Yes, state freq, EIRP, call sign, and physical location for each antenna and service

                        Qualification Information
Coverage objective obtained (Y/N):
% of area site will cover versus design objective:
IMD study complete:
EMF study complete:
Site particular comments:




                                        Site Acceptance Form
                                        Document #: _____    Date: _____
                                        Design engineer: _____
                                        Reviewed by: _____
```

Figure 4.8 Site acceptance form (SAF).

```
┌─────────────────────────────────────────────────────────────────────┐
│                       Site Rejection Form                           │
│                          RF Engineering                             │
│                                                                     │
│  Search area code:                                                  │
│                                                                     │
│  The (name of test location) was visited on (date of test) and did not meet the design criteria │
│  for the search area defined.                                       │
│                                                                     │
│  The location did not meet the design criteria for the following reasons (state reasons). │
│                                                                     │
│                                                                     │
│  RF Engineer: _____                               │
│  Engineering Manager: _____                           │
└─────────────────────────────────────────────────────────────────────┘
```

Figure 4.9 Site rejection (SR) form.

It is recommended that the site rejection process include a design review with a sign-off by the manager. This is to ensure that the reasons for rejecting the site are truly valid and the issues are properly communicated. The SR form needs to be distributed to the same parties that the SAF would be sent to. The reason is that just because a site does not meet the design criteria specified at this time in the network design does not mean it will always be unsuitable. Therefore, it is imperative that the SQT information collected for this site be stored in the search area's master file.

4.7 FAA Guidelines

The Federal Aviation Administration (FAA) compliance is mandatory for all the sites within a system. The verification of whether the site is within FAA compliance should be covered during the design review process. If a site does not conform to the FAA guidelines, then a redesign might be in order to ensure FAA compliance.

The overall key elements that need to be followed for compliance are

1. Height
2. Glide slope
3. Alarming
4. Marking and lighting

The verification of the height and glide slope calculations is needed for every site. It is recommended that every site have the FAA compliance checked and included in the master site reference document. If there is no documented record for a site regarding FAA compliance, it is strongly recommended that one be completed immediately. An FAA compliance check takes relatively little time and effort. It can be completed within a week for a system with several hundred cells.

4.8 EMF Compliance

Electromagnetic field (EMF) compliance needs to be factored into the design process and continued operation of the communication facility. The use of an EMF budget is strongly recommended to ensure personnel safety and government compliance. A simple source for the EMF compliance issue should be the company's EMF policy.

Most of the current operators tend to take a reactive role in dealing with the EMF topic. Specifically, the unwritten rule tends to be, do not discuss it and it might go away. However, this issue is not going to disappear even with the advent in the United States of the 1996 Telecom Act.

Therefore, during the site acquisition phase there are many aspects that can be accomplished in the beginning steps to minimize the controversy associated with electromagnetic fields and the installation. Currently towers or monopoles tend to draw the most responses; however, more structures in the urban and suburban areas are being located on top of existing building structures.

Since speed is the primary driving force in building out a network, properly conducting the real estate acquisition phase can expedite the build program through selecting locations that do not require any variances prior to pulling the permit. However, no matter how well you perform your site acquisition process, you will need to apply for variances at some of your locations.

Everyone who has had the pleasure of participating in the site acquisition process can tell you there are several key items that constantly arise after the lease negotiation phase: aesthetics, property devaluation, and electromagnetic radiation. The first two items are probably the easiest to address since they are the most tangible. However, the third item, electromagnetic radiation, will always invoke an emotional response from the landlord, local residents, and/or the board members, since it is so frequently associated with cancer.

Prequalifying a potential site is one of the most important ways to help move the process along. Explaining to the landlord what your intentions are prior to conducting a transmitter test is essential for determining if the landlord will eventually give you difficulties. Problems with the landlord have at times kept sites from being constructed even after they have passed all the design acceptance phases and a large amount of capital dollars have been expended. Some of the problems experienced are caused by the landlord, and others are self-inflicted during the initial site selection phase due to miscommunication or deliberately leaving out key items.

One necessity for prequalifying a site on top of a building is to walk the rooftop, where possible, and determine, prior to any transmitter test, where the antennas would potentially be placed. By conducting this pretransmitter survey you can avoid having the antennas placed right above someone's balcony or roof patio or next to or right at someone's window. If the design calls for a monopole or a tower to be constructed, it is important to have some flexibility in where the structure will be placed on the property. Most of the time a properly placed antenna or tower not only meets the design condition for the location but also minimizes or eliminates any negative responses.

One of the more expedient methods for site acquisition involves colocating with another service provider on a tower, monopole, or building. Primarily if a tower already exists, it is a target of opportunity, provided the location meets the engineering design criteria established for that area.

Whether you are a sole occupant or one of many at a communication site there are a few rules that must be followed that pertain to electromagnetic fields. The intermodulation issue is usually the primary focus of the RF engineer in a colocation situation. However, there is another critical issue that tends to play a background role in the site selection process. Cellular system operators utilize part 22 of the Code of Federal Regulations, Part 47 (CFR 47) rules for determining the allowable electromagnetic emission levels from their facilities. PCS providers will be utilizing part 24 of the CFR 47 rules for determining the allowable electromagnetic emission levels. One key common point is the fact that they both reference part 1 of CFR 47 with specific reference to 1.1307. The importance of this is that they both must follow C95.1 as the basic reference for determining the allowable electromagnetic field strength.

The C95.1-1992 specification has two basic sets of criteria that must be followed in determining if a communication facility meets the regulations. The two basic criteria pertain to controlled and uncontrolled environments. *Controlled* pertains to the environment that workers, including RF technicians, are exposed to. *Uncontrolled* pertains to the environment that the general public is exposed to. The primary difference between the two is that there is a difference in the electromagnetic power levels that a person is allowed to be exposed to. As a general rule a person in an uncontrolled environment is exposed to about one-fifth the power level as someone in a controlled environment.

Table 4.1 is a simple reference chart showing controlled and uncontrolled levels for the 800-MHz to 2.0-GHz band. At first glance you will notice that the specification not only differs for controlled and uncontrolled environments, but

TABLE 4.1 Power Density Chart*

	Power density, mW/cm^2		
		C95.1-1992	
Frequency f, MHz	C95.1-1982†	Controlled†	Uncontrolled‡
880	2.933	2.933	0.586
900	3.0	3.0	0.6
1800	6.0	6.0	1.2
1900	6.333	6.333	1.266
2000	6.666	6.666	1.333

*Power density levels are the maximum permissible exposure levels.
†Derived from the formula $f/300$.
‡Derived from the formula $f/1500$.
SOURCE: Values taken from MPE tables for controlled and uncontrolled environments (C95.1).

it also varies with frequency. Currently cellular base stations transmit at between 869 to 894 MHz. The new PCS systems will now be operating between 1850 to 1990 MHz. There are several caveats with this chart when you reference the C95.1 specification, in particular the time averaging that this measurement is to take place over.

One other item that is rarely covered or discussed is the composite power issue associated with the communication site. Generally the particular company that is petitioning to get on the facility will point out that its emission levels are well below the specification levels and that there is no real issue here. The change involves when you are colocated with another operator, which is becoming more common. The total power from the transmitters must be factored into the equation at that facility. This, however, is often not done due to the fact that most of the colocation situations are run by a site manager. The degree of documentation from one facility to another varies greatly, and calculating the composite power for a facility becomes very difficult. The alternative is to take a spectrum analyzer and/or a wideband power density probe to measure the power density in the area around the facility. While this might meet with some success on the superficial level, it still makes the overall determination of composite power difficult since full loading of cellular, paging, and two-way systems is not normal. However, conducting a field measurement is a viable method for determining what the overall power density levels are at the location. The biggest disadvantage to using the physical testing methodology is the time and effort spent collecting data when an analytical method could suffice for over 90 percent of the locations.

During the site acceptance phase of the project, it is strongly advisable to keep the antennas away from common access areas in residential buildings. Although this should be common sense, most of us can recall situations where this simple first step did not take place resulting in much damage control. When you are conducting the initial site survey, a simple chart with distances and power levels comparing them to the chart in Fig. 4.10 should be used as an initial step in the process. Once the actual antenna locations are decided upon during a structural review, the distance calculations could be checked again to ensure that everything is all right at this stage of the process. We would suggest as an up-front process to utilize the free-space path loss equation as the first round since it is more brutal and can be trimmed down later with an actual field measurement. The equation utilizes a 20 log scale which seriously overemphasizes the propagation characteristics. The equation also does not take into account the air interface loss, antenna pattern correction, dynamic power control, modulation techniques, or the fact that the site is not fully loaded all the time.

A quick glance at the chart in Fig. 4.10 tells you that for a tower or monopole situation where the height is at least 100 ft above the ground there is no issue with the emission levels. However in a rooftop situation with apartments or offices across the street or right below the antenna structure, the situation could be different especially in a colocation environment.

Cell:
Date:
Sector 1

# Channels	19
Setup Channel	1
ERP/Channel	100W
Total Power	2000W

Distance, ft	Total Power, W	Power Density, mW/cm^2	Max for Band, mW/cm	% Budget
10	2000	1.713998474	0.586666667	292
25	2000	0.274239756	0.586666667	47
100	2000	0.017139985	0.586666667	3

Figure 4.10 EMF power budget.

There are a variety of things you can do to keep in compliance with the specification if the calculations show you could possibly exceed the levels. One obvious method is to reduce the power, but this has a negative impact on the coverage area you are trying to satisfy with this site. Another suggestion is to relocate the antennas so you get more vertical isolation; however, this might lead to aesthetic problems, installation complications, and cost increases for the facility. A third suggestion is to lower the maximum amount of channels available at this facility, but this would eventually make this a limiting site in your system design. A fourth suggestion is to utilize a wideband probe and a spectrum analyzer and conduct a series of measurements to document the actual levels that would be there and make adjustments from there.

Therefore, the establishment of an EMF power budget should be incorporated into the master source documents for the site and stored on the site itself identifying the transmitters used, power, who calculated the numbers, and when it was last done. As a regular part of the preventative maintenance process the site should be checked for compliance and changes to the fundamental budget calculation.

The EMF power budget should be signed off by the manager for the department and shared with the operations department of the company. An EMF budget needs to be completed for every cell site in operation and also for those proposed. Additionally since EMF compliance is a safety issue, some states are now requiring that a professional engineer licensed in that state or a health physicist certify the compliance.

4.9 Planning and Zoning Board

Depending on the land use acquisition process needed, preparing for a presentation to a zoning or planning board could and should be part of the design review process. Not only is the presentation important, it might be possible with a modification to the original site design to eliminate this process entirely. There have been many actual cases where a modification to the site design would have eliminated the need to request a variance from the town and thus would have prevented massive delays in the site build program. While this step seems obvious, it is rare that local ordinances are checked and incorporated into the design process.

It is recommended that the local ordinances for the site be included with the site source documents. However, when it comes time to present the case of why the site is needed to the local planning and zoning boards, a well-rehearsed presentation is needed. It is recommended that the program be rehearsed prior to the meeting night to ensure that everyone knows what to say and when.

Engineering's role in the process tends to focus on why the site is needed and health and safety issues associated with electromagnetic fields. The items that should be presented or prepared include, as a minimum,

1. Description of why the site is needed
2. Description of how the site will improve the network

3. Drawing of what the site will look like
4. Views from local residences
5. EMF compliance chart
6. EMF information sheets and handouts for the audience

Before the meeting it is essential that the local concerns be identified so they can be specifically addressed before or at the meeting. It is also recommended that the presentation be focused at both the public and the board members.

As in every case it is imperative that all the issues needed to launch a successful appeal for a negative ruling by the council be covered. The overall preparation for the meeting is essential since the comments made by the company employees or consultants is a matter of public record and will be used solely for the appeal process.

4.10 Design Guidelines

The actual design guidelines that should be utilized by the RF engineers need to be well documented and distributed. The design guidelines, however, do not need to consist of voluminous amounts of data. They should consist of a few pages of information that can be used as a quick reference sheet by engineering. The design guideline sheet has to be based on the system design goals and objectives set forth in the CDR.

Additionally the advent of multiple technology platforms being used by wireless operators either in the same market or in affiliate markets has led to the use of multiple design guidelines.

The actual content of the design guideline can vary from operator to operator. It is essential that a list of design guidelines be put together and distributed. This will ensure there is a minimum level of RF design specifications in the network.

A generic RF design guideline is shown in Fig. 4.11. It is structured for an AMPS analog system, but the concept and general format can easily be adapted to include any one of the many wireless access technologies that exist today or will in the near future. We have used an AMPS system to illustrate what an RF design guideline should look like. Obviously if the service includes code-division multiple access (CDMA), then Eb/No needs to be factored into the mix. Depending on the system configuration, more specifically the amount of wireless technologies being deployed, it might be best to have some general guidelines with technology-specific ones that follow.

4.10.1 Performance criteria

A subset of the design criteria is the performance requirements for a wireless system. The performance criteria are often overlooked in the design process, since they are expense-driven and not capital-driven. The performance criteria are listed separately since they are very technology specific. Again if there

```
                    AMPS RF Design Guideline
System name:
Date:
          RSSI        ERP          Cell area       Antenna type

Urban     −80 dBm     16 W         3.14 km²        12 dBd, 90H/14E
Suburban  −85 dBm     40 W         19.5 km²        12 dBd, 90H/14E
Rural     −90 dBm     100 W        78.5 km²        10 dBd, 110H/18E

Voice channel:                      17 dB (90th percentile)
Frequency reuse:                    N = 7
Maximum # transmitters/sector:      19

Antenna System
Sector cell orientation:            0,120,240
Antenna height:                     100 ft or 30 m
Antenna pass band:                  825–894 MHz
Antenna feedline loss:              2 dB
Antenna system return loss:         20–25 dB
Diversity spacing:                  d = h/11 (d = receive antenna spacing, h = antenna AGL)
Receive antennas per sector:        2
Transmit antennas per sector:       1
Roof height offset:                 h = x/5  (h = height of antenna from roof,
                                              x = distance from roof edge)
```

Figure 4.11 RF design guideline example.

are multiple technologies deployed in a network, or subsystem, then a separate set of criteria need to be defined for each technology base.

We have included some performance criteria recommendations in Tables 4.2 through 4.6, which may or may not meet your particular market requirements. However, the performance criteria are an integral part of the design process and need to be addressed by the design team for initial designing, expansion, and upgrades. The suggested performance criteria should be used to help define the nominal operating ranges for the system. Once the nominal operating ranges for the system are defined, then the quest to improve upon them can take place. Performance improvement recommendations or rather some troubleshooting guidelines, will need the performance criteria defined in the design process for establishing the benchmark.

We did not put into the performance criteria any issues associated with customer care (service) for responding to trouble tickets (specific customer complaints requiring technical responses). However, it is recommended that the trouble ticket response process be incorporated into the performance criteria with the requisite escalation procedures.

4.10.2 AMPS

The advanced mobile phone system (AMPS) performance criteria are probably the most straightforward. Since many operators that occupy the cellular band have AMPS, the performance criteria associated with the analog part of the system need to be well defined. With the push to move the operation to a digital platform, either 2G or 2.5G, AMPS cannot be forgotten about. Table 4.2 is a list of proposed AMPS performance criteria.

4.10.3 IS-136

The IS-136 performance criteria shown in Table 4.3 are very similar to those of AMPS with the exception of the bit error rate (BER) requirements. Please note that for the cellular operators the existence of the AMPS is a likely factor to contend with. Therefore, when establishing or refining the performance criteria for this specific technology, it is strongly suggested that the common issues between AMPS and IS-136 be the same. For example, access failures should be set the same provided the ERP for the digital AMPS (D-AMPS) and AMPS are the same. Alternatively you may want to relax the AMPS radio blocking criteria, say to 5 percent, with IS-136.

4.10.4 IS-95/CDMA2000 (1XRTT)

The IS-95/CDMA2000 performance criteria are depicted in Table 4.4. Please note that for the cellular operators the existence of AMPS is a likely factor to contend with and the level of radio blocking will likely be an issue besides the mutual interference issues. When establishing or refining the performance criteria for this specific technology, it is strongly suggested that the common issues between AMPS and CDMA be the same, but this is not necessarily a prerequisite.

4.10.5 iDEN

For an integrated dispatch enhanced network (iDEN) system the performance criteria are shown in Table 4.5.

TABLE 4.2 AMPS Performance Criteria

Metric	Range
Lost call rate	<1.5%
Attempt failure rate	<1.0%
Handoff failures	<0.5%
Radio blocking (erlang B)	1%> and <2%
Radio utilization rate	60–80%
Usage minutes/lost call	50
% area > −85 dBm (RSSI design threshold is market dependent)	95%
% area > 18 dB C/I	90%

Note: RSSI = reverse signal strength indicator.

TABLE 4.3 IS-136 Performance Criteria

Metric	Range
Lost call rate	<1.5%
Attempt failure rate	<1.0%
Handoff failures inbound	<0.25%
Handoff failures outbound	<0.25%
RF radio blocking (erlang B)	1%> and <2%
TDMA radio utilization rate	60–80%
Usage minutes/lost call	50
BER	<1%
% area <2% BER	95%
% area > −85 dBm (RSSI design threshold is market dependent)	95%
% area > 18 dB C/I	90%

Note: TDMA = time-division multiple access, BER = bit error rate, RSSI = reverse signal strength indicator.

TABLE 4.4 CDMA Performance Criteria

Metric	Range
Lost call rate	<1.5%
Access failure rate	<1.0%
Handoff failures inbound	<0.25%
Handoff failures outbound	<0.25%
RF radio blocking (erlang B)	1%> and <2%
CDMA radio resource utilization rate (voice + data)	60–80%
Usage minutes/lost call	50
FER	<1%
% area <2% FER	95%
% area > −85 dBm (RSSI design threshold is market dependent)	95%
% area > 7 dB Eb/Io	90%
Data services	
Attach failure rate	<1%
Session drops	<2%
Packets/session drop	30 Mbit
% sessions downgrades at handoff	<5%
% packet retransmissions	<10%
% area > 56 kbit/s	90%
% area > 100 kbit/s	20%

Note: FER = frame error rate.

TABLE 4.5 iDEN Performance Criteria

Metric	Range
6:1 blocking %	1%> and <2%
3:1 blocking %	1%> and <2%
Dispatch que %	1%> and <5%
Packet que %	5%> and <10%
BR utilization rate	60–80%
DCCH fail %	<0.4%
Interconnect dropped call %	<1.5%
Total handoff drop %	<0.5%
Bad SQE measurements	<5%
HO threshold fail %	<5%
RF completion ratio	>95%
Usage minutes/lost call	50
PCCH % outbound	<80%
PCCH % inbound	<13%
% area > −85 dBm (RSSI design threshold is market dependent)	95%
% area < 22 SQE	90%

Note: BR = base radio, DCCH = digital control channel, SQE = signal quality estimate, HO = handoff, PCCH = packet paging channel, RSSI = reverse signal strength indicator.

4.10.6 GSM/GPRS

A set of suggested global system for mobile communications (GSM) and general packet radio services (GPRS) performance criteria are proposed in Table 4.6. Therefore, it will be necessary to examine the corporate-defined proper key performance indicators and then expand upon them.

Please note that for some PCS operators the existence of IS-136 is a likely factor to contend with. Therefore, when establishing or refining the performance criteria for this specific technology it is strongly suggested that the common issues between IS-136 and GSM be the same. Using access failures as an example, the antenna systems might be shared and coverage areas might be designed to be the same. Alternatively you may want to relax the IS-136 radio blocking criteria, say to 5 percent, with GSM's introduction.

4.11 Link Budgets

The establishment of a link budget is one of the first things the RF engineer needs to perform when beginning the design process. The establishment of the link budget, of course, can only be done after a decision as to which technology platforms to use. When introducing say a 2.5G platform into a 2G system, it

TABLE 4.6 GSM/GPRS Performance Criteria

Metric	Range
Lost call rate	<1.5%
Access failure rate	<1.0%
Handoff failures inbound	<0.25%
Handoff failures outbound	<0.25%
Radio blocking (erlang B)	1%> and <2%
GSM radio resource utilization rate (voice + data)	60–80%
Usage minutes/lost call	50
BER	<1%
% area < 3 RXQUAL	95%
% area > −85 dBm (RSSI design threshold is market dependent)	95%
% area > 14 dB C/I	90%
Data services	
Attach failure rate	<1%
Session drops	<2%
Packets/session drop	30 Mbit
% sessions downgrades at handoff	<5%
% packet retransmissions	<10%
% area > 56 kbit/s	90%
% area > 100 kbit/s	20%

Note: BER = bit error rate, RXQUAL = receive quality, RSSI = reverse signal strength indicator.

will be necessary to have a link budget established for each of the individual technology platforms involved. In addition with the introduction of packet data the higher data rates have a direct influence on the range of the site and on its capacity.

The link budget is a power budget that is one of the fundamental elements of radio system design. It is the part of the RF system design where all the issues associated with propagation are included. Put simply, the link budget can either be forward- or reverse-oriented and must account for all the gains and losses that the radio wave will experience as it goes from the transmitter to the receiver.

The link budget is the primary thing an RF engineer must first determine in order to ascertain if a valid communication link can and does exist between the sender and the recipient of the information content. It incorporates many elements of the communication path. Unless the actual path loss is measured empirically, the RF engineer has to predict just how well the RF path itself will perform. The many elements involved in the communication path incorporate assumptions made regarding various path impairments.

Figure 4.12 shows what part of the radio communication path the link budget tries to account for. There are two portions to the link budget, the uplink and downlink. The uplink path is the path from the subscriber unit to the base station. The downlink path is the path from the base station to the subscriber unit. Both the uplink path and the downlink path are reciprocal, provided they are close enough in frequency. However, the actual paths should be the same with the exception of a few key elements that are hardware related. The actual path loss associated with the path the radiowave traverses from antenna to antenna is the same whether it is uplink or downlink directed.

The maximum path loss, or limiting path, for any communication system determines the effective range of the system. The example shown in Table 4.7 involves a simplistic calculation of a link budget associated with a 1G system, AMPS, and is used for the determination of which path is the limiting case to design from. In this example the receiver sensitivity value has the thermal noise, bandwidth, and noise figures factored into the final value presented.

The uplink path, defined as mobile to base, is the limiting path case. From the example just given, the talk-back path is 6 dB less than the talk-out path. The limiting path loss is then used to determine the range for the site using the propagation model for the network.

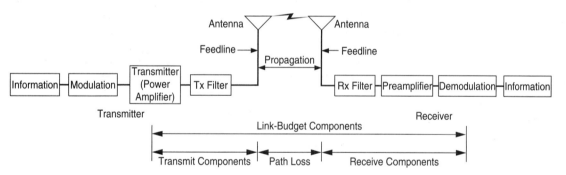

Figure 4.12 Radio path components.

TABLE 4.7 1G Link Budget

	Downlink	Uplink
Transmit (ERP)	50 dBm	36 dBm
Receiver antenna gain	3 dBd	12 dBd
Cable loss	2 dB	3 dB
Receiver sensitivity	−116 dBm	−116 dBm
C/N	17 dB	17 dB
Maximum path loss	150 dB	144 dB

Note: C/N = carrier-to-noise ratio.

With the introduction of a 2G and or 2.5G platform into the wireless system the issue of the components with a link budget become more complicated. The complications arise due to differing the modulation technique and bandwidth, as well as the process gain and finally the Eb/No or carrier-to-noise ratio (C/N) values required for a proper BER or FER. The differences lie in the usable signal level for a given type of service; obviously data throughput may have a more stringent link budget than, say, voice communication, to mention one possibility.

It will be common to have more then one link budget based on the morphology and, of course, the technology platform used. However, the morphology variation in the actual link budget is included in the propagation analysis for the particular site and is a varying value depending on the local particulars. The link budget itself is an establishment of the maximum path loss, in either the uplink or downlink, that the signal is allowed to attenuate while still meeting the system design requirements for a quality signal.

It is recommended that the items in Tables 4.8 and 4.9 be included in the calculation of the actual link budget. Not all of the items need to be utilized in the physical system being installed, but all of the items can impact the link budget. It is also highly possible that other devices can be added in the path to either enhance or potentially degrade the performance of the network.

Our example helps define the forward and reverse radio path components which comprise the forward and reverse link budgets. One final note on the path is that certain wireless access technologies utilize different modulation formats on both the uplink and downlink paths. If this is the case, then some of the reciprocity may not be applicable. Since it is such an integral part of the RF design process, the link budget used for the system design needs to be documented and made available for the design community to utilize.

4.12 Frequency Planning

Frequency planning (also called frequency management) is an integral part of the system design. It is a critical function for all wireless communication systems. How much frequency planning there is in a network is largely determined by the technology platform that is chosen by the operator. There are many variants to frequency planning ranging from coordination of a single transmit channel to orchestrating the manipulation of hundreds of radio channels. Within the cellular and PCS arenas the amount of frequency planning can range from segmentation of the available spectrum to defining the different pseudorandom number (PN) short codes for CDMA.

The advent of more sophisticated computer modeling tools has fostered in the era of automatic, or computer-driven, frequency plans. The use of computer model frequency planning has the promise of facilitating the design effort and shortening the design loop significantly. The use of a software tool to perform the frequency plan requires the obvious issue of a complete and accurate database of site configuration information, assuming, of course, that the software program does not have any bugs or other such glitches. The age-old saying "garbage in equals garbage out" is very relevant here for obvious reasons.

TABLE 4.8 Generic Downlink Link Budget

Downlink path	Units
Base station parameters:	
Transmitter PA output power	dBm
Transmitter combiner loss	dB
Transmitter duplexer loss/filter	dB
Jumper and connector loss	dB
Lightning arrestor loss	dB
Feedline loss	dB
Jumper and connector loss	dB
Tower top amplifier transmitter gain or loss	dB
Antenna gain	dBd or dBi
Total power transmitted (ERP/EIRP)	W or dBm
Environmental margins:	
Transmitter diversity gain	dB
Fading margin	dB
Environmental attenuation (building, car, pedestrian)	dB
Cell overlap	dB
Total environmental margin	dB
Subscriber unit parameters:	
Antenna gain	dBd or dBi
Receiver diversity gain	dB
Processing gain	dB
Antenna cable loss	dB
C/I or Eb/No	dB
Receiver sensitivity	dB
Effective subscriber sensitivity	dBm

Some programs are advocating the ability to recommend or define tilt angles and orientations for antenna systems to augment the frequency plan. We would shy away from physical alterations of the antenna system to support a one-off frequency plan. The site's configurations should be structured to provide the best performance overall and not for one specific frequency plan. On this same topic the engineers responsible for the frequency plan need to be fluent in the ability to frequency plan in order to properly review and possibly alter the recommended plan put forth by the computer modeling program.

Frequency management is critical to the success of a communication system. Often it is the frequency planner who sets the direction for the performance of a communication system. The frequency planning process needs to be rigorously checked on a continuous basis to always refine the system. As a minimum

226   Chapter Four

TABLE 4.9 Generic Uplink Link Budget

Uplink	Units
Subscriber unit parameters:	
Transmitter PA output	dBm
Cable and jumper loss	dB
Antenna gain	dBd or dBi
Subscriber unit total transmitter power (ERP, EIRP)	W or dBm
Environmental margins:	
Transmitter diversity gain	dB
Fading margin	dB
Environmental attenuation (building, car, pedestrian)	dB
Total environmental margin	dB
Base station parameters:	
Receiver antenna gain	dBd or dBi
Tower top amplifier net gain	dB
Jumper and connector loss	dB
Feedline loss	dB
Lightning arrestor loss	dB
Jumper and connector loss	dB
Duplexer/receiver filter loss	dB
Receiver diversity gain	dB
C/I Eb/No	dB
Processing gain	dB
Receiver sensitivity	dBm
Base station effective sensitivity	dBm

for frequency planning, the designs, no matter how minor they seem, need to be passed through a design review process, even when using computer modeling software. Until the software tools have fully matured, providing a frequency plan that will change based on traffic patterns, humans will still be needed in the process to generate the frequency plan.

Continuing frequency planning as a general rule is more of an art form than a defined science, even with automatic frequency planning since it still needs to be checked. Therefore, the use of C/I and whether you use an $N = 7$, $N = 3$, $N = 4$, or $N = 12$ pattern are more clinical in nature. How you go about defining what the frequency plan design tradeoffs are for an area is an art. Some engineers are better artists than others, but the fundamental issue of controlling interference and how well it is done falls on the design review process.

There are numerous technical books and articles that go over how to frequency plan a network, from a theoretical standpoint. It is very important to understand the fundamental principles of frequency planning in order to design a frequency management plan for any network. Failure to adhere to a defined frequency design guideline will limit the system's expansion capability.

The rationale behind defining this RF design process as an art is based on the multitude of perturbations available for any given frequency management plan. There are several methods available for use in defining the frequency management of a network. The method chosen by the mobile carrier needs to be factored into the frequency management scheme capacity requirements, capital outlays, and adjacent market integration issues to mention a few. Obviously the method that is used for the frequency plan also has to ensure that the best possible C/I is obtained for both cochannel and adjacent channel RF interference.

The use of a grid is essential for initial planning, but the notion is academic in nature. The reason is that the site acquisition process tends to drive the system configuration and not the other way around. It is dealing with the irregularities of the site coverage, traffic loading, and configurations that requires continued maintenance of the network frequency plan.

With the introduction of other wireless technologies within the same network frequency planning becomes more complicated. The complication arises due to sharing of the limited spectrum, which requires excellent spectrum management and mutual interference issues.

As mentioned before there are several different methods of assigning frequencies in a network. The most common methods utilized in cellular are $N = 12$, $N = 7$, and $N = 4$. Figures 4.13 to 4.15 are meant to assist in laying out an initial plan using each of the methods mentioned.

The $N = 12$ method (Fig 4.13) is usually deployed in an omni-configuration system which is in areas that do not require the high-traffic-density carrying capacity as provided in an $N = 7$ or $N = 4$ pattern, and is, therefore, the most efficient for trunking (having more erlang carrying capability with the same physical number of radios).

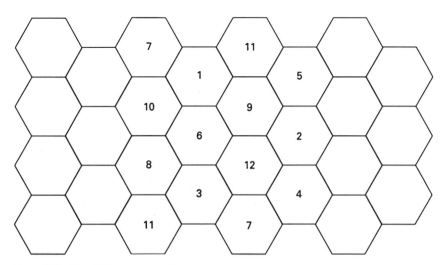

Figure 4.13 $N = 12$ frequency grid.

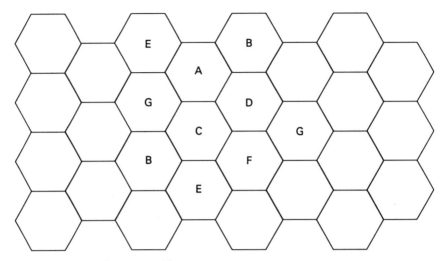

Figure 4.14 $N = 7$ frequency grid.

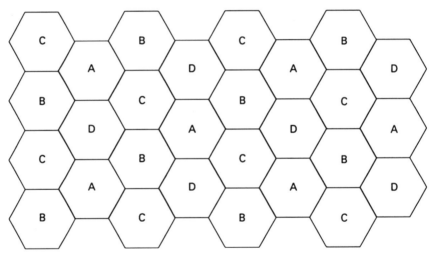

Figure 4.15 $N = 4$ frequency grid.

The $N = 7$ method (Fig. 4.14) of channel assignments is one of the most popular methods for assigning frequencies. The advantages to using the $N = 7$ pattern lie in its ability to provide a high level of trunking efficiency, increased flexibility for the placement of the RF channels in the network, and excellent traffic capacity. It utilizes a 120° sectored cell design that equates to three sectors per cell site. This design facilitates the reuse of the same channels in close proximity to another cell site.

The $N = 4$ method (Fig. 4.15) is another popular method for assigning frequencies. The key advantage with using an $N = 4$ pattern is the high traffic capacity handling ability obtained. It delivers the most channels per square mile or kilometer out of all the channel assignment methods. The disadvantage with this method is that it is not as flexible for cell site placement and does not have the trunking efficiency under average traffic loads that the $N = 7$ pattern does. The $N = 4$ pattern utilizes 60° sectors, six sectors per cell site for AMPS and IS-136 systems, but it is also used with 120° sectors with GSM.

A series of frequency planning checklists which are technology specific are included in Chap. 7. The inclusion of the checklist is both a design and performance issue.

Regardless of the technology platform used, the main idea to keep in mind when frequency managing a system is to maximize the distance between reusers. Whatever method or process is used when maximizing the reuse distance, it will be wrought with problems. When selecting a frequency set for a cell site, whether it is a new site, an expansion issue, or a correction of an interference problem, the current and future configurations of the network need to be factored into the equation.

Many of us can recall the simple question, "Who selected those frequencies and why?" It is recommended that a regular program of retuning the network take place when adding sites and radios to the network. The objective behind implementing regular retunes is to help define a schedule whereby planning for channel and site expansion can take place. The use of regular retunes also enables a review of the compromises implemented into the network and establishment of a plan for their correction in a regulated fashion. For example, you can divide your system up into three or four sections and either proceed in a clockwise or counterclockwise pattern for retuning the network. It is recommended that one region be retuned every 3 or 4 months depending on the size of the network. The retune or revamping process suggested is highlighted in Chap. 7 as part of the continuous process to improve system performance.

Rules for assigning radio channels

1. Do not assign cochannels or adjacent channels at the same cell site (not applicable for CDMA).
2. Do not assign cochannels in adjacent cell sites (not applicable for CDMA).
3. Do not mix and match frequency assignment groups in a cell or sector.
4. Avoid adjacent channel assignments in adjacent cell sites (not applicable for CDMA).
5. Maintain proper channel spacing for any channel assignments for a sector or site.
6. Maximize the distance between reusing cell sites.

Obviously the assignment rules can and should be expanded upon, but the concept is to establish a set of rules, usually available from the infrastructure vendor, and follow them.

When putting together a frequency plan for a site or a system, it is essential that a process exists that will ensure the major items are always reviewed. Since there is a large amount of artistic leeway involved with frequency planning, i.e., more than one design can exist for a given situation, the use of a checklist will expedite the review process and ensure the accuracy of the plan.

Even for a simple, or perceived simple, channel alteration the design needs to be reviewed by the RF design engineer, the performance engineer, and the engineering managers. When making any changes to the frequency plan, whether small or large, the following major items need to be identified.

1. Reason for change
2. Number of radio channels predicted
3. Cell site antenna orientation, standard or nonstandard
4. Proposed ERP levels by sector
5. Coverage prediction plots generated
6. C/I, Eb/No prediction plots generated
7. Cochannel reusers for the next three rings identified (PN code)
8. Adjacent channel cell sites identified
9. PN and color code assignments checked for cochannel and adjacent channel
10. Link budget balance checked
11. Adjacent band conflicts checked

The final requirement of any frequency plan proposal must include a written document issued to the engineering department. It is recommended that the written document take on the form shown at the top of the next page to ensure a paper trail is started and maintained. Every frequency change, channel assignment, and alteration needs to have this document attached.

4.12.1 Frequency plan and alteration test plans

When implementing a frequency plan, it is essential that a test plan be developed. For example, assistance from the operations department is usually required to actually alter the frequencies at a site. In the event that additional radios and potentially spans are needed, there is another logistics piece that must be overcome. The test plan generally needs to have several key elements. Some of these are

1. Design objective
2. Design reviews
3. Coordination meeting and time lines
4. Method of procedure

```
                    Frequency Assignment Addition/Change Form
    Document number:

    Date:

    Subject:



    The following alterations are required to be implemented into the network:

    Cell         Current        Future         Required date




    Design engineer _____
    Engineering manager _____
```

5. Implementation
6. Postanalysis

The test plan items identified are used for a system or regional retune. When making individual channel or localized retunes, a similar process needs to be followed but on a lesser scale.

The design objective for the project can be crafted as a single paragraph which informs the company what the engineering department is trying to accomplish. If the frequency plan involves a major system alteration, then additional design reviews should take place with other departments within the company that are affected by this plan. It is essential in a major system retune that all the personnel in the engineering, operations, and implementation departments be involved with the process since they will be affected. The reviews should utilize the checklist identified above as a basis for discussion during the meeting.

Once a design is agreed upon, it will be necessary to call a general meeting to review the initial time tables focusing on the deliverables of each organization. The objective with this meeting is to identify where the critical paths are and if the schedule proposed is realistic or not. The coordination with adjacent markets might be an essential element for the time table if they have to alter their system or if they have not reviewed the plan.

The next step that needs to be done is the generation of the method of procedure (MOP). The MOP will be the source document from this point forward identifying who does what and when. It is recommended that the MOP be as detailed as possible and include an escalation procedure to be followed for notifying key company personnel of the project's status. The MOP needs to have a back-out procedure in the event that things go significantly wrong.

The implementation of the plan is probably the most exciting and nerve-racking part of the process because this is where the true design review takes place. The actual implementation of the plan is covered in the MOP, and in all likelihood is carried out by the operations department. How you implement the program is dependent upon whether you have manual or auto tuning combiners or linear amplifier combiners (LACs). Several approaches can be taken, ranging from a phased approach of altering one channel set at a time to flash cutting the network and letting the interference fall where it will during the transition process. The particular approach is dependent upon the design and logistics factors involved with the process.

Identification of the drive routes that will be utilized as part of the post-retune process is essential and should be included in the design review and implementation process. The drive routes should be sectioned into three categories. Category 1 is the critical routes where a problem would most likely occur. Routes in this category need to be focused on immediately after the retune takes place. Category 2 includes all the other roads that need to be driven as part of the process. Category 3 roads and routes are those that arise as a result of the previous postanalysis phases 1 and 2 where problems are identified and restoration action takes place.

Postanalysis of the retune is one way that all involved can know if their efforts were successful or not. When conducting a retune, it is essential that a final report comparing preanalysis and postanalysis change data be completed. It is also recommended that for the first week after the retune a daily status report be issued identifying all the problems found and the fixes implemented to date. No retune takes place that does not have some level of problems associated with it, so it is best to plan for these with proper documentation.

The first week's data should be monitored on an hourly basis and then this can be relaxed to a daily basis. The data should be compared with the previous weeks data, day by day and hour by hour, for comparison. The small amount of data, i.e., time frame, will tend to make things confusing, but it will identify the worst sites in the network and expedite the troubleshooting process.

The data reviewed should include the following items:

1. Lost call
2. Blocked or sealed channels
3. Access failures
4. Out of service (OOS) channels
5. Customer complaints
6. Drive test data
7. Usage/lost calls
8. FER, BER, SQE
9. Handoff failures

Finally, since the data extracted will most likely come from various support systems in the network, it is essential that management information systems (MIS) know about the system retune to prepare any data required by the engineering department and to be aware of the possible changes in other company reports.

4.12.2 System radio channel expansion

Radio expansion is an ongoing process of a growing wireless communication system. There is always the need to determine how much blocking in the system is acceptable. Radio expansions should take place on a 6-month basis for all the sites in the network. The rationale behind utilizing a 6-month time frame is that this schedule allows for better utilization of company capital equipment and minimizes the ongoing expense of leased T1 lines for cell sites and port expanse with switches.

The frequency plan should incorporate a yearly projection of additional voice channels for planning purposes and should correspond with a rotational retune program. However, for better channel management any new channels should be implemented in a logical fashion with the operations department 1 month before they are needed by the system.

The guidelines to follow for this effort involve monitoring of the voice channel blocking levels, percent utilization, capital, and lease line expenses. The first step that must be followed is to determine the current system growth rate that is taking place. This data can be secured from marketing or you can do a linear approximation following system growth. One caveat is that the system traffic growth is seasonal and should be incorporated into the modeling data, but conducting a 1-year projection almost negates this approach.

The issue of peak versus average traffic needs to be resolved before the system growth projection can be completed. One method that works involves utilizing an average of the 10 busiest days of the month for the peak month in that year during the system busy hour. This method has been used with much success and will remove peak traffic spikes from overinflating your capital deployment of radios.

The seasonal traffic adjustments can be used to facilitate frequency planning by minimizing the amount of reused channels in the network. The benefit here is reduced overall network interference.

The utilization rate of the voice channels used in a network can range from 80 to 100 percent, depending upon your budgetary and marketing requirements. It is recommended that the 80 percent utilization level be used as the trigger point for determining when additional radios will need to be placed in service in the network.

The blocking level used for a network tends to vary according to an operator's design philosophy. Some standard blocking levels range from a maximum of 2 percent to a minimum of 1 percent. The use of the 2 percent maximum follows traditional cellular engineering and may be adjusted downward for competitive reasons. The type of blocking table used should be checked and noted. Most companies utilize an erlang B table, while some make use of a Poisson table. Be aware that there are a few traffic charts available that promote themselves as erlang B but are in fact some hybrid approach that follows no logical pattern except to add channels.

The following is a brief radio expansion procedure to be used as a guide for what to do when adding radios to the system. The process can be led either by the manager for network or by performance engineering.

On a 6-month basis

1. Determine, based on growth levels, the system needs on a 1-, 3-, 6-, and 12-month period.

2. Factor into the process expected sites available from the build program, accounting for deloading issues.

3. Establish the number of radios needed to be added or removed from each sector in the network.

4. Modify the quarterly plan previously issued.

5. Determine the facilities and equipment bays needed to support channel expansions.

6. Inform the frequency planners, performance engineers, equipment engineers, and operations personnel of their requirements.

7. Issue a tracking report showing the status of the sites requiring action. This report should contain as a minimum the following items:
   - Radios currently at every site
   - Net change in radios
   - When exhaustion, 2 percent blocking, is expected to occur
   - Radios ordered (if needed)
   - Facilities ordered (if required)
   - Radios secured
   - Frequencies issues
   - Cell site translations completed
   - Facilities secured (if needed)
   - Radios installed or removed
   - Activation date planned for radios

8. On a quarterly basis conduct a brief 1-h meeting to discuss the provisioning requirements and arrangements for workers.

9. Perform a biweekly traffic analysis report to validate the quarterly plan and issue the status report of the progress at the same time.

The last step in the system tracking process involves the activation date planned for the radios. This particular piece will enable you to pre-position the radios in the network without actually activating them. Specifically if the radios are needed for the middle of March but are installed in January, you can keep the interference levels to a minimum by having them remain out of service.

The postimplementation of the channels and the task of tracking what stage you are in is essential and included as part of the given checklist. It is recommended that a MOP be generated and followed to ensure that the radios are properly added into the network.

4.13 Antenna Systems

See Sec. 2.7 for a review of basic antenna systems.

The choice of which antenna to use will directly impact the performance of either the cell or the overall network. The radio engineer is primarily concerned in the design phase with the base station antenna since this is the fixed location and there is some degree of control over the performance criteria that the engineer can exert on the location.

The correct antenna for the design can overcome coverage problems or other issues that are trying to be prevented or resolved. The antenna chosen for the application must take into account a multitude of design issues as well as adhere to the system RF design objectives.

4.13.1 Base station antennas

There are a multitude of antennas that can be used at a base station. However, the specifics of what comprise a base station antenna, or rather antenna system, are determined by the design objectives for the site coupled with real-world installation issues. For wireless radio systems, most, if not all, of the antenna design decisions are determined by the type of base station they will be deployed at. For instance the antenna system for a macrocell will most likely be different than that used for a microcell and definitely different than that used for a picocell.

Base station antennas are either omnidirectional, referred to as omni, or directional antennas. The antenna selected for the application should be one that at a minimum meets the following major points:

1. Elevation and azimuth patterns meet requirements.
2. Antenna exhibits the proper gain desired.
3. Antenna is available from common stock, company inventory.
4. Antenna can be mounted properly at the location.
5. Antennas will not adversely affect the tower, wind, and ice loading for the installation.
6. Negative visual impact has been minimized in the design and selection phase.
7. Antenna meets the required performance specifications.

This section will restrict itself to collinear, log periodic, folded dipole, yagi, and microstrip antennas with respect to passive, i.e., no active electronics in the antenna system itself. Of the antenna classifications mentioned, collinear and log periodic antennas are the more common.

4.13.2 Diversity

Diversity as it applies to an antenna system refers to a method for comparing signal fading in the environment. *Diversity gain* is defined as the ratio of the gain that occurs when diversity is used and the gain that occurs when diversity is not used. In the case of a two-branch diversity system if the received signal into both antennas is not of an equal signal strength, then there cannot be any diversity gain. This is an interesting point considering most link budget calculations incorporate diversity gain as a positive attribute. The only way diversity gain can be incorporated into a link budget is if a fade margin is included in the link budget and the diversity scheme chosen attempts to reduce the fade margin that is included there. Where the actual gain is achieved is in aperture gain, i.e., the larger cross-sectional area of the antenna intercepts more of the radio energy.

There are several types of diversity which need to be accounted for in both the legacy systems as well as the 2.5G and 3G platforms. For example, iDEN uses three-branch diversity, while for some picocells no diversity is deployed.

When discussing diversity, the radio engineer is usually focused on the receive path, the uplink from the mobile to the base station. With the introduction of 2.5G (and soon 3G) platforms, transmit diversity has been introduced but is implemented in a fashion where the subscriber does not need a second antenna.

The type of antenna diversity used can and is often augmented with another type of diversity which is accomplished at the radio level. There are several types of radio diversity that are used, including spatial, horizontal, vertical, polarization, frequency, time, and angle.

For most wireless systems two antennas separated by a physical, horizontal distance is used. As mentioned, some 2.5G platforms like iDEN utilize a three-branch diversity receive scheme, but they are the exception, and the usual method is to deploy only two antennas per sector for diversity reception.

The spacing associated with antennas located in the same sector is normally a design requirement that is stipulated from RF engineering. Diversity spacing is a physical separation between the receive antennas which is needed to ensure that the proper fade margin protection is designed into the system. As mentioned earlier, horizontal space diversity is the most common scheme used in wireless communication systems.

The following equation is used to determine the required horizontal diversity requirements for a site as displayed in Fig. 4.16.

$$n = \frac{h}{d} = 11$$

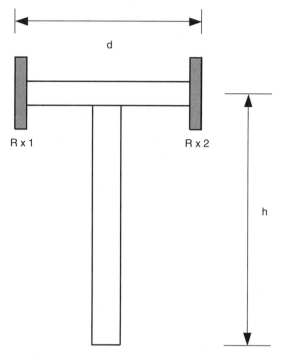

Figure 4.16 Two branch diversity spacing.

where h = height and d = distance between antennas, both in feet. This equation was derived for cellular systems operating in the 800-MHz band but has been successfully applied for the other wireless bands in the 1800- and 1900-MHz regions.

With the introduction of CDMA2000 the application for transmit diversity needs to be factored into the antenna design. The use of two different transmit diversity schemes is driven by the practical issue of antenna installation concerns. The two possible transmit diversity schemes are space transmit diversity (STD) and orthogonal transmit diversity (OTD). The preferred transmit diversity scheme, when implemented, is the STD method.

4.13.3 Installation issues

In any wireless communication system there is always a host of antenna installation problems ranging from space restrictions to tower compression or shearing loading factors or even the physical ports available to be used. The more common problems are associated with physically mounting the antennas.

With the implementation of alternative wireless platforms to an existing operator the antenna installation issues are only compounded due to the lack of the number of physical antennas available. As with the introduction of any new radio access platform, each technology has its own special issues.

Figure 4.17 shows a typical situation with CDMA and AMPS where there are two or three antennas per sector available for use. Sometimes there is only one antenna if it is a cross pole antenna. With an AMPS system as the underlying legacy system the use of an STD transmit diversity scheme is possible with a configuration as shown in Fig. 4.17, with the exception that only one carrier is used for CDMA. If a second carrier is added, then OTD is utilized and the configuration shown in Fig. 4.17a is used. Now if the operator has been able to secure more antennas per sector, e.g., 5, then the configuration shown in Fig. 4.17b is the desired method where the AMPS and CDMA systems are bifurcated. The use of STD or OTD is again dependent upon the number of carriers required at the site.

The deployment of GPRS into an existing GSM network is rather straightforward from an antenna aspect since the carriers and fundamental infrastructure issues remain the same. The only difference lies in the amount of antennas that may need to be added due to transmitter combing losses. However, there is no unique antenna configuration issue that need to be adhered to other then standard GSM deployment schemes.

When implementing a GPRS system over an IS-136 system, there are antenna issues which need to be thought about prior to acquiring the cell site or installing antennas. One such issue is the fundamental problem that GPRS or IS-136 relies on a different modulation scheme and, therefore, has different performance parameters and design guidelines.

The lesson learned with IS-95 deployment into a AMPS environment is that for performance and optimization reasons a set of separate antennas were possible. It is not that the technologies cannot share the same antenna, but the optimization techniques for GPRS or GSM are different than that for IS-136. Therefore, if the antenna system is not separated, performance compromises will be experienced in both the new and legacy systems.

The determination of where to place antennas or the methodology that is used to place the antennas is often encountered when not utilizing a monopole

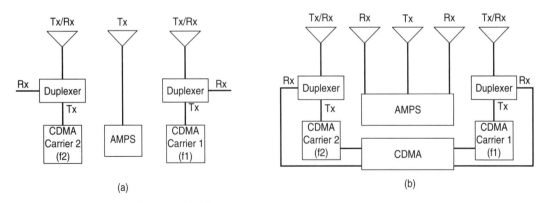

Figure 4.17 CDMA 2000/IS-95 and AMPS systems sector antenna configurations.

or tower installation. Figure 4.18 shows an example of a omnisite or single sector involving a transmit and two receive antennas. The transmit antenna is installed in the center, while the receive antennas are aligned as best as reasonably possible to provide maximal diversity reception for the major road shown in the figure. Obviously, the example is more relevant for the small stretch of highway, and different installation schemes can be implemented.

The diagram shown in Fig. 4.19 is a slight modification from that shown in Fig. 4.18. The change addresses the issue of when the antennas cannot be installed at the edge of the buildings roof and need to be installed on the penthouse of the building.

When installing an antenna system on a penthouse, or on any building installation where the antennas are not installed at the edge of the building for either visual or structural reasons, a setback rule needs to be followed. The setback rule involves the relationship between the antenna's installation point, its height above the rooftop, and, of course, the distance between the antenna and the roof edge.

Figure 4.20 shows the relationship in a simplistic drawing of the antenna placement to roof edge when installing on a roof. The concept is to avoid violating the first Fresnel zone for the antenna; however, since each antenna has a different pattern and there are different operating frequencies, the following relationship will provide the necessary clearance:

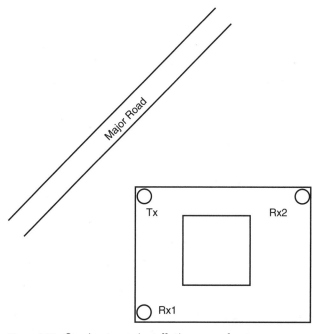

Figure 4.18 Omni antenna installation example.

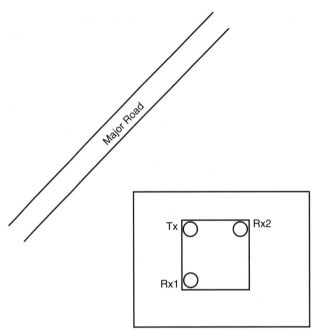

Figure 4.19 Penthouse antenna installation.

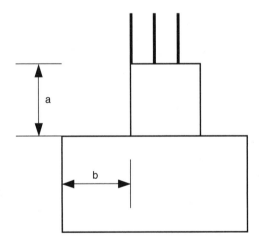

Figure 4.20 Building installation.

$$a = 5 \times b$$

where a = height, in feet or meters, and b = offset from edge of building, in feet.

From this equation we can draw the conclusion that the farther the antenna is from the roof edge, the higher it will need to be installed to obtain the necessary clearance.

4.13.4 Wall mounting

For many building installations it may not be possible to install the antennas above the penthouse or other structures for the building. Often it is necessary to install the antennas onto the penthouse or water tank of an existing building. Rarely has the building architect factored in the potential installation of antennas at the onset of the building design. Therefore, as shown in Fig. 4.21, for a three-sector configuration, the building walls may meet one orientation needed for the system but rarely all three. It is necessary to determine what the offset from the wall of the building structure needs to be. Figure 4.22 illustrates the wall-mounting offset that is required to ensure proper orientation for each sector.

Figure 4.22 shows that in order to obtain the directionality of the sector a structure is installed that will meet that requirement. The method for determining the wall offset is shown in the next equation.

$$\alpha = d \cdot \sin \phi$$

where α = distance from wall
d = diversity separation for two-branch system
ϕ = angle from wall

4.13.5 Antenna installation tolerances

When designing or installing antenna systems for a wireless communication facility, there will be some variance to the design. Just what variances are allowed needs to be stipulated from the onset of the design process. The antenna installation tolerances apply directly to the physical orientation and plumbness of the antenna installation itself. There are usually two separate considerations: how accurate should the antenna orientation be and how plumb should the antenna installation be. The obvious issue here is not only the design requirements from engineering, but also the practical implementation of the antennas for cost reasons. Use the guidelines in Table 4.10 to determine antenna installation tolerances.

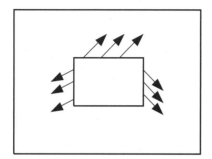

Figure 4.21 Sector building installation.

242 Chapter Four

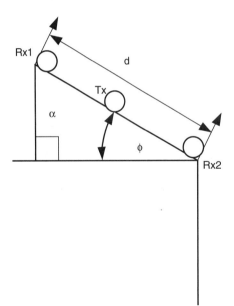

Figure 4.22 Wall offset.

TABLE 4.10 Antenna Installation Tolerances

Type	Tolerance
Orientation	±5% or antenna's horizontal pattern
Plumbness	±1° (critical)

The antenna orientation tolerance is a function of the antenna pattern and can be unique for each type of cell site. Obviously for an omni cell site there are no orientation requirements since the site is meant to cover 360°. However, for a sector or directional cell site the orientation tolerance becomes a critical issue. The orientation tolerance should be specified from RF engineering, but in the absence of this the guideline is to be within 5 percent of the antenna's horizontal pattern. Table 4.11 illustrates the use of some of the more standard types of antenna patterns. The obvious goal is to have no error associated with the orientation of the antenna, but this is rather impractical.

As the antenna pattern becomes more tight, the tolerance for the orientation error is reduced. The objective defined here is ±5 percent, but the number can be either relaxed or tightened depending on your particular system requirements. The 5 percent number should also factor in any potential building sway which can and does occur; this is usually not an issue due to the height of the buildings used for wireless installations.

4.13.6 Cross-pole antennas

The use of cross-pole antennas has proved to be very beneficial in wireless mobile applications when installation restrictions prevail. There are two types

TABLE 4.11 Horizontal Antenna Tolerances

Antenna horizontal pattern, degrees	Tolerance from boresite, degrees
110	±5.5
92	±4.6
90	±4.5
60	±3.0
40	±2.0

of cross-pole antennas that have been used. The first type can and is used for diversity reception. The second type is meant to facilitate either two transmit antennas in one housing or two different technology platforms which cannot share the same antenna for performance reasons.

Although the use of cross-pole antennas has advantages, there are disadvantages depending on the application implemented within. Typically a cross-pole antenna consists of two unique antennas that occupy the same general antenna housing, called a *radome*. There are two main variants to the cross-pole antenna. One variant has a vertical antenna array and a separate horizontal antenna array, used primarily for diversity gain. The other main variant of cross-pole antennas uses two antennas that each have a 45° polarization, one at +45° and the other at −45° referenced to vertical, used primarily for dual technology deployment. Figure 4.23 provides a visual representation of the two cross-pole variants. The cross-pole antenna consists of two separate antenna arrays which are oriented so both arrays have an angular separation of 90°.

Regardless of which trade journal, antenna manufacturer, or technical paper you read regarding the advantages of using cross-pole antennas there are a few things to always consider. First, the cross-pole antenna has a lower correlation between branches than ordinary special diversity. On the surface this would appear to be desirable; however, for diversity you want the antennas to be correlated, not 100 percent, but 70 percent. Therefore, the lack of correlation does not provide as good a protection for fades as space diversity.

The other item to consider is that the use of cross-pole antennas does not reduce the amount of cables (feeders) which need to be installed for a communication site. Referring to Fig. 4.23, each antenna array has its own feed, meaning that the installation advantage is appearance and some wind loading reduction. The wind loading reduction is not 50 percent due to the physical housing requirements associated with two antenna arrays.

If you use a cross-pole antenna, you should utilize the ±45° variant for several reasons. The first reason addresses the correlation issue and that the elements oriented ±45° from vertical will provide the best correlation coefficient possible, assuming diversity is a requirement with only one antenna.

The second reason addresses the possibility of having more than one technology type colocated at the same wireless facility, e.g., IS-136 and GSM, to mention one obvious combination. The use of the ±45° cross-pole antenna would

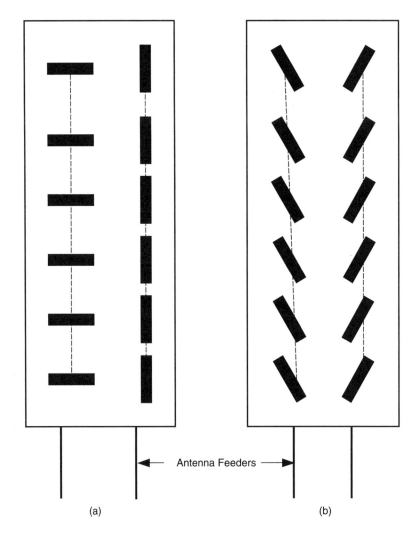

Figure 4.23 Cross-pole antennas. (*a*) Vertical and horizontal poles; (*b*) ±45° poles.

afford the possibility of having a IS-136 system on pole 1 and GSM on pole 2. If there was a second cross-pole antenna, then special diversity would also be possible using each technology on a separate antenna array.

When looking at the antenna array shown in Fig. 4.23, the rationale for using the 45° antenna should be apparent; e.g., which system goes on the horizontal leg if the ±45° antenna is not used?

Obviously if there is more than one technology used by the operator, the possibility exists that they will be in different frequency bands. When this occurs, the typical mode is to use one cross-pole antenna for technology A and another separate cross-pole antenna for technology B, again using the ±45° antenna for each cross-pole antenna. There are also some dual-band antennas that are being used. If this is available, then the use of the ±45° cross-pole antenna is once again the viable option.

Keep in mind that since the polarization for wireless mobility is vertical, special diversity using vertically polarized antennas is still the preferred method for aperture gain as well as cross correlation of the diversity signals thereby providing the best fade margin.

4.13.7 Antenna change or alteration

The alteration of an antenna system, whether it is in orientation, degree of inclination, or antenna type, can have a large impact, both positive and negative, upon a network. The need to ensure that any antenna system change is done with a design review process is essential to ensure the health of a communication network.

The first step in the process is to define the purpose for this change and the benefits expected. Often this simple step is overlooked, and a change takes place with much capital dollars begin expended without truly knowing what is the desired end result.

The criteria used for defining the antenna system change need to be well defined. For example, if the intent is to improve coverage by increasing the gain of the antennas, a link budget analysis needs to be performed. If the desire is to utilize a narrower horizontal antenna pattern, the impact to the coverage in the area currently handled by the site needs to be evaluated as a minimum. A third example would be for downtilting of the antenna where the objective would be to reduce interference and/or current cell coverage, and alternative methods of achieving the same result should be investigated. As with all antenna changes the internal Federal Communications Commission (FCC) process needs to be followed to ensure the proper documentation is maintained and the site source documents are updated.

After a design change has been completed, it is essential that some form of posttesting take place to ensure that the cure is not more problematic than the illness you were trying to correct. The need for pretesting and posttesting is essential for closing the design loop process. Drive testing of the affected area before and after these changes needs to take place.

Statistical analysis will also need to be done to ensure that adjacent sites are not adversely affected. The following is a simple checklist for important metrics that need to be reviewed:

1. Reasons for changing
2. Criteria for selecting antenna system
3. Desired results
4. Pretesting plan
5. Implementation plan
6. Internal and external coordination

7. FCC internal process
8. Posttesting
9. Conclusion report
10. Cell site source file updated

4.14 Site Types

There are numerous types of communication sites which comprise the 1G, 2G, 2.5G, and future 3G configurations associated with wireless mobile systems. There are also a plethora of other communication sites that the design engineer may encounter in the design process including existing mobile systems: local multipoint distribution system (LMDS); point to multipoint (PMP); multichannel, multipoint distribution system (MMDS); specialized mobile radio (SMR); enhanced specialized mobile radio (ESMR); paging; broadcasting; frequency modulation (FM), and amplitude modulation (AM). Each of these different types of wireless sites, depending on its proximity, may need to be included in the design phase.

The usual colocation considerations are

1. Antenna placement
2. Frequency of operation [adjacent channel and cochannel (adjacent market)]
3. Intermodulation—third- and fifth-order intermodulation distortion (IMD) products along with spectral regrowth
4. Site maintenance obstructions—window washing equipment, sand blasting, etc.

The most common types of sites considered for 2G and 2.5G implementation are macro-, micro-, and picocell sites. How these terms are defined is dependent upon the service area the base station will cover. For instance, if the site is to cover 25 square miles (mi^2), it is considered a macrocell. If the site is to cover 0.25 mi, it is usually referred to as a microcell. A site that is meant to cover a meeting room is often referred to as a picocell. Since there is no specification that defines the service area and the name for the particular communication site, the definitions of macro-, micro-, and picocells remain somewhat vague.

A typical cell site consists of the following components, which are referenced in Fig. 4.24. The piece components are the same for all cell site types. The chief difference lies in the form factor which impacts the overall capacity carrying capability for the site and, of course, the power.

4.15 Reradiators

Operators have effectively used reradiators to solve communication problems. Reradiators take on three distinct forms: (1) high power, (2) low power, translating, and (3) nontranslating or bidirectional.

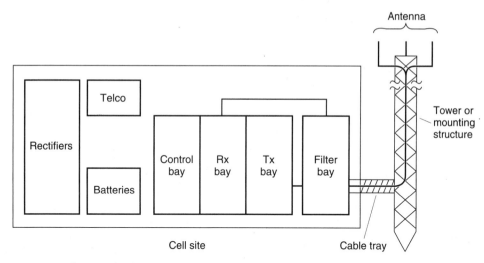

Figure 4.24 Communication site.

The only true objective for utilizing a reradiator is to increase coverage since this technology platform will not add capacity to the network. The reradiator, or repeater, redistributes capacity from one cell site to another area. The repeater can alter the capacity distribution of the network through the selection of the donor cell from which the capacity is drawn.

The primary negative attribute of a repeater is that it can add both mobile-to-base and base-to-mobile interference into the network. The interference is a result of the repeater extending the actual coverage of the donor cell beyond what it was designed to operate over. Which type of reradiator to utilize depends on the design objective and the configuration options available for the situation.

4.16 Inbuilding and Tunnel Systems

Designing an interior wireless system to support an inbuilding, subway, or tunnel application can and should be a rewarding situation. Some of the applications include improving coverage for a convention center or large client, disaster recovery, or a wireless private branch exchange (PBX). The issues that are encountered for interior communication system applications are very unique to that associated with a macrosystem.

There has been a lot of focus on inbuilding applications over the years with surges in effort taking place. The fundamental problem which keeps arising is the inherent cost of the facility. The cost of the inbuilding system typically far outweighs any potential for increased wireless revenue for the wireless operator. The cable and antenna installation, whether distributed antennas or leaky feeder, are the real driving costs of these installations. The inclusion of 802.11g wireless local area networks (LANs) has proved exceptional for configuration and installation ease but has provided little in the way of increased revenues for the wireless companies. Regardless of the application and the solutions sought, we have yet to get involved with any interior communication system which involves less then three visits due to all the issues which constantly arise.

Typically the inbuilding design approach has been the "surround and drown" method, which should be evident in the link budget calculations associated with urban environments. However, the propagation of the RF energy takes on unique characteristics in an interior or confined application as compared to an outdoor environment.

The primary differences in propagation characteristics for interior versus outdoor are the fading, shadowing, and interference. A deeper fade occurs in interior situations because the wave valleys are spatially closer than normally encountered in exterior applications. Shadowing is also quite different due to the lower antenna heights; excessive losses through floors, walls, and cubicles; or vehicle blockage as in the case of a tunnel. The shadowing effects limit the effective coverage area to almost line of site (LOS) for wireless communication whether it is in the cellular or PCS band. The interference issue with an interior communication system can actually be a benefit since the interference is primarily noise driven and not cochannel interference, assuming no reuse is involved; obviously this is not a CDMA issue. Interior communication systems are primarily noise driven because of the attenuation experienced by external cell sites as they enter into the buildings and various structures. This is not to say, however, that there cannot be IMD or out-of-band emission problems which disrupt the communication system.

There are some unique considerations that need to be factored into the communication system design:

1. Base-to–subscriber unit power
2. Subscriber unit–to-base power
3. Link budget
4. Coverage area
5. Antenna system type and placement
6. Frequency planning

The base-to-subscriber power needs to be carefully considered to ensure that the desired coverage is met, deep fades are mitigated in the area of concern, the amplifier is not being overdriven or potentially underdrive, and mobile overload does not take place. The desired coverage that the interior system is to provide might require several transmitters because of the limited output power available from the units themselves. For example, if the desired coverage area required 1-watt (W) ERP to provide the desired result, a 10-W amplifier would not be able to perform the task if you needed to deliver a total of 40 channels to that location, meaning only 25 mW of power per channel was really available. The power limitation can, and often does, make the limiting path in the communication system for an inbuilding system the forward link.

The forward link power problem is further complicated by the fact that portable and potential mobile units will be operating in very close proximity to the interior system's antenna. If the forward energy is not properly set, a subscriber unit could easily go into gain compression causing the radio to be desensitized.

The subscriber-to-base power also needs to be factored into the interior design. If the power window's dynamic power control is not set properly, then imbalances could exist in the talk-out and talk-back paths. Usually the reverse link in any interior system is not the limiting factor, but the subscriber-to-base path should be set so that there is a balanced path between the talk-out and talk-back paths.

Most interior systems have the ability to utilize diversity receive, but for a variety of reasons it is often not utilized. The primary reason for not utilizing diversity receive in an interior system is the need to place two distinct antenna systems in the same area, which is rarely possible due to installation and cost restrictions.

The link budget for the communication system needs to be calculated in advance to ensure that both the forward and reverse links are set properly. The link budget analysis plays a very important role in determining where to place the antenna system, distributed or leaky feeder, and the amount of microcell and picocell systems required to meet the coverage. The link budget associated with an inbuilding or tunnel system is for all intents and purposes a line of sight model. The simple rule is: if you can see the antenna, you have coverage. Interior fading and attenuation is very severe, and rounding a corner in an office will usually result in a signal loss which will deteriorate the call or have it terminate prematurely.

There are several methods available for a wireless operator for interior coverage.

Macrosystem (surround and drown)

Micocell or picocells

Reradiators

Please note that the frequency planning for an interior system needs to be coordinated with the external cellular network. The coordination is needed since most interior communication systems are designed to facilitate handoffs with the macrosystem. If the inbuilding system is utilizing a microcell with its own dedicated channels assigned to it, then it is imperative that the interior system be integrated into the macronetwork.

Reiterating, the concepts for inbuilding or tunnel coverage are very similar in that they both rely on line of sight (LOS) and not multipath to ensure a reliable communication link. This is a fundamental change from the macrosystem design which by default relied on the use of multipath to ensure the communication link.

4.16.1 Antenna system

The antenna system selected for the interior application is directly related to the uniformity of the coverage and quality of the system. The antenna system primarily provides LOS coverage to most of the areas desired in the defined coverage area. Based on the link budget requirements the antenna system can either by passive or active. The antenna system for an inbuilding system may have passive and active components in different parts of the system to satisfy the design requirement.

Figure 4.25 shows a distributed antenna system (DAS) which can be augmented by additional active components if the design warrants it. Typically a passive antenna system is made up of a single or distributed antenna system or can also utilize a leaky coaxial system. The inbuilding system shown in the figure utilizes a distributed antenna system for delivering the service. A leaky coaxial system could also be deployed within the same building to provide coverage for the elevator in the building.

The advantage a leaky coaxial system has over a distributed antenna is that it provides a more uniform coverage to the same area over a distributed antenna system. However, the leaky coaxial system does not lend itself for an aesthetic installation in a building. The use of a distributed antenna system for providing coverage in an inbuilding system makes the communication system stealthy. But for providing coverage for elevators the only method we have found to be successful is the use of the leaky feeder due to the metal box the elevator makes when the doors close.

If the antenna system requires the use of active devices in the communication path, the level of complexity increases. This is because the active devices require alternating current (ac) or direct current (dc) power and introduce another failure point in the communication system. However, the use of active devices in the inbuilding system can make the system work in a more cost-effective fashion. The most common active device used in an inbuilding antenna system is a bidirectional amplifier.

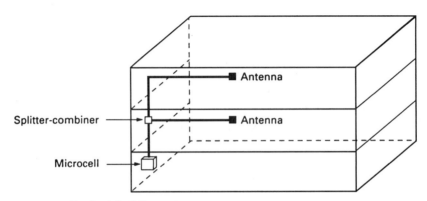

Figure 4.25 Passive inbuilding system.

4.16.2 Inbuilding application

For covering a simple room or meeting areas within a hotel, the use of a bidirectional system, whether permanent or temporary, may prove the most viable. Figure 4.26 is an example of an inbuilding bidirectional system. Please note the reliance on the external system and the need to provide isolation between the external, directional antenna and the DAS antenna shown in the figure. To prevent feedback, 70 dB of port-to-port isolation is typically required, which can be achieved by antenna placement or use of an attenuator for minor adjustments.

4.16.3 Tunnel applications

Tunnel systems are unique, and there have been many papers written on the topic of tunnel coverage. What we have found is that external illumination provides coverage until the first bend in the tunnel. Therefore, unless it is a small tunnel or short underpass, some form of cell site or reamplication of the macrosystem needs to take place to provide reliable coverage.

For tunnel coverage we have used both the distribute antenna system, using bidirectional amplifiers, and also the leaky feeder approach. An example of a bidirectional amplifier method is shown in Fig. 4.27, which could be done with a microcell or picocell. The antennas should be yagi antennas and installed after the first bend or as deep in the tunnel as possible.

The leaky feeder approach provides more uniform coverage and accounts for bends, etc. The location of the leaky feeder has been on the top of one of the sidewalls placing the feeder against the roof, with the appropriate standoffs provided by the leaky feeder vendor. An example of a leaky feeder install for a tunnel is shown in Fig. 4.28, where a microcell is used as the antenna system feeder instead of relying on a reradiation technique.

At both ends of the leaky feeder a 50-Ω antenna is installed, which provides both a load for balancing as well as a method for providing better overlap with

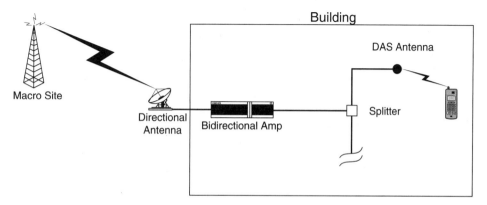

Figure 4.26 Bidirectional antenna system.

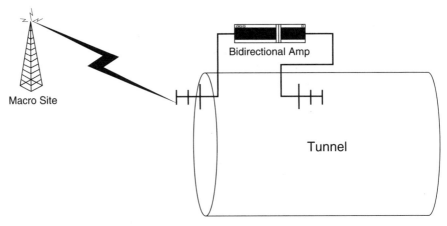

Figure 4.27 Tunnel coverage using rerad approach.

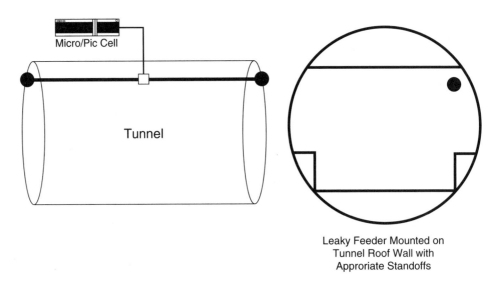

Leaky Feeder Mounted on
Tunnel Roof Wall with
Approriate Standoffs

Figure 4.28 Leaky feeder system for a tunnel.

the macrosystem. Additionally if the tunnel is excessively long, which you can determine by evaluating the link budget, it is necessary to use two micro cells or bidirectional amplifiers. The middle point of the tunnel has a termination on both ends to allow for a handoff zone (see Fig. 4.29).

Please note that the leaky feeder installation is effectively a single-antenna, nondiversity site. This is not an issue since in most cases line of sight is maintained. However, installing a second leaky feeder for diversity will not net you any benefit and will escalate the installation cost needlessly.

Anyone who has ventured into a tunnel system knows some of the unique installation concerns. One of the issues is access for maintenance. That is why

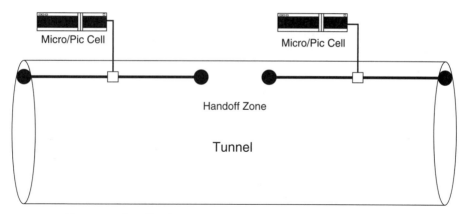

Figure 4.29 Long tunnel application.

we are strong proponents of placing the equipment near the entrance of the tunnels and having the RF distribution system remain passive.

Lastly, there is the question of what to do when several operators want to use the same leaky feeder. The obvious issue is a combining technique for the forward energy and a receiver multiplex system for the receive. However, this will prove problematic since someone has to maintain the receiver multiplexer if it is active, as most are. Therefore, a crossband coupling is recommended which allows all the operators to use a duplexer for transmitting and receiving the signal. You must pay attention to the operating band of the coupler port, the losses of the combing system, which can get excessive, and the overall power being delivered into the unit. Once you understand all the losses, you need to redo the link budget and ensure that you can still have reliable communication.

4.16.4 Planning

The issues associated with designing an inbuilding or tunnel system for a wireless system are listed here. The time durations that accompany each of the steps are not included since they depend directly upon the size of the system as well as the time to market requirements.

Project kickoff meeting

Implementation
- Antenna mount and installation
- Equipment installation
- Uninterruptible power supply (UPS) system installation
- Equipment rack installation
- Equipment installation
- Microcell installation
- Cable installation to microcell or picocell
- Radiax cable and/or distributed antenna system
- Installation constraints

RF engineering project design
- System requirements
- Design criteria defined
- Establishment of responsibility centers
- Performance criteria
- Review system designs
- Microcell and picocell system review
- Macrocell system review
- Coverage requirements
- Handoff requirements
- Cable design
- Tunnel operator's system design review
- Link budget
- Intermodulation
- Noise levels
- Filter requirements
- Microcell system review
- Radiax system review
- RF plumbing design
- Path analysis
- Antenna system
- Antenna selection
- Lightning protection
- Transmitter combing
- Frequency plan
- Translations
- Parameter adjustments

Establishment of equipment list and ordering

Acceptance of test plans
- Criteria establishment
- Generate test plan
- Issue ATP
- Conduct tests (facilities, fiber testing, power acceptance test)
- ATP signoff

4.17 Isolation

Many wireless mobile operators approach colocation requirements largely driven by rules of thumb which may or may not be entirely applicable considering the application where it is applied. The utilization of rules of thumb for defining the isolation requirements can and has made communication sites less than optimal or potentially undesirable from either a RF, permitting, or construction aspect. With the added pressure to utilize existing structures for wireless communications, there is a strong need to establish mutually acceptable colocation guidelines for all the operators in the given market.

Besides physically ensuring that there is space and that there are no structural problems for the installation, the following issues need to be addressed.

1. Isolation
2. Intermodulation
3. Grounding
4. EMF compliance

Antenna colocation requirements stem from the need for isolation from one service provider's transmitter (source) to another service provider's receiver (victim). However, when colocating antennas for your own service and when working with another service provider, there is a need to maximize the performance for the site from both perspectives.

Other sections of this book address intermodulation and EMF compliance, and these topics will not be repeated here. However, grounding is an entire discipline by itself and each vendor and wireless operator has its own particular requirements for grounding. The common issues revolve around ground loops which result in voltage differences which in turn induce current to flow between the loops raising the noise floor and compromise the grounding system integrity. The specific requirements for grounding requirements at a tower or building application should be secured from your infrastructure vendor.

The antenna colocation guidelines are centered around the issue of required isolation between the source and victim. For the isolation study several critical assumptions are made that form the basis of this guideline. The critical assumptions are listed here for reference and will be utilized throughout this guideline.

1. The equipment utilized by both the source and victim meet or exceed the FCC requirements and are fundamentally well designed.
2. The grounding system at each of the sites is adequate.
3. There are no intermodulation products caused by mixing of frequencies.
4. Receiver blocking will not be the driving issue for colocation isolation requirements since receiver desensitization will take place before blocking occurs.
5. The isolation requirements will be based on a rise in the usable noise floor by 0.5 dB which has the source of the out-of-band interference being 10 dB below the victim's usable receiver sensitivity.
6. Relaxation of the isolation requirements can take place if the margin between the usable noise floor and the operating signal for the service has enough headroom to relax the minimal threshold.
7. All radio transmitters have low levels of emissions outside of their intended channel and band.
8. The isolation requirements were compiled from the operator's and manufacturer's input.

4.17.1 Isolation requirements

The isolation requirements for each of the operators is dependent upon a multitude of variables. The isolation requirements are best defined by referencing the antenna input port of the equipment for the base station. This normalization can then be utilized for further refinement of the requirements.

Some of the items that need to be factored into the isolation requirements concern the receiver and performance requirements for each operator. The isolation requirements for each of the services can either be defined at the 1-dB compression point for out-of-band energy or at the level of inband energy which begins to degrade the receiver's sensitivity. For colocation situations with other wireless operators the recommended method for determining the isolation requirements is based on the impact to the receiver's sensitivity.

The starting point for determining the receiver desensitization would be first to establish what the effective sensitivity of the receiver is and then from there determine what level of receiver desensitization would be acceptable in a colocation environment. For example, if the operating signal is -80 dBm and the receiver sensitivity is -115 dBm with a Eb/No requirement of 20 dB, then there is almost 15 dB of headroom still available for colocation situations. Specifically in this case the isolation requirements could be relaxed by 15 dB to enable colocation to take place.

Figure 4.30 visually depicts the inband interference issues caused by a colocated transmitter in another band. For this example the source is a cellular operator utilizing AMPS technology and the victim is a PCS1900 service provider in the PCS block A. The opposite scenario is depicted in Fig. 4.31 where the source is now the PCS1900 system and the victim is a cellular operator.

Figures 4.30 and 4.31 attempt to highlight the fact that isolation requirements apply in both directions. However, the amount of isolation needed is directly dependent upon the technology chosen, the operational service issues, and the physical separation of the antennas. It is, of course, possible to improve the isolation requirements by further filtering the transmitter of the source to remove more out-of-band energy making colocation less onerous. However, one

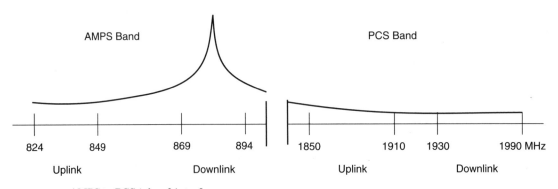

Figure 4.30 AMPS to PCS inband interference.

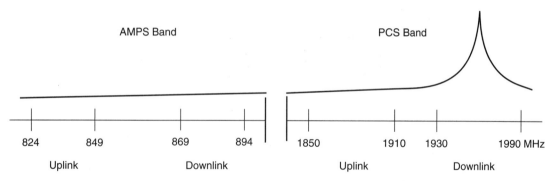

Figure 4.31 PCS to AMPS inband interference.

of the drawbacks is that the inclusion of additional filtering at the transmitter comes with a cost in terms of insertion loss.

4.17.2 Calculating needed isolation

The method used for calculating the required isolation can be extremely complicated. The complication for determining the isolation requirements is driven by the fact that not all the issues are know for any location. Also there are different infrastructure vintages and models available for an operator to utilize at any location. The infrastructure utilized for location A may not be the same for location B. It is strongly suggested that you include in your initial site design the isolation requirements for both your system and the other operators so as to avoid redesigns.

The method of procedure that should be utilized to determine the isolation required for inband interference is as follows.

1. Determine if the site is a colocation site (if it is not, then the following is not applicable).
2. Determine who the other service providers are at the facility (including future ones).
3. Determine the optimal antenna locations for your installation.
4. Compare the desired locations with the ones already allocated.
5. Reposition the antennas for your service if the location has already been allocated for another.
6. Determine the required isolation needed for your system from each of the other operators.
7. Determine the required isolation for each of the other operators from your system.
8. Reconfigure the antenna system if there is an isolation violation in steps 6 and 7 and redo the calculations.

9. Generate a lease exhibit drawing showing the antenna locations chosen.
10. Submit the drawing for review to the other operators involved in your antenna placement scheme.
11. Allow 5 business days for feedback on the design.
12. The operator initials the drawing for approval; if no feedback occurs, utilize the escalation procedure.
13. If there is a discrepancy in the design, utilize reasonable efforts to resolve the issue to the parties' mutual satisfaction.
14. Generate a revised lease exhibit drawing and resubmit to operators for their records.

The method for calculating the out-of-band interference allowed into any system is given here.

$$\text{ES (effective receiver sensitivity, in } -\text{dBm}) = \text{TN} + \text{BW} + \text{NF}$$

Y (maximum allowable in-band signal permitted, in $-$dBm)
$$= \text{ES} + 10 \text{ dB}$$

where TN = thermal noise floor = -174 dBm/Hz
  BW = bandwidth = $10 \log B$
  B = bandwidth of signal in Hz
  NF = front-end noise figure (includes cable loss and preamplifier NF), in dB

Table 4.12 illustrates individual system requirements based on the technology that each system utilizes. The noise floor for your market as well as the receiver sensitivity should be verified. The values included here are more nominal. In addition, depending on the services and particular RF plan, the E_b/N_o or C/I requirements may also need to be adjusted.

TABLE 4.12 Isolation Calculation

Technology	Bandwidth	Thermal noise floor, dBm	Receiver sensitivity, dBm	Eb/No or C/I, dB	Maximum inband signal, dBm
iDEN	25 kHz	−130	−104	20	−134
AMPS	30 kHz	−129	−116	18	−126
IS-136	30 kHz	−129	−103	17	−130
IS-95	1.23 MHz	−113	−108	7	−118
PCS1900 (GSM)	200 kHz	−120	−104	9	−114

4.17.3 Isolation requirements

Now that we have determined how much isolation is required, the next step is to determine just how this can be achieved. The determination of how much isolation is afforded due to the physical separation of the antennas themselves is indicated here. There are several components to physical separation.

1. Free space loss
2. Antenna patterns of source and victim at the victim frequency
3. Cable and connector losses
4. Physical isolation: horizontal, vertical, and slant
5. Transmit filtering of the out-of-band energy from the source
6. Antenna "filtering" at the victim frequency band

4.17.4 Free space

For a given spatial separation the application of the free-space path loss is considered to be a reasonable assumption. The free-space path loss (in decibels) is shown for both the 800- and 1900-MHz bands in the following equations. It is important to note that the free-space equation utilized for the isolation requirements needs to be at the victim's frequency.

$$\text{Cellular/ESMR: } L = 20 \log f_{\text{MHz}} + 20 \log l_m - 31.02$$

$$\text{PCS: } L = 20 \log f_{\text{MHz}} + 20 \log l_m - 38.02$$

where f_{MHz} is the frequency in megahertz and l_m is the distance in feet. Also, 10 ft is equal to 3.048 m.

The free-space calculation assumes that the antennas, both source and victim, are in the far field, when in fact most of the time they are actually in the near field. This is a valid assumption when trying to determine the expected field strength in a given area. The free-space path loss will, however, be less or equal to the actual value at the site. FCC Office of Science and Technology Report 65 (OST 65) and antenna theory books indicate why this is a valid assumption when trying to estimate the field strength. Additionally, an alpha of 2, $1/r^2$, is utilized for the slope, and this is a standard free-space loss model that is again applicable for the situation.

The frequency of operation that is chosen for the free-space calculation should involve the lower portion of the receive band you are utilizing as the victim. The rationale behind this method is the fact that the propagation characteristics are better at lower frequencies and when comparing inband interference the utilization of a transmitter frequency is not relative, unless it is an on-channel hit.

4.17.5 Antenna patterns

A critical part of the isolation determination has to do with the antenna patterns for both the source and victim. The particular pattern used for both needs to be evaluated at the victim's frequency band of operation.

The full gain of the antenna at the victim's receive frequency band will have to be factored into the antenna pattern since a far field assumption is being made. This again is a valid assumption for modeling estimates since the actual gain for both the source and victim will vary both positively and negatively until the far field region is achieved. Therefore, for the modeling aspects, the maximum gain obtainable at any one point is determined by the antenna pattern itself and is entered into the equation. The antenna gain can be reduced from maximum depending on the angle of attack from the source and victim antennas. Both antenna patterns will need to be corrected for the situation at hand.

It is assumed that antenna 1 is for the existing radio system (source) and antenna 2 is for the colocating equipment (victim). The combined antenna gain in decibels is

$$G = G_{MAX,1} - G_1(\theta_1, \phi_1) + G_{MAX,2} - G_2(\theta_2, \phi_2)$$

where $G_{MAX,1}$ and $G_{MAX,2}$ are the maximum gains of the respective antennas. The antenna gain at a particular azimuth θ and elevation ϕ is $G(\theta, \phi)$.

4.17.6 Vertical separation

Vertically separating the source and victim antennas affords the greatest amount of isolation. The utilization of vertical separation may not be practical for certain applications. An example of what constitutes vertical separation is depicted in Fig. 4.32. Note that the vertical separation is defined from the base on the top antenna to the top of the bottom antenna.

The method used for determining the isolation requirements is defined in the following two equations:

$$I_V = L - G \qquad (4.1)$$

$$= 28 + 40 \log \frac{S_V}{\lambda} \qquad (4.2)$$

where I_V = vertical isolation, dB; the maximum value for I_V is 70 dB; L = free-space loss, dB; G = antenna gain, dB; S_V = vertical separation, ft; λ = wavelength, ft. Equation (4.1) is utilized when the antenna information is available, and Eq. (4.2) is utilized when no information is obtainable for the other service provider's antenna system.

Example 4.1, for determining the vertical isolation where the source is an AMPS and the victim is a PCS1900 (GSM) system, is presented next.

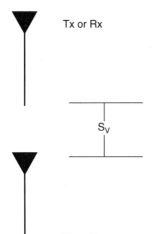

Figure 4.32 Vertical separation.

Example 4.1 Several assumptions are made regarding the antenna system: AMPS = ALP_9214 (14 dBd) and GSM = DB980H90-M (15 dBd).

Therefore, when the antennas are vertically separated, the antenna factors are

$$G = 14 - 25 + 15 - 20 = -16 \text{ dB}$$

Similarly, at 1865 MHz and 10 ft of separation, the path loss is

$$L = 65.41 + 20 - 38.02 = 47.39 \text{ dB}$$

The total isolation is

$$I = 47.39 + 16 = 63.39$$

Since the isolation required is 30 dB, a vertical separation of 10 ft gives an additional *isolation margin* of 33.39 dB (63.39 − 30 dB). This is even without factoring in the cable losses for the source and the victim and the source antenna frequency attenuation. Therefore, the physical separation requirements are limited to practical installation requirements, e.g., avoid touching the antennas.

4.17.7 Horizontal separation

Horizontal separation of antennas is not by itself a practical method of obtaining large amounts of isolation through physical separation. The method for determining the isolation is shown in the following two equations. If the particulars of the antenna system are known, use the following equation:

$$I_H = L - G \qquad (4.3)$$

where I_H = horizontal isolation, dB. If the particulars of the antenna system are not known, then use this equation:

$$I_H = 22 + 20 \log \left[\frac{S_H}{\lambda} - (G_T + G_R) \right] \quad \frac{S_H}{\lambda} > 10 \quad (4.4)$$

where S_H = horizontal separation, ft; G_T = transmitter antenna gain, dB; G_R = receiver antenna gain, dB.

The distances between the antennas utilized for horizontal separation are shown in Fig. 4.33. The reference point for both the source and victim antennas are the base or middle of the antenna itself. For Eqs. (4.3) and (4.4) it is assumed that both the source and the victim antennas are on the same horizontal plane and that no variation for slight changes in antenna height due to gain and frequency issues need to be factored into the equations.

Utilizing the same methodology as for vertical separation, Example 4.2 illustrates a horizontal isolation example.

Example 4.2 Two antenna systems are pointing straight at each other and are at the same height:

$$G = 14 + 15 = 29 \text{ dB}$$

Similarly, at 1865 MHz with 20 ft of horizontal separation, the path loss is

$$L = 65.41 + 26.02 - 38.02 = 53.41 \text{ dB}$$

and the total isolation is

$$I = 53.41 - 29 = 24.41 \text{ dB}$$

Thus the antenna systems are directly pointing at each, which is a possibility for adjacent building applications. This illustrates that the required isolation of 30 dB is not met for this application by close to 6 dB (30 − 24.41).

The frequency correction for the antenna system used by the source at the victim's frequency band was not factored into the numeric calculation in

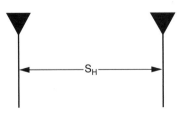

Figure 4.33 Horizontal separation.

Example 4.2, nor were the cable or connector losses. Assuming that the victim cable and connector loss (C_L) is 2 dB and the source antenna correction (SAC) is 10 dB, a refinement to the isolation situation can be made.

$$I_{adj} = I + C_L + \text{SAC}$$
$$= 24.41 + 2 + 10$$
$$= 36.41 \text{ dB}$$

where I_{adj} = adjusted isolation, dB. Note that the source antenna correction factor needs to be provided by the manufacturer of the antenna in question. However, if this is not available, a minimum of 10 dB can be assumed when comparing frequency bands that are significantly separated.

The isolation provided by cables, connectors, and the source antenna provide the additional isolation needed for Example 4.2. However, if say the distance was reduced to 10 ft, then the isolation requirements with all the correction factors added would just meet the requirements, with no safety margin. If the isolation required for reliable communication cannot be achieved with the proposed antenna configuration, the following steps can be taken to try and achieve the required isolation.

1. Add vertical separation isolation.
2. Change the orientation of the antenna systems between the source and the victim.
3. Add additional transmit filtering to the source transmit path to reduce out-of-band energy emissions.

If the directional antennas are mounted side by side, then the antenna pattern attenuation can be included to reflect the situation thus improving the isolation greatly. This type of installation is shown in Fig. 4.34. It is important to note that the separation between the mounting configurations is shown as I_{SX}, and this information needs to be delivered to the construction team associated with the implementation of the site itself.

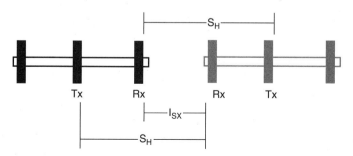

Figure 4.34 Horizontal separation installation guideline. S_H = horizontal separation; I_{SX} = installation separation.

264   Chapter Four

4.17.8 Slant separation

As is often the case in wireless installations the antennas for the source and the victim are neither purely horizontally or vertically separated. When the exact parameters for both the source and victim are known, corrections for horizontal and vertical issues are accounted for in the antenna pattern factors. The distance that is utilized for free-space calculations is the shortest distance between the source and the victim as depicted in Fig. 4.35.

However, when the particulars for the location are not known about the other antenna system, the following equation can be utilized.

$$I = (I_V + I_H) \cdot \frac{\theta}{90} + I_H$$

where θ = angle of 0 to 90° and where

$$I_V = 28 + 40 \log \frac{S_V}{\lambda}$$

$$I_H = 22 + 20 \log \frac{S_H}{\lambda} - (G_T + G_R)$$

$$I_{V,\max} = 70 \text{ dB}$$

$$\frac{S_H}{\lambda} > 10$$

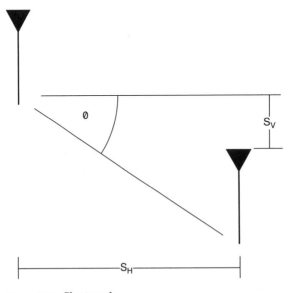

Figure 4.35 Slant angle.

4.18 Base Station Site Checklist

There can be a multitude of installation issues encountered at a site since each building or installation has some problem or problems associated with it. The coordination of the various contractors used for the installation process is always an issue with installation deals. In many instances the dependence is serial in that project B cannot begin until project A is completed.

The following is a brief listing of some of the more common installation issues that are encountered:

1. Access to the site
2. Hours allowed for construction (usually a tenant improvement issue)
3. Base station equipment delivery
4. Telco acceptance or installation of a point-to-point (PtP) radio link
5. Securing the necessary permits
6. Architectural engineering drawing approvals from internal groups as well as the landlord
7. Antenna system installation
8. Landlord claims of damage to the building as a result of the installation work
9. Power system upgrade
10. Floor loading (primarily for batteries)
11. Heating, ventilation, and air conditioning (HVAC) venting and installation
12. Noise abatement for the cell site and HVAC system
13. Alternative power requirements (e.g., generator hookup)
14. Parking and bathroom facilities

Table 4.13 on pp. 266 and 267 is a generic checklist of the major items that should be checked prior to or during the commissioning of a communication site. Following this checklist will help ensure that everything is accounted for prior to the communication site going commercial, and it can be tailored for your particular application.

TABLE 4.13 Cell Site Checklist

Topic	Received	Open
Site location issues		
1. 24-hour access		
2. Parking		
3. Direction to site		
4. Keys issued		
5. Entry/access restrictions		
6. Elevator operation hours		
7. Copy of lease		
8. Copy of building permits		
9. Obtainment of lean releases		
10. Certificate of occupancy		
Utilities		
1. Separate meter installed		
2. Auxiliary power (generator)		
3. Rectifiers installed and balanced		
4. Batteries installed		
5. Batteries charged		
6. Safety gear installed		
7. Fan/venting supplied		
Facilities		
1. Copper, fiber, PtP		
2. Power for fiber hookup (if applicable)		
3. PtP radios aligned		
4. POTS lines for operations		
5. Number of facilities identified by engineering		
6. Spans load-tested for 24 hours		
HVAC		
1. Installation completed		
2. HVAC tested		
3. HVAC system accepted		
Antenna system		
1. FAA requirements met		
2. Antennas mounted correctly		
3. Antenna azimuth checked		
4. Antenna plumbness check		

TABLE 4.13 Cell Site Checklist (*Continued*)

Topic	Received	Open
Antenna system (*cont.*)		
5. Antenna inclination verified		
6. SWR check of antenna system		
7. SWR record given to operations and engineering		
8. Feedline connections sealed		
9. Feedline grounds completed		
Operations		
1. User alarms defined		
Engineering		
1. Site parameters defined		
2. Interference check completed		
3. Installation MOP generated		
4. FCC requirements document filled out		
5. Optimization complete		
6. Performance package completed		
Radio infrastructure		
1. Bays installed		
2. Equipment installed according to plans		
3. Radio equipment subjected to acceptance test procedure		
4. Transceiver output measured and correct		
5. Grounding complete		
6. Equipment bar coded		

Chapter 5

RF System Performance and Troubleshooting

Radio frequency system performance and troubleshooting remains one of the most challenging and rewarding aspects of working on a wireless network. Many wireless systems implement wireless data in addition to voice services, so there is pressure for continuous improvements in the radio environment. *RF system performance and troubleshooting* implies many things to both technical and managerial individuals within the wireless technology community. However, there are some basic concepts of RF system performance, also referred to as optimization, that transcend technology and system configurations. The concepts of performance improvement in wireless systems are based on the same fundamental principles regardless of whether the service is voice or data or whether the technology is advanced mobile phone system (AMPS), global system for mobile (GMS), IS-136, code-division multiple access (CDMA), or a fixed wireless technology like a local multipoint distribution system (LMDS), or a multichannel, multipoint distribution system (MMDS).

These central concepts involve the following three basic principles. First, you must define the key goals and objectives for the performance of the system. Second, you must define how these goals and objectives will be measured or monitored. Third, you must define the actions to be taken on a daily, weekly, and monthly basis to improve system performance. Therefore, system performance for any wireless network involves a continuous series of adjustments and refinements.

Regarding the first principle, the key goals and objectives need to be broken down into meaningful and objective criteria, and a defined time frame needs to be associated with each criteria. Very frequently a time frame of 1 or 2 years may be necessary to reach a *key* goal or objective. The goal or objective must be broken down into quarterly or monthly time scales, with a current benchmark for the foundation or starting point.

The goals and objectives also need to be defined and/or apportioned to each of the managers and engineers involved, so their specific contributions can be factored into the process. For example, when defining the specific contribution a performance engineer will make to the overall network, his or her specific area of responsibility, region, and/or cell sites need to be defined. The goals and objectives for each performance engineer must be crafted to reflect his or her area of responsibility.

Monitoring of the goals and objectives needs to be done in such a fashion that management can allocate resources to resolve issues. The monitoring, or reporting, needs to be performed on a daily, weekly, and monthly basis, and it needs to both incorporate the key factors and indicate a variance from the norm or goal. Reports should be done on a regional basis, in addition to on the overall system, with a comparison to the norm or goal for each geographic area. Both exceptions and tracking reports are required.

The third principle involves what to do after you have defined what the goals are and how they are measured. The steps to take are

1. Establish a weekly and monthly overall plan of attack.
2. Review the daily exception report and take action if needed.
3. Examine system parameter settings for errors and correct on a daily basis.
4. Examine key metrics on a weekly basis for progress and/or possible changes in plans.
5. Document changes and communicate findings to fellow engineers.

As implied earlier, many aspects exist to the role of the system performance engineer. One essential fact is that, no matter how thorough the design work is, the system performance engineer has to make the equipment really work, from an engineering point of view. Therefore, good engineering practices during system performance troubleshooting are imperative. It is the performance engineer who ensures that the lost call rate and other quality factors are at their best, thus ensuring maximum customer satisfaction and revenue potential.

System performance and troubleshooting involve applying a set of critical techniques that have passed the test of time. Revisiting the three main criteria, a sequence of steps, methods, or techniques can be defined which, when applied, will greatly improve the chances of finding a true solution to any performance problem.

Step 1. Identify your objective with the effort you are about to partake in and document it.

Step 2. Isolate the item you are working on from the other variable parameters involved with the mission statement.

Step 3. Identify what aspect of the system you are trying to work on: switch, telco, cell site, mobile, or RF environment.

Step 4. Establish a battle plan. Write down what you want to accomplish, how you will accomplish it, and what the expected results will be.

Step 5. Communicate your objective—what you want to do and why; usually this is called a test plan or resolution plan.

Step 6. Conduct the work or troubleshooting that is identified in your objective.

Step 7. Perform a postanalysis of your work and issue a closing document either supporting or refuting your initial conclusions and identifying what are the next actionable items.

It needs to be stressed that the performance and troubleshooting techniques referenced in this chapter are not the be-all and end-all of techniques. Technologies change over time, and wireless operators either replace technologies or overlay new technologies into their existing networks. This chapter provides guidance to some technology-specific issues; however, it is the concepts and troubleshooting techniques that are emphasized.

The intention of this chapter is to discuss problems that have happened in the past so that they may be avoided in the future. Topics include how to monitor the network and implement fixes that will be expedient and cost effective. Pertinent equations and why they should be used will also be covered.

5.1 Key Factors

The performance of a wireless network, from the RF perspective, has many dynamic aspects to it. The fact that the system is dynamic and many issues are interrelated makes the task of improving the performance of a system daunting. Handling issues related to the addition of new sites, changes to existing sites, new features, maintenance, customer service, and a host of other topics taxes the performance engineer's time.

Since everything in a commercial system is time- and money-related, the performance engineer needs to allocate his or her time appropriately, meeting both internal and external objectives. To maximize the RF system performance and expedite the troubleshooting process it is exceptionally important to determine the critical system metrics, or key factors, that need to be monitored and the frequency and level of detail needed.

All too often there is either an information overload or underload problem in engineering. Information overload occurs when everyone in the organization is receiving all the reports. Information underload occurs when there are too few reports being distributed. As systems continue to grow in size and complexity, the use of statistics for determining the health and well-being of the network becomes more and more crucial. Therefore, when you determine what reports and information you want to see regarding the network, it is imperative that a support system is installed. The support system will ensure that the report generation, report distribution, and analysis of the data are done in a timely and accurate manner and on a continuous basis.

Most system operators have several key metrics they utilize for monitoring the performance of their networks. The metrics are used for both day-to-day operation and for upper management reports. The particular system metrics utilized by the operator are dependent upon the actual infrastructure manufacturer being used and the software loads. The metrics that are common to all operators are lost calls, blocking, and access failures. The metrics mentioned are very important to monitor and act upon, but how you measure or calculate them is subject to multiple interpretations. The fundamental problem with having multiple methods for measuring a network is that there is no standard procedure to use and follow.

There are numerous metrics that need to be monitored, tracked, reported, and ultimately acted upon in a wireless network for both voice and data. The choice of which metrics to use, their frequency of reporting, who gets the reports, and the actual information content largely determines the degree of success an operator has in maintaining and improving the existing system quality. With the proper use of metrics a service provider can be proactive with respect to system performance issues. But when a service-affecting problem occurs in the network, it is better for the engineering and operations departments to already be aware of the problem and have a solution that they are implementing.

This brings up the issue of what metrics you want to monitor, the frequency with which you look at them, and who receives the information. The key metrics that we have found to be most effective, regardless of the wireless technology, infrastructure vendor, or software load currently being used and regardless of whether the service is voice or data, are listed here:

1. Lost calls
2. Blocking
3. Access failures
4. Bit error rate (BER), frame error rate (FER), and signal quality estimate (SQE)
5. Customer complaints
6. Usage and RF loss
7. Handoff failures
8. RF call completion ratio
9. Equipment out of service (such as radios or cell site spans)
10. Technician trouble reports

Focusing on these key parameters for the RF environment will net the largest benefit to any system operator, regardless of the infrastructure currently being used. When operating a multiple-vendor system for a wireless mobile infrastructure, you can cross map the individual metrics reported from one vendor and find a corollary to it with another vendor. The objective with cross mapping the metrics enables everyone to operate on a level playing field regarding system performance for a company, and ultimately the industry as a whole. The individual equations for each of the key metrics will be discussed in later sections of this chapter.

The key metrics identified are relatively useless unless you marry them to the goals and objectives for the department, division, and company. For example, knowing that you are operating at an access failure level of 2.1 percent, depending on how you calculate it, does not bode well when your objective is 1.0 percent. The decision to use bouncing busy hour versus busy hour for the evaluation also is an important aspect in measuring the system's performance. The use of the system-defined busy hour, however, is far better for evaluating trends and identifying problematic areas of the network.

When reporting metrics, you need to address both what metrics you are monitoring and how you report on the metrics. It is very important to produce a regular summary report for various levels of management to see so that they know how the system is operating. When crafting a metrics report, you should determine in advance who *needs* to see the information versus who *wants* to see it. More times than not there are many individuals in an organization that request to see large volumes of data, with valid intentions of acting on them, but in the process become so overcome with data input that they enter the "paralysis of analysis" phase.

Establishing regular, periodic action plans is very effective in helping maintain and improve system performance in a network. System performance requires constant and vigorous attention. Establishing a quarterly and monthly action plan for improving the network is essential in ensuring its health. In particular every 3 months, once a quarter, you should identify the worst 10 percent of your system following the list of metrics. The focus should not only be on the cell site but also on the sector, or face, of a particular cell site so that problems can be cross correlated. The quarterly action plan should be used as the driving force for establishing the monthly plans.

Coupled into long-term action plans are the short-term action plans which help drive the success or failure of the overall mission statement for the company. The key to ensuring that the long- and short-term goals are being maintained is through the requirements of periodic reports. The periodic reporting issues are covered in another chapter; however, they are essential, if conducted properly, for ensuring the company's success.

Some reports that facilitate focusing on the performance of the system are listed here. This list should be made available to essential personnel on a daily and weekly basis. It is recommended that the following key items of information be included in the reporting structure on a weekly basis.

1. Weekly statistics report for the network and the area of responsibility of the engineer trended over the last 3 months.
2. Current top-5 worst-performing cells and top-10 worst-performing sectors in the network and each region using the statistics metrics.

3. Listing of the cells and sectors which were reported on the last weekly report with a brief description of the action taken toward each, and the problem resolved, if any.
4. List of the cells on the current poor-performing list, including a brief description as to the possible cause for the poor performance and the action plan to correct the situation.
5. Number of radio channels [channel elements (CEs) for CDMA] in the network, by region point in time indicating the total number of
   - Channels
   - Radios out of service for frequency conflicts (non-CDMA)
   - Radios out of service for maintenance
6. Status of the technical trouble reports (weekly).

Of course, merely presenting the information in a timely and useful fashion is not enough—it has to be used to have an effect. For example, you identify the worst 10 sectors of a system, and they may be all physically pointing in the same area of the network indicating a possible common problem. By overlaying the other metrics, patterns may appear, which will enable the performance department to focus its limited resources on a given area and net the largest benefit.

5.2 Performance Analysis Methodology

The methodology that should be used by the performance engineer and his or her immediate management involves a layered approach to problem resolution. Remembering the seven major steps in any problem identification and resolution, we will attempt to address some basic system problems. The discussion will focus on an AMPS cellular system using lost calls, access failures, and blocked calls. All wireless technology access platforms have these three fundamental concepts, or performance issues, in common. For wireless data not only is there the contentious issue of blocked calls but also the potential degradation in throughput due to the RF environment, the radio or network system configuration, or off-net problems.

The concepts put forth in trying to identify and formulate an action plan, whether for an AMPS, CDMA, time-division multiple access (TDMA), or GSM technology platform, follow the same fundamental methodology. What changes when focusing on the individual technology platforms is the vendor-specific metrics and unique protocol issues. In a later part of this chapter some troubleshooting guidelines and flowcharts will be presented which are technology platform specific. However, we cannot overstress the importance of understanding a fundamental methodology for problem identification and resolution.

- Identify objective
- Remove variables
- Isolate system components

- Test plan
- Communicate
- Action
- Postanalysis

The chart in Fig. 5.1 represents a portion of a weekly system statistics report for a network. The chart has only four cell sites listed on it so that the example presented here is clearer. The performance criteria for the system include a lost call rate of 2 percent, attempt failure of 1 percent, and radio blocking between 1 to 2 percent.

The chart is interesting by itself, but a simple review of the data indicates that there are a few sectors which potentially need investigation. The chart can be converted to be more of a visual aid in the troubleshooting analysis. The analysis methods presented here show a step-by-step approach. The method you choose to use can combine several of the steps presented here, but for clarification they are presented separately here.

1. Sort the chart by lost call percent focusing on the worst performers.
2. Sort the chart by the raw number of lost calls focusing on the highest raw number.

The reason for the two sorts for the lost calls pertains to how the metrics are calculated. Sorting by the lost call percent alone might not net the largest system performance improvement. The sorting method needs also to incorporate the raw number of lost calls. The lost call percent number can be misleading if there is little usage on the site or there is a large volume of traffic on the site. If there is little usage on the site, one lost call can potentially represent a 10 percent lost call rate. If the site has a large amount of traffic, the site might be operating within the performance criteria, 1.9 percent, but this represents 10 percent of the entire system's lost calls for the sample period. The resulting display for the lost call percent and raw number of lost calls is shown in Fig. 5.2. There are several sectors all pointing to a general location using the data from the chart in Fig. 5.1.

The next step is to produce a similar chart for the radio blocking statistics. The radio blocking statistics need to be sorted by radio blocking percent and also the raw number of radio blocks. The rationale behind these two sorting methods is exactly the same as that used for the lost call method.

3. Sort the chart by radio blocking percent focusing on the worst performers.
4. Sort the chart by the raw number of radio blocks focusing on the highest raw number.

The resulting display for the radio blocking percent and raw radio blocking numbers is shown in Fig. 5.3. The information displayed in the figure does not indicate any system-level problems.

Sample Busy Hour System Report

Date:

Cell Site	Time	Usage	O&T	LC %	# LC	% AF	# AF	% Block	# Block	Usage/LC
1A	1700	262	238	2.1	5	1.26	3	1.26	3	52.38
1B	1700	393	357	7	25	2.52	9	0.84	3	15.71
1C	1700	183	167	1.8	3	4.20	7	1.20	2	61.11
2A	1700	770	700	1	7	0.43	3	—	0	110.00
2B	1700	147	133	1.5	2	1.50	2	1.50	2	73.33
2C	1700	770	700	2	14	1.29	9	0.14	1	55.00
3A	1700	367	333	1.2	4	0.30	1	2.70	9	91.67
3B	1700	419	381	2.1	8	2.10	8	1.05	4	52.38
3C	1700	3438	3125	4	125	0.45	14	1.38	43	27.50
4A	1700	500	455	11	50	0.44	2	0.22	1	10.00
4B	1700	592	667	1.5	10	1.50	10	0.30	2	59.20
4C	1700	183	167	3	5	3.60	6	0.60	1	36.67

Figure 5.1 Partial example of weekly statistics report for an AMPS cellular system.

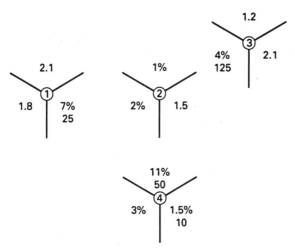

Figure 5.2 Visual display of percent lost calls and number of lost calls by sector.

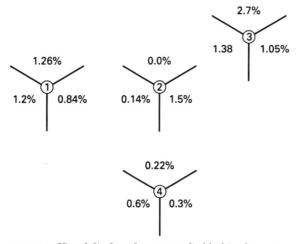

Figure 5.3 Visual display of percent radio blocking by sector.

The next step is to produce a similar chart for the attempt failure statistics. The attempt failure statistics need to be sorted by attempt failure percent and also the raw number of attempt failures. The rationale behind these two sorting methods is exactly the same as that useed for the lost call method.

5. Sort the chart by attempt failure percent focusing on the worst performers.
6. Sort the chart by the raw number of attempt failures focusing on the highest raw number.

The resulting display for the attempt failure percent and raw attempt failure numbers is shown in Fig. 5.4. The information displayed in the figure does not indicate any system-level problems.

The next step is to produce a similar chart for usage and RF losses. The usage and RF losses need to be sorted by the worst performers. The usage and RF loss worst performers are those with the lowest amount of usage between a lost call.

7. Sort the chart by usage and RF loss focusing on the poorest performers.

The resulting display for the usage and RF loss is shown in Fig. 5.5. The information displayed in the figure indicates a potential problem focusing on the same area as with the lost calls. Figure 5.6 is a composite view of the metrics evaluation.

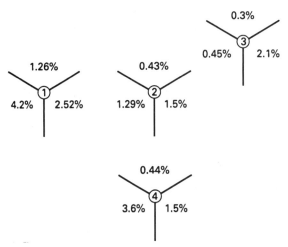

Figure 5.4 Visual display of percent attempt failures (access) by sector.

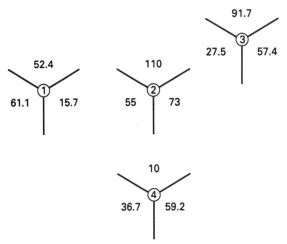

Figure 5.5 Visual display of usage/RF loss by sector.

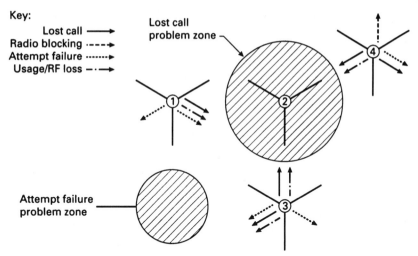

Figure 5.6 Visual display of compilation.

Obviously this procedure can and should be used with the other metrics for the network. The visual display of the information can and should be coupled into the metrics reporting mechanisms used for the network. All the key performance metrics should be checked for any correlation issues since this is one of the best methods available to identify performance-related trends.

The reports getting to upper management must be able to tell a story that is both factual and brief. Upper management needs to know that the system is running at a particular level but also that you are in control and do not require their intervention. Many times a senior-level manager that has a technical background will generate many questions and inadvertently misdirect the limited resources when given too much data. The simple rule for dissemination of reports is to minimize the information flow to only those people who really need to know the material. This is not meant to keep other people in the dark but to help ensure that the group which needs to focus on improving the network continues doing just that.

However, it is very important to let members of the technical staff know what the current health of the network really is on a regular basis. One very effective method that has been used in various forms is to have the key metrics displayed on a wall so everyone can see the network's performance. The metrics displayed should be uniform in time scale, and it is recommended that they trend over at least a year so everyone can see how well you are doing over time.

The wall chart in Fig. 5.7, if done correctly, will foster competition between fellow engineers working on the performance of the network. For example, if the data are displayed by network region, the engineer who has the worst performance in an area will feel compelled, through peer pressure, not to be on the bottom of the heap for the next reporting period. If this is handled correctly, the efforts of the various engineers will ensure that the overall

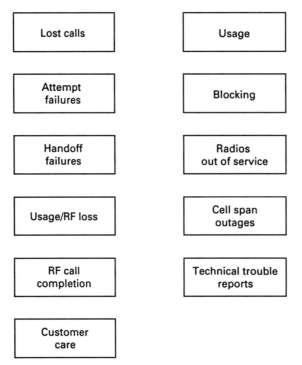

Figure 5.7 Wailing wall.

performance of the network continues to improve. The chart can, of course, be expanded upon to incorporate wireless data services.

The other item which needs to be addressed is the need to give regular presentations to the upper management of the technical arena. This is not a classic empowerment method but is meant to stress to the upper management that engineering is performing a good job. This will arm upper management with key information so that when they are confronted with irate customers who complain about the lack of good system performance they have some personal knowledge of what is happening.

Upper management reports should take place every 6 months and focus on what you have done and what you will do over the next 6 months. The critical issue here is that if you tell your superiors you are going to do something, make sure you actually do it and report on it in the next presentation, including what you did, when, and the results. It is also important that the meeting not take more than 1 hour. Based on the size of the department, i.e., the number of presenting engineers, the time frame should be well rehearsed and time minimized. Focus on three to five items which you can talk about quickly and have the answers for. It is imperative that when you bring up a problem you have a solution that goes hand in hand. The key issue, though, in presenting to upper management is to be truthful and never offer up information or proposed solutions to problems that exist when you do not have direct knowledge or control of all the issues.

RF System Performance and Troubleshooting

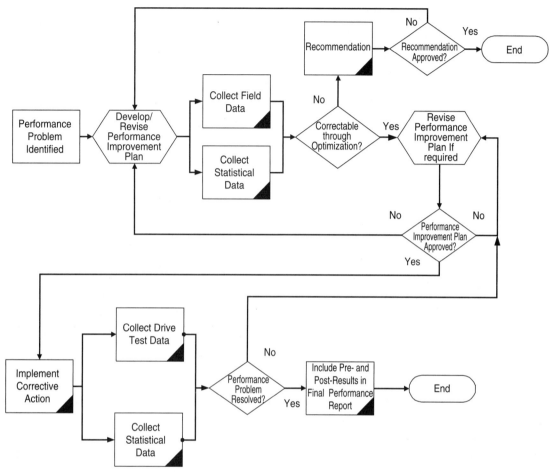

Figure 5.8 Performance improvement flow chart.

Figure 5.8 is a simple flowchart which indicates the major process associated with performance improvements. Please note that the flowchart is independent of the radio-access technology used.

5.3 Lost Calls

The lost call metric is probably one of the most discussed and focused-on parameters in the wireless industry, short of blocking and soon quality of service (QOS) for wireless data. The lost call metric is personalized because everyone who has a wireless piece of subscriber equipment has experienced a lost call and is personally aware of the frustration and aggravation it causes. Unfortunately, lost calls are a fact of life in the wireless industry and will remain so for the foreseeable future.

There have been major improvements in detecting and isolating lost calls since the inception of wireless mobile systems. However, the proliferation of wireless and its prevalence in everyday life has placed great expectations on the operators by the very subscribers who pay their bills. Additionally for many system operators a key component to engineers' and upper management's incentive is based on the lost call performance numbers. Therefore, a key step in the road to focusing on lost calls and reducing their occurrence is the setting of goals and the calculation method used.

You might be asking why goal setting is important. There are several reasons, and they are technology platform independent. The primary reason is that this is the first step in the seven steps toward problem resolution. The second reason is it prevents the "what 'is' is" syndrome. In addition the calculation method that is used needs to be established. It is important to establish a consistent lost call objective and value to strive for system performance since it is the paying customers who are having their calls dropped.

Regardless of the technology platform used when focusing on a system objective, setting the goal on a raw percentage has merit since it is easily traceable and definable. However, when you equate it to the raw volume of traffic on a network, a 2 percent lost call rate for a large network can mean over 10,000 physically dropped calls in any given system busy hour. If you had said I have 10,000 lost calls in the busy hour versus 2 percent, the system performance metrics would reflect a different picture but with the same quality impact.

The percentage lost call rate is a valid figure of merit for a system that is growing and has coverage gaps of significant size. The lost call percentage is an extremely effective tool for helping to pinpoint problems and monitor the overall health of any network. However, using the raw lost call percentage number for a large, growing network may in fact be counterproductive since more actual problems in raw volume may occur than the number represented in the percentage indicates.

For example at a 40 percent system growth rate the 10,000 lost calls at the 2 percent lost call design rate would possibly increase to 14,000 lost calls the next year. Assuming all the parameters increase with the same rate, the lost call rate would remain at 2 percent, but the raw number would increase by 40 percent. Obviously this is not the trend that you would want to take place on your own system. An alternative to using the percentage value would be to utilize an additional metric to help define the quality of the network.

Additionally when operating multiple-technology platforms in a system, it would be advisable to have different goals for each platform. For example, if there are AMPS, IS-136, and GSM platforms being used in the same market, the goal might be to have the subscribers migrate to the GSM platform. Therefore the tightest lost call goal may be placed on GSM, with IS-136 next, and the most lax goal placed on AMPS.

The percentage of lost calls should be set so that every year the percentage of lost calls is decreasing as a function of overall usage and increased subscriber penetration levels. Specifically the lost call rates need to be set so that the ultimate goal is 0 percent lost calls in a network. While we do not believe this is feasible at the present state of technology and capital investment, it is still the proper ultimate goal to set. Anything less should not be acceptable.

But reality dictates that setting of reasonable and realistic goals needs to be done in such a fashion that the real lost call rates are reduced and at the same time the methods utilized are sustainable. How to set the actual lost call rate is an interesting task since one lost call is too many. The suggested method is to provide at the end of the third quarter of every year the plan for the lost call rate to be striven for in the next year. The goals should be set so that you have a realistic reduction in the lost call rates for the coming year, trying to factor in where you might be at the end of the current year.

The following is a few examples on how to set the lost call objective for your network. Obviously you and your management should be comfortable with the set values. The lost call goal should not be set in the vacuum of an office and then downward directed. The goal should be driven to improve the overall performance of the network factoring into it the growth rate expected, budget constraints, personnel, and the overall network build program.

One example of lost call goal setting involves the situation where several wireless systems were owned by the same company through acquisitions. The interesting issue is that the lost call rates used for goals were different for each wireless system. The difference in the lost call rate was not necessarily the goal of, say, 2 percent, but rather the methodology of calculating the equation. This has led to many interesting situations where the individual wireless system was able to pick the method which would present the individual company in the best light to the parent company.

Specifically, one system used the lost call method of overall lost call, and another used the call segment approach. The difference between the two is dramatic and needs to be watched for. We chose the call segment method for reporting to the parent company and used the overall lost call rate for an internal method. The difference between the two techniques is best represented by a simple example using the same system performance statistics.

$$\text{Originations and terminations (O\&T)} = 10{,}000$$

$$\text{Handoffs (HO)} = 10{,}000$$

$$\text{Lost calls (LC)} = 200$$

$$\text{Overall lost call} = \frac{\text{LC}}{\text{O\&T}} = 2\%$$

$$\text{Call segment} = \frac{\text{LC}}{\text{O\&T} + \text{HO}} = 1\%$$

The difference between the two methods is rather obvious using the simplistic numbers listed for the example. Therefore, when you hear or see a lost call percent value, the underlying equation and methodology used needs to be known to best understand the relevance of the numbers presented.

Resolution of the different methods of reporting was never attempted. However, the overall method calculation is not necessarily the best method to use when analyzing system performance since handoffs are part of the overall situation. Some infrastructure vendors do not have the ability to record originations and terminations on a per-sector level.

Whether you use the overall method or the call segment method, it is important to be consistent. The relative health of the network can be determined by either method using simple trending methods that compare past and present performance.

When reviewing the example, the concept of the technology platform was not used. The reason the technology platform was not introduced is that it is not relevant. All wireless mobile systems require subscribers to either place or receive a call or data session with the possibility that during the service delivery the connection will be prematurely terminated. Understanding the fundamental implications of overall versus call segment methods of presenting lost calls is essential in the drive to improve system performance.

When reporting the lost call rates to upper management, the method chosen should be visual. Visual methods are extremely useful for conveying a story quickly. However, charts and graphs can be extremely deceiving especially if the x-axis and y-axis scales and legends are not defined.

For example, one situation of scaling involved two engineering divisions in the same company. Both of the engineering groups were using the same equations and time frames to calculate their lost call rates. However, both divisions utilized different y-axes to display the data. The division with the poorer lost call rate used a larger y-axis with the lost call rate placed in the middle of the chart. The division with the better lost call rate chose scales which exemplified the lost call rate. The difference in y scales between the divisions was close to a 2:1 margin.

The most interesting point with this example is that the group using the more granular y axis had a lower lost call rate with the higher system usage. The perception of upper management was that the division using the more granular reporting scheme was performing the worst since the line on the chart was higher. There were many lessons learned with this example, and one of them was that perception is very important.

The time used for conducting lost call analysis should also be clearly delineated. For example, if you use a bounding busy hour method for determining your lost call rate, no matter which equation type you use, it will be difficult to trend. Therefore, when monitoring the trend of the system, it is important to establish a standard time to use from day to day and month to month. The standard time to use is the busy hour of the system on weekdays, Monday to Friday, excluding holidays. The same hour should be used for each of the days for the entire sampling period.

The system busy hour, barring fraud, is usually between 4 to 6 P.M. in the United States. The establishment of the actual system busy hour can be done through a simple analysis of the system traffic usage broken out by hour and day for several months to establish a comfort level that the time picked is valid. It is important to take a snapshot of what the system busy hour is on a regular monthly basis to verify that traffic patterns are not changing.

It is important to set yearly and, at a minimum, quarterly lost call goals. Setting overall and interval goals serves two primary purposes. The first purpose is that short-term goals help direct the efforts of the company. The second purpose is that interval goals ensure that the overall goal is being met, thereby avoiding an end-of-year surprise of not meeting the goal.

Figure 5.9 shows one way to set the lost call goal for the network. The lost call rate at the beginning of the year for this example is 2.2 percent, and the desired goal is 2 percent. The chart is divided into quarters, so a gradual improvement objective can be set. Please note the year on the chart and that if the trend line was to be followed, then calls would be awarded to subscribers by this time frame. The following examples are presented to help drive the methodology of establishing a lost call goal for a wireless system.

The first example involves a situation where the lost call rate was set based on what the overall corporate goal for the lost call rate was, in terms of a quality figure of merit. The lost call rate was 1.50 percent for the network. The network's current performance, however, was a 2.1 percent lost call rate. Obviously moving from a 2.1 to a 1.5 percent lost call rate in 1 year while experiencing a 40 percent growth rate may not be a realistic, obtainable goal. Achieving the final number immediately within the single management cycle would require over 28 percent reduction in the current lost call rate, regardless of the method chosen for calculation.

In establishing the lost call objective for the network, the goals may be unobtainable if additional resources are not allocated by management, for a sustained level of time, in order to facilitate the reduction effort.

For the example chosen management did not permit additional resources in terms of the work force or equipment to facilitate the reduction. The objective chosen was to do more with less, which is a common methodology used by many who have never done any of the work. Instead the value chosen was 1.9 percent which involved a 10 percent improvement as a minimum and an overall stretch of 1.8 percent for the final lost call rate. The final numbers were going to be the last month's lost call rate which was to be the total lost calls during the busy hours of all the days in the month.

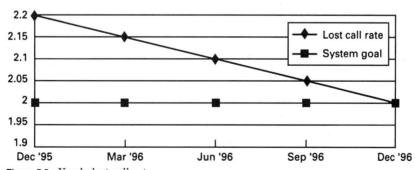

Figure 5.9 Yearly lost call rate.

Another example of setting goals for the department involves the upper management picking a value. A senior director picked a value for the lost call rate to commit to corporate for the next year's goals and objectives. The goal set was not discussed with any of the management of the engineering department nor with operations to receive their input as to the feasibility of the number being set. The end result was that there was a serious problem with meeting the number, and in fact the goal was altered during the midpoint of the year so it could be easily met. Surprisingly no one was held accountable for altering the performance numbers that determined the yearly bonuses. The revised number used was the one determined by the engineering department working level managers.

While it is important to set aggressive goals to work toward, it is also equally important to involve members of the staff whose job it is to ensure that the mission statement is met. In particular a meeting should have been established to arrive at a realistic value to submit to upper management. The setting of the value without subordinate involvement left the final goal as being the senior director's number and not the number that the department believed in, creating a fundamental friction point within the organization.

However, when you set a value, there is a midpoint between the examples discussed here. To arrive at a midpoint goal you need to know what the lost call rate trend is and determine what percentage improvement is needed for the coming year, factoring in the network subscribers, usage volume, and cell growth.

For example, setting a goal to reduce the lost call rate by 25 percent from 2 percent, i.e., a goal of 1.5 percent, is not realistic if you are growing the network at a 40 percent rate and allocate no additional resources to help in the effort. Instead the goal should be to maintain or improve the network by, say, a minor increment ensuring that the rapid growth rate does not negatively impact the network. This seems simple in concept, but the reality of determining this value is very difficult.

The ability to focus on an incremental improvement is based on the fundamental premise that the lost call rate has been successfully managed. If the lost call rate was say at 10 percent, regardless of calculation method, then focusing on a small improvement may prove futile, and in this instance the current resources, methodology, and skills used need to be revisited in a very hasty manner.

Obviously, arriving at the methodology for calculating the lost call rate is exceptionally important. In the process of determining the method to use it is equally imperative to arrive at the intervals to use also. For example do you use a monthly average, weekly average, or yearly average to determine the end result? We recommend using a quarterly average using the system busy hour, most likely 4 to 5 P.M. for weekday traffic.

This recommendation will eliminate weekly and monthly fluctuations in the lost call rate. It will also compensate for the first few months of the year when the lost call rate is normally lower due to demographic patterns that seem to always take place. The methods will also compensate for a bad cell site software load that is potentially put into the system.

We also recommend using the overall method for deterring the lost call rate of the network since this is the value the subscriber sees. In particular, if the subscriber hands off 3 times during a call and then experiences a lost call, the lost call rate is only 33 percent using the call segment method. But in the example presented the customer really experiences a lost call rate of 100 percent for that call.

After arriving at the time frame and methodology to use, the actual value will need to be set. The best method for setting the value to be arrived at is dependent upon what the current level of performance is. If your current overall lost call rate is 3 percent during the busy hour, then setting the goal of a 33 percent reduction in the lost call rate is reasonable. But if your lost call rate is 2 percent, currently setting a goal of reducing the rate by 33 percent for the coming year means a 1.34 percent level.

This brief discussion has only focused on the percentage method of determining the lost call rate. If you use the raw number of lost calls in a network, based on the system growth projection, then be very careful to look at what the relative performance improvements will be versus the resources available to combat this issue.

Specifically if you are currently operating at a lost call rate of 2 percent, but the raw number of lost calls during the busy hour is 200, reducing the raw lost calls by 10 percent from the initial number might be a larger percentage when you factor in the actual growth of the network.

At a 40 percent growth rate the O&T level would be 14,000 in the busy hour at the end of the next year, up from 10,000. Reducing the overall lost calls from, say, 200 to 180 would mean going from a 2 percent lost call rate to a 1.28 percent rate, which is a very aggressive goal. A more realistic goal would be to tie the lost call rate to the system growth level by having the raw lost call rate not exceed an overall percentage number but at the same time not increase, and the raw number not increase by more than 50 percent of the system growth rate or stay the same, netting a real improvement. Specifically the goal using this method should be a raw number of 240 as the minimum goal and 220 as the stretch goal to work for.

Whatever the value picked for use as the benchmark for determining the system health of the network, it is imperative that all the groups involved with achieving the objective help set the goals. However, you need to watch out for paralysis of the committee deciding the actual value. The end result is that there is no real magical solution to setting the value for lost calls. However, you must know what your objective is before trying to set the value, i.e., step 1 in the system performance and troubleshooting process.

Once you set your goals for the lost call rate for the system, the next issue you must face is how to identify the poor performers in the network. There are many techniques used for deterring the poor performers in the network, all of which have a certain level of success in trying to reduce the lost call rates.

One technique is to use the same report used for reporting the health of the network to your upper management and determining what sites are the poorest performers from this set of data. There are two fundamental ways to focus on problems when looking to rate the lost call numbers. The first method is to sort the list by poorest performance, determined by the percentage of lost calls reported on a per-sector or per-cell basis. This list will include the whole system, but you should focus on, at most, the top 10 site for targeting action plans. The second method is to focus on the raw number of lost calls. Both techniques should be used when determining which cell sites to focus attention on first.

For example, using the first percentage method, the poor performers will be identified regardless of the traffic load. The percentage method will help identify if there is a fundamental problem with the site. However, if the cell site has virtually no usage, say 10 calls and has one lost call, the percentage calculation is 10 percent indicating that there is a serious problem at this site. However, a site operating at a 1.9 percent lost call rate may be contributing 10 percent of the overall lost call rate to the network, but since it is such a high-volume cell, it is showing a lower overall percentage issue by itself. Obviously the focus of attention should be on the site contributing the largest volume of lost calls in this case, not the cell which has the highest percentage of lost calls.

The primary point with these two examples is that you must think in several dimensions when targeting poor-performing cell sites. The individual lost call rates should be looked at, plus the overall impact to the network as a whole needs to be addressed when focusing on what sites to deploy resources at.

Another successful technique used in lost call troubleshooting is to utilize another parameter. Use of the parameter of usage per number of lost calls on a per-sector and per-cell basis has met with great success in system troubleshooting. This parameter is exceptionally useful for identifying the worst performers in a network, regardless of the vendor or software loads used.

The parameter of usage per number of lost calls will also give you a figure of merit for determining the level of problems experienced at a site. For example, if you have five lost calls with 50 usage minutes, this equates to one lost call every ten min. When you are troubleshooting a system, the interval between the lost calls themselves is very important, since the shorter the interval the more problematic the problem is and the higher the probability of finding the root cause in a shorter period of time.

When looking at these three methods of lost call analysis, it is imperative that you try to find a pattern. Use the data in Fig. 5.10 for the sample system. The pattern search is best achieved through a three-step method. The first step is to sort the worst performer by lost call percent (Fig. 5.11). The second step involves generating a sort by raw lost call numbers (Fig. 5.12). The third step in the process is to sort by the usage/RF value (Fig. 5.13). You then take each of the worst 10 or 15 sites and put them onto a map or other visual method and look for a pattern as shown in Fig. 5.14.

More times than not there are several sites that focus on a cluster occurring in a given area. As shown in Fig. 5.14, the identification of the worst performers will in most instances show a pattern of an area that is experiencing a problem.

The key issue here is that you need to focus on a given area, besides the individual sites involved. The root cause of the problems could be as simple as a handoff table adjustment to a frequency plan problem. There are, of course,

Sample Busy Hour System Report						
Date:						
Cell Site	Time	Usage	O&T	LC %	# LC	Usage/LC
1A	1700	262	238	2.1	5	52.38
1B	1700	393	357	7	25	15.71
1C	1700	183	167	1.8	3	61.11
2A	1700	770	700	1	7	110.00
2B	1700	147	133	1.5	2	73.33
2C	1700	770	700	2	14	55.00
3A	1700	367	333	1.2	4	91.67
3B	1700	419	381	2.1	8	52.38
3C	1700	3438	3125	4	125	27.50
4A	1700	500	455	11	50	10.00
4B	1700	592	667	1.5	10	59.20
4C	1700	183	167	3	5	36.67

Figure 5.10 Lost call statistics.

Sample Busy Hour System Report						
Date:						
Cell Site	Time	Usage	O&T	LC %	# LC	Usage/LC
4A	1700	500	455	11	50	10.00
1B	1700	393	357	7	25	15.71
3C	1700	3438	3125	4	125	27.50
4C	1700	183	167	3	5	36.67
1A	1700	262	238	2.1	5	52.38
3B	1700	419	381	2.1	8	52.38
2C	1700	770	700	2	14	55.00
1C	1700	183	167	1.8	3	61.11
2B	1700	147	133	1.5	2	73.33
4B	1700	592	667	1.5	10	59.20
3A	1700	367	333	1.2	4	91.67
2A	1700	770	700	1	7	110.00

Figure 5.11 Lost calls sorted by percent lost calls.

situations where there is no real solution to the problem. However, there are always methods available to minimize the problem at hand. Sometimes the problem at hand may not be the site producing the poor statistics.

To illustrate the concept of one site causing another site to perform poorly, consider a system retune. Shortly after a system retune took place, a series of problems were reported at one site involved in the retune effort. Looking at Fig. 5.15, site 1 was reporting that it was producing a high volume of lost calls. Analysis of the data at hand indicated that this was the primary culprit to the problem. The drive team dispatched to investigate the situation confirmed that there was a major problem with the problematic site. Analysis of the cell parameter and neighbor lists of the site itself and the surrounding sites indicated no database or cell site parameter problems.

	Sample Busy Hour System Report					
Date:						
Cell Site	Time	Usage	O&T	LC%	# LC	Usage/LC
3C	1700	3438	3125	4	125	27.50
4A	1700	500	455	11	50	10.00
1B	1700	393	357	7	25	15.71
2C	1700	770	700	2	14	55.00
4B	1700	592	667	1.5	10	59.20
3B	1700	419	381	2.1	8	52.38
2A	1700	770	700	1	7	110.00
4C	1700	183	167	3	5	36.67
1A	1700	262	238	2.1	5	52.38
3A	1700	367	333	1.2	4	91.67
1C	1700	183	167	1.8	3	61.11
2B	1700	147	133	1.5	2	73.33

Figure 5.12 Lost calls sorted by number of lost calls.

	Sample Busy Hour System Report					
Date:						
Cell Site	Time	Usage	O&T	LC%	# LC	Usage/LC
4A	1700	500	455	11	50	10.00
1B	1700	393	357	7	25	15.71
3C	1700	3438	3125	4	125	27.50
4C	1700	183	167	3	5	36.67
3B	1700	419	381	2.1	8	52.38
1A	1700	262	238	2.1	5	52.38
2C	1700	770	700	2	14	55.00
4B	1700	592	667	1.5	10	59.20
1C	1700	183	167	1.8	3	61.11
2B	1700	147	133	1.5	2	73.33
3A	1700	367	333	1.2	4	91.67
2A	1700	770	700	1	7	110.00

Figure 5.13 Lost calls sorted by usage per lost call.

There was one aberration, the low usage on one sector of an adjacent cell site, number 2. The site technician was contacted and indicated that the problem being experienced was always there. However, a historic plot of the site, Fig. 5.16, showed a dramatic reduction of the usage on that sector over the same period that the problems appeared at the problematic site.

A site visit was performed by the site technician, who found that a receive antenna was physically disconnected at the antenna input on the tower. This problem was corrected with the aid of an antenna rigging crew which was immediately dispatched to the site, and the problematic site returned to normal operation and traffic levels resumed on the other site.

This example simply highlights that the problems at an area might not be as directly apparent when just using statistics. Adjacent sites performing

RF System Performance and Troubleshooting

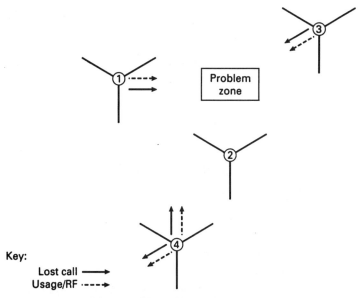

Figure 5.14 Visual display of lost call data from Fig. 5.9.

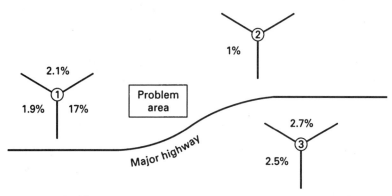

Figure 5.15 Problematic area.

poorly, either through low usage or parameter misadjustments, can directly affect the lost call rate.

Another successful technique for helping identify areas to focus on for system troubleshooting is to utilize input from customer service. Based on the sophistication that exists between the technical group and customer service the level and volume of information exchanged can range from great to sporadic.

It is recommended that customer service information regarding reported lost calls be identified and mapped to identify cluster areas. The identified problem areas should cross correlate with the data collected from the statistics. However, this customer service data should be used to check the data coming

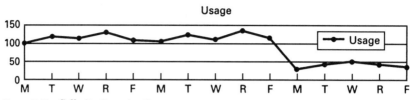

Figure 5.16 Cell site 2, sector 3 usage.

from the metric system and identify the volume of problems for a given area. For example, if 20 percent of customer complaints were coming from one major roadway and the static data did not corroborate the same story, this could be an indication that other problems are occurring in the network.

The basic questions to consider when identifying the worst performers in the network are

1. What changes were made to the network recently in that area?
2. Are there any customer complaints regarding this area?
3. Who are the reusers, what is the co-channel, and what is the adjacent channel for the area, two or three rings out? (Not applicable for CDMA)
4. What is the site's configuration, hardware, antennas, etc.?
5. Do the topology handoff tables indicate one-way handoffs, either into or out of the site?
6. Are there any unique settings in the cell site parameters?
7. Does an access failure problem also exist for the same area?
8. What is the signal-level distribution for the mobiles using the site?
9. Is there a maintenance problem with the site?
10. Does one individual radio cause most of the problems?
11. Is there a software problem associated with the cell site load?

If you use this list as a general reminder when initially looking at a poorly performing site, it will expedite your efforts. Most of the time the problems leap out at you, though, of course, at times they do not. It is when the problems are not obvious that it is imperative to utilize a checklist. There are many causes of lost calls in a network, and the effort to minimize or eliminate them is an ongoing process.

5.4 Access Failures

Access failures, also known as access denied levels and attempt failures, are another key metric to monitor and continuously work on improving. The attempt failure level is important to monitor and act on since it is directly related to revenue. Attempt failures can occur either as a result of poor coverage, maintenance problems, parameter settings, or software problems. Regardless of the exact cause for the attempt failure, when you deny a customer access on the network due to its received signal level, this is lost revenue. It is important to note that from the customer's aspect, whether the call or data session is denied because of access problems or because of radio congestion, the frustration level is the same.

The value and methodology used for setting and troubleshooting is very important for determining and improving the health of a network. Most vendors have a software parameter that can be set for establishing the actual received signal strength, or access threshold, level for denying a subscriber service on the network. One constant theme from the ranks of engineering is that the value needs to be set at a level that will ensure a good-quality call because subscribers would rather experience no service than marginal service. However, what constitutes marginal service versus good service is very subjective. We will argue that the ultimate litmus test for determining what the correct access level is can be derived by monitoring actual system usage and customer care complaints.

The other argument that constantly arises is that if the access parameter is set too low, this will increase the lost call rate of the network. Conceptually these two items appear to be directly related. However, in reality the parameter setting and the lost call rate are not as strongly related as initially believed. Specifically in a dense urban environment using three different vendors' equipment the lost call rate is not coupled to the access denied level on a one-to-one basis. In fact access values were set to almost correspond to the noise floor of the system with an improvement in the lost call rate still taking place at the same time.

It must be cautioned that just setting the threshold parameter to near the noise floor will not necessarily result in reduction in the lost call rate. If the access threshold settings are just changed with no other proactive action taken, the net result could easily be an increase in the lost call rate. However, as part of a dedicated program of system performance improvements the attempt failure rate can be successfully reduced at the same time the lost call rate is being reduced.

There are several methods for measuring the attempt failure level in a network. Two of the methods used are identified here:

Method 1: $\quad$ Attempt failures $= \dfrac{\text{no. of access denied}}{\text{total seizures}} \times 100$

Method 2:

$$\text{Attempt failures} = \dfrac{\text{no. of access denied} - \text{directed retry}}{\text{total seizures} - \text{directed retries}} \times 100$$

Using some simplistic numbers, a simple comparison of these two methods for calculating attempt failures can be accomplished.

$$\text{No. of attempt failures} = 1000$$
$$\text{No. of directed retries} = 500$$
$$\text{Total seizures} = 50{,}000$$

Method 1:   Attempt failures $= \dfrac{1000}{50{,}000} \times 100 = 2\%$

Method 2:   Attempt failures $= \dfrac{1000 - 500}{50{,}000 - 500} \times 100 = 1.01\%$

The first method of calculating the access failure rate for the network will indicate the true level of problems associated with access issues. However, if you utilize method 2 for your attempted failure calculation, the actual level of problems being experienced could easily be misleading.

Another comment about method number 2 pertains to the directed retry value reported by the network. Specifically the directed retry will cause another attempt failure at the new target cell site and also add to the total seizure count for the network. The use of directed retry for any reason has to be tightly controlled and monitored to ensure adverse system performance does not take place. The reason the directed retry parameter needs to be tightly controlled is that using it can easily mask system problems. The primary lesson you can derive from this simplistic example is that once again it is exceptionally important to understand the equation that is being used in the metric calculation.

The next logical issue concerns how you define the system goals and how you monitor your progress and report it to upper management. Like the lost call rate there are several methods from which to calculate attempt failures. However, it is important to keep in mind that it is important to have any system performance report reflect what the subscriber experiences.

Two methods were shown for calculating the attempt failure levels in a network. Method 1 is referred to as the overall method for determining the access failures on a network. Method 2 is referred to as a diluted method for determining access failures onto a network.

Whether you use the overall method or the diluted method, it is important to be consistent. The relative health of the network can be determined by either method using simple trending models that compare past and present performances. However, the diluted method will mask actual problems in the network. If your goal is to improve the overall quality of the network and the revenue potential for the company, the overall method is best.

Several equipment manufacturers have the ability to report attempt failures on a per-sector basis. An obvious issue is that many vendors have each sector populated as a separate cell. Another issue is determining the proper level of granularity to utilize for monitoring this important parameter. Obviously the goal ultimately desired is to have no denied access to the network as a result of signal strength levels received. However, until there are no coverage or double-access problems in a network, an interim value must be used as part of the path for improving the network.

A similar approach should be used for setting the value for access failures as that used for the lost call value method discussed previously. The time frame used for the attempt failure rate should be identical for the lost call metrics because you should use these parameters together as a method for qualifying the site's overall performance. Also, as for the lost call goals, it is important to set yearly and quarterly access failure goals and to have all the groups involved with achieving the objective help set the goals.

Now that you have determined the access failure level to use for the system, the next step is the same as for the lost call goals: You need to identify the poor performers in the network. The methods used are the same as that for the lost calls, percentage and raw numbers. For example, using the percentage number method, the poor performers will be identified, regardless of the traffic load. Using the percentage method will help identify if there is a fundamental problem with the site for attempt failures. However, if the site is co-setup, co-DCC with another site, the problem might not be with the site itself but with the actual frequency assignment for the setup channel.

With a co-setup, co-DCC situation an attempt failure will be recorded by the system for the call being placed, but the subscriber will gain access to the network. The system will either report an attempt failure due to the signal level being inadequate or a facility assignment failure due to the mobile not arriving on the target channel. This situation is best represented in a series of diagrams.

The mobile is near cell site 7 in Fig. 5.17 and originates a call on the network. The mobile scans all the setup channels and determines which cell it is receiving the most forward energy on. The mobile selects the strongest channel and responds with a request to access the system on the reverse control channel, 321 for this example (see Fig. 5.18).

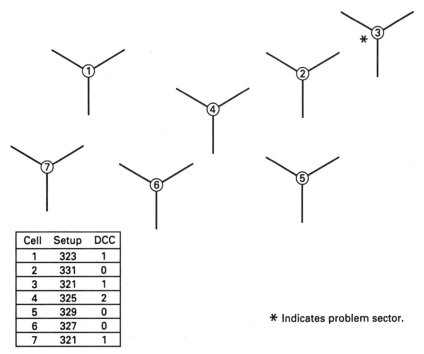

Cell	Setup	DCC
1	323	1
2	331	0
3	321	1
4	325	2
5	329	0
6	327	0
7	321	1

✶ Indicates problem sector.

Figure 5.17 Access failure example.

296 Chapter Five

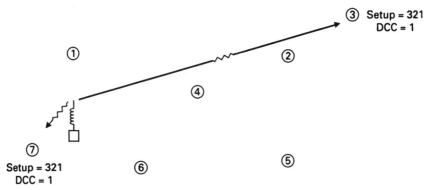

Figure 5.18 Access failure example.

When the subscriber unit responds on the reverse control channel, cell 7 evaluates the mobile for access on the network as part of its call-processing algorithm. However, cell 3 is using the same setup channel and DCC as cell 7 is. The mobile's signal, which is sent on the reverse control channel 321, is not only detected by cell 7 but also by cell 3.

In this situation cell 3 will attempt to assign the mobile a voice channel at the same time cell 7 is doing the same. Depending on the system configuration and software, a channel could be assigned or an attempt failure would be recorded at cell 3. If cell 3 tries to assign the subscriber a voice channel, it will ultimately record a facilities failure for this assignment since the subscriber will have really arrived on cell 7.

In this example the subscriber will have been assigned channels at several potential cell sites at the same time but will successfully arrive at the closer site and be assigned a channel, while the problem site identified is reporting problems, but this does not really impact on the subscriber. This situation described could be simply resolved by assigning a different DCC value, possibly DCC = 2, for site 10.

An example of another attempt failure problem occurred due to a serious imbalance in the link budget for a site. The effective radiated power (ERP) for the site was incorrectly specified by engineering for the location. In particular the ERP for the site was in excess of 200-W ERP when the site should have been operating at a maximum of 50-W ERP. The disparity in signal level was driven by the miscalculation of feedline loss. The operations department set the value for transmit power according to the engineering department's recommended value. The real ERP from the site set up a major disparity in talk-out versus talk-back paths. The additional 6 dB of talk-out power than the site was designed for resulted in mobiles originating on the site outside of the receive path's radius. The imbalance in talk-out versus talk-back paths resulted in numerous attempt failures due to the sign levels received at the site. The situation was corrected after a site visit to the area with the technician when the feedline miscalculation was noticed.

Access failures can also point to a bad receive antenna where the access failure levels increase right after rainstorms and heavy moisture. Since all the sites utilize switching or maximum ratio combining for diversity receive, a bad leg of the receive path can adversely affect the access failure levels of a site.

These examples point out again that just altering a parameter by itself is not necessarily the solution. There are many aspects that can cause an access failure, and it is only through isolation of the variables that the real problem can be uncovered.

When looking to help minimize the access failure levels in a network, it is imperative that you look for a pattern. The pattern search is best achieved through a two-step method. The first step is a pattern search that involves sorting the worst performer by access failure percentage number. The next step in the pattern search involves sorting the list by the highest raw access failure. You then take each of the worst 10 or 15 sites and put them on a map or other visual method and look for a pattern.

The basic questions to consider when identifying the worst performers in the network are

1. Does a lost call rate problem also exist for the same area?
2. What are the co-setup, co-DCC sites [digital control channel (DCCH), pilot pollution, etc.]?
3. What is the signal level distribution for the originating signals on the site?
4. Are there any customer complaints regarding this area?
5. Is there a maintenance problem with the site?
6. Does one individual radio cause most of the problems?
7. Is there a software problem associated with the cell site load?
8. Were most of the problems caused by one mobile or a class of mobiles?

One of the major problems associated with attempt failures is the lack of a dominant server for the area. This is often referred to as a random origination location. The lack of a dominant server for CDMA systems is referred to as *pilot pollution*. The primary problem with random originations is that there is no one setup, or control, channel that dominates the area where the problem occurs. The lack of a dominant server leads to mobiles originating on distant cells creating interference problems in both directions, uplink and downlink.

The uplink problem occurs since the mobile is in the wrong geographic area and is now in a position to spew interference into a reusing cell site's receiver. The downlink problem occurs simply because the frequency set designed for the area is not being used, or potentially not being used. The downlink problem is more pervasive since the subscriber unit is the victim due to the transmit power reusing cell site.

One key concept to keep in mind is that any reduction or increase in ERP has a dramatic impact on a cell site's coverage. While the concept of the relationship between ERP and the cell's coverage area seems trivial and obvious, the reality behind it is staggering. For example, reducing a cell's ERP by 3 dB will significantly reduce the overall coverage area involved. The relationship is easier to follow if you look at the equation for the area of a cell.

$$\text{Area of cell} = \pi R^2 \tag{5.1}$$

where R = radius of cell. This concept eluded two design engineers.

When you adjust the ERP of a cell, the overall area it serves is significantly altered. The usual attention placed on ERP is the forward portion of the link budget, i.e., base to mobile. However, the reverse link needs to also be accounted for since a significant imbalance can easily impact the system's performance in a negative fashion.

The most important point of this is that when you reduce the coverage of one cell a new dominant server is created, expanding the coverage area of another. Using the single issue design concept could lead to more attempt failures and lost calls through focusing only on an individual cell site and not on the system.

5.5 Radio Blocking (Congestion)

Another key element in system performance and troubleshooting involves radio blocking levels of the system. Radio blocking has a direct impact on the system's performance from a revenue and service quality aspect. Determining just what are the appropriate blocking levels for a network has been a subject of many debates. Radio blocking affects both voice and data services.

Depending on the fundamental design for the wireless system, when introduced, the data services may need to share existing radio facilities with the voice network or may occupy their own unique spectrum and facilities. Presently there is no system that will, for example, have IS-136 operational and then allow a subscriber who is denied access due to radio blocking to be passed onto the GSM network, and vice versa. This ability is available, however, for CDMA systems and even between PCS and cellular bands. It is a matter of time before the introduction of multitechnology customer premise equipment (CPE) is prevalent. Until that time radio blocking will be treated separately for each technology base.

It is also very important to address the QOS aspect with wireless data. Regardless of pundits' comments, wireless data for mobility cannot replace the digital subscriber line (DSL) or cable always-on service. The reason for this bold statement is the simple fact that wireless mobile data are contention-based, i.e., subscribers need to negotiate a link every time they initiate services.

When addressing radio blocking itself, there are primarily three schools of thought. The first philosophy is that any blocking in a network is too much and will result in lost revenue. The second philosophy is that a system should operate at a 2 percent blocking level on the macrolevel. The third philosophy is that the network should be operated within a band, or range, of blocking and, as much as possible, kept within that band.

The first philosophy has its merit for when you want to ensure the most amount of network capacity at a given time. However, this philosophy will lead to overprovisioning of the network in terms of infrastructure equipment, radios, and facilities. The end result is that the inherent operating costs to the network have been substantially increased.

One serious downside to this approach, besides the inherent costs associated with the method, is the impact it has on frequency planning and other associated resources. One of the key concepts with frequency planning is to minimize the amount of reusers in a given area; this also applies to adjacent channels and to CDMA since the level of mutual interference increases. If you constantly overprovision the network, the fundamental interference levels will naturally increase since you are using more channels. While there are many techniques for controlling interference, the overprovisioning of channels only makes frequency management more difficult.

One additional comment on this approach is that if you desire no blocking on the network, this effort involves more than just the RF portion; it will be necessary to modify the public switched telephone network (PSTN) blocking level design. For example, if you have a PSTN blocking level of 1 percent, it will be very difficult to ensure a no-blocking-level approach. When establishing a radio blocking level for the network to operate at, a complete evaluation of all the components in the call-processing chain needs to be factored into the solution. Additionally with the introduction of data services the packet data serving node (PDSN) path, whether it is for connectivity to the Internet or a virtual private network (VPN), needs to have sufficient resources to meet the service-level agreement (SLA) with the customer.

The second school of thought is the more common method employed where a top-end number is used. The top-end number is a ceiling which is the level not to be exceeded. This method is a valid approach when initially deploying a system since the system is experiencing massive growth and expansion. It is also effective when there is very limited work force to monitor and adjust a system's blocking level.

The ceiling approach has been successfully used by many operators for designing a network since it is simple and straightforward. The only real issues with the ceiling-level approach is defining the blocking levels, which equations to use, the utilization rate, and the time intervals. The most common blocking level used for designing is the 2 percent blocking level for the busy hour. The other particulars associated with the ceiling-level approach which need to be focused on will be discussed later.

The third approach (banding) to blocking level designs is more relevant to a mature system and requires constant monitoring and adjustments. It is a very efficient way to utilize a network's resources. The banding method involves setting top-end and lower-end blocking levels to operate at. It is customary to set the upper range at 2 percent and the lower range at 1 percent (Fig. 5.19).

Based on the traffic and growth of the network the adjustments needed for the network might seem daunting. However, the banding method can and has been used to keep the amount of fluctuations in network provisioning requirements

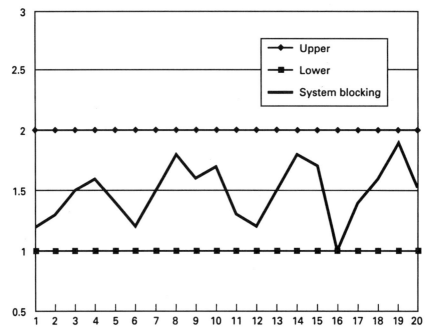

Figure 5.19 Banding method.

to a minimum. Implementing the banding approach will require a dedicated and focused approach to keep the system operating within the limits set.

The banding approach will also focus on removing channels from cell sites that are not exhibiting high usage. The advantage with removing unneeded radio channels is that this will make additional channels available for reuse or will simply reduce the aggregate interference levels in the network.

Regardless of the method utilized, ceiling or banding, for setting the blocking level of the network there are several similar methods of operation that need to be adhered to. The items to focus on for keeping system blocking levels within some design guidelines are which equations, utilization rate, time intervals, and logistics management are used.

Blocking calculations take on several forms; however, there has and will be a constant running debate between whether to use erlang B or Poisson for voice services and erlang B or C for mobile data. The equations for erlang B and Poisson are listed below for reference. Erlang B is the normal industry standard and should be applied for determining the trigger points to act from for voice services.

Erlang C is used for data services, e.g., packet, and PSTN facility dimensioning. However, erlang C does not lend itself to spreadsheet entry, i.e., a straightforward equation like erlang B. With that said, it is possible to use erlang B to approximate erlang C; e.g., erlang B 5 percent is about the same as erlang C 2 percent. The relationship between erlang B and erlang C is not linear, and you should consider whether you want to use this approach.

A key concept in all traffic engineering is the definition of an erlang. The *erlang* is a dimensionless unit since the numerator and denominator are of the same units and thus cancel out. In the wireless industry an erlang is simply defined as 3600 s of usage in a 1-h time period. Therefore, if you have two radios and each is occupied for 1 h, 3600 s, then this represents 2 erlangs of usage.

The erlang B equation is as follows:

$$\text{Grade of service} = \sum_{n=0}^{n} \frac{E^n}{n!} \quad (5.2)$$

where E = erlang traffic
n = number of trunks (voice channels) in group

There are numerous books and technical articles regarding erlang B's statistical nature and accuracy. However, the primary driving element with erlang B is that it estimates, statistically, the probability that all circuits will be busy when you try and make a call. The issue with the erlang B calculation is that it assumes the call, if not assigned a circuit, is lost from the network permanently.

The Poisson equation

$$\text{Grade of service} = e^{-a} \sum_{n=c}^{\infty} \frac{a^n}{n!} \quad (5.3)$$

where a = erlang traffic
c = required number of servers
n = index of number of arriving cells

should be compared against the erlang B value arrived at for the same amount of circuits and grade of service. The fundamental difference between erlang B and Poisson is that in Poisson you are assuming blocked calls get put into a queue instead of being discarded.

Obviously neither erlang B nor Poisson represents the real world as it pertains to wireless since neither are exact models for subscriber and traffic patterns in wireless. However, erlang B is used as the predominant statistical method for traffic calculations and growth predictions for voice and many packet services. Using erlang B as the foundation for projecting radio blocking values is the recommended method to use, but the grade of service (GOS) needs to be company and market specific.

It is very important to know what method of blocking statistics your company is using when projecting radio growth for a network. The reason is that knowing which method and equation to use has lead to some interesting discoveries on how blocking levels are determined. Two primary situations occurred which seriously affected the capital deployment of the networks involved due to homemade equations being promoted as erlang B.

In one situation the operator was using a traffic table called erlang B. The traffic table was used for defining how many radios needed to be added into a given cell site. A problem was uncovered when putting together a growth study and with the desire by engineering to use a spreadsheet for calculating channel growth requirements. Since the traffic table was touted as erlang B, a spreadsheet was structured using the erlang B equation (5.2). The error in the traffic table was uncovered when spreadsheet values were compared to the traffic table provided. The net result was that overprovisioning of radios was being done with the supplied traffic table. After many internal debates, the incorrect table was used since it inflated the system requirements to upper management for budget purposes.

The second situation occurred where another operator was using a table, again call erlang B, that was crafted by a traffic engineer. The current traffic engineer when challenged over the table inconsistencies was unaware of what the erlang B equation was and was simply using the traffic table provided. In investigating this issue it was found that there was a deliberate attempt by the engineering manager to overprovision equipment at a cell site to account for a maintenance buffer. The correct erlang B equation was then used by traffic engineering, and the amount of radios needed for the network was significantly reduced.

The primary point with these two examples is that it is very important to know the baseline equation being used for provisioning the network. An incorrect equation can lead to over- or under-provisioning of the network.

The utilization rate for the radio equipment is another key parameter that needs to be focused on and defined. The utilization rate of the radio's channels for a cell site's sector determines at what point in the process it will be necessary to start provisioning for expansion. When you are at 80 percent utilization rate for a sector, it is time to ensure that additional radios are being planned for the site. When you are at 100 percent utilization rate for a cell, based on the derived blocking level used, the turnaround time for restoration is too short.

A low utilization rate is indicative of overprovisioning, and there should be a lower limit placed on this value to keep the system from being underutilized. The utilization rate should be used as a trigger point. The rate picked by you is entirely dependent upon your internal work force and financial resources.

Once you have arrived at a utilization rate to operate from, you then perform a trend analysis. The trend analysis needs to focus on the next period of growth with the purpose of determining the amount of radios you will require.

The next step in the process is to define the time used for monitoring the network. The same system busy hour used for reporting lost calls and attempt failures should be used. However, it is important that a bouncing busy hour traffic report gets generated on a biweekly basis to ensure that there is not a fundamental shift in the traffic pattern. The bouncing busy hour report will also help identify fraudulent usage on the network or unique demographics that can help marketing.

This leads to the question of how to go about provisioning the network. The step-by-step generic process is listed here. For this process we will assume the banding method uses a 1 to 2 percent erlang B blocking level range. In addition the system busy hour and an 80 percent utilization rate are the two other key parameters. The process presented here can either be led by the manager for network or performance engineering.

On a 6-month basis

1. Determine, based on growth levels, the system needs for 1, 3, 6, and 12 months.
2. Factor into the process expected sites available from the build program, accounting for deloading issues.
3. Establish the number of radios needed to be added or removed from each sector in the network.
4. Modify the quarterly plan that is already issued.
5. Determine the facilities and equipment bays needed to support channel expansions.
6. Inform the frequency planners, performance engineers, equipment engineers, and operations of the requirements.
7. Issue a tracking report showing the status of the sites requiring action. This report should have in it as a minimum:
   - Radios currently at the site
   - Net change in radios
   - When radio exhaustion, percent blocking meeting design goals, is expected
   - When other radio element exhaustions [e.g., channel elements (CE) for CDMA] are expected
   - Radios ordered (if needed)
   - Facilities ordered (if required)
   - Radios and other related resources secured
   - Frequency issues (technology specific)
   - Cell site translations completed
   - Facilities secured (if needed)
   - Radios installed or removed
   - Activation date planned for radios
8. On a quarterly basis conduct a brief 1-h meeting to discuss the provisioning requirements and arrange for the work force.
9. Perform a biweekly traffic analysis report to validate the quarterly plan and issue the tracking status report at the same time.

The last step in the tracking system process involves the activation date planned for the radios. This particular piece will enable you to pre-position the radios in the network without actually activating them. Specifically if the radios are needed for the midpoint of March and then are installed in January, you can keep the interference levels to a minimum by having them remain out of service.

Blocking problems in a network can occur as a result of a multitude of reasons ranging from normal network growth, a localized traffic jam, changes in traffic patterns due to construction detours, hardware problems, or software problems.

5.6 Technology-Specific Troubleshooting Guides

In this section are several troubleshooting guidelines which are meant to help facilitate the identification and resolution of performance problems in a wireless network. Obviously there are numerous types of problems that can and do arise in a wireless system. However, procedures put forth should net some benefits. As with any technology, the guidelines may need to be modified to reflect vendor- and network-specific issues.

The guidelines here do not address multiple-technology platforms within any market because of the magnitude of perturbations and the simple fact that each technology platform, within a wireless network, will be configured differently starting with parameter settings.

Additionally there are some common troubleshooting issues that are relevant regardless of the technology platform used. The issues are simple in nature, but how they manifest themselves in a system is unique based on the underlying technology and, of course, infrastructure vendor and software load currently being used.

1. Neighbor list problems
2. Coverage problems
3. Frequency plan problems
4. Capacity problems
5. Base station controller (BSC) and mobile switching center (MSC) boundary problems
6. Site-specific parameters and configuration problems
7. Maintenance and equipment outage problems

Any performance engineer will be familiar with these topics. Therefore, when investigating and ultimately resolving any technical performance problem, the fundamental seven universal issues should always be considered.

With this said, we have attempted to define some of the more specific trouble-shooting issues in the next sections. As one would expect, the checklists are not all-inclusive but should prove useful.

5.7 IS-136

The troubleshooting guidelines that should be followed for any IS-136 system need to factor in whether the system uses IS-136 channels only, e.g., a U.S. personal communication service (PCS) system, or is a cellular operator that has AMPS and IS-136 channels in operation. With the advent to third-generation (3G) services most IS-136 operators at this time have chosen to migrate some or all of their systems to GSM, at least the portion of the system which is in the PCS band.

The methodology set forth in this section will focus on using the seven main performance issues. Throughout this section for troubleshooting it will be assumed, to avoid too much redundancy, that the seven-step performance troubleshooting process, which was discussed at the beginning of this chapter, is followed.

The main problems that will be covered for IS-136 include

1. Lost calls
2. Static
3. Access
4. Handoff failures

Since these four main performance issues are the more pertinent to any operation and are evident to the customer, they will be focused on. The items do not necessarily need to be checked in this order.

5.7.1 Lost calls

Typically a lost call for an IS-136 system occurs when there is a digital voice color code (DVCC) timeout. This happens when a call that is on a digital traffic channel loses the DVCC. The loss of the DVCC can take place at the mobile or cell site, and the tolerable amount of time for the loss of the DVCC is set by the wireless operator. Typically this value is a default value which is established by the infrastructure vendor.

The three main causes of lost calls for IS-136 are

1. Coverage problems
2. Interference, both cochannel and adjacent channel
3. Neighbor list problems

When reviewing data, if the site is experiencing a high lost call rate, or any lost call rate which requires focusing resources on, the following items need to be checked:

1. Determine what changes were made to the network recently in that area (if the area worked well before the changes).

2. Verify that there are no coverage problems with the prediction and field data. You need to verify that the prediction and field data (if available) confirm or deny that there is a problem. If there is a coverage problem, then there exists the distinct possibility that this may be the leading, but not the only, source of the problem. Coverage problems could be caused by terrain or the existing system configuration. Sometimes the adjacent site configuration is the cause or can be altered to resolve the issue.

Many of the infrastructure vendors have internal diagnostics available which are able to plot the nominal signal level, the reverse signal strength indicator (RSSI), for the subscriber unit from the cell site's aspect at a minimum. What you want to check is that there is sufficient C/N to ensure the call is sustainable for 90 to 95 percent of the reported signals.

3. *Review the neighbor list for the primary site.* Check the site's neighbor list to verify that there are no missing or incorrect neighbors, and check and verify that there is not an excessive number of neighbors. You should also check for reciprocity for the neighbor list, but this will not manifest itself in a high lost call rate for this site, only the neighboring site which may not be reciprocal. Also check the hysteresis value for the handoff. At times the hysteresis value could be set to preclude the handoff from taking place.

Most vendors have built into their operations and maintenance center (OMC), or like product, the ability to look for one-way handoffs and to define the amount of handoffs that occur on a per-candidate basis. For example, handing off from cell A to B might occur 60 percent of the time, while from cell A to C it might occur 5 percent of the time. What you want to look for is the proximity of the sites and if the candidate selection process makes sense from your experience. Keep in mind that time of day plays a major role in mobility issues, but stationary coverage, e.g., building, may be worth checking. For the stationary situation the frequency of handoffs may be an indication of a ping-pong effect and in this case attention to hysteresis may be appropriate.

4. *Check the frequency plan.* You need to review the frequency plan that is in place for the site and sector involved. Things to look for involve cochannel and adjacent channel interference, including when the adjacent channel is in the same or adjacent sector or when the cell with that sector overlaps the problem sector. What you want to do is check the frequency plan on a per-sector and per-channel basis for any obvious issues with cochannel reusers. If there is an interference issue, it might be resolved temporarily by placing the offending channel out of service, provided there is sufficient capacity to warrant this approach. However, depending on the severity of the problem, rendering a channel out of service can cause additional problems at the sites adjacent to the site with the channel taken out of service.

5. *Check traffic loading.* Next review the traffic load at the surrounding sites and ensure that each site has sufficient capacity so that a mobile can hand off to it. If the neighbor site is blocking, this could be a cause of the lost call rate. Also directed retry may be activated resulting in subscriber units (SUs) operating outside of their designed coverage area. Things to look for involve the amount of DTCs in use and where any of the messaging channels have to discard packets due to loading issues. Also the amount of measurement requests initiated by the cell site need to be reviewed to determine if the frequency of them could be reduced.

6. *Check cell site parameters.* This part of the check involves reviewing the site's specific configuration and software settings. Things to check involve topics beyond the neighbor list and possibly the ERP for the site. For example, you need to review the power control settings for the site and existing software load. For any of the parameters checked relative to the site or sector you need to compare them to the design standard which is used in the system. If the parameters are not within nominal range, then they need to be checked to ensure that they are not the root source of the problem.

One mentioned item to look for besides the dynamic power control (DPC) and fade time is the digital mobile attenuation code (DMAC) value for the sector. Additionally the site's software configuration needs to be checked, i.e., the channels and DVCC associated with each channel. In other words check the datafill for anything out of the norm.

7. *Check the MSC borders.* Verify the location of the MSC borders and ensure that high traffic areas are not on an MSC border. If they are, then the source of the problem may be the border itself where due to the time required for the handoff process to occur, the SU dropped because the fade timer initiated the premature termination.

8. *Check for maintenance issues.* When checking for maintenance issues, you need to determine if the problem can be isolated to a single channel. If it can, then all DTCs will be affected. You will also want to verify that there were no T1/E1 outages for the site at that time or the adjacent sites since this would negatively affect the performance of the site. One other item to look for is the possibility of an antenna system problem, which would warrant a site visit, if no other problems could be found. Some vendors have the ability to monitor the antenna system for voltage standing-wave ratio (VSWR) which would indicate whether a problem might exist or not. Also if historic antenna data are available, it might be worth it to check and see if there is any degradation and if the problems appear more prevalent after a rainstorm indicating water intrusion.

5.7.2 Handoff failures

Handoff failures should not be an issue with IS-136 with mobile assisted handoff (MAHO). However, there are still handoff failures which occur. The following is a brief listing of items to check when investigating a high handoff failure rate for a site or sector.

1. *Review the neighbor list for the primary site.* See part 3 of the list in Sec. 5.7.1. You should check for reciprocity for the neighbor list, but this will not manifest itself in a high handoff failure rate for this site, only the neighboring site which may not be reciprocal. In addition, what you want to check, besides the actual neighbor list, is the time between handoffs. There is a systemwide number which you should use as the benchmark for performance. If the time between handoffs is very small, as compared to the norm, then either the cell site is shedding traffic or it may not be the correct candidate to use from the beginning. However, coverage and terrain issues may justify a low time to handoff ratio.

2. *Check cell site parameters.* This part of the check involves reviewing the site's specific configuration and software settings. Things to check involve ERP imbalances for the site where the DCCH is operating at one power level and the other channels are operating at a lower power level, leading to possible handoff failures, and, of course, dropped calls. You should also review the power control settings for the site and existing software load. You need to compare any of the parameters checked relative to the site or sector to the design

standard which is used in the system. If the parameters are not within nominal range, then they need to be checked to ensure that they are not the root source of the problem. This check should also be done at the potential source sites to ensure that they are set properly. Again for this part of the investigation all aspects of the datafill need to be looked at for both the source and target cells and sectors.

3. *Check the MSC borders.* See part 7 of the list in Sec. 5.7.1.

4. *Check for maintenance issues.* See part 8 of the list in Sec. 5.7.1. If historic antenna data are available, it might be worth it to check and see if there is any degradation and if the handoff failures appear more prevalent after a rainstorm indicating water intrusion.

5.7.3 All servers busy (ASB)

The problems with access to the wireless system involve both blocking, i.e., resources not available, and an insufficient signal. In the situation where resources are not sufficient to address the current traffic, the obvious response would be to just simply add radio channels. However, the traffic increase may be caused by incorrect parameter settings or a deliberate attempt to reallocate traffic. There are several items to look for in verifying that indeed you need to add channels or possibly a cell site to relieve the traffic.

1. *Determine the source of the blockage.* It is important to know the source of the blockage since it will tell you how to possibly relieve the situation both on short-term and long-term bases. You will need to determine the volume of traffic type for the cell or sector and if it is origination based or handoff based. For instance, a normally balanced cell for a typical mobile system will have one-third originations and two-thirds hand-ins; your numbers may vary, but the illustration is important. If this cell were to have two-thirds originations and one-third hand-ins, you might indicate that the blockage problem is near the site itself. If the majority of the traffic was hand-in, then the source of the traffic would be from an adjacent site.

The source is important in determining the possible solutions. For traffic originating at the site you might want to relax the MAHO hysteresis for target cells and hand off more traffic shedding the load. Or you might want to increase the hysteresis for this site since it is the target cell, reducing incoming traffic if the problem is inbound. You should also check for a directed retry situation where the cell site is the target of the directed retry.

2. *Check for maintenance issues.* When checking for maintenance issues, you need to determine if the problem can be isolated to a single channel. If it can, then all DTCs will be affected. You will also want to verify that there were no T1/E1 outages for the site at that time or the adjacent sites since this would negatively affect the performance of the site.

3. *Check the cell overlap.* If there is sufficient cell overlap, you might want to visit the issues of shedding traffic, origination, and possibly handoff, by reducing the site's ERP. Additionally or separately the DMAC values can be

reduced, provided, of course, that the adjacent sites have sufficient capacity and the frequency plan supports the configuration change.

5.7.4 Insufficient signal strength (IS)

Insufficient signal strength is caused by either lack of coverage or parameter manipulation by the operator. Therefore, there are a few items to check.

1. *Verify there are no coverage problems with the prediction and field data.* You need to verify that the prediction and field data (if available) confirm or deny that there is a problem. If there is a coverage problem, then there exists the distinct possibility that this may be the leading, but not the only, source of the insufficient signal level. Coverage problems could be caused by terrain or the existing system configuration. Sometimes the adjacent site configuration is the cause or can be altered to resolve the issue.

Many of the infrastructure vendors have internal diagnostics available which are able to plot the nominal signal level, RSSI, for the subscriber units which access the system. This is an important diagnostic tool to use, if available. For instance, if the IS percent goal is less than 2 percent but the diagnostics indicate that 10 percent of the SUs that access the site are below the required RSSI, prior to DPC adjustment, then this leads you to check coverage or to the possibility that the site is a directed retry candidate.

2. *Verify the dominant server.* You will need to check if there is an area the cell covers that has no dominant server. The lack of a dominant server may either lead to originations occurring on the wrong cell and being denied due the IS level setting at the potential target cell or manifest in high lost calls and handoff failures.

3. *Check cell site parameters.* See part 2 of the list in Sec. 5.7.2. In addition, the particular IS parameter setting needs to be checked, this is often the source of the problem if there is sufficient coverage for the area.

4. *Check the MSC borders.* Verify the location of the MSC borders and ensure that double originations are not taking place resulting in glare or denials due to double access caused by the location of the MSC border. If they are, then the source of the problem may be associated with the DCCH assignment methodology used.

5. *Check for maintenance issues.* When checking for maintenance issues, you need to determine if the problem can be isolated to a single channel; if it can, then all DTCs will be affected. IS problems will not manifest themselves with a T1/E1 outage for the site but will be caused by a possible bad antenna system.

5.7.5 Static

The need for an investigation of static for a cell will not necessarily appear on the statistics but will in all likelihood be born from customer service or internal reports. However, static is usually a result of poor voice quality which is a result of reduced BER and FER.

Identifying the source of the static problem is a combination of reviewing coverage, handoff, and possible access problems to the network.

1. *Determine what changes were made to the network recently in that area (if the area worked well before the changes were made).*
2. *Verify that there are no coverage problems (prediction and field data).* See part 2 of the list in Sec. 5.7.1.
3. *Review the neighbor list for the primary site.* Check the site's neighbor list to verify that there are no missing or incorrect neighbors and check and verify that there is not an excessive number of neighbors. You should also check for reciprocity for the neighbor list. In addition, check the hysteresis value for the handoff. At times the hysteresis value could be set to preclude the handoff from taking place.
4. *Check the frequency plan.* See part 4 of the list in Sec. 5.7.1.
5. *Check traffic loading.* See part 5 of the list in Sec. 5.7.1.
6. *Check cell site parameters.* See part 6 of the list in Sec. 5.7.1.
7. *Check for maintenance issues.* See part 8 of the list in Sec. 5.7.1.

5.8 iDEN

An iDEN system is unique in the RF performance environment since it is really two systems within one. There is an interconnect and a dispatch system which share some similar resources. For example, if the interconnect to a particular tandem is not dimensioned correctly, i.e., it has too few circuits, then the customer can experience blocking. The customer is not aware of where the problem occurs, only that it does occur. Another example is if the registration borders for both the interconnection location area (ILA) and dispatch location area (DLA) are not properly designed. This could lead to either an increase in PCCH activity or the potential for a subscriber to experience a registration problem.

Motorola has a wealth of specific troubleshooting and performance-related guidelines for the iDEN system. These guidelines should be used since the list of issues that have a direct impact on the system performance of a network is vast. In addition, as more features and equipment are implemented into the network, the complexity of the situation only increases.

We cannot realistically cover every possible performance situation that might occur in an iDEN system. However, we will attempt to list some of the more salient issues following the seven-step performance troubleshooting process which was discussed at the beginning of this chapter.

For reference it is important to note that there are 3:1 and 6:1 or 12:1 interconnect and voice calls, as well the dispatch which uses the 1:6 process in addition to any packet service. At this writing there are several enhancements being pursued for iDEN which should greatly improve its traffic-handling capability while not sacrificing voice quality.

5.8.1 Lost calls

Typically a lost call for an iDEN system can be traced to one of the following three culprits (assuming there is no maintenance issue which can also exacerbate the problem):

1. Coverage problems
2. Both cochannel and adjacent channel interference, either self-induced through reuse or from legacy specialized mobile radio (SMR) users.
3. Neighbor list problems

It is important to note that lost calls are not necessarily associated with dispatch calls because the very nature of dispatch involves push to talk (PTT) service and is by default short in duration. Therefore, when reviewing data, if the site is experiencing a high lost call rate for interconnect, or any lost call rate which requires focusing resources on, the following items need to be checked.

1. *Determine what changes were made to the network recently in that area (if the area worked well before the changes).*
2. *Verify the 3:1 and 6:1 or 12:1 relationship.* Determine if the lost call rate is associated with the 6:1 and 3:1 or 12:1 interconnects. If the problem is 3:1 and not 6:1 or 12:1, then there might be a capacity problem or other network-related issue. If the problem is 12:1 or 6:1 and not 3:1, it could be an SQE or raw coverage problem.
3. *Verify that there are no coverage problems with the prediction and field data.* You need to verify that the prediction and field data (if available) confirm or deny that there is a problem. If there is a coverage problem, then there exists the distinct possibility that this may be the leading, but not the only, source of the lost calls. Coverage problems could be caused by terrain or the existing system configuration. Sometimes the adjacent site configuration is the cause or can be altered to resolve the issue.
4. *Review the neighbor list for the primary site.* Check the site's neighbor list to verify that there are no missing or incorrect neighbors. You also need to check and verify that there is not an excessive number of neighbors. You should also check for reciprocity for the neighbor list, but this will not manifest itself in a high lost call rate for this site, only the neighboring site which may not be reciprocal. Also check the hysteresis value for the handoff. At times the hysteresis value could be set to preclude the handoff from taking place. Review the time between handoffs, i.e., the usage or minutes between handoffs. If the number is really low, then it is likely that the cell, as a source, is incorrect, and this might be the nature of the problem. Other parameters that fit this condition are requests for handoff, successful handoffs, and intracell handoffs. Also check that the color code for all the neighbor sites is correct in the datafill. Determine if the lost call occurred during handoff and, if so, whether it was a cochannel or adjacent channel base radio (BR).

5. *Check the frequency plan.* See part 4 of the list in Sec. 5.7.1. You need to also check for SMR interference by reviewing all known SMR operators that could be using the same channel or adjacent channels in the area of concern.

6. *Check traffic loading.* Next review the traffic load at the surrounding sites and ensure that each site has sufficient capacity so that a mobile can hand off to it. If the neighbor site is blocking, this could be a cause of the lost call rate. Also reselection and handoff classes may be nondefault resulting in SUs operating outside of their intended service area.

Make sure that the 6:1, 3:1, or 12:1 relationship is also checked for capacity-related issues. More specifically check the utilization of the time slots for 12:1, 6:1, and 3:1 usage. Also check the amount of loading that is due to dispatch for target cells to see if this could contribute to the problem.

7. *Check cell site parameters.* This part of the check involves reviewing the site's specific configuration and software settings. Things to check involve topics beyond the neighbor list and possibly the ERP for the site. For example, you need to review the power control settings for the site and existing software load. For any of the parameters checked relative to the site or sector, you need to compare them to the design standard which is used in the system. If the parameters are not within nominal range, then they need to be checked to ensure that they are not the root source of the problem. In other words check the datafill for anything out of the norm.

Some additional issues to look for involve verifying that the transmit and receive frequencies are indeed 45 MHz in separation, depending on the software load. You should also verify that the carrier number matches the frequency for the site on a BR level.

8. *Check ILA borders.* Verify the location of the ILA borders and ensure that high traffic areas are not on an ILA border. If they are, then the source of the problem may be the border itself.

9. *Check for maintenance issues.* When checking for maintenance issues, you need to determine if the problem can be isolated to a single BR. If it can, then all traffic on that particular BR will be affected. You will also want to verify that there were no T1/E1 outages for the site at that time or the adjacent sites since this would negatively affect the performance of the site. One other item to look for is the possibility of an antenna system problem, which would warrant a site visit, if no other problems could be found. Some vendors have the ability to monitor the antenna system for VSWR which would indicate whether a problem might exist or not. Also if historic antenna data are available, it might be worth it to check and see if there is any degradation and if the lost calls appear more prevalent after a rainstorm indicating water intrusion. In many situations the BR could have an unterminated antenna port on the receiver leading to ingress noise.

5.8.2 Access problems

Access problems for an iDEN system can be associated with the interconnect as well as the dispatch parts of the system. There is, however, a distinct difference associated with access problems between the dispatch and interconnect parts of the system which relates to the provisioning aspects. Most of the access problems associated with dispatch are provisioning related, provided there are facilities available.

For this section a dispatch access problem followed by an interconnect access problem will be addressed. In both cases the capacity at the site will need to be reviewed, but although dispatch is packet-driven and interconnect is circuit-switched, both share the same radio resource.

Dispatch access problems. For dispatch access problems the issues involve both the network configuration and provisioning. When a dispatch problem occurs, usually it is via a trouble ticket and is somewhat reactive in nature unlike the interconnect part which can be addressed in a more proactive fashion. This is not to say that the dispatch problems are all reactive issues. Interference, cochannel or co-color code, DLA location updating, PCCH collisions, and subrate blocking all disrupt customer access and can be monitored for performance degradations.

1. *Perform a provisioning check.* Verify that the subscriber records indicate that they are in the allowed service. This is a simple concept but very important. You also need to check to see if the subscriber is associated with the correct fleet or group. Also determine if the problem is private or group dispatch–related, and if the subscriber trying to be raised is also allowed to be dispatched to and is provisioned in the system properly.

2. *Determine if the problem is outbound or inbound.* If the problem is outbound, the issue may be provisioning which may have been overlooked or missed in the previous step. Additionally it is possible to increase the zones the subscriber is allowed to dispatch in, and again the location where this occurs is important.

Problem messages associated with constrained outbound-PCCH would be "target not available" or "PVT-in-use, PLS try again." Outbound problems related to reuse or cochannel or co-color code contentions would create "network trouble" or "PVT-in-use" problem messages for the sending SU and/or multiple alerts, or dispatch paging chirps, muting, or received alert and mute on the receiving SU. Insufficient digital signal level 0 (DS0) allocation, at times, will allow a successful page, yet will be mute one way or both ways. On occasion, it will also generate "network trouble" messages for the sending SU. Loss of dispatch paging due to constrained outbound-PCCH will also cause a loss of SMS paging and voicemail alerts.)

If the problem is inbound, e.g., the subscriber does not get the dispatch page (or alert) on the receiving end, the problem could be the DLA border, assuming the subscriber is provisioned correctly, or PCCH collisions due to reuse, interference, or cochannel or co-color code conditions. The DLA border could be the issue if the subscriber registered as being in, say, DLA#2 and the subscriber is now in DLA#3 but the system has not updated the registration information. This is often the case when either the DLA borders are incorrectly established or cells on the borders experience excessive DLA location updates; which are characterized by high inbound-PCCH utilization.

3. *Determine if the dispatch problem is in one area or multiple areas.* If it is in multiple areas, it is most likely a provisioning or faulty subscriber unit (poor RF performance) issue, but if it is in one area, then it is likely a DLA configuration or enhanced base transceiver (EBTS) problem which is inhibiting access to the PCCH or traffic channel (TCH).

4. *Determine if the subscriber can make interconnect calls.* If the subscriber can make interconnect calls, then the subscriber is registered on the system. However, the problem still could be provisioning related, so the dispatch application processor (DAP) and EBTS datafill needs to be checked in addition to the site or sites where the problem is reported to have happened.

5. *Check maintenance issues.* Verify that dispatching operates on the site and the surrounding sites. If it doesn't, then there is a potential maintenance problem with the EBTS hardware, subrate allocations, or DAP and EBTS datafill. Also verify that the DAP and MOBIS (Motorola's GSM A-bis) datalinks are active and there are no other network-related problems, in addition to checking the frame relay link to the sites.

Interconnect access problems. Interconnect access problems are typically due to blockage, traffic, or provisioning. Blockage associated with the TCH is driven by proper traffic engineering and allocation of the TCHs for interconnect and dispatch usage. There are several items to look for in verifying that indeed you need to add channels or possibly a cell site to relieve the traffic.

1. *Determine the source of the blockage.* See part 1 of the list in Sec. 5.7.3.

2. *Determine what changes were made to the network recently in that area (if the area worked well before the changes).*

3. *Verify that there are no coverage problems with the prediction and field data.* See part 3 of the list in Sec. 5.8.1. What you want to check is that there is sufficient SQE to ensure the call is sustainable for 90 to 95 percent of the reported signals.

4. *Check cell site parameters.* This part of the check involves reviewing the site's specific configuration and software settings. Things to check involve ERP imbalances for the site where the beacon channel is operating at one power level and the other channels are operating at a lower power level. You should also review the power control settings for the site and existing software load. You need to compare any of the parameters checked relative to the site or sector to the design standard which is used in the system. If the parameters are not within nominal range, then they need to be checked to ensure that they are not the root source of the problem. Again for this part of the investigation all aspects of the EBTS and DAP datafill need to be looked at for both the source and target cells/sectors.

5. *Check the ILA borders.* Verify the location of the ILA borders and ensure that there are no potential location update problems as a result of the border location.

6. *Check for maintenance issues.* When checking for maintenance issues, you need to determine if the problem can be isolated to a single channel; if it can, then all TCHs will be affected. You will also want to verify that there were no T1/E1 outages for the site or the adjacent sites at that time since this would negatively affect the performance of the site.

Proper T1/E1 subrate allocations must also be verified. T1s which are groomed and/or have microwave links between sites could have their DS0 allocations reset to default levels when BSCs, IGX, IPX, or BPX are reset or reprogrammed.

7. *Check the cell overlay.* If there is sufficient cell overlap, you might want to visit the issue of shedding traffic, origination, and possibly handoff, by reducing the site's ERP. Additionally or separately the subscriber power level values can be reduced. This is done indirectly by modifying the serving cell's contour and lowering the ERP in undesired areas and maintaining path balance on the handheld unit. This is, of course, provided the adjacent sites have sufficient capacity, and the frequency plan supports the configuration change. Additionally the subscriber power level values can be reduced at the site.

5.9 CDMA

The guidelines for troubleshooting a CDMA-based system need to follow, as with all troubleshooting techniques, the seven-step process defined earlier in Sec. 5.7.1. With CDMA systems, as with other wireless mobile systems, the troubleshooting methodology needs to include other RF technology platforms, based on your system configuration. The other technology platforms to be included for CDMA include AMPS, which can receive CDMA handoffs; and possibly even TDMA/GSM which could be interfering with the CDMA system. Additionally the use of cellular and PCS frequency bands in addition to the existence of different CDMA versions need to be factored into the trouble-shooting process.

The version differences apply to CDMAOne and CDMA2000 systems when you are migrating to a CDMA2000 system. The inclusion of legacy issues, i.e., the existing infrastructure, also should be considered, plus the possible separation of service class, for voice and mobile IP, based on the CDMA carrier.

The six main performance issues for any CDMA system, whether it is CDMAOne or CDMA2000, include the following topics.

1. Lost calls
2. Access failures
3. Handoff failures
4. Packet session access
5. Packet session throughput
6. Capacity and blocking

These six issues are pertinent to any operation and are evident to the customer, so we will focus on them. Obviously, packet services are only related to CDMA2000 systems which have a PDSN associated with them. The items do not need to necessarily be checked in the order presented. Based on your particular system configuration and the RF access platforms being used, the performance questions, or checks, can be merged with those for another access technology, since many have similar results but are manifested with differing symptoms, and also can be merged with the list of seven performance topics common to all technologies, as mentioned earlier in Sec. 5.7.1.

5.9.1 Lost calls

Typically a lost call for a CDMA system occurs when the BTS can no longer communicate with the SU. This relatively simple concept has a multitude of possible causes. Because of the greater reliability of calls inside the core coverage area of a wireless system where more coverage and handoff options exist, some difference in lost calls performance is expected between sectors which cover the core and sectors which cover areas outside of the core.

The main issues which cause failure of BTS to MS communication in CDMA systems are

1. Coverage problems
2. Neighbor list problems
3. Hard handoff problems

One possible culprit to look for is the location of the transition zone or area where hard handoffs take place, either between CDMA carriers, or when handing off to AMPS. If the lost call is neither in a hard handoff area, nor associated with an attempted hard handoff, then CDMA system lost calls are usually associated with a high FER in either the uplink or downlink direction, leading to the breakdown in communication, and hence coordination of action, between the BTS and the SU. Knowing the first direction of high FER, immediately before control failed, can give some critical clues as to the cause of the high FER, which ultimately led to the lost call.

Therefore, when reviewing data, if the site is experiencing a high lost call rate, or any lost call rate which requires focusing resources, the following items should be checked:

1. Determine what changes were made to the network recently in that area (if the area worked well before the changes).

2. Verify that there are no coverage problems with the predictions, field data, and sectors or spans out of service. See part 3 of the list in Sec. 5.8.1. Coverage issues could also be caused by the power distribution within the CDMA carrier, in that it has to dedicate a large percentage of the power budget to keep distant subscribers on the cell or sector. This, of course, leads to the breathing cell phenomenon, where problems are intermittent. Therefore, it is important to review the predicted coverage, with the designed load, and then look for possible pilot pollution locations, where the problem may be too much signal from too many sources rather than lack of signal. If the strongest of these multiple pilots is still relatively weak, especially in relation to the other pilots being received by the SU, then this coverage problem is often described as lack of a *dominant pilot.*

You want to ensure that there is a sufficient Eb/Io for the call to be sustainable with an acceptable FER for 90 to 95 percent of the reported signals, in both the uplink and downlink directions, where possible. Also when a sector is out of service, part of the area it normally covers can revert to a pilot pollution situation, if no neighbor is sufficiently dominant to fill the coverage gap.

3. *Review the neighbor list for the primary site.* Check the site's neighbor list to verify that there are no clearly missing or incorrect neighbors. You also need to check that there are not an excessive number of neighbors. In addition, you should check for reciprocity of the neighbor list. However, lack of reciprocity will not manifest itself in a high lost call rate for this site, but rather for the neighboring sites which may not be reciprocal.

Check the T_ADD, T_DROP, and T_COMP parameter values to see if they are too lax or too tight, depending on the location of the neighbor sites. Clearly, in general,

- Lowering T_ADD will make a neighbor active sooner.
- Lowering T_DROP will hold on to an active neighbor longer.
- Lowering T_COMP will make it easier to replace an active neighbor with some other neighbor having a stronger pilot, and the reverse if the thresholds were to be raised instead of lowered.

Verify that on average there is not an excessive number of active-list neighbors (high proportion of time to be spent in soft handoff) for the sector, resulting in capacity restrictions for other subscribers, or in Walsh code and/or power budget–related problems for the sector. You may want to increase the T_COMP value if many calls for the sector have more than three active neighbors.

Additionally, CDMA systems report all pilots in the pilot signal measurement message (PSMM). You should use these values to verify that the neighbor cells that are active for the lost call have an acceptable signal level.

4. *Check for pilot pollution.* You need to review the existing system plan and field data for the site and sector involved. Check for lack of a dominant server in the problem area. Issues to contend with involve determining *why* there is not a dominant server and *if* one of the sites covering the area can be altered to provide the dominant server for the problem area.

5. *Check the round-trip delay (RTD).* Depending on the specific infrastructure vendor you may need to verify that the RTD parameter is set correctly. If it is set too long, calls can drag on border zones. Alternatively if it is set too short, this could lead to lost calls occurring on another site. Therefore, you need to review the RTD for the surrounding sites as well.

6. *Check traffic loading.* Next review the traffic load at the surrounding sites and ensure that they have sufficient capacity to accept a mobile handoff.

Lack of capacity could be caused by excessive soft and softer handoffs, or by a lack of channel elements or Walsh codes. A lack of Walsh codes is not common, except for a very small and close-in coverage area with very high traffic. If the neighbor site is blocking, this could be a cause of the lost calls. Also, directed retry, through service redirection, may be activated; resulting in SUs operating outside of their designed coverage area. Check the mobile data usage for the sector and surrounding sites to ensure that the data portion of the network is not a culprit, and vice versa, that voice traffic is not blocking data calls.

7. *Check the global positioning system (GPS).* Check to ensure that there was not a GPS problem with the source site or sector or any of the sites on the neighbor list. If a site has a faulty GPS, its local time base may be out of synchronization with the rest of the network, leading to the "island cell" phenomenon where calls may be able to hand into the island cell successfully, but probably will not be able to hand out successfully, due to the lack of synchronization's effect on the SU's pilot search windows and on the SU's decoding of other signals.

8. *Check BTS parameters.* This check involves reviewing the site's specific configuration and software settings. Things to check include topics beyond the neighbor list and possibly the RTD for the site. For example, you need to review the power control settings for the site and existing software load. All site or sector parameters should be compared against the design standard for the system. Therefore, you will need to check the site's software and hardware configurations. A prime area to consider is the pilot parameters. If the parameters are not within nominal range, then they need to be investigated to ensure that they are not the root source of the problem. Sometimes it is hard to verify the site's hardware configuration due to documentation problems. If the configuration is required, a site visit may be in order, especially if the system is trifurcated with other services.

9. *Check BSC borders.* Verify the location of the BSC borders and ensure that high traffic areas do not fall on a border. If they do, then the source of the problem may be the border itself, due to the time required for the handoff process to occur in the SU.

10. *Check for maintenance issues.* When checking for maintenance issues, you need to determine if the problem can be isolated to a single frequency allocation (FA) or channel element. You will also want to verify that there were no T1/E1 outages for the site at that time, or for the adjacent sites, since this would negatively affect the performance of the site. A review of the BTS and BSC alarms may uncover the cause of the problem as well.

Another possibility is an antenna system problem, which would warrant a site visit, if no other problems could be found. Some vendors have the ability to monitor the antenna system VSWR, which could indicate the presence of a problem, but not the absence of all problems. Also, if historic antenna data are available, check and see if there is any degradation over time, and if the lost calls appear more prevalent after a rainstorm, indicating water intrusion in a cable or jumper.

If your system is newly installed or recently modified, be sure to consider the many ways in which antennas can be misconnected. For example, the alpha and beta sector transmit antenna cables (only) might be cross connected, resulting in failed call attempts in major areas of both the alpha and beta sectors, due to apparently weak uplink signal levels received from the MS.

These 10 areas of inquiry may seem like a lot to check for a "simple" problem of a few lost calls. However, the depth and diversity of these 10 areas simply reflects the many possible reasons for a spate of lost calls.

5.9.2 Handoff failures (problems)

Specific handoff failures in a CDMA system are not as prevalent as with other wireless access systems. However, handoff failures and problems usually manifest themselves as a lost call. The handoff failures, of course, can be both from the source or target. Since CDMA systems have soft, softer, and hard handoffs, the types of failures and their root causes cover a wider range of possible issues than with other technologies. Handoff failures can occur between sectors of a cell, between cells, between CDMA carriers (i.e., an f1 to f2 handoff), and in a transition zone with a handoff from CDMA to AMPS. Obviously, the hard handoff when you hand down from CDMA to AMPS is only applicable to the cellular band operators, where AMPS may exist, and not to the PCS operators. Table 5.1 may assist in determining the type of handoff and supporting services which can and does occur with a handoff.

Another important point to raise regarding handoffs is the radio configuration (RC) differences that could and most likely do exist in a network. For example, depending on the RC for the call in progress the active neighbors will also need to have the same RC. Why this is important to note is that a RC1/2 to RC3 is not supported which means that you either have to go from RC3 to RC3, RC3 to RC1/2 or RC1/2 to RC1/2. Therefore, the configuration of the neighbor sites focusing on the CDMA version available or active needs to be verified to ensure that this is not a contributing cause of the problem.

TABLE 5.1 Handoff Compatibility Table

Source	Target	Destination traffic channel type
AMPS	IS-95/CDMA 2000	NA
IS-95	AMPS	AMPS
	IS-95	IS-95
	IS-2000	IS-95
IS-2000	AMPS	AMPS
	IS-95	IS-95
	IS-2000	IS-2000

Regardless, the largest source of handoff failures is driven by pilot pollution or errors in the neighbor list. Therefore, here is a list of items to check when investigating a high handoff failure rate for a sector.

1. *Determine what changes were made to the network recently in that area (if the area worked well before the changes).*

2. *Review the neighbor list for the primary site.* Check the site's neighbor list to verify that there are no missing or incorrect neighbors. Check and verify that there is not an excessive number of neighbors. You should also check for reciprocity for the neighbor list, but this will not manifest itself in a high lost call rate for this site, only the neighboring site which may not be reciprocal.

Verify that there are not an excess number of active neighbors for a call resulting in capacity restrictions for other subscribers, Walsh code, or power budget related problems. You may want to increase the T_COMP if there are more than three active neighbors typically for the site.

Usually excess active list sectors are measured as a percentage of time in the test area that the call is in three-way, four-way, etc. For example, you may not want your percentage of four-way or more soft handoffs (SHOs) to be more than, say, 10 percent.

Additionally CDMA systems report all the PSMM values. You should use these values to verify that the neighbor cells that are active are of an acceptable level.

A high FER in both the uplink and downlink directions is symptomatic of a neighbor list problem, which is not coverage- or resource-related. For a high FER in the uplink direction only, check the RTD values. For a high FER in the downlink direction, check the power budget.

3. *Check the handoff parameters.* Sometimes the handoff parameters used to select the active neighbor list may need to be adjusted. Check the T_ADD, T_DROP and T_COMP parameter values to see if they are too lax or too tight depending on the location.

The search window for the neighbor, SRCH_WIN_R, and remaining sets, SRCH_WIN_N, could be tightened or relaxed. They should be tightened if there are too many neighbors being included for consideration. However, note that an excluded neighbor will interfere with the active sectors.

Alternatively, the search window used for the active and candidate sets, SRCH_WIN_A, could be adjusted, thereby either relaxing or restricting the selection process. However, the search windows are not often changed within a system from the default values recommended by the infrastructure vendor

4. *Check handoff types.* You will need to check what percentage of the handoffs into and out of the site are soft or hard handoffs. Of specific interest is what triggers a hard handoff. Is it because of a transition zone between CDMA/AMPS, or because the FA being used has a limited footprint in the system?

If the problem is related to hard handoffs, then check the RTD setting and determine if the handoff should be moved closer or farther away from the source site or sector. This remedy also applies if the site is experiencing a lot of target handoff problems.

One of the most common boundary handoff failure reasons is that the datafill on one or both sides is wrong. If the handoff failures are inbound, target handoff problems for the site, then you should also review the power budget for the sector to determine if it is possibly causing pilot pollution resulting in the site being included for handoff selection when it should not be.

5. *Check traffic loading.* Next review the traffic load of the surrounding sites and ensure that they have sufficient capacity to accept a mobile handoff. Lack of capacity could be caused by excessive soft and softer handoffs, or by a lack of channel elements or Walsh codes. Check the mobile data usage for the sector and surrounding sites to ensure that the data portion of the network is not using too much capacity, or vice versa.

6. *Check the GPS.* Check to ensure that there was not a GPS problem with the source site or sector or any of the neighboring sites that are on the neighbor list. This could lead to call dragging, effectively causing the SU to use a nonoptimal candidate. Additionally a GPS failure should cause a major alarm to be tripped and calls on all sectors for that site and the surround sites would be affected.

7. *Check BTS parameters.* See part 8 of the list in Sec. 5.9.1.

8. *Check BSC borders.* Verify the location of the BSC borders and ensure that high traffic areas are not on a border. If they are, then the source of the problem may be the border itself, due to the time required for the handoff process to occur in the SU.

9. *Check for maintenance issues.* When checking for maintenance issues you need to determine if the problem can be isolated to a single FA or channel element. You will also want to verify that there were no T1/E1 outages for the site at that time, or the adjacent sites, since this would negatively affect the performance of the site. A review of the BTS and BSC alarms may produce the cause of the problem as well.

One other item to look for is the possibility of an antenna system problem, which would warrant a site visit, if no other problems could be found. Some vendors have the ability to monitor the antenna system for VSWR which would indicate whether a problem might exist or not. Also if historic antenna data are available, it might be worth it to check and see if there is any degradation and if the lost calls appear more prevalent after a rainstorm indicating water intrusion.

5.9.3 All servers busy

Problems with access to a wireless system involve both blocking, i.e., no resources available, and insufficient signal levels. If the resources are not sufficient to address the current traffic, the obvious response would be to simply add a new FA or more channel elements. However, the traffic increase may be caused by incorrect parameter settings or a deliberate attempt to reallocate traffic.

There are several items to look for in verifying that you truly need to add channels, or possibly even a cell site, to relieve the traffic.

1. *Determine if anything has changed in the system.* Did you or someone else alter another site's coverage, or add or remove a site or sector from the system?

2. *Determine the source of the blockage.* It is important to know the source of the blockage since it will tell you how to possibly relieve the situation both on short-term and long-term bases. You will need to determine the volume of traffic type for the cell or sector and if it is origination based or handoff based. For instance, a normally balanced cell for a typical mobile system will have one-third originations and two-thirds hand-ins; your numbers may vary, but the illustration is important. If this typical cell were to have two-thirds originations and one-third hand-ins, it might indicate that the blockage problem is near the site itself. If the majority of the traffic was hand-in, then the source of the traffic would be from an adjacent site.

The source is important in determining the possible solutions. For traffic originating at the site you might want to relax the T_COMP for target cells and hand off more traffic, thus shedding the load. Or, you might want to increase the T_COMP value for this site being the target cell, thus reducing incoming traffic if the problem is inbound. The T_COMP value in this case would be adjusted at the neighboring sites or sectors of the blocking site or sector. You should also check to see if the cause may be service redirection activated in the cluster of sites where the blocking cell site is the target of the traffic shedding.

3. *Check Walsh codes.* With CDMA, the Walsh codes can, depending on your aspect, be a limited resource. Therefore, you should check for any blockages due to the lack of Walsh codes. Lack of Walsh codes is probably more problematic if the same FA is being used for packet data services.

If there is indeed a Walsh code shortage, then review of whether the traffic requesting the codes is originating at the site or is soft handoff driven. If packet services are being used and this is a CDMA2000 channel, then another issue to check is the packet data usage. The more data being used on a sector, the more Walsh codes are consumed. Usually, Walsh code limitations also manifest themselves with handoff failures and lack of packet data throughput.

4. *Check for maintenance issues.* When checking for maintenance issues you need to determine if the problem can be isolated to a single FA or channel element card. You will also want to verify that there were no T1/E1 outages for the site at that time, or the adjacent sites, since this would negatively affect the performance of the site. Additionally, you should check for loss of GPS.

5. *Check for cell overlap.* If there is sufficient cell overlap, you might want to shed traffic, both origination and possibly handoff traffic, by reducing the site's coverage footprint. The limitation of the site's footprint is done typically by altering the physical configuration of the antenna system. If the antenna alteration is not possible or does not produce the desired result, then adjusting the site's pilot power could be attempted. However, care must be exercised in adjusting the pilot power to avoid unbalancing the receive side of the system. Additionally, the coverage area or footprints of the adjacent sites or sectors could be reduced depending on what the desired goal is.

Most operators prefer it the other way around. They prefer to limit coverage with the antenna pattern and keep all the pilot powers the same at the link budget value. Changing the pilot power, except on the analog border of the system, is risky as it unbalances the sector on the receive side.

5.9.4 Access problems

CDMA systems can exhibit access problems. Access problems manifest themselves as no page responses, sometimes causing a large volume of calls to go to voice mail. In addition, the access problems also result in call failed messages. The major reasons for access failures are a few common culprits, pilot pollution, or possibly registration problems.

Therefore, here are some items to check.

1. *Determine if anything has changed in the system.* Did you or someone else alter another site's coverage or add or remove a site or sector from the system? Was a new BSC added, or was there a HLR failure, etc.?

2. *Verify that there are no coverage problems (prediction and field data).* See part 3 of the list in Sec. 5.8.1. Many of the infrastructure vendors have internal diagnostics available which are able to plot the Eb/Io for the subscriber units which access the system. This is an important diagnostic tool to use, if available.

3. *Pilot pollution.* You will need to verify if the cell covers areas that have no dominant server. Lack of a dominant server may lead to originations occurring on the wrong cell and also manifest itself in high lost calls and handoff failures, due to pilot pollution. Some issues involve determining *why* there is not a dominant server, and *if* one of the sites covering the area can be altered to provide the dominant pilot channel.

4. *Check the registration area.* If the access problem occurs in a border area, either internal to the network or between systems, then the issue could be caused by the registration process itself. If the SU is on a border between CDMA and AMPS, where there is intermittent coverage from both systems, then the SU may be changing modes between digital and analog and causing call delivery problems. At heart of the matter is the updating of the home location register (HLR) or visitor location register (VLR), causing flip-flop registrations. The solution is to improve the coverage or shorten the registration interval.

Now if the access problem is internal to the network, and it is between MSC or registration borders, then either revisit the location of the border and alter it, shorten the registration interval, or global page the subscriber if not found with the first page. The last suggestion has obvious performance concerns associated with it. Registration problems will show up when reviewing the statistics for the number of registrations divided by the sum of originations plus terminations.

5. *Check cell site parameters.* This check involves reviewing the site's specific configuration and software settings. Things to check include the pilot database focusing on the RTD, sector and radio type, the FA and any restrictions or unique configuration issues associated with it. If the site or sector is on a BSC or other border, the cell ID and target sector(s) datafill, either analog or CDMA,

needs to be checked as well. A power budget imbalance between the pilot database and TCH needs to be checked, as well as all other site parameter settings.

You need to compare each of the site or sector parameters to the design standard used in the system. If the parameters are not within nominal range, then they need to be checked to ensure that they are not the source of the problem. Again, for this part of the investigation, all aspects of the datafill, or system parameters, need to be considered for both the source and target cells or sectors.

6. *Check for maintenance issues.* When checking for maintenance issues, you need to determine if the problem can be isolated to a single FA or channel element. A review of the BTS and BSC alarms may produce the cause of the problem also. Lastly, access problems will not manifest themselves with a T1/E1 outages for the site, but can be caused by a defective antenna system.

5.9.5 Packet session access

CDMA2000 introduces the real use of packet data for wireless mobile systems. When troubleshooting a performance problem with the packet data network, issues need to encompass both the radio as well as the packet network, including both on-net and off-net applications. However, besides the radio environment, and along with the subscriber unit being the correct vintage, the key to the packet network is the PDSN.

Since the PDSN connects to the CDMA radio network on the access side of the system, an important issue in the troubleshooting process is that, with the exception of the radio resources, the PDSN does not interact with the voice elements directly. In general, the CDMA2000 data and voice networks are separated as much as possible. This is important since you could easily have a voice network problem and not a packet problem, or vice versa, depending, of course, on where the problem is located.

Packet data network performance problems can also be caused by the subscriber configuration, either in the authentication, authorization, and accounting (AAA) system, the network subscribers are trying to connect to, or the mobile PC's network interface card (NIC) settings. Therefore it cannot be overemphasized that to expedite the problem resolution, try to determine what elements of the system could be involved with the problem.

This list of actionable items are for a 1XRTT packet enabled system. It must be assumed that packet services are active in the entire system. However, if not, then the obvious question concerns the packet system boundaries as a possible cause for the problem.

The investigation of packet access can be initiated by a customer complaint or by monitoring the system for traffic packets. Therefore, in the proactive sense, you can begin investigating if there is no 1XRTT packet activity on the site or sector, but there was previously, i.e., one can monitor for negative activity.

The following fundamental areas need to be checked.

1. *Determine what changes were made to the network recently in that area (if the area worked well before the changes).* Try and determine where the problem took place or if it is an individual subscriber.

2. *In the case of an individual subscriber, determine if the subscriber is active in the HLR and AAA for packet services.*

3. *Packet and other traffic loads.* Verify that no blocking or other facility-related issues for the site, sector, BSC, or PDSN are involved with the initiation of the packet session. The availability of channel elements should also be checked.

With CDMA the use of Walsh codes can be a limited resource. Therefore, you should check to see if there were any blockages due to the lack of Walsh codes. Walsh codes could be a problematic issue if the same FA is being used to support voice services.

If indeed a Walsh code shortage exists, then review whether the traffic requesting the codes is originating at the site itself or is soft handoff driven. Another issue to check, if packet services are being used, is what the packet data usage is, since the more data being used on a sector, the more Walsh codes are consumed. Usually Walsh code limitations should also manifest themselves with handoff failures and lack of packet data throughput.

4. *Check cell site parameters.* This check involves reviewing the BTS and BSC specific configuration and software settings. Things to check include determining whether packet services have been activated for the sector and radio type, the FA settings, and any restrictions or unique configuration issues associated with it. If the site or sector is on a BSC border, or another border, the cell ID needs to be checked as well. A power budget imbalance between the pilot database and TCH needs to be checked, as well as all the other site parameter settings.

For any of the site or sector parameters checked, you need to compare them to the design standard used in the system. If the parameters are not within nominal range, then they need to be checked to ensure that they are not the source of the problem. Again, for this part of the investigation, all aspects of the datafill or system parameters need to be reviewed for both the source and target cells/sectors.

5. *Voice processing.* It is very important to verify if there is voice traffic on the site. Depending on whether the FA is separate or combined usage, this might help determine the source of the problem. If there is voice usage on the site and there are no FA problems, then the problem could be configuration- or PDSN-related.

6. *Check the PDSN.* For this part of the investigation you will need to verify that there have been no major core network service disruptions which could preclude the initiation of service for the subscriber. Usually, however, if there is a major outage or disruption, other sites and services will also be affected and this will not be an isolated situation.

Additionally the PDSN configuration needs to be checked as well. This is separate from the BTS and BSC configuration checks.

- Ensure that the packet zone is activated in the data fill; this is easy to overlook.
- Check for a valid range and setting in the PDSN IP address.

- Ensure that the L2TP tunnel is set up. Verify that the number of retries is not exceeded. A possible cause is the physical link, the processing capacity, or that the retries are set too low.
- Check the time and rate between which the minimum data rate forces the session to go from active to dormant. Similarly, check the values for the transition from dormant to active. These parameters should be set to prevent unnecessarily frequent transitions between the dormant and active modes for the subscriber.
- Ensure the routers used for the PDSN are not experiencing any congestion on their ports, their CPU, or their buffers.

7. *Check handoff during a session.* If a 1XRTT-capable SU is in a dormant data session, and meanwhile moves to a nonpacket system, the point-to-point (PPP) session between the PDSN and the SU will time out and be torn down. The SU will not be able to reestablish a packet data call until it is in a 1XRTT system again.

8. *Check for maintenance issues.* When checking for maintenance issues, you need to determine if the problem can be isolated to a single FA, BTS, or BSC. Therefore, you need to review the BTS and BSC alarms for a clue as to the cause of the problem. You also need to review the PDSN network, again focusing on any service disruptions or alarms that may have been produced during the problematic time period.

5.9.6 Packet session throughput problems

Packet session throughput is determined by the radio environment as well as by the site's configuration, overall fixed network capacity, the service being offered, the service's destination, and, of course, what SLA has been assigned to the packet user. Other factors could be the subscriber's configuration, the service configurations, and their throughput.

Throughput problems can be summed up in two basic issues:

1. Insufficient amount of kilobits per second

2. Session terminated prematurely

It should be obvious then that many issues play a role in the ability to deliver and sustain 1XRTT sessions. Since the session is packet based, the FA's Walsh codes are and should be a shared resource. Obviously, you need to determine what the available resources are for the packet session; if the desired throughput level, say 144 kbit/s, is really achievable; or if the design (or desired) level is really more like 40 kbit/s.

However, when a packet session issue arises, it will most likely be in the form of a customer complaint. Using system statistics to determine packet session throughput will only indicate usage, since each session is unique in what it is accessing and in terms of bandwidth required, both for the uplink and downlink.

With this said, here is a starting point to begin an investigation. Often, if the system is not experiencing problems, the issue is the subscriber's configuration or perception of a problem. For example, the session speed may be 15 kbit/s, but the subscriber thought the speed was 30 kbit/s. However, the additional bandwidth was not needed, based on the negotiated rate with the off-net server. Perception is difficult to overcome and should not be presumed in the investigation, since the customer is paying for the service.

Lastly, the troubleshooting process, after verifying that the subscriber is allowed packet service, requires some specificity about the location of the problem, since assuming the entire system is at fault would imply a PDSN problem that should be evident from other alarms.

The following fundamental issues need to be checked.

1. *Identify the specific problem and, if possible, the geographic area where the problem occurred.*

2. *Determine what changes were made to the network recently in that area (if the area worked well before the changes).* Also try to determine where the problem took place.

3. *In the case of an individual subscriber, determine if the subscriber is active in the AAA for packet services and if the SLA matches what is desired or claimed by the subscriber.*

4. *Determine the type of services experiencing the problem.* If they are having a file transfer protocol (FTP) problem, but can gain access via hypertext transfer protocol (HTTP), then it might be their configuration. If possible, determine if the server or PDSN router can be pinged, that is, respond with an acknowledgment after receiving a message. The fundamental objective is to determine what service is affected and whether it can be isolated. Additionally, we'll need to know if the problem is on-net or off-net.

If the problem is on-net, then just a ping may not be sufficient, since the PDSN may be configured to deny a particular service or set of services based on the port number. In addition, if the service includes a VPN with 802.11g interoperability, then you need to determine if the link connecting the PDSN is able to communicate with the external service (including the Internet).

5. *Verify traffic usage.* Verify that the traffic usage, i.e., Walsh codes, channel elements, and other resources allocated for 1XRTT packet services, is sufficient for the traffic load. FA channels can be shared, but there is a limit to this and it is user definable. Therefore, check to see if the throughput problem could really be a voice services–related blocking problem.

6. *Review the neighbor lists.* Handoff for CDMA packet services works the same as for voice services. Therefore, if the packet session is terminated prematurely, many of the issues previously associated with handoff failures are also applicable to packet data. As part of the investigation, we would compare the voice network to the packet network and determine if there are similar problems, or determine if a setting was made to favor one service over the other and this was an unintended consequence.

Therefore, as part of the troubleshooting, check the site's neighbor list to verify that there are no missing or incorrect neighbors and check and verify that there is not an excessive number of neighbors. You should also check for reciprocity for the neighbor list, but this will not manifest itself in a high lost call rate for this site, but rather for the neighboring site which may not be reciprocal.

Check the T_ADD, T_DROP, and T_COMP parameter values to see if they are too lax or too tight depending on the site locations. Verify that there is not an excessive number of active neighbors, resulting in capacity restrictions for other subscribers, or in Walsh Code or power budget–related problems. You may want to increase the T_COMP if there are more then three active neighbors typically for the site. Another issue to verify is whether a hard handoff took place, or even whether a handoff to a more packet restricted site took place.

7. *Active or dormant data session.* If a 1XRTT capable SU is involved in an active data session and moves to a nonpacket system, then a hard handoff is denied. If the SU is in a dormant data session, the PPP session between the PDSN and the SU will time out and be torn down. The SU will not be able to reestablish a packet data call until it is in a 1XRTT system again.

8. *Check cell site parameters.* You need to verify that 1XRTT packet service is enabled for the site, sectors, and for every FA involved. Also, verify that the neighbor sites have 1XRTT packet services enabled. The allocation of available system resources, i.e., Walsh codes and power that should be associated with packet-based services, need to be checked as well.

The detach timer should be checked to ensure that it is not set too short, causing a reattachment, and ensuring that the VLR knows where the mobile is when the packet session is idle.

- *Idle timeout setting.* The time, usually in minutes, that the PDSN will wait for the PPP connection to receive traffic before being disabled.
- *Session timeout setting.* The time a PPP connection may be active before being turned off, with or without receiving traffic.

Again, for this part of the investigation, all aspects of the datafill need to be considered both for the source and for the target cells and sectors.

9. *Check voice processing.* It is very important to check whether there is TCH traffic for voice calls on the site. If there is, then the data problem could be configuration or PDSN-related issues.

10. *Check for maintenance issues.* For this part of the investigation, you will need to verify that there have been and currently are no major core network service disruptions, which could preclude the initiation of service for the subscriber. Usually, however, if there is a major outage or disruption, other sites and services will also be affected, and this problem will not be an isolated situation.

When checking for maintenance issues, you need to determine if the problem can be isolated to a single FA, BTS, or BSC. Therefore, you need to review the BTS, and BSC alarms for a clue as to the cause of the problem. You also need to review the PDSN network again, focusing on any service disruptions or alarms that may have been produced during the problematic time period.

5.10 GSM

The troubleshooting guidelines for any GSM system should follow the seven main performance techniques presented at the beginning of this chapter. Additionally you need to factor into the troubleshooting process whether the system is a pure stand-alone GSM system or is sharing spectrum with IS-136 or CDMA, U.S. PCS specific. With the advent of 3G services, GSM operators at this time have chosen to introduce general packet radio services (GPRS) into their systems enabling IP packet data services to be made available instead of using circuit-switched techniques.

The main problems for any GSM and GPRS system involve

1. Lost calls
2. Access
3. Handoff failures
4. Packet session access
5. Packet session throughput

Since these five main performance issues are the more pertinent to any operation and are evident to the customer, they will be focused on. The items do not necessarily need to be checked in this order.

5.10.1 Lost calls

Typically a lost call for a GSM system occurs when the BTS can no longer communicate with the subscriber. Although this is a simple concept, there are many reasons behind this type of occurrence. The three main issues, which cause lost calls for GSM, are

1. Coverage problems
2. Interference, both cochannel and adjacent channel
3. Neighbor list problems

When reviewing data, if the site is experiencing a high lost call rate, or any lost call rate which requires focusing resources on, the following items need to be checked:

 1. *Determine what changes were made to the network recently in that area (if the area worked well before the changes).*
 2. *Verify that there are no coverage problems with the prediction and field data.* You need to verify that the prediction and field data (if available) confirm or deny that there is a problem. If there is a coverage problem, then there exists the distinct possibility that this may be the leading, but not the only, source of the lost calls. Coverage problems could be caused by terrain or the existing system configuration. Sometimes the adjacent site configuration is the cause or can be altered to resolve the issue.

Many of the infrastructure vendors have internal diagnostics available which are able to plot the nominal signal level, RSSI, and RXQUAL in addition to the RSSI for the various neighbor BTSs on the MAHO list. This information is available for both the uplink and downlink paths, with the exception of the neighbor BCCH RSSI values which are downlink only. This information could also indicate that there might be an out-of-band or in-band interferer.

You will want to ensure that there is sufficient C/N so that the call is sustainable for 90 to 95 percent of the reported signals in both the uplink and downlink directions, where possible.

3. *Review the neighbor list for the primary site.* Check the site's neighbor list to verify that

- *There are no missing or incorrect neighbors.* Typically, the common missed or incorrect neighbors are those within the site. Referring to Fig. 5.20 site 1, cell A is a neighbor of cells B and C, but cell B is not a neighbor of cells A and C, and this is all within the same site. In addition, a whole site could be missing or overlooked from the neighbor list. For example, Fig. 5.21 illustrates how site 1 (cell B) is a neighbor to site 2 (cell D) and site 3 (cell I),

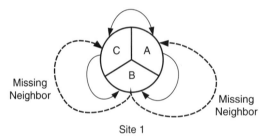

Figure 5.20 Intracell missing neighbor.

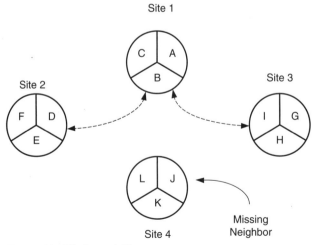

Figure 5.21 Missing neighbor.

but not to site 4 (cells J and L). Site 4 (cells J and L), then, is missing neighbors, which could cause a dropped call depending upon the mobile's selection, signal strength, handoff, etc.
- *There is not an excessive number of neighbors.* If a cell has an excessive number of neighbors, then those extra neighbors may cause the mobile to make the wrong decision in a handoff situation and cause a dropped call when routing a call to a cell further than the nearest most logical cell to the mobile. This is best illustrated through Fig. 5.22 where site 1 (cell B) has site 3 (cell G), site 7 (cell U), site 2 (cell F), and site 8 (cell V) as unwarranted and excessive neighbors.
- *All cells on the neighbor list have reciprocal or two-way neighbors.* This allows a call from one cell to another to occur in both directions. From cell A to B and from cell B to A. For example, from Fig. 5.23, site 1 (cell B) is a neighbor of site 2 (cell D), and site 2 (cell D) is a neighbor of site 1 (cell B). Please note that this will not manifest itself in a high lost call rate for this site; it is only the neighboring site that may not be reciprocal.
- *The hysteresis value for the handoff does not preclude the handoff from taking place.*

Most vendors have built in to their OMC, or like product, the ability to look for one-way handoffs and to define the amount of handoffs that occur on a per-candidate basis. For example, handing off from cell A to B might occur 60

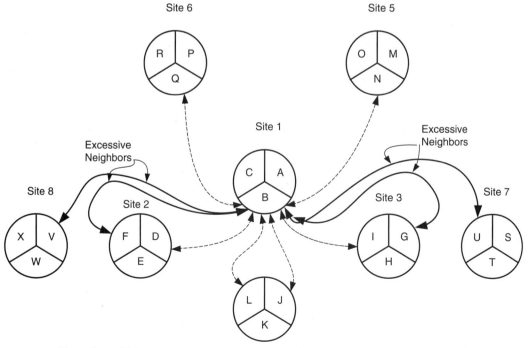

Figure 5.22 Excessive neighbors.

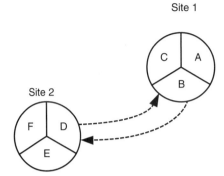

Figure 5.23 Reciprocal neighbors.

percent of the time, while from cell A to C it might occur 5 percent of the time. What you want to look for is the proximity of the sites and if the candidate selection process makes sense from your experience. Keep in mind that time of day plays a major role in mobility issues, but stationary coverage, e.g., building, may be worth checking.

For the stationary situation the frequency of handoffs may be an indication of a ping-pong effect, and in this case attention to hysteresis may be appropriate. Additionally GSM systems through the MAHO process report all the RSSI values for the respective BCCHs. You should use these values to verify that the neighbor cell chosen best matches the RSSI ranking, i.e., if cell 2 typically reports −80 dBm and cell 3 reports −85 dBm, you would expect to see more handoffs to cell 2 than cell 3.

4. *Check the frequency plan.* You need to review the frequency plan that is in place for the site and sector involved. Things to look for involve cochannel and adjacent channel interference, including when the adjacent channel is in the same or adjacent sector and when the cell with that sector overlaps the problem sector.

What you want to do is check the frequency plan on a per-sector and per-channel basis for any obvious issues with cochannel reusers. If a frequency plan is implemented and an obvious cochannel interferer is missed, this could cause a numerous amount of dropped calls within the direction that the cochannel sector is pointing and, depending upon the power setting, a number of sites away. For example Fig. 5.24 uses a $N = 5$ plan, $1 + 1 + 1$ site configuration for all sites. Site 1 (cell B) has channel 591 and site 6 (cell Q) has channel 591, which, based upon the direction of the cells, causes a cochannel interference problem.

In many cases the cochannel interferer may not be on the first TRx (a GSM radio channel), but may be present when the second or third TRx comes on-line, depending upon the traffic load. If the second or third TRx's are not hopping and have the same channel as an adjacent site's cell, depending upon the direction that the adjacent cell is pointing, and its power setting, you could have a cochannel interference problem.

If there is an interference issue, it might be resolved temporarily by placing the offending channel out of service, provided there is sufficient capacity to warrant this approach. However, in most GSM systems removing a full TRx,

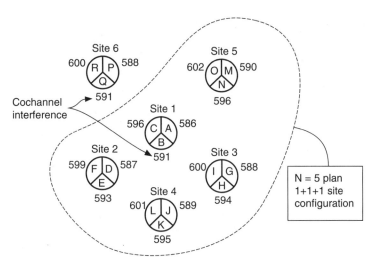

Figure 5.24 Cochannel interference.

8 potential TCHs, is not usually a viable option. Therefore, an alternative frequency assignment needs to be pursued.

You also need to check if frequency hopping is being used. There are many different frequency hopping methods being used throughout the industry, and most are driven by the amount of channels. However, when reviewing and possibly modifying the frequency hopping pattern, no one specific configuration is better than the other, provided, of course, it is properly engineered.

To help illustrate a possible frequency hopping method, Table 5.2 is an implementation of a method which would also simplify the assignment of the hopping sequence number (HSN), for each site, by use of when the BSIC [network color code (NCC) and BSC color code (BCC)] is assigned. This is a simple implementation, due to the fact that the HSNs range from 0 to 63 and the BSIC in decimal format ranges from 0 to 63. A table could be made to cross-reference and combine the two.

Therefore, if a frequency hopping method is implemented and enabled, then the following items should be checked and verified:

- Ensure that the proper mobile allocation list (MaList) is implemented per site for all second, third, and fourth, etc., TRx's. Remember that for now, the first TRx, the BCCH transmission, does not hop (Fig. 5.25).
- Ensure that the proper channel allocation within the MaList is used. Each MaList has a set of unique channels allocated to it that does not repeat within another MaList. (Table 5.3 and Fig. 5.25).
- Ensure that the proper mobile allocation index offset (MAIO) is utilized within the MaList, which allocates a starting channel within the MaList, for each TRx, of the second, third, and fourth, etc., TRx's per sector (Table 5.3 and Fig. 5.25).
- Ensure that the proper HSN is implemented per site for all second, third, and fourth, etc., TRx's (Table 5.2 and Figure 5.26).

TABLE 5.2 HSN = BSIC Combined Table Example

		\multicolumn{8}{c}{NCC}	
		0	1	2	3	4	5	6	7	
B	0	0	8	16	24	32	40	48	56	
C	1	1	9	17	25	33	41	49	57	H
C	2	2	10	18	26	34	42	50	58	S
	3	3	11	19	27	35	43	51	59	N
	4	4	12	20	28	36	44	52	60	
	5	5	13	21	29	37	45	53	61	
	6	6	14	22	30	38	46	54	62	
	7	7	15	23	31	39	47	55	63	
		\multicolumn{8}{c}{HSN}	

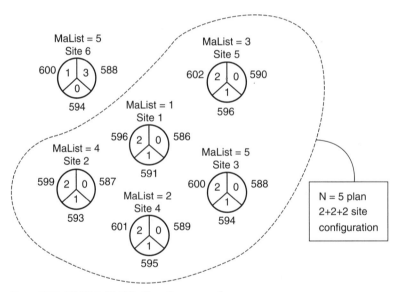

Figure 5.25 MaList–MAIO allocation example.

Performing the items listed here will ensure that the correct hopping method, table, and sequence are entered for each sector, for each site, and for the entire network.

5. *Check traffic loading.* Next review the traffic load at the surrounding sites and ensure that each site has sufficient capacity so that a mobile can hand off to it. If the neighbor site is blocking, this could be a cause of the lost call rate. Also directed retry may be activated resulting in SUs operating outside of their designed coverage area.

RF System Performance and Troubleshooting

TABLE 5.3 MaList–MAIO Table Example

MAIO	MaList					
	1	2	3	4	5	6
0	713	714	715	716	717	
1	718	719	720	721	722	
2	723	724	725	726	727	
3	728	729	730	731	732	
4						
5						

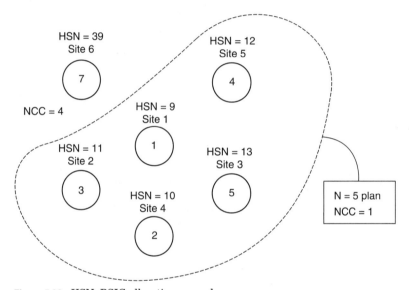

Figure 5.26 HSN–BSIC allocation example.

Things to look for involve the amount of TCHs in use and where any of the messaging channels have to discard packets due to loading issues. The inclusion of GPRS in the system will also preclude TCHs from being used, so you need to verify the GPRS allocation methodology, if applicable.

6. *Check BTS parameters.* This part of the check involves reviewing the site's specific configuration and software settings. Things to check involve topics beyond the neighbor list and possibly the ERP for the site's BCCH. For example, you need to review the power control settings for the site and existing software load. For any of the parameters checked relative to the site or sector you need to compare them to the design standard, which is used in the system. If the parameters are not within nominal range, then they need to be checked to ensure that they are not the root source of the problem. You also need to ensure that the other transmitter channels are operating at the right power

level, after combining. Additionally, the site's software configuration needs to be checked, i.e., the channels and BSIC associated with each channel. In other words check the datafill for anything out of the norm.

7. *Check LA/RA borders.* Verify the location of the location area/routing area (LA/RA) borders and ensure that high traffic areas are not on a LA/RA border. If they are, then the source of the problem may be the border itself where due to the time required for the handoff process to occur the SU could drop.

8. *Check for maintenance issues.* When checking for maintenance issues, you need to determine if the problem can be isolated to a single TCH or TRx channel. If it can, then all TCHs will be affected. You will also want to verify that there were no T1/E1 outages for the site at that time or the adjacent sites since this would negatively affect the performance of the site. Review of the BTS and BSC alarms may also produce the cause of the problem.

One other item to look for is the possibility of an antenna system problem, which would warrant a site visit, if no other problems could be found. Some vendors have the ability to monitor the antenna system for VSWR, which would indicate whether a problem might exist, or not. Also if historic antenna data are available, it might be worth it to check and see if there is any degradation and if the lost calls appear more prevalent after a rainstorm indicating water intrusion.

5.10.2 Handoff failures

Handoff failures should not be an issue with GSM due to MAHO. However, there are still handoff failures, which occur. The following is a brief listing of items to check when investigating a high handoff failure rate for a site or sector.

1. *Review the neighbor list for the primary site.*
   - Check the site's neighbor list to verify that there are no missing or incorrect neighbors (Fig. 5.21).
   - Check and verify that there is not an excessive number of neighbors (Fig. 5.22).
   - Check for reciprocity for the neighbor list (Fig. 5.23), but this will not manifest itself in a high handoff failure rate for this site, only the neighboring site which may not be reciprocal.
   - Check the hysteresis value for the handoff, at times the hysteresis value could be set to encourage handoffs to occur when, in fact, they should not due to a terrain issue.

Most vendors have built into their OMC, or like product, the ability to look for one-way handoffs and to define the amount of handoffs that occur on a per-candidate basis. For example, handing off from cell A to B might occur 60 percent of the time, while from cell A to C it might occur 5 percent of the time. What you want to look for is the proximity of the sites and if the candidate selection process makes sense from your experience. Keep in mind that time of day plays a major role in mobility issues, but stationary coverage, e.g., building, may be worth checking.

Reiterating a previous issue for a stationary situation, the high occurrence of handoffs may be an indication of a ping-pong effect and in this case attention to hysteresis may be appropriate.

What you also want to check, besides the actual neighbor list, is the time between handoffs. There is a systemwide number which you should use as the benchmark for performance. If the time between handoffs is very small, as compared to the norm, then either the cell site is shedding traffic or it may not be the correct candidate to use from the beginning. However, coverage and terrain issues may justify a low time to handoff ratio.

Many of the infrastructure vendors have internal diagnostics available which are able to plot the nominal signal level, RSSI, and RxQual in addition to the RSSI for the various neighbor BTSs on the MAHO list. This information is available for both the uplink and downlink paths, with the exception of the neighbor BCCH RSSI values which are downlink only.

You will want to ensure that there is sufficient C/N so that the call is sustainable for 90 to 95 percent of the reported signals in both the uplink and downlink directions, where possible.

2. *Check the frequency plan.* You need to review the frequency plan that is in place for the site and sector involved. Things to look for involve cochannel co-BSIC resulting in wrong handoff decisions. Interference, whether it is cochannel or adjacent channel, needs to be checked to ensure that this is not the root cause of the problem (Fig. 5.24). However, if it is an interference problem, lost calls will usually be prevalent for the site also.

3. *Check handoff types.* You will need to check if the handoffs into and out of the site are synchronous or nonsynchronous. The handoffs should be synchronous, i.e., using the previous or present timing advance from the cell for handoff when on the same BTS or when handing off to adjacent cells (other sectors at the same BTS). The synchronized handoff leads to a short communication interruption of about 100 ms (Fig. 5.27). The nonsynchronous handoffs, i.e., those obtaining new timing advance information, when the SU is handing into the BTS or to another BTS, which is not collocated with itself could be the source of the handoff problems. The nonsynchronized handoff (Fig. 5.27), leads to a long communication interruption of about 200 ms.

Also if the handoff failures are inbound problems, then the possible cause is the BTS itself, i.e., the target. However, if the handoff failures are outbound, then the possible cause might be the target cell.

4. *Check cell site parameters.* This part of the check involves reviewing the site's specific configuration and software settings. Things to check involve topics beyond the neighbor list and possibly the ERP for the site's BCCH. For example,

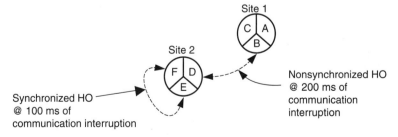

Figure 5.27 HO synch example.

you need to review the power control settings for the site and existing software load. You need to compare any of the parameters checked relative to the site or sector to the design standard, which is used in the system. If the parameters are not within nominal range, then they need to be checked to ensure that they are not the root source of the problem. You also need to ensure that the other transmitter channels are operating at the right power level, after combining. Additionally, the site's software configuration needs to be checked, i.e., the channels and BSIC associated with each channel. In other words check the datafill for anything out of the norm. This check should also be done at the potential source sites to ensure that they are set properly. Again for this part of the investigation all aspects of the datafill need to look for both the source and target cells and sectors.

The final source to check, which may not be so obvious, and is not commonly your first line to check, is if the two sites are between a BSC boundary. If so, you should check to see if the MSC parameters for those cell sites are in sync with the BSC parameters for those cell sites. If an error in the MSC and BSC comparison is found, it is normally the wrong cell identity (CI), the wrong BSC identity (BSCid), or even the wrong location area code (LAC). These are not commonly found errors and should be one of the last troubleshooting checks performed.

5. *Check the LA/RA border.* Verify the location of the LA/RA borders and ensure that high traffic areas are not on an LA/RA border. If they are, then the source of the problem may be the border itself where due to the time required for the handoff process to occur the SU could drop.

6. *Check for maintenance issues.* When checking for maintenance issues, you need to determine if the problem can be isolated to a single TCH or TRx channel. If it can, then all TCHs will be affected. You will also want to verify that there were no T1/E1 outages for the site at that time or the adjacent sites since this would negatively affect the performance of the site. Review of the BTS and BSC alarms may also produce the cause of the problem.

One other item to look for is the possibility of an antenna system problem, which would warrant a site visit, if no other problems could be found. Some vendors have the ability to monitor the antenna system for VSWR, which would indicate whether a problem might exist, or not. Also, if historic antenna data are available, it might be worth it to check and see if there is any degradation and if the handoff failures appear more prevalent after a rainstorm indicating water intrusion.

5.10.3 All servers busy

The problems with access to the wireless system involve both blocking, i.e., resources not available, and also insufficient signal. In the situation of resources not being sufficient to address the current traffic, the obvious response would be to just simply add radio channels. However, the traffic increase may be caused by incorrect parameter settings or a deliberate attempt to reallocate traffic.

There are several items to look for in verifying that indeed you need to add channels or possibly a cell site to relieve the traffic.

1. *Determine the source of the blockage.* See part 1 of the list in Sec. 5.7.3. If GPRS is in use for the site, check the GPRS usage and the TCH allocation to see if this might be over- or underallocated.

2. *Check for maintenance issues.* When checking for maintenance issues, you need to determine if the problem can be isolated to a single TCH or TRx channel. If it can, then all TCHs will be affected. You will also want to verify that there were no T1/E1 outages for the site at that time, or the adjacent sites since this would negatively affect the performance of the site.

3. *Check the cell overlay.* If there is sufficient cell overlap, you might want to visit the issue of shedding traffic, origination, and possibly handoff, by reducing the site's ERP for the BCCH and the other associated TRx channels for that sector. Additionally or separately the power level (PL) values can be reduced, provided, of course, that the adjacent sites have sufficient capacity and the frequency plan supports the configuration change. Additionally, the PL values can be reduced at the site thereby, or the adjacent cells can have their BCCH and associated TRx channels reduced in power.

5.10.4 Insufficient signal strength

Insufficient signal strength is caused by lack of coverage, which is determined by access attempts which failed when there were sufficient TCHs available to handle the traffic load. Therefore, there are a few items to check.

1. *Verify that there are no coverage problems with the prediction and field data.* You need to verify that the prediction and field data (if available) confirm or deny that there is a problem. If there is a coverage problem, then there exists the distinct possibility that this may be the leading, but not the only, source of the insufficient signal level. Coverage problems could be caused by terrain or the existing system configuration. Sometimes the adjacent site configuration is the cause or can be altered to resolve the issue.

Many of the infrastructure vendors have internal diagnostics available which are able to plot the nominal signal level, RSSI, for the subscriber units which access the system. This is an important diagnostic tool to use, if available.

2. *Verify the dominant server.* You will need to check if there is an area the cell covers that has no dominant server. The lack of a dominant server may lead to originations occurring on the wrong cell and also manifest in high lost calls and handoff failures.

3. *Check cell site parameters.* This part of the check involves reviewing the site's specific configuration and software settings. Things to check involve ERP imbalances for the site where the BCCH channel is operating at one power level and the other channels are operating at a lower power level. You should also review the power control settings for the site and existing software load. You need to compare any of the parameters checked relative to the site or sector to the design standard, which is used in the system. If the parameters are not

within nominal range, then they need to be checked to ensure that they are not the root source of the problem. Again for this part of the investigation all aspects of the datafill need to be looked at for both the source and target cells and sectors.

4. *Check LA/RA borders.* Verify the location of the LA/RA borders and ensure that double originations are not taking place resulting in glare or denials caused by double access due to the location of the LA/RA border. If they are, then the source of the problem may be associated with the BCCH assignment methodology used.

5. *Check the frequency plan.* You need to review the frequency plan that is in place for the site and sector involved. Things to look for involve cochannel co-BSIC resulting in ghosts or double access.

6. *Check for maintenance issues.* When checking for maintenance issues, you need to determine if the problem can be isolated to a single TCH or TRx channel. If it can, then all TCHs will be affected. Review of the BTS and BSC alarms may produce the cause of the problem also.

IS problems will not manifest themselves with T1/E1 outages for the site, but will be caused by a possible bad antenna system.

5.10.5 Packet session access

This brief list of actionable items is for a GRPS-enabled system. It has to be assumed that GPRS is active in the entire system; however, if it is not, then the obvious question arises as to the location and if it is in the GRPS network boundaries.

Packet session access requires both the wireless system elements and the SU to be configured correctly. There are state changes, modes, that the SU goes through when it is GRPS enabled. The modes are ready, standby, and idle (Fig. 5.28). The ready mode is when the SU location is known to the serving sector, site, and system and the MS is transmitting or has just finished transmitting. The standby mode is when the MS location is known to RA and the SU is capable of being paged data. The idle mode is when SU location is not known and subscriber is not reachable by the GPRS network.

The investigation of GPRS access can be initiated by a customer complaint or by monitoring the system for traffic (packets). Therefore, in the proactive sense you can begin investigating the first few steps if there is no GPRS activity on the site or sector.

Therefore, the following fundamental issues need to be checked.

1. *Determine what changes were made to the network recently in that area (if the area worked well before the changes).* Try and determine where the problem took place, if it is an individual subscriber.

2. *In the case of an individual subscriber, determine if the subscriber is active in the HLR for GPRS services.*

3. *Check the traffic load.* Verify that the traffic usage, i.e., TCHs allocated for GPRS, is sufficient for the traffic load. GPRS channels can be shared, but

RF System Performance and Troubleshooting 341

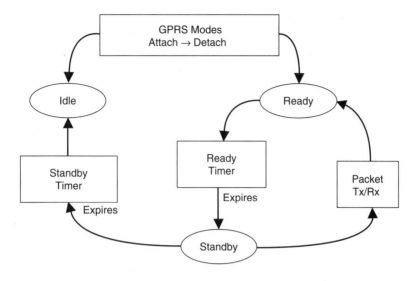

Ready Mode
 1. SU location is known to the serving sector, site, and system.
 2. SU is transmitting or has just finished transmitting.

Standby Mode
 1. SU location is known to RA.
 2. SU is capable of being paged data.

Idle Mode
 1. SU location is not known.
 2. Subscriber is not reachable by the GPRS network.

Figure 5.28 GPRS modes.

there is a user-definable limit to this. Therefore, check to see if the access problem could be really a GPRS blocking problem.

4. *Check site parameters.* You need to verify that GRPS is enabled for the site, sectors, and for every TRx involved that should carry GRPS traffic. In addition, the allocation of how many TCHs that should be associated with GRPS service needs to be checked as well. Again for this part of the investigation all aspects of the datafill need to be looked at for both the source and target cells and sectors.

5. *Check voice processing.* It is very important to verify if there is TCH traffic for voice calls on the site. If there is, then the problem could be configuration or PDSN related.

6. *Check the PDSN.* For this part of the investigation you will need to verify that there have been no major core network service disruption which could preclude the initiation of service for the subscriber. Usually, however, if there is a major outage or disruption, other sites and services will also be affected and this will not be an isolated situation.

7. *Check the frequency plan.* You need to review the frequency plan that is in place for the site and sector involved. Things to look for involve cochannel and adjacent channel interference, including when the adjacent channel is in the same or adjacent sector and when the cell with that sector overlaps the problem sector.

What you want to do is check the frequency plan on a per-sector and per-channel basis for any obvious issues with cochannel reusers. Therefore, an alternative frequency assignment needs to be pursued. However, if this is an interference-born problem, then LC and HO failures should be prevalent also.

8. *Check for maintenance issues.* When checking for maintenance issues, you need to determine if the problem can be isolated to a single TCH or TRx channel. If it can, then all TCHs will be affected. Review of the BTS and BSC alarms may produce the cause of the problem also.

A field visit may be warranted to validate suspected or determined problems. If so deemed, the field visit will need to be performed on a sector-by-sector basis for those sites in question.

Listed here is a set of procedures for testing the attaching and detaching of a sector.

- Each sector of a site should be tested for GPRS capabilities and should be performed with a mobile that has engineering test software loaded.
- Switch the mobile off until you have reached the target sector. Then by checking the CELLid and LAC on the mobile's test menu, you can ensure that you are camped on the right sector.
- Perform attaching and detaching at each sector of the site. Each mobile, depending upon the vendor, indicates this in a simple and noticeable way, such as "Attached," for when you have attached to the system and "Detach" or (the word *Attached* goes away) when you are detached from the system.
- If you were unable to attach to the sector, then perform the following subtests.
  - Switch the mobile off, and then back on.
  - Check if a voice call is possible in that sector, and then check the signal level, while the voice call is up, as shown on the mobile.
  - Check to see if problem is only in that sector or in the whole site, by checking all sectors.
  - Check to see if problem is only in that site, in that BSC, or in a different site of another BSC.
  - Remove and reinsert the subscriber identity module (SIM) to verify that the test unit or rather the SU is not the root cause or adding to the problem.
  - If availability is the issue, then swap out the GPRS mobile.

Repeat these steps for all sectors of the site and for each site in question. The sole purpose of the field visit is to validate your suspicions, and is sometime not warranted. However, once your suspicions are validated, they need to be reported to the appropriate maintenance department for rectification.

5.10.6 Packet session throughput

The issue of packet session throughput is determined by the radio environment as well as by the site's configuration, overall fixed network capacity, the service being offered, the service's destination and, of course, what SLA has been assigned to the packet user, let alone the subscribers' configurations and the server or services' configurations and throughput.

The determination of whether a throughput is good or acceptable, and to know whether there is a problem or not, varies depending on the level of service provided to the customers and the costs involved. A temporary block flow (TBF) is a one-way session for packet data transfer between MS and BSC (PCU). It uses either uplink or downlink, but not both, except for associated signaling. It can use one or more time slots and takes two TBFs (uplink and downlink) to allow bidirectional communication.

With the above mentioned, a completion success rate is one way that a client can determine if the throughput is acceptable or not. A normal rage for a TBF completion success rate is between 95 and 100 percent, per BSC. In order to get a thorough and reliable statistic, there should be a minimum of 5000 TBF establishments in an uplink for each BSC. The formula for the TBF completion success ratio is

TBF completion % =

$$\frac{1 - (\text{TBF establishments} - \text{normal TBF releases} - \text{releases due to flush} - \text{releases due to suspend})}{\text{TBF establishments} - \text{releases due to flush} - \text{releases due to suspend}}$$

The TBF completion success rate is one network monitoring method that should be used to determine if there might be a packet throughput problem.

From this discussion, it is obvious that many issues can and do play a role in the ability to delivery and sustain GRPS sessions. Since the session is packet based, the TCHs are and should be a shared resource. When this question or issue arises, it will most likely be in the form of a customer complaint. Using system statistics to determine packet session throughput will only indicate usage since each session is unique in what it accesses and needs in terms of bandwidth, both uplink and downlink.

When troubleshooting GPRS throughput issues the coding scheme needs to be factored into the analysis. GPRS has four coding schemes which are used to determine the rate of the throughput for data in addition to C/N, coverage, and signal strength. The four current coding schemes and their data rates are shown in Table 5.4.

Figure 5.29 shows the relationship between coding schemes and the radio environment. The example is only meant for illustrative purposes, but the concept tries to show the relationship between RF coverage and changing data rates.

The following is a brief starting point to begin the investigation. Often, if the system is not experiencing problems, the issue is the subscriber's configuration

344 Chapter Five

TABLE 5.4 GRPS Coding Schemes

Coding scheme	Data rate, kbit/s
CS1	9.06
CS2	13.4
CS3	15.6
CS4	21.4

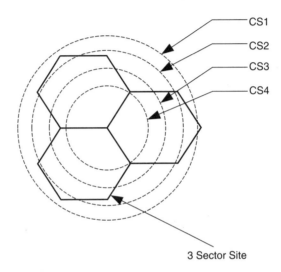

CS	QOS	C/N Nonhopping (dB)	C/N Frequency Hopping (dB)
1	BLER <10%	10.3	7.5
2	BLER <10%	12.6	11.1
3	BLER <10%	14.0	13.3
4	BLER <10%	18.3	20.6

Figure 5.29 GRPS coding scheme illustration.

or perception of a problem. For example, the session speed for accessing a server may be 9.06 kbit/s, but the subscriber thought the speed was 13.4 kbit/s, using only one TCH. The important concept here is that additional bandwidth is not needed based on the negotiated rate. The perception is difficult to overcome and should not be presumed in the investigation since the customer is paying for the service.

Lastly the troubleshooting process, after verifying the subscriber is allowed GPRS service, requires some specificity about the location of the problem, since assuming the entire system is at fault would imply a PDSN problem and should be evident on other alarms.

The following fundamental issues need to be checked.

1. *Determine what changes were made to the network recently in that area (if the area worked well before the changes).* Try and determine where the problem took place.

2. *In the case of an individual subscriber, determine if the subscriber is active in the HLR for GPRS services and the SLA.*

3. *Determine the type of services.* See part 4 of the list in Sec. 5.9.6.

4. *Verify traffic usage.* Verify that the traffic usage, i.e., TCHs allocated for GPRS are sufficient for the traffic load. GPRS channels can be shared, but there is a limit to this, and it is user definable. Therefore, check to see if the throughput problem could really be a GPRS blocking problem.

5. *Check RA borders.* Ensure that there are no apparent issues or problems with the RA borders. Routing areas are used for GPRS mobility management and are each served by only one serving GPRS support node (SGSN). The size of the RA is dependent upon the capacity loads, size of the service area, etc., that the carrier is trying to provide to the customer. A small RA increases RA updates, while a large RA increases the paging load and, depending upon the particular vendor, has a limit to the number of active subscribers per RA (Fig. 5.30).

In addition verify that the locations of the RA borders do not coincide with any high traffic areas. If they do, then the source of the problem may be the

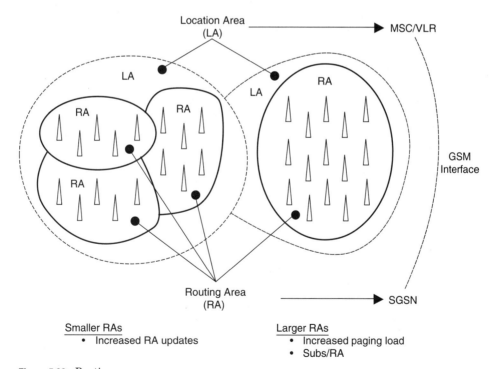

Figure 5.30 Routing areas.

border itself where due to the time required for the handoff process to occur the SU causes the session to slow down or be prematurely terminated.

6. *Check site parameters.* You need to verify that GRPS is enabled for the site, sectors, and every TRx involved that should carry GRPS traffic. In addition the allocation of how many TCHs that should be associated with GRPS service needs to be checked as well. Also verify that the neighbor sites are GRPS enabled.

The detach timer should also be checked to ensure that it is not set to short causing a reattachment and ensuring the VLR knows where the mobile is when the packet session is idle. Again for this part of the investigation all aspects of the datafill need to be looked at for both the source and target cells and sectors.

7. *Voice processing.* It is very important to verify if there is TCH traffic for voice calls on the site. If there is, then the problem could be configuration or PDSN related.

8. *PDSN.* For this part of the investigation you will need to verify that there have been and are currently no major core network service disruptions, which could preclude the initiation of service for the subscriber. Usually, however, if there is a major outage or disruption, other sites and services will also be affected and this will not be an isolated situation.

9. *Check the frequency plan.* You need to review the frequency plan that is in place for the site and sector involved. Things to look for involve cochannel and adjacent channel interference, including when the adjacent channel is in the same or adjacent sector, and when the cell with that sector overlaps the problem sector.

What you want to do is check the frequency plan on a per-sector and per-channel basis for any obvious issues with cochannel reusers. Therefore, an alternative frequency assignment needs to be pursued. However, if this is an interference-born problem, then LC and HO failures should be prevalent also.

10. *Check for maintenance issues.* When checking for maintenance issues, you need to determine if the problem can be isolated to a single TCH or TRx channel. If it can, then all TCHs will be affected. Review of the BTS and BSC alarms may produce the cause of the problem also.

5.11 Retunes

Any wireless mobile system will at some time require a frequency adjustment, commonly referred to as a retune. The retune can either be channel related, as in the case with GSM, IS-136, iDEN, and AMPs, or be PN code driven, as in the case with CDMA.

Most mature systems experience different levels of retunes on a regular basis. How successful your frequency plan is depends directly upon your approach to retunes. The level and scope of the interference control in a network needs constant attention especially with a rapid cell and radio expansion program.

There are several approaches to dealing with system retunes, based upon your configuration and experience level. Retunes can and do take on many shapes and forms. Depending on your own perspective a retune can be viewed as a fundamental design flaw or as part of the ongoing system improvement process. We firmly believe that retunes are part of an ongoing system improvement process.

The rationale behind this philosophical approach is that there is no real grid, contrary to popular belief. The primary driving point of the lack of a true grid focuses on real estate acquisition. There is rarely a site that is selected and built in a network where some level of compromise is achieved with it from the RF point of view alone.

The typical retunes that take place, which most people associate with cellular, involve adjustment to the surrounding sites when a new site is introduced into the network. The other type of retunes that occur are the result of a problem found in the network where there is a cochannel or adjacent channel interference problem.

Based on the volume of new sites being introduced into the network, the frequency of problems will most likely track with them. With every site added into the network the adjustments which are made can either help or hinder future expansions for the area, either new cells or radio.

There are several methods used for retunes and each has its pros and cons. The main retune methods used are

1. Systemwide flash cuts
2. Cell-by-cell retunes
3. Sectional retunes

The systemwide cuts were probably the most favorite at the beginning of cellular due to the size of the networks at the time. The usual comment made during a systemwide flash cut is, This systemwide cut will put the system on a grid and will eliminate the need for more retunes. However, as long as there is system growth the aspects of no more retunes will never materialize no matter what the underlying technology is.

The method of performing individual sites for retunes is valid for a rural statistical area (RSA) and basic trading area (BTA) type environment. However, for a metropolitan trading area (MTA) and metropolitan statistical area (MSA) type environment utilizing individual site retunes as the primary method for resolving many of the system's frequency management problems is not viable. Specifically the time frame and effort needed to retune an area that, say, has 50 cells will take multiple visits to the same sites as the adjacent sites or subsequent rings are worked on.

There are two key parameters that need to be looked at for any retune, besides improved service, and they are time, opportunity cost, and personnel logistics. When putting together a retune, the need for completing it in a timely fashion is important. In addition when you are focusing resources on retunes, the same resources are not working on other projects for the network. The personnel logistics, which tie into lost opportunities, will be strained and general maintenance will suffer which will impact the system performance of the network.

The specific method recommended for use involves sectioning the network into major quadrants. The shape of the quadrants is directly dependent on the size and scope of the system and personnel available. For example, if your system has 300 cell sites, it would be advisable to partition the network into three, possibly four, sections for retunes. The direction, i.e., clockwise or counterclockwise, is more dependent upon where you think it is more applicable to begin the efforts.

In addition it is necessary to limit the scale of the retune itself. The scale of the retune can be contained by predefining the specific items you are desiring to change for this retune. During this design process, there is a tendency to continually increase the size and scope of the retune due to a variety of reasons. To prevent the size of the retune from increasing to a level which is unmanageable, it is advisable that the manager or director for the engineering department establish a line in the sand after where no changes can take place. Ultimately the line established will probably be compromised, but the extent of that compromise will be less and the size and scope of the retune will be within the logistics available to be successful.

An example of how to set the line is shown in Figs. 5.31 and 5.32. The retune regions are defined as well as the do-not-exceed line (DNEL), which is different in area from the quadrants set up for the network. The amount of cell sites within the DNEL are more than for the retune section itself. The actual starting point for the retune sequence and the rotation pattern is a function of several issues. The first issue is the amount of problems, adjacent system interaction, growth plans, and a swag.

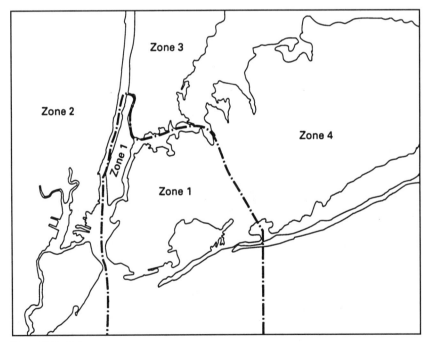

Figure 5.31 Retune zones.

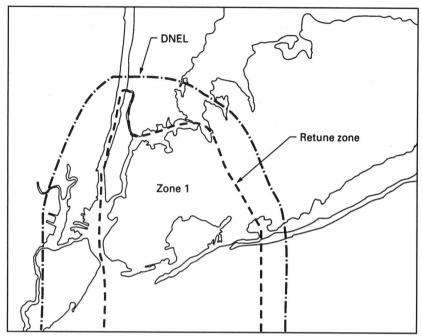

Figure 5.32 Do not exceed line (DNEL).

When setting up a retune, it is important to ensure that everyone you need for the retune effort is involved with this. Failure to ensure that all the groups are involved with the up-front planning will only complicate things at a later stage. The rationale for involvement lies in the fact that the frequency plan is always a point of contention for every department. Since retunes are a point of contention, having all the major groups in the process involved will eliminate finger pointing and keep the focus on improving the network.

To ensure that the retune process moves as smoothly as possible it is recommended that the following procedure be followed.

Pre-retune process

X-X-XX	Retune area defined.
X-X-XX	Project leader(s) defined and time tables specified as well as the scope of work associated with the project.
X-X-XX	Traffic engineering provides radio channel count.
X-X-XX	Frequency planning begins design.
X-X-XX	Phase 1 of design review (frequency planning only).
X-X-XX	Phase 2 of design review (all engineering).
X-X-XX	Phase 3 of design review (operations and engineering).
X-X-XX	Phase 4 of design review (adjacent markets if applicable).
X-X-XX	Frequency assignment sheets given to operations.

X-X-XX   Retune integration procedure meeting.
X-X-XX   Executive decision to proceed with retune.
X-X-XX   Adjacent markets contacted and informed of decision.
X-X-XX   Secure post-retune war room area.
X-X-XX   Briefing meeting with implementers of retune.
X-X-XX   MIS support group confirms readiness for postprocessing efforts.
X-X-XX   Customer care and sales notified of impending actions.

Retune process (begins X-X-XX at time XXXX)

X-X-XX   Operations informs key personnel of retune results.

Operations personnel conduct brief post-retune test to ensure call processing is working on every channel changed.

Operations manager notifies key personnel of testing results.

Post-retune process (begins X-X-XX at time XXXX)

Voice mail message left from engineering indicating status of retune (time).

Begin post-retune drive testing phase 1 (time).

Database check takes place.

Statistics analysis takes place.

Voice mail message left from RF engineering indicating status of post-retune effort (time).

Phase 2 of post-retune drive testing begins.

Commit decision made with directors for retune (time).

Phase 3 of post-retune drive testing begins.

X-X-XX   Continue drive testing network.

Statistics analysis.

Conduct post-retune analysis and corrections where required.

X-X-XX   Post-retune closure report produced.

During the design reviews it is important to ensure that the frequency planning checklist used for regular frequency planning is used for the entire process. The following are some checklists that need to be used for a regional-wide or systemwide retune and are somewhat technology platform specific.

These checklists present the minimum required to ensure that a proper design review takes place. It is suggested that several reviews take place at different stages of the design process to ensure a smooth integration process. Also the method of procedure (MOP) defined for the retune needs to have internal and external coordination delineated. In addition to all the coordination, the actual people responsible for performing the various tasks need to be clearly identified at the beginning of the process, not in the middle of it.

The following is the checklist that needs to be used for a regional or systemwide retune for AMPS and should be used as a starting point for a checklist.

Voice channel assignments
1. Reason for change
2. Number of radio channels predicted for all sites
3. New sites expected to be added
4. Proposed ERP levels by sector for all sites
5. Coverage prediction plots generated
6. C/I prediction plots generated
7. Cochannel reusers identified by channel and supervisory audio tone (SAT)
8. Adjacent channel cell sites identified by channel and SAT
9. SAT assignments checked for cochannel and adjacent channel
10. Link budget balance checked

Control channel assignments
1. Reason for change
2. Coverage prediction plots generated
3. Cochannel C/I plots generated
4. Proposed ERP levels by sector
5. Cocontrol channel reusers identified by channel and digital color code (DCC)
6. Adjacent control channels reusers identified by channel
7. DCC assignments checked for dual originations
8. 333/334 potential conflict checked

Frequency design reviewed by
1. RF design engineer
2. Performance engineer
3. Engineering managers
4. Adjacent markets (if required)

The following is the checklist that needs to be used for a regional or system-wide retune for an IS-136 system. The checklist assumes a dual-mode situation for cellular and can be easily modified to reflect IS-136 for PCS alone and should be used as a starting point for a checklist.

Voice channel assignments (AMPS)
1. Reason for change
2. Number of radio channels predicted for all sites for analog
3. Number of time slots and, therefore, radios predicted for all sites for digital
4. Digital and analog spectrum allocations
5. Guard band defined
6. New sites expected to be added
7. Proposed ERP levels by sector for all sites
8. Coverage prediction plots generated
9. C/I prediction plots generated
10. Differences noted between analog and digital reuse patterns and groups
11. Cochannel reusers identified by channel, SAT, and DVCC

12. Adjacent channel cell sites identified by channel, SAT, DVCC
13. SAT assignments checked for cochannel and adjacent channel
14. Link budget balance checked

Control channel assignments
1. Reason for change
2. Coverage prediction plots generated
3. Cochannel C/I plots generated
4. Proposed ERP levels by sector
5. Cocontrol channel reusers identified by channel and DCC
6. Adjacent control channels reusers identified by channel
7. DCC assignments checked for dual originations
8. 333/334 potential conflict checked

Digital control channel assignments
1. Reason for change
2. Coverage prediction plots generated
3. Digital control channel assignments matched to preferred channel list
4. Cochannel C/I plots generated
5. Proposed ERP levels by sector
6. Codigital control channel reusers identified by channel and DCC
7. Adjacent digital control channel reusers identified by channel
8. DCC assignments checked for dual originations

Frequency design reviewed by
1. RF design engineer
2. Performance engineer
3. Engineering managers
4. Adjacent markets (if required)

The following is the checklist that needs to be used for a regional or system-wide retune for a GSM system and should be used as a starting point for a checklist.

Channel assignments
1. Reason for change
2. Number of radio channels predicted for all sites
3. Number of time slots and therefore radios predicted for all sites
4. Spectrum allocation restrictions
5. Guard band defined
6. New sites expected to be added
7. Proposed ERP levels by sector for all sites
8. Coverage prediction plots generated
9. C/I prediction plots generated
10. Cochannel reusers identified by channel and BSIC
11. Adjacent channel cell sites identified by channel and BSIC
12. Link budget balance checked

Frequency design reviewed by
1. RF design engineer
2. Performance engineer
3. Engineering managers
4. Adjacent markets (if required)

The following is the checklist that needs to be used for a regional or system-wide retune for a iDEN system and should be used as a starting point for a checklist.

Radio channel assignments
1. Reason for change
2. Number of time slots predicted for all sites for 3:1 interconnect traffic
3. Number of time slots predicted for all sites for 6:1 and/or 12:1 interconnect traffic
4. Number of time slots predicted for dispatch traffic
5. Number of time slots predicted for DCCH traffic
6. Total number of radios, BRs, defined from time slots required
7. Spectrum allocation available for the market
8. Guard band defined
9. New sites expected to be added
10. Proposed ERP levels by sector for all sites
11. Coverage prediction plots generated
12. C/I prediction plots generated
13. Cochannel reusers identified by channel and DVCC
14. Adjacent channel cell sites identified by channel, and DVCC
15. Link budget balance checked

Frequency design reviewed by
1. RF design engineer
2. Performance engineer
3. Engineering managers
4. Adjacent markets (if required)

The following is the checklist that needs to be used for a regional or system-wide retune for a CDMA system. The retune here is based on the PN codes and not the RF allocations. However, there is a check for microwave clearance which is included for reference and should be used as a starting point for a checklist.

CDMA carrier assignments
1. Reason for change
2. Number of traffic channels predicted for all sites for analog
3. Spectrum allocation restrictions
4. CDMA channels available defined
5. Guard band defined
6. New sites expected to be added
7. Proposed pilot power level distribution defined

8. Coverage prediction plots generated
9. Pilot pollution problems identified
10. PN codes defined
11. PN reusers identified, including shift
12. Link budget balance checked

Frequency design reviewed by
1. RF design engineer
2. Performance engineer
3. Engineering managers
4. Adjacent markets (if required)

The following is the checklist that needs to be used for a regional or system-wide retune for a cellular CDMA system. The checklist assumes a dual-mode situation for cellular and should be used as a starting point for a checklist.

Voice channel assignments
1. Reason for change
2. Number of radio channels predicted for all sites
3. Number of traffic channels predicted for CDMA carrier
4. Digital and analog spectrum allocations
5. Guard band defined
6. New sites expected to be added
7. Proposed ERP levels by sector for all sites
8. Coverage prediction plots generated
9. C/I prediction plots generated
10. Cochannel reusers identified by channel and SAT
11. Adjacent channel cell sites identified by channel and SAT
12. SAT assignments checked for cochannel and adjacent channel
13. Proposed pilot power level distribution defined
14. Pilot pollution problems identified
15. PN codes defined
16. PN reusers identified, including shift
17. Hard handoff sites identified
18. Link budget balance checked

Control channel assignments
1. Reason for change
2. Coverage prediction plots generated
3. Cochannel C/I plots generated
4. Proposed ERP levels by sector
5. Cocontrol channel reusers identified by channel and DCC
6. Adjacent control channels reusers identified by channel
7. DCC assignments checked for dual originations
8. 333/334 potential conflict checked

Frequency design reviewed by
1. RF design engineer
2. Performance engineer

3. Engineering managers
4. Adjacent markets (if required)

Reiterating the retune method, which is recommended regardless of the technology platform used, is the regional retune process since it takes a systematic approach to frequency management. The systematic approach stresses that there is no perfect frequency plan, and the dynamics of the network in terms of channel and cell site growth necessitate a regular program for correcting the system compromises that are introduced. We have found this method to be very successful in improving the network's performance since this ensures that the system compromises introduced into the network over the year can be eliminated or simply improved upon.

The systematic retune process is advantageous but at the same time fraught with many potential downsides. It will enable a dedicated group to focus on an area of the network and optimize it to the best of their and the system's ability. Frequency reassignments are not the only factor looked for in a retune. Some additional issues are the current and future channel capacity, cell site growth, handoff, and cell site parameters. Basically in a retune you are scrubbing a section of the network involving many aspects besides channel assignments.

The reason you need to look at a multitude of additional parameters for the retune lies in the inherent fact that frequency management is the central cog for cellular engineering. Handoffs, cell site parameter setting, and overall performance of the network are directly impacted by the frequency plan. Periodically scrubbing a section of the network on a continuous basis will ensure that the compromises made during cell introductions and temporary retunes will be rectified at a predetermined time in the near future. This simple concept will enable the designers to seek more short-term fixes to the multitude of problems they face. Since the knowledge of the systematic retune date for the region is known in advance, the designers can use this future date as design completion date.

The stop point is important in this process quite simply because how you approach a problem and propose a solution is entirely dependent upon whether it is for 6 months or 5 years. If it is a 5-year solution, the time to bring the solution and the volume of unknown variables makes the problem all the more daunting and unmanageable. While it is important to design certain items for 10 to 20 years of useful life, frequency management should be considered short term.

When conducting region retunes, the scope of the project can easily be expanded to an unmanageable level. For example, a period retune used to help introduce close to two dozen cells occurred. The initial scope of the project involved retuning some 70 cells and introducing into the network the new cells at the same time. During the course of the efforts management lost sight of the actual work being done by the engineers, and the overall plan resulted in over 130 cells being returned and over 100 new channels being added to the existing sites, besides the new cells.

The expansion of the size of the project created a severe strain on both operations and implementation, not to mention the rest of engineering. The strain was caused largely by the frequency planners repeatedly altering their project scope and management not stepping in to stop it. The end result was operations, as usually is the case, pulled engineering bacon out of the fire. However, the level of system problems introduced due to the strain of the retune took over a month to correct.

Regardless of the method used for establishing retunes, it is important to always conduct a pretest and posttest. The pretest level for the retune is usually a few drive tests of selected road and a large volume of statistics analysis prior to the switch being thrown. The post-retune analysis is one of the most important aspects of the retune process because the accuracy of the design is done. It is also important because without it you will miss many opportunities for additional system performance improvements. The most important aspect is that you will never know if your design efforts were successful and how you can improve upon them unless you perform the post-retune analysis. Continuous refinement is the only way you can improve the network on a sustained basis.

One key element that was listed in the retune MOP is the time that the retune will take place over. It is strongly suggested that when you retune a network, the subscriber impact be considered. While this seems rather elementary, the person who is actually paying your salary, the customer, may be forgotten in the heat of the battle to get the task done.

It is recommended that all retunes take place in the maintenance window and over a weekend period. This will enable the least amount of negative impact on customers and will allow for maximum time for the groups, primarily engineering, to correct any issues.

The recommended post-retune process is listed below:

1. *Identification of the key objectives and desired results prior to the retune taking place.*   For example, if the goal is to facilitate adding five new cells into the network, then this is *the* goal. Identifying key metrics and anticipating their relative change and direction is very important. If you are operating at a 2 percent lost call rate and you are aiming for a reduction in lost calls by 10 percent to say 1.8 percent with this retune, this might prove to be an unrealistic goal. This would be a difficult proposition if your before and after channel count remained the same, meaning the overall channel reuse in the network would stay about the same. A better objective in this case would be to position the network for future growth without degrading the service levels already there and aiming for improvements in selected zones.

2. *Statistics analysis for 2 weeks prior to the effort, using the same time frames and reference points.*   One week is the minimum, but the more weeks you have in the analysis, the better it is to identify a trend. Obviously more than a couple of months worth of data is not relevant for this effort since traffic, system configuration, software, and seasonable adjustments makes comparison very difficult.

3. *Full cooperation of operations, implementation, customer service, MIS, and, of course, engineering for staffing levels.*   Support from each of these groups is critical for the mission to succeed.

4. *Staffing during changes.* It is important that the crew which is on duty during the retune document all the problems which occurred during the process and list what they have checked to prevent the need to reinvent the wheel.

5. *Post-retune statistics analysis.* While the drive tests are initially collecting the first tier of data for analysis, it is important for the engineers to validate that the entire system is configured in a fashion per the design. There has never been a case where problems have not been found during this stage for a multitude of reasons ranging from data entry mistakes to outright design flaws.

6. *Identification of the most problematic areas in the network.* Initial statistic analysis is done at this time during or right after the configuration is checked for the network. The problem sites need to be identified by following the key metrics listed before, which are
- Lost calls
- Attempt failures
- Blocks
- BER, FER, and SQE
- Channel failures
- Usage and RF loss
- Customer complaints
- Field reports from the drive test teams

The initial statistics will only focus on an hourly basis due to the freshness of the retune itself. The data are then checked against the expected problem areas identified before the retune itself and are also plotted on a map of the system so that patterns can be identified.

This process is repeated every 4 h for the first 2 days after the retune and then daily for the next 2 weeks. Obviously the degree of detail employed is relevant to the scope of work and the ultimate level of problems encountered.

7. *The initial drive test data are then analyzed for the key potential problem.* The focus is on the nature of the lost calls and any other problems reported, including dragged calls, interference but no drop, and dropped calls. The nature of the problems are then prioritized according to the severity perceived and cross checking with the statistic and anticipated problem areas. An action plan is then put together for each of the problems identified. Sometimes the problem is straightforward, like a missing handoff table entry, or nothing is determinable which requires additional testing as part of phase 3 of the drive testing.

8. *Drive test data are then analyzed for the general runs for the rest of the retune area.* This involves again focusing on any problems that occurred in phase 2.

9. The third phase of testing analysis involves the follow-up tests and post-parameter change corrections needed to the network, and, if required, additional tests are then performed.

10. Over the next 2 weeks a daily statistics and action report is generated showing the level of changes and activities associated with the effort. This effort is concluded by issuance of a final report.

5.12 Drive Testing

The concept of drive testing a network is usually well understood in terms of its importance and relevance for determining many design issues. Drive testing is used to help define the location of potential cells for the network, integrate new cells into the network, improve the existing network through pre- and postparameter changes, and retune support to mention but a few.

It is always interesting to note that it is usually the drive test team which sets up the test for collecting the data to qualify a potential new site. A well-trained drive test team is exceptionally critical for the success of a network. As most engineers know, any test can be set up to fail or succeed if the right set of conditions are introduced. Therefore, it is very important to have a strong interaction between various engineering departments and the drive test teams.

There are several types of drive tests which take place, for example,

1. Pre site qualification
2. SQT
3. Performance testing
4. Pre- and postchange testing
5. Competition evaluations
6. Post cell turn-on
7. Post retune efforts

Obviously these items need to be performed exceptionally well. To ensure that the testing is done well a defined test plan needs to be generated and reviewed with the testers so that they understand what the overall objective really is. Often there are many alterations to a test plan that are left to the test team. If the test team understands what the desired goal and/or results are for the test, they have a better chance of ensuring that the alterations and observations they make during the test are beneficial and not detrimental.

One of the key critical elements which needs to be monitored and checked on a periodic basis is the maintenance and calibration of the equipment. Vehicle maintenance needs to be adhered to in order to ensure the fleet is at key operational readiness at all times. Since major problems in the network requiring full deployment of resources is never a planned event, having the fleet in top operational condition is a high priority. It would be a tragedy if half the fleet, assuming more than two vehicles, were in for some level of repair when a major system problem occurred.

The one area which requires continued attention is the calibration of the field equipment. For the SQT equipment the transmitters used need to be stress tested on a periodic basis to ensure that they are functioning properly. It is important to ensure that a transmitter used for an SQT that will be operational for, say, 6 h maintains the same output power for the duration of the testing. One simple way to check this fact is to ensure that a wattmeter is in line for the test and before and after test readings are done to ensure that there were no ERP alteration issues that took place during the test.

The antennas and cables used for the SQT need to be validated on a regular basis. A full depot check on all the equipment needs to be performed on a 6-month basis. However, spot checks can be done for each test and recorded in the test log to ensure that these components are operating properly.

The cables and antennas used need to be swept at 3-month intervals to ensure that nothing abnormal has occurred with them. The test should be done on a more regular basis but with the per-test snapshot and the quarterly SQT equipment integrity check, the level of confidence with the test equipment functionality should be high.

Equipment used for measuring power should be calibrated yearly as should the spectrum analyzer, network analyzer, and service monitors. Since so much reliance is placed on the accuracy of the data collected by the test equipment, it is only logical that the test equipment be checked on a routine basis to ensure its integrity.

The test vehicle measurement equipment itself needs to be validated on a regular basis due to continued problems which happen to all drive test vehicle equipment. It is strongly recommended that the equipment be tested on a monthly basis using a full calibration test which checks out, among other items, the functionality of the antennas, receiver sensitivity, RSSI accuracy, adjacent channel selectivity, transmit power, and data deviation.

The calibration of the test equipment needs to be recorded and stored in a central book that is available for quick inspection by all in the department. The calibration of the field measurement equipment and other SQT pieces needs to be listed on the test forms as "in calibration." The calibration records should also include the equipment serial number. In the event that the equipment turns up missing, you will then have a source to track it from.

Whatever the field measurement equipment used in the drive test vehicle is, it is exceptionally important that you know the accuracy of the equipment. This involves the adjacent channel selectivity, which is an important value to know because when you are monitoring, say, the control channels of the cell sites, you obviously are trying to measure an adjacent channel signal level from the dominant server in the area. The ability of the receiver to reject adjacent channel signals is imperative for making rational decisions on problems or potential problems. For example, if you do a single channel plot and notice that it is rather hot in an area near another site, the cause might be that the adjacent channel selectivity is not sufficient to isolate the desired signal from the undesired signal.

The receiver sensitivity is also an important value to know for the equipment since a receiver sensitivity of -102 dBm is not sufficient to measure signals at, say, -110 dBm. The difference in the two is an 8-dB signal and in most cases if you are designing for a 17- or 18-dB C/N level, the desired serving level might be mistaken for a -85 instead of a -93 dBm which has a major impact on the design criteria for the area.

There are many other variables for just the test equipment in terms of how it collects the signals which are imperative to understand. There are many source documents which help determine the sampling intervals required for an accurate RSSI measurement to take place.

The postprocessing of the data is a key area which many people overlook. With the large amount of data collected many times, there is a significant amount of postprocessing done to reduce the amount of data displayed at the end for the engineer to see. Obviously if you average many points and are performing an interference analysis, then a problem area might be masked due to the peak interfering signal being averaged with many other bins of data to come out with some nominal level. An example of this would be when driving a large area and going over a bridge or a high elevation on a highway and the interference is only a two-block area compared to the 20-mile data collection run. It is imperative that all postprocessing steps be defined in advance of any data reduction process so the tradeoffs made are understood before they take place.

The SQT role of the drive team is critical in the capital deployment process for any company. As stated before, if the SQT is not done properly, a wrong decision can easily be made either to build or not to build a cell site at that location. To ensure the SQT is performed properly it is necessary to establish and follow a test plan.

As mentioned in the design guidelines for the SQT it is imperative that the test plan be made in advance of the site visit by the testing team. The objective for this effort is to ensure that the mounting of the antenna and the other ancillary pieces is done according to the design. The drive routes used for the testing must be sufficiently clear to enable the team to follow the desired direction.

In addition to the SQT is performance testing which we have labeled pre- and postchange testing. This is a very important aspect of the drive team, and the feedback the group gives the engineers is critical for determining the nature and cause of the problem at hand.

Therefore, to ensure maximum output for the drive team the engineer requesting the test must explain the test's objective exactly to the team. Knowing the objective of the test beforehand will ensure critical feedback regarding particular issues that arose during the testing is collected. An example of feedback might be that when the test team was helping to determine why an area was experiencing a high lost call rate, part of the major road they drove on was almost below grade and would have had a significant influence on the lack of signal for the test area.

For performance drive tests there are functionally two general classifications of tests, internal and external. The two test objectives both focus on trying to identify and help correct a real or perceived problem in the network. Both internal and external performance testing follow the same general testing format. The objective as with all optimization techniques is defining what the objective is and then what you are going to do prior to any action taking place.

There are several types of performance drive tests that can be conducted. The following is a listing of the major items.

1. Interference testing (cochannel and adjacent channel)
2. Coverage problem identification
3. Customer complaint validation

4. Cell site parameter adjustments
5. Cell site design problems
6. Software change testing
7. Postchange testing

When scheduling the testing, it is important to identify a priority level for these testing types. The objective of defining a priority level is to have a brief procedure in place when a problem occurs, which it always will, so that the highest priority level item will be taken care of first. This will prevent the first in, first out (FIFO) effect of drive test scheduling being the only criteria for scheduling tests.

Regarding all performance changes made to the network, the need to have a pre- and postchange test conducted can never be overstressed. Pre- and postchange tests ensure that the problem is truly identified. The prechange test data should be used as part of the design review process. Often many good ideas seem great, but when you actually put them down on paper and submit them for peer review they may not be as valid as initially thought.

Once a change is implemented into the network, it is equally important to conduct a postchange test. The postchange test is meant to verify that the design change worked. It is also meant to verify that the changes made to the network do not cause any unanticipated problems to appear. The postchange test plan also needs to be presented at the design review.

Regarding competitive evaluations the nature of how this is done is more of an art than a science. Competitive evaluations are an art since the methods used and the benchmarking calculations are primarily nebulous. The reason why they are so nebulous is that the variables involved are subjective and they define quality. Our personal definition of quality is probably different from your definition of quality and since there is no specification to truly benchmark against, this quality figure is left to the determination of an outside consulting firm or the senior management office.

Several methods for attempting to evaluate the current quality of the network and the ranking against the competition have been devised. A few companies offer a device which measures the audio levels on both the uplink and downlink paths. The audio level measurements are made quantitatively, which enables the evaluation to be done the same way time and time again. However, the parameters used in the testing and postprocessing are user defined leaving quite a lot open to interpretation. Using this method will produce the best results since it gets rid of many of the subjective items associated with quality testing.

Another method used is to hire people to drive sections of the network and make a series of calls on your system and the competitor's systems. Presumably the calls placed are similar enough in time of day, location, and duration to make a direct comparison. However, the people on the landline or the mobile are left to describe the call by listing it as good, poor, or excellent. Obviously the skills and subjectivity of this type of test leaves a lot to be desired.

An example of some subjectivity problems occurred once when a quality test came back saying an area of the system was performing exceptionally when engineering and operations knew better. Specifically the area reported to be performing great was a known trouble area which during this time was receiving a lot of attention. The good report was hailed as a success. However, on the next competitive test that was done for the same area the report came back that the area was performing poorly. Upper management was shown the previous test and the current test which caused much teeth gnashing in engineering and operations. There was an exceptional amount of effort then placed in trying to refute the report which previously was hailed as a very good report. The fundamental problem in this case was not the report itself but the fact that when good news is presented, but it is incorrect, it rarely gets challenged. What should have happened here is that the test should have been challenged when it said things were very good when it was known that in fact it was not.

The post-turn-on, or activation, testing is exceptionally critical to ensure that the system is not degraded by the entrance of a new cell site. However, there has never been a site introduced into a network that has not had some level of problem with it. The argument here is that when there are no problems found, no one is really looking.

The key to the post-turn-on testing involves two simple principles. The first principle is that you have to define the test prior to the site becoming commercial. The second principle is that you need to have the post-turn-on testing done immediately after the new site goes commercial to ensure that the problems the site has will be found quickly and rectified.

The problems associated with the new site should not be impossible to overcome, provided the design guidelines listed in a previous chapter are followed. Instead the issue is the handoff table adjustment, power level setting, or bias adjustment which needs to take place right after turn-on.

An example of the post-turn-on test request is shown in Fig. 5.33. The post-turn-on test needs to be an integral part of the post-turn-on MOP discussed later in this chapter. The actual form utilized as a trigger point for requesting resources should be well defined. A suggested format to use is proposed, but like all the other forms and procedures presented it is essential that they are crafted to reflect internal organization structures.

The last major area of drive testing involves post-retune efforts, which are similar to many of the other tests done, except that these are done in a tiered approach. Specifically it is recommended that these are done in a three-phase approach where each phase has a unique mission statement. An example of the retune zone is shown in Fig. 5.34.

Phase 1 is the identification and characterization of the most highly probable problem areas for the design which need to be validated first. Generally this involves bridges, the major roadways, and areas on C/I plots which show anything 19 dB C/I or less for AMPS, or where the testor suspects a problem (Fig. 5.35) based on the wireless technology used. Once the potential problem areas are identified, a drive route is designed corresponding to the potential problem areas previously identified (Fig. 5.36).

RF System Performance and Troubleshooting   363

```
Cell site code: _____

MOP: _____

Requester: _____

Expected drive test start date: _____

Expected drive test start time: _____

Number of test vehicles needed: _____

Estimated drive test duration: _____

Data to be collected: _____

Report any problems to: _____

Special comments: _____

Drive map attached (Y/N)
```

Figure 5.33 Cell site activation post-turn-on test form.

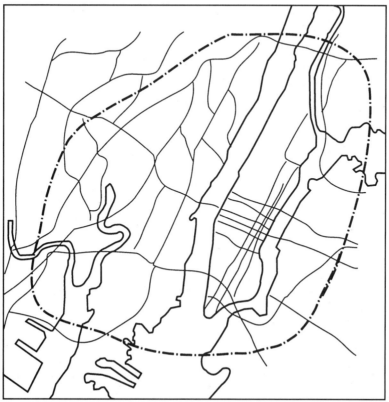

Figure 5.34 Retune zone.

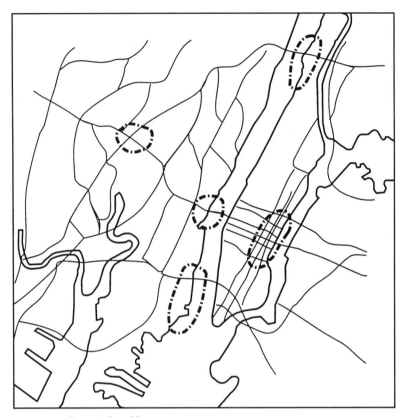

Figure 5.35 Potential problem areas.

Phase 2 involves the characterization of class 1 and 2 roads in the retune zone (Fig. 5.37). This is meant to identify any problems along the major arteries in the network. The objective is to ensure that the major throughways do not have any apparent problems with them prior to a full system load.

Phase 3 involves testing all the areas which showed up as problems in phases 1 and 2 for either further clarification of the problem or validation that the fixes implemented worked.

This tiered method enables resources to be focused on resolving the problems in a timely and efficient method. A key to all the posttesting activities is that they are conducted immediately after the action takes place. This immediate time frame is important so that problems can be found by the testing team before subscribers discover them.

5.13 Site Activation

The philosophy of site turn-on varies from company to company. Some of the philosophies are driven by engineering; others are driven by financial objectives. There are several philosophies used in the wireless industry. The first

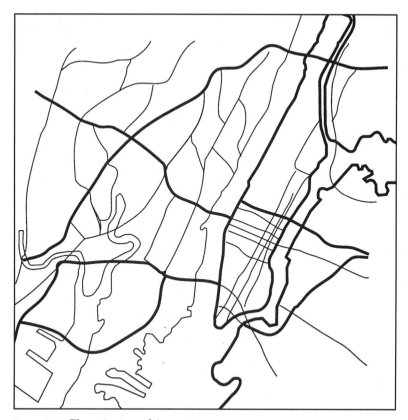

Figure 5.36 Phase 1 retune drive routes.

philosophy is that when a site is finished being constructed, it should be activated into the network. The second primary philosophy is that the site's depreciation should be minimized or maximized, depending on the accounting method employed by the company. The third method is where the sites are not activated until the implementation plan put forth by engineering dictates the timing of the new cells.

The first philosophy of cell site activation (turn it on immediately) has an emotional and upper management appeal. The appeal is that the site is being constructed to resolve some system problem, and the sooner it is put into service, the sooner the problem the site is designed for will be resolved. While this simple philosophy has its direct merit, it also has a few key drawbacks which if done incorrectly can create more system problems than it was intended to fix.

Specifically the drawbacks involve timing and coordination of the engineering plan to bring the site into the network. If site A requires handoff changes and is activated when the last ATP function is completed, the possibility of the handoff changes being implemented at this time is low. One alternative to this is to have the topology changes done in advance of the site's activation, but if it is done incorrectly, then handoff problems and lost calls could result. Another situation

366　Chapter Five

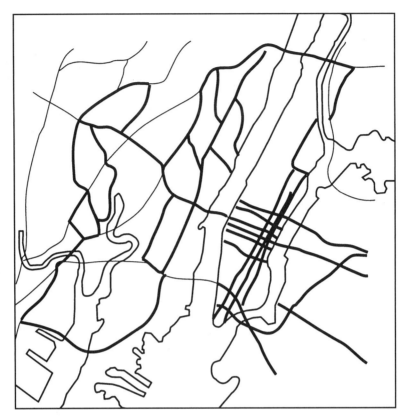

Figure 5.37　Phase 2 retune drive routes.

could occur where there are too many handoff candidates in the topology tables of the sites complicating the frequency plan for the area.

Another major problem with this activation philosophy is the coordination of resources. As in most cases a site needs some level of intrasystem and intersystem coordination. If the site is activated at a seemingly random time, there is no guarantee that all the required coordination has been completed.

The third disadvantage with this method is the queuing of post-turnoff resources for the system troubleshooting phase of a site turn-on. The post-turn-on efforts to be fully effective require coordination in terms of timing. If a site is expected to be turned on anywhere within a 3-day window based on implementation problems, it is difficult to ensure that post-turn-on testing will begin right after turn-on. If you, however, want to ensure that post-turn-on testing begins right after turn-on of the cell, then there will be additional opportunity costs associated with this effort since resources will be significantly mismanaged.

The fourth disadvantage with this effort is the most important aspect, and it is the customer impact. Turning on the site when the implementation process is finished will most likely occur during the day. It is strongly advised that any site or major system action be conducted in the system maintenance window. The simple objective here is to minimize any negative impact the subscriber might experience and try and allow enough turnaround time for the engineering and operations teams to correct any problems before the subscribers find it.

The second philosophy of activating (to maximize or minimize depreciation costs) is largely driven by the financial requirements of the company. To maximize depreciation costs, the operator usually scrambles to activate as many sites as possible by the end of the fiscal year, usually corresponding with the calendar year. The objective here is to maximize the potential depreciation expense the company will have in any fiscal year.

The minimization of depreciation philosophy involves attempts to defer the depreciation of the new cell site into the next fiscal year. What is typically done here is that a site is prepared for activation into the network but will not be turned on until the next fiscal year. The objective here is to minimize the amount of expenses reported by the subsidiary to its parent company.

Usually the philosophy to maximize depreciation expenses has the activation philosophy of "turn it on now." The minimization of depreciation philosophy usually involves a plan issued by engineering which matches the financial goals of the company.

The third philosophy for site activation (no new cell site or system change without a plan being issued and approved by engineering) ensures that the introduction of the new cell into the network has been sufficiently thought out, resources are planned and staged, and all the coordination required has or will be done in concert with the activation. This philosophy is essential to ensure that new cells are introduced into a mature system gracefully.

For this process usually more than one cell site is activated at the same time. Typically, if logistically possible, a series of cell sites in a region are activated at the same time. This philosophy enables a maximization of post-turn-on resources to focus on the issues at hand and also ensures the minimization of system alterations required. The map in Fig. 5.38 shows how combining several cell sites into a single turn-on for the system will facilitate maximizing resources and minimize the amount of changes required to the network.

The largest problem associated with trying to utilize this philosophy is getting upper management approval. The opposition occurs when a site may have to wait several weeks for activation due to configuration changes needed in the network to ensure its smooth transition. The issue of having a site ready for service and not activating it immediately is the hardest obstacle to overcome when presenting the case.

However, there are several key advantages that need to be stressed with this philosophy which might or might not be apparent.

1. Reduced system problems
   - Interference
   - Handoffs
   - Parameter settings
2. Coordinated efforts between all departments
3. Design reviews of integration plan take place
4. Intrasystem and intersystem coordination is more fluid

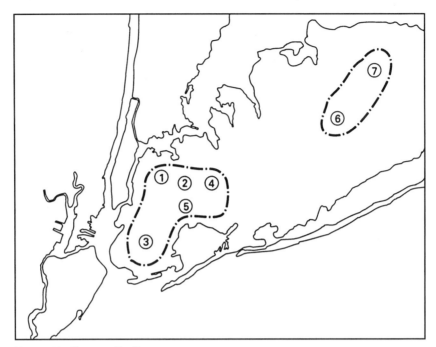

Figure 5.38 Group activation.

5. Pre-turn-on testing is conducted
6. Minimized negative customer impact through activation in maintenance window

There are obviously more positive attributes; however, the key issues are reduced system problems, pre-turn-on testing, and minimization of negative customer impacting issues as a result of following a plan. It is essential that for every cell site brought into a network that a plan is generated for its introduction and then carried out.

The design reviews necessary for a new site activation into the network need to be conducted by several parties. There are several levels of design reviews for this process. The first level of design reviews involve the RF engineer and the performance engineer discussing the activation plans and reviewing the plan of action put forth. The second level of design reviews involves having the manager of the RF engineering group sign off on the implementation design with full concurrence with the performance manager. The third level of design review involves reviewing the plan with the director for engineering and operations personnel to ensure that all the pieces are in place and that something has not been left out, like who will do the actual work.

After the design reviews are completed, the MOP for the activation is released. It should be noted that during the design phases the MOP should have been crafted and all the parties involved informed of their roles. A sample MOP is listed in Fig. 5.39 for comparison; obviously the exact MOP for the situation is different and needs to be individually crafted.

It is essential to always include a back-out procedure for cell site activation in the event of a major disaster taking place. The escalation procedure should be defined in the MOP and the decision to go or not to go needs to be at the director level, usually the engineering director or the operations director.

After the MOP is released and the design reviews are completed, it is essential that the potential new cell be visited by the RF and performance engineers

---

Date

Preactivation process

X-X-XX New cell sites to be activated defined
X-X-XX Project leader(s) named and timetables specified, as well as the scope of work associated with the project
X-X-XX Phase 1 design review (frequency planning only and RF engineer for site)
X-X-XX Phase 2 design review (all engineering)
X-X-XX Phase 3 design review (operations and engineering)
X-X-XX Phase 4 design review (adjacent markets if applicable)
X-X-XX Frequency assignment and handoff topology sheets given to operations
X-X-XX New cell site integration procedure meeting
X-X-XX Performance evaluation test completed
X-X-XX Executive decision to proceed with new cell site integration
X-X-XX Adjacent markets contacted and informed of decision
X-X-XX Secure post-cell-site activation war room area
X-X-XX Briefing meeting with drive test teams
X-X-XX MIS support group confirms readiness for postprocessing efforts
X-X-XX Customer care and sales notified of impending actions

New cell site activation process (begins X-X-XX at time XXXX)

X-X-XX Operations informs key personnel of new cell site activation results
      Operations personnel conduct brief post-turn-on test to ensure call processing is working on every channel and that handoff and handins are occurring with the new cell site
      Operations manager notifies key personnel of testing results

Post-turn-on process (begins X-X-XX at time XXXX)

      Voice mail message left from engineering indicating status of new cell sites (time)
      Begin post-turn-on drive testing, phase 1 (time)
      Database check takes place
      Statistics analysis takes place
      Voice mail message left from RF engineering indicating status of postretune effort (time)
      Phase 2 of post-turn-on drive testing begins
      Commit decision made with directors for new cell site (time)
      Phase 3 of post-turn-on drive testing begins

X-X-XX
      Continue drive testing areas affected
      Statistics analysis
      Conduct post-turn-on analysis and corrections where required

X-X-XX
      Post-turn-on closure report produced
      New site files updated and all relevant information about the site transferred to performance engineering

Figure 5.39 Method of procedure for new cell site integration.

at various stages of the construction period. However, prior to activation it is essential that a pre-turn-on procedure take place. The pre-turn-on procedure is meant to ensure that the site is configured and installed properly so that when the site is activated into the network, the basic integrity of the site is known. The pre-turn-on procedure that should be followed is listed in a later part of this chapter.

Internal coordination involving new sites being introduced into the network is essential. The MOP listed above focuses on voice mail notifications to many groups inside and outside the company. However, it is essential that the activation of new cells and major system activities be announced to other departments in the company to inform them of the positive efforts being put forth by engineering and operations.

The primary groups to be informed to ensure some level of notification takes place are

1. Sales
2. Marketing
3. Customer service
4. Operations, real estate, and engineering
5. Corporate communications
6. Legal and regulatory

Basically, the entire company needs to be notified of the positive events that take place. One of the most effective methods is through the company's internal voice mail system. However, not everyone will have an individual voice mail account.

To ensure that all the people are notified of the new site's activation into the network, a series of communications can be accomplished. One method is to issue an electronic mail message to all the employees notifying them of the new sites and any particulars about the intended improvements, if any, to the network. Another method is to issue a memo to everyone in the company declaring the activation of the sites and the improvements that have been made.

External coordination for new sites is as essential as internal coordination. Specifically the neighboring systems should know when you are bringing new sites into the network and other major activities. The reason behind this effort is that your actions may have an unintended consequence on them, either positive or negative, which they need to know. In the same light, by providing your neighboring systems with new site activation information, you will receive reciprocal communications.

After the sites are activated into the network, it is essential that post-turn-on testing begins immediately. There has never been a site activated into a network, that we are aware of, which did not have some type of problem with it. Therefore, it is essential that the efforts put forth in this stage of the site activation process receive as much attention as the design phases did.

The key parameters which need to be checked as part of the post-turn-on activities are site configuration checks, metrics analysis, and drive test analysis.

1. *Site configurations from the switches point of view.* The objective here is to check all the cell site parameters as reported by the switch to those intended for the initial design. What you are looking for here is a possible entry mistake or even a design mistake made during the design process. Usually a data entry mistake is found in this process or an entry is left out. It is imperative that the neighbor cell sites also be checked in this stage of the process.

2. *Metrics analysis.* The objective with this part of the post-turn-on activities is to help identify and isolate for problem resolution problems reported in the network by the system statistics. This process requires continued attention to detail and an overall view of the network at the same time. The metrics that you should focus on involve the following items:
- Lost calls
- Blocking
- Usage
- Access failures
- BER, FER, and SQE
- Customer complaints
- Usage and RF loss
- Handoff failures
- RF call completion ratio
- Radios out of service
- Cell site span outage
- Reported field problems called in by technicians or the drive test team

The statistics monitored should be the primary sites activated and their neighboring cells. The issue here is to not only look at the sites being brought into service but also to ensure that their introduction did not negatively impact the system.

The actual numbers to use for comparison need to be at least 1 week's data for benchmarking, if possible. In addition the numbers used for the new cell should be compared against the design objective to ensure that the site is meeting the stated design objectives.

3. *Drive testing.* The post-turn-on drive test data analysis needs to take place here. This effort usually begins at the specified time after turn-on, usually early in the morning or late at night depending on the activation schedule. The drive tests are broken down into three main categories.
- *Phase 1.* Involves focusing on areas where there is the highest probability of experiencing a system design problem. The identification of these areas can be through use of prior experience, C/I plots, or SWAG.
- *Phase 2.* Involves targeting the rest of the areas involved with the site activation activities, usually the remaining class 1 and class 2 roads not already driven.
- *Phase 3.* Involves driving areas that were uncovered as problems in phases 1 and 2. This level of testing either verifies that the problem identified previously is still in existence or the change introduced into the network did its job.

4. *New site performance report.* The last stage in the new site activation process is the issuance of the new site performance report. The performance report will have in it all the key design documents associated with the new site. The key design documents associated with this new site should be stored in a central location instead of a collection of people's cubes. The information contained in the report is critical for the next stage of the site's life. The next stage of the site's life involves ongoing performance and maintenance issues.

To ensure that poor designs do not continue in the network it is essential that the new site meet or exceed the performance goals set forth for the network. If the site does not meet the requirements set forth, then it should remain in the design phase and not the ongoing system operation phase. The concept of not letting the design group pass system problems over to another group is essential if your goal is to improve the network.

The new site performance report needs to have the following items included in it as the minimum set of criteria.

1. Search area request form
2. Site acceptance report
3. New cell site integration MOP
4. Cell site configuration drawing
5. Frequency plan for site
6. Handoff and cell site parameters
7. System performance report indicating the following parameters 1 week after site activation
    a. Lost calls
    b. Blocking
    c. Access failures
    d. Customer complaints
    e. Usage/RF loss
    f. BER, FER, and SQE
    g. Handoff failures
    h. RF call completion ratio
    i. Radios out of service
    j. Cell site span outage
    k. Technician trouble reports
8. FCC site information
9. FAA clearance analysis
10. EMF power budget
11. Copy of lease
12. Copy of any special planning or zoning board requirements for the site

The new cell site performance report is an essential step in the continued process for system improvements. Only once a site is performing at its predetermined performance criteria should the site transition from the design phase to the maintenance phase.

Lastly as part of the presite activation a site checklist should be completed, which is given in Table 5.5.

TABLE 5.5 Cell Site Checklist

Topic	Received	Open

Site location issues
  1. 24-hour access
  2. Parking
  3. Direction to site
  4. Keys issued
  5. Entry/access restrictions
  6. Elevator operation hours
  7. Copy of lease
  8. Copy of building permits
  9. Obtainment of lien releases
  10. Certificate of occupancy

Utilities
  1. Separate meter installed
  2. Auxiliary power (generator)
  3. Rectifiers installed and balanced
  4. Batteries installed
  5. Batteries charged
  6. Safety gear installed
  7. Fan/venting supplied

Facilities
  1. Copper or fiber
  2. Power for fiber hookup (if applicable)
  3. POTS lines for operations
  4. Number of facilities identified by engineering
  5. Spans load tested over 24 hours

HVAC
  1. Installation completed
  2. HVAC tested
  3. HVAC system accepted

Antenna systems
  1. FAA requirements met
  2. Antennas mounted correctly
  3. Antenna azimuth checked
  4. Antenna plumbness check
  5. Antenna inclination verified
  6. SWR check of antenna system
  7. SWR record given to operations and engineering
  8. Feedline connections sealed
  9. Feedline grounds completed

Operations
  1. User alarms defined

Engineering
  1. Site parameters defined
  2. Interference check completed
  3. Installation MOP generated
  4. FCC requirements document filled out
  5. Drive test complete
  6. Optimization complete
  7. Performance package completed

TABLE 5.5 Cell Site Checklist (*Continued*)

Topic	Received	Open
Radio infrastructure		
1. Bays installed		
2. Equipment installed according to plans		
3. Receiver and transmitter filters tested		
4. Radio equipment completes acceptance test procedure (ATP)		
5. Transmitter output measured and correct		
6. Grounding complete		

5.14 Site Investigations

Cell site investigations can be equated to a hunting trip. If you are a good tracker and understand the game, your chances of success are greatly improved. There are two primary types of site investigations, new sites and existing sites. The new sites referred to here are those which have not begun to process commercial traffic and thus are not activated into the network when visited. The existing cell sites, however, are currently active sites in the network. The existing cell site usually has some particular problem associated with it that now warrants an engineering investigation.

5.14.1 New sites

For a new cell site the site investigation is essential to ensure that there are no new problems introduced into the network as a result of the physical configuration of the site. The objective of a new site investigation is to validate that site is built to the design specifications put forth by engineering. Some of the key areas to validate are

Antenna system orientation

Antenna system integrity

Radio power settings

Cell site parameters

Hardware configuration

Grounding system

This list of major items needs to be expanded upon since they are primarily engineering issues and do not focus on the radio and cell site commissioning aspects required by operations.

The antenna system orientation is essential to have validated prior to activation. It has a direct impact on how the cell site will interact with the network. For example, if the orientation is off by say 30°, then the C/I levels designed for it cannot be met. When inspecting the orientation of the antennas, it is essential to validate their location and installation versus the AE drawings for the site which were approved by engineering. Lastly here, if there is an obstruction to the antenna system itself, this needs to be corrected quickly.

The orientation of the site can be validated through several methods depending on whether it is a rooftop or tower installation. Ordinarily, whoever installs the antennas for the site are required to validate the orientation of the antennas through some visual proof provided to the operator.

If the installation is on a rooftop, you should use a 7.5-minute map with some type of optical alignment method, usually a transit. By referencing a point on the 7.5-minute map, it is easy to validate the orientation of the antennas with a high degree of accuracy.

When installation is on a tower or monopole, the orientation check is a little more difficult. However, the site drawings should reference the orientation of the legs of the tower itself. Once you have the orientation, you can verify that the antennas and mounts used are installed in the right locations. However, validating the orientation is more difficult; therefore, you will need to utilize a 7.5-minute map and locate three points at a distance from the tower. When traveling to each of these points, it will be necessary to establish your bearings; use an optical device (transit) and site the antennas for the sector you are in and validate that the bore site for the antenna appears to be correct.

This discussion has addressed physically checking sectored sites, but obviously for an omni-cell site orientation is not the issue but rather it is the plumpness of the antenna. It is necessary, where applicable, to also validate the plumpness of every antenna at a site. This can be accomplished through use of a digital level or visual inspection where the application of a level is not practical or safe.

In validating the physical aspects of the antenna system it is necessary to check if downtilt is employed as part of the design and if so what angle is utilized. This can be checked again through use of a set of simple measurements made on the antenna itself. One word of caution, Do not just utilize peg holes for validating the downtilt angle unless the exact angle versus peg hole count is known for that antenna and installation kit used. Figure 5.40 is an example of how to calculate the degree of downtilt employed at a site.

An additional step in the inspection of the antenna system involves validating that the correct antennas are in fact installed at the site. The simplest method is to verify the make and model number by reading it right off the antenna, where applicable.

One key element to note is the physical mounting of the antenna. On several occasions an antenna that has electrical downtilt employed as an omni-antenna has been installed upside down since this is a standard installation on a monopole site. The additional interesting aspect to the antenna inverting situation was that the drain hole was now on the top and the antenna was effectively becoming a rain-level indicator, an interesting sideline but not the intended purpose. Additionally if you are using electrical downtilt and invert the antenna, the pattern result is significantly altered from the desired uptilt situation.

After the physical installation characteristics are checked for an antenna system, it is necessary to perform an S11 test on the antenna system.

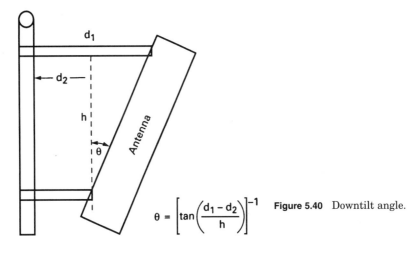

Figure 5.40 Downtilt angle.

$$\theta = \left[\tan\left(\frac{d_1 - d_2}{h}\right)\right]^{-1}$$

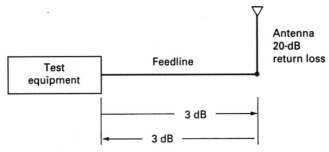

Figure 5.41 Antenna system.

When checking the antenna system, it is important to predetermine the values anticipated prior to actually making the measurements (see Fig. 5.41). For example, if you expect to get a value of 26 dB return loss

$$\begin{aligned}
\text{Antenna retune loss} &= 20 \text{ dB} \\
\text{Feedline loss up} &= 3 \text{ dB} \\
\text{Feedline loss back} &= \underline{3 \text{ dB}} \\
\text{System return loss} &= 26 \text{ dB}
\end{aligned}$$

you would pass the antenna system with a return loss of anywhere from 24 to 26 dB. However, if you got a 14-dB return loss, then this would indicate a major system problem since

$$\begin{aligned}
\text{System return loss} &= 14 \text{ dB} \\
\text{Feedline loss up} &= 3 \text{ dB} \\
\text{Feedline loss back} &= \underline{3 \text{ dB}} \\
\text{Antenna return loss} &= 9 \text{ dB}
\end{aligned}$$

The interesting point here is that most operators will pass an antenna system with a standing-wave ratio (SWR) of 1.5:1 which is a 14-dB return loss. However, when you factor in the cable loss, a SWR of 1.5:1 really means that your antenna is experiencing a real SWR of greater than 2.0:1. A SWR of 2.0:1 would have any cellular operator demanding action. Therefore, it is important to remember that the feedline can and will mask the real problem if you are not cognizant of the ramifications.

The last step in the antenna system validation involves validating that the antennas are indeed connected as specified. The real issue is to check if the antenna 1 feedline in the cell site is really connected to the actual antenna 1 for the sector.

The simple test involves using a mobile that is keyed on a particular channel and driving to the bore site of every sector, in a directional cell, and making sure that maximum smoke is measured on the antennas assigned for that sector. The key point here is that even the transmit antenna should be checked at this time, provided it is designed to pass energy in the mobile receive band also.

Figure 5.42 depicts a mobile located in the bore site of sector 1 for the cell site. The mobile operator keys the mobile's transmitter and sends energy on a cell site receive frequency. Figure 5.43 is an illustration of what the person at the cell site would be observing with either a spectrum analyzer or a service monitor.

One additional test associated with any cell site involves checking the filtering system used for the site. It is imperative that the actual filter characteristics

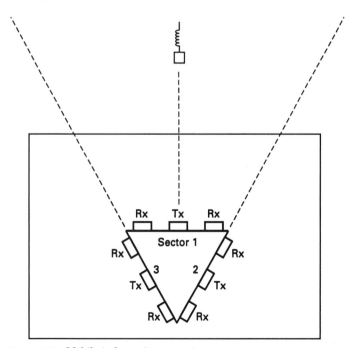

Figure 5.42 Mobile in bore site sector 1.

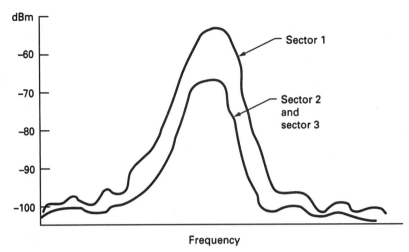

Figure 5.43 dBm display of antenna alignment test.

are checked at this time to prevent additional problems from being introduced into the network. The most effective method for performing this effort is through a S21 test or rather a through test.

The following is a brief listing of items which need to be checked for a new cell site. Some of the items in this list have previously been talked about. When reviewing the new cell site checklist, it is important to also obtain operation's input into the proposed form to ensure there is one form and not several. This list should be used as a punch list for correcting identified problems prior to the site going commercial and issues after it has.

It is essential that the checklist provided be used as a minimum and that it should be modified and expanded upon based on your own particular system requirements.

Cell site performance checklist

1. Antenna system
    a. Installation completed
    b. Installation orientation and mounting verified
    c. Feedline measurements made and recorded
    d. Return loss measurements made and acceptable
    e. Feedlines grounded and waterproofed
2. FAA
    a. Lighting and marking completed (if required)
    b. Alarming system installed and operational
3. Receive and transmit filter system
    a. Bandpass filters performance validated
    b. Notch filter performance validated (if required)
    c. Transmit filter performance validated
4. EMF power budget provided

5. Cell site parameters
   a. Frequency assignments validated
   b. Handoff topology lists checked
   c. Cell site parameters checked
   d. Cell site software load for all devices validated
6. Spectrum check
   a. Sweep of transmit and receive spectrum for potential problems
   b. Colocated transmitter identified

Many of the common problems associated with a new site investigation involve antenna orientation and obstruction issues. It is not that uncommon on a rooftop installation to have an antenna obstructed by either a neighboring building or the air-conditioning unit on the roof. The rectification to this situation is to try and relocate the antenna itself or see if possibly using the antenna for another orientation is a better application.

Regarding orientation this situation often involves someone reading a 7.5-minute map incorrectly and orienting the site according to the bad reference point. That is why it is important to have the contractor provide the orientation reference as part of the completion specifications.

5.14.2 Existing cell sites

When trying to improve a system's performance, it is often necessary to physically visit an existing cell site or location in order to try and determine firsthand the exact nature and cause of the problem for the area. As with all engineering efforts it is essential to have a battle plan laid out prior to performing the mission. It is strongly suggested that the battle plan put forth include possible problems and suggested solutions prior to conducting the field visit itself. With a predefined hypothesis as to the exact nature of the problem, you are in a position of either validating or refuting the initial hypothesis.

Before conducting a site visit to an existing cell site, it is recommended that the following checklist of items be followed to help expedite the efforts.

1. Review of the statistics for the cell site in question and the surrounding cells
2. Review of the site's frequency plan and the surrounding cells
3. Handoff topology review of the site and its neighbors
4. Expected problems to find and possible recommended action items
5. Review of the cell site's hardware configuration
6. Review of the maintenance issues for the site over the last month
7. Site access secured for the location
8. Maps of the area and directions to the site secured
9. Test equipment needed for the investigation secured

When defining the site to be investigated, it is imperative that a test plan be formulated for the effort. The methods used for defining the area or cell in question is a result of analysis of the key factors that are monitored on a continuous basis.

Regardless of the method picked and the nature of the investigation the following steps need to be taken.

1. Define the objective and specifically what the test or site visit is meant to check for.
2. Identify the area to investigate. Although this is an obvious point the geographic area for investigation is critical to the success or failure of any field test or trip.
3. Ensure that the site picked for the investigation is related to the area defined in step 2.
4. Define prior to the investigation the time you are allocating to investigate the problem, e.g., 2 weeks.
5. Define the physical resources required for the testing, equipment, and personnel.
6. Issue a status report and concluding report on the investigation.
7. Review and update the site-specific books.

There are many types of tests which can be done for a site, and each requires a different approach to uncover the real problem and determine recommended fixes.

Interference. One test that could be done is to investigate potential interference at a location. In defining the test it is essential that you identify the potential problems at this location by looking at the following items in addition to the items listed above which are part of the normal site investigation list.

1. Current frequency plan
2. Coverage plots for the area (voice and setup)
3. C/I plots for the site and the surrounding area
4. Drive tests

Design issue. Another type of test involving an existing cell site could involve investigating potential design problems overlooked in the initial site's deployment or later modifications. Some of the items to check for in this part of the investigation are

1. Receive configuration
2. Equipment provisioning aspects for the site
3. IF settings, if applicable
4. Antenna tilt angles, looking for overtilt or unbalanced tilting between sectors
5. Antenna elevations, too high or too low
6. Antenna types used

7. Cell site firmware
8. Cell site equipment vintages
9. Setup and voice channel ERP
10. Cell site parameters

Coverage. Another site investigation test involves identifying and qualifying coverage problems. For this type of testing the objective is to determine if there is insufficient coverage in an area that will promote the possibility of interference caused by the lack of coverage in the area in question.

Coverage testing is similar to interference testing in that the basic issues investigated are the same. The recommended use for a coverage investigation involves verifying if the addition of a new cell or changing the antenna system at the site, increasing the ERP, or doing nothing at all is needed.

For cell site parameter alterations and problems there is a different slant taken to the site investigation. The testing for the site focuses on evaluation of the handoff window and call-processing parameters for the site. A drive test is essential for the pre- and postchange testing. The pre- and postchange testing methodology was covered previously in this chapter.

Regardless of the type of site investigation taking place, some level of documentation needs to take place detailing the findings and recommended actions. A site improvement plan is presented here. The format used can be followed or altered based on the particular situation encountered. The situation presented here was an investigation of a site which had its antenna system redesigned. The redesigned antenna system began having network-related problems and an investigation into the nature of the problems commenced.

Site improvement plan

1. Generate propagation plots for the site at its current antenna height and at the previous antenna height.
2. Compare the current propagation predictions with the data actually measured.
3. Evaluate data to determine if the coverage presently produced by the site is the desired result for the area.
4. Conduct a S11 test of the antenna system.
5. Conduct a test utilizing the previous antenna system as a comparison.
6. Based on data collected in step 5 determine if further action is required. If further action is required, then determine if the site's previous antenna height should be used employing downtilt or another height picked.
7. Implement changes to the antenna system, and conduct a S11 test of the new antenna system.
8. Conduct additional field tests to evaluate if the desired results are achieved and make corrections if needed or possible.

> **Field Test Report**
>
> Date: 7-24-95
>
> Subject site: Cell X
>
> *Reason for conducting site visit.* The site was chosen to be visited as part of the ongoing process to improve area 3 of the network.
>
> *Purpose.* The purpose of the site visit was to try to quantify specifically the reasons for the poor performance of sectors 1 and 2.
>
> *Site configuration.* The site configuration is shown as figure 1, attached. The site consists of four bays of radio equipment and 41 physical radios. There are a total of nine antennas at the site. The site parameters and software load were validated to be correct.
>
> *Observations.* During the course of the investigation into the site it was noted that there was a serious obstruction to sector 1 and partially for sector 2. Pictures of this situation were taken and are included for reference. It was also found that there was a defective transmit filter for sector 2 of the site, limiting its power output. The filter was acting as a load and therefore did not set off any SWR alarms.
>
> *Recommendations.* It is recommended that the antenna design for sectors 1 and 2 be reworked to avoid the obstruction. The proposed configuration is shown as figure 2. The defective transmit filter was replaced the day of the site investigation, and no further action is required for this issue.
>
> Engineer: ─────────────
>
> Operations: ─────────────

Figure 5.44 Field test report.

9. Evaluate the cell site and its neighbors statistics performance.

10. Issue a closing report on the engineering activities by a specified date.

An example of a site visit report is given in Fig. 5.44. The cell site investigated was one of the worst-five performing cell sites in the network. Several tests were conducted with field personnel regarding the site netting marginal results. It was, therefore, decided to conduct a physical investigation of the site by engineering with operation's support. The key concept to always remember when conducting a site investigation and improvement process is to document what you have and will do. Every cell site can undergo some level of improvement, no matter how small it may seem.

5.15 Orientation

The orientation of the sectors in a wireless network directly determines the effectiveness of the frequency management scheme employed by the operator regardless of the technology platform used. The consistent orientation of the sectors is critical for getting the maximum C/I available in frequency assignment for TDMA-based systems. However, for CDMA the orientation criteria is relaxed and not required.

Frequency reuse is maximized though controlling where the potential interference will be. Using different orientations in a network will lead to increases in lost call rates and poor performance due to increased interference. There are numerous technical articles pertaining to the use of orientations and frequency management. The use of standard orientations also facilitates the system performance troubleshooting through the elimination of variables.

The orientation of the cell sites is critical not only for frequency management but also for handoffs. Orientation is not that critical a factor in a young system because most of the performance problems are directly related to coverage issues. However, coverage issues can be caused by incorrectly orienting antennas at a cell site.

5.16 Downtilting

Downtilting or altering the antenna inclination of a cell site is one of the techniques available to a radio engineer for altering a site's coverage. The rationale used to alter the concatenation of the cell site's antenna system can be varied. Some of the rationale used for altering involves reducing interference, improving inbuilding penetration, improving coverage, or limiting a site's coverage area. The concatenation of the antenna system can have a major impact on the actual performance of the cell site itself.

The alteration of the tilt angle for a cell site should be done with extreme care since this can have a major impact, both positive and negative, on the performance of the network. To maximize the benefits of altering the tilt angle for a cell site while minimizing your exposure to problems, it is suggested that you follow a test plan.

There have been several articles and papers written regarding the use of downtilting antenna systems. Generally, the reports written stress the use of tilting the antenna system, usually half the vertical beam width, to achieve a 3-dB reduction signal level at the periphery of the sector's coverage. The tilt angles are usually specified, or requested, by the frequency planners for the intention of improving the C/I at another cell site.

I have found that taking the terrain aspects near the cell site can provide large signal attenuation with just a minor tilt angle. The use of terrain to assist in the attenuation of the signal is based on diffraction of the signal. The use of diffraction on attenuating the signal is a very effective tool when trying to maximize the signal near the cell site and attenuate it near the horizon.

For example, if your objective is to improve the coverage near the cell site but contain the coverage of the cell so it does not create an interference problem to the network, altering of the antenna system's tilt angle can be considered as one possible solution, assuming the antenna system employs an antenna with a 5° downtilt. Figure 5.45 shows where the main lobe of the antenna pattern

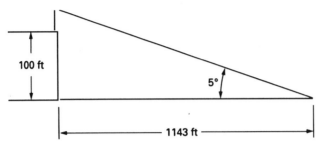

Figure 5.45 Downtilt.

strikes the ground. The main lobe of the antenna system using a 100-ft-high cell site at 5° hits the ground at 1143 ft. The example in the figure, however, assumes a flat-earth situation which is not realistic.

The tilt angle for the antenna system also has a direct impact on the coverage for the cell site. The example shown in Fig. 5.45 indicates that if a cell radius is 2800 ft, another cell site might now be required for the area to provide coverage. The coverage loss due to downtilting is often an overlooked aspect when this technique is employed as a solution.

The coverage loss to the network could be reduced or eliminated by utilizing diffraction. The use of diffraction on attenuating the signal for a cell site has a greater impact than merely tilting the antenna by half its vertical beam width.

Figure 5.46 is an example of using diffraction to attenuate the signal. The figure shows an obstruction 3000 ft away from the bore site for the sector. The desired goal is to have the signal level near the cell site but minimize the negative effects of interference at the reusing cell site, several sites away. The antenna tilt angle needed to achieve over 21 dB of attenuation in less than 2°. The use of terrain to assist in the attenuation of the signal should be exploited in order to maximize the coverage of the cell site and achieve the necessary attenuation of the signal to facilitate frequency reuse.

When planning on altering an antenna system's inclination, it is strongly advised that a test plan or MOP be used. The key element for the plan which must be done prior to altering the antenna pattern for a cell site is to define what your objective is before you begin. Defining the objective beforehand seems simplistic in nature, but it is imperative that this is done.

The following is a procedure for altering the tilt angle of an antenna system, usually a sector, for a cell site. The procedure here, as with other procedures, should be modified to reflect the particulars of the situation. The individual procedure listed here is meant for a trisector cell site. If you have a situation where the site has a different amount of sectors, a modification to this procedure involves simply increasing the number of data points to test.

An important aspect is that altering the tilt angle for a cell site can involve increasing and decreasing the angle of inclination currently used there. Many improvements to a network's performance have been achieved simply by reducing the tilt angles currently employed at a cell site.

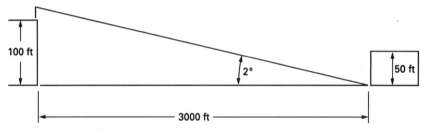

Figure 5.46 Downtilt.

Downtilt procedure

1. Identify the problem.
2. Check operations.
3. Make physical observations.
4. Perform drive testing.
5. Define cell's coverage area.
6. Perform design review. The items which need to be reviewed at the design review meeting are
   - The objective for the antenna alteration
   - Desired coverage area for cell site and sectors
   - Preliminary tilt angles desired based on the previous two items
   - Pre- and posttest plan
   - Pass/fail criteria
7. Perform test plan.
8. Perform closure activities.

5.17 Intermodulation

Intermodulation situations will present themselves to any radio engineer at various stages of one's career. The cause of intermodulation and how to remedy the situation has employed many talented engineers and will continue to do so. In order to find and ultimately resolve the intermodulation problem, it is important to know the basic concepts of just what is intermodulation.

Intermodulation is the mixing of two or more signals that produce a third or fourth frequency which is undesired. All cell sites produce intermodulation since there is more than one channel at the site. However, the fact that there are intermodulation products does not mean there is a problem.

Various intermodulation products are shown below for reference. The values used are simplistic in nature to facilitate the examples. In each of the examples, $A = 880$ MHz, $B = 45$ MHz, $C = 931$ MHz, and D is the intermodulation product. This example does not represent all the perturbations possible.

Second order: $A + B = D$ (925 MHz)
                     $A - B = D$ (835 MHz)

Third order: $A + 2B = D$ (970 MHz)
                     $A - 2B = D$ (790 MHz)
                     $A + B + C = D$ (1856 MHz)
                     $A - B + C = D$ (1766 MHz)

Fifth order: $2A - 2B - C = D$ (739 MHz)

The various products that make up the mixing equation determine the order of the potential intermodulation. When troubleshooting an intermodulation problem, it is important to prepare for the encounter in advance.

All too often when you conduct an intermodulation study for a cell site there are numerous potential problems identified in the report. The key concept to remember is that the intermodulation report you are most likely looking at does not take into account power, modulation, or physical separation between the source and the victim, to mention a few. Therefore, the intermodulation report should be used as a prerequisite for any site visit so you have some potential candidates to investigate.

Intermodulation can also be caused by your own equipment through bad connectors, antennas, or faulty grounding systems. However, the majority of the intermodulation problems encountered were a result of a problem in the antenna system for the site and well within the control of the operator to fix.

Just how you go about isolating an intermodulation problem is part art and part science. We prefer the scientific approach since it is consistent and methodical in nature. If you utilize the seven-step approach to troubleshooting listed at the beginning of the chapter, you will expedite the time it takes to isolate and resolve the problem. The biggest step is identifying the actual problem, and after this the rest of the steps will fall in line. Therefore, it is recommended that the following procedure be utilized for intermodulation site investigations.

Previsit work

1. Talk to the cell site technician and have him or her go over the nature of the problem and all the steps taken to correct the problem.

2. Examine the site-specific records for this location and see if a previous problem was investigated and if there were any changes made recently to the site.

3. Determine if there are any colocated transmitters at this facility and conduct an intermodulation report looking for hits in your own band or in another band based on the nature of the problem.

4. Collect statistic information on the site to try to determine any problem patterns.
5. Review maintenance logs for the site.
6. Formulate a hypothesis for the cause of the problem and generate a test plan to follow.
7. Secure the necessary test equipment and operations support for the site investigation.
8. Allocate sufficient time to troubleshoot the problem.

Site work

1. Perform initial test plan.
2. Isolate the problem by determining if the problem is internal or external to the cell site.
3. Verify all connectors are secure and tight.
4. Monitor the spectrum for potential intermodulation products determined from the report.
5. When intermodulation products appear, determine common elements which caused the situation.

Based on the actual problem encountered, the resolution can take on many forms.

If the problem is a stray paging transmitter, the recommended course of action is to notify the paging company and request, first, that they resolve the situation immediately. For this situation you will need to conduct a posttest to validate if the change took place and netted the desired result.

If the problem is a bad connector producing wideband interference, the situation is corrected by replacement of the connector itself.

If the problem is a bad duplexer or antenna, again the situation is rectified through replacement of the equipment itself.

If the intermodulation product is caused by the frequency assignment at the cell site, then it will be necessary to alter the frequency plan for the site, but first remove the offending channels from service.

If the intermodulation problem is due to receiver overload, the situation can be resolved by placing a notch filter in the receive path if it is caused by a discrete frequency. If the overload is caused by cellular mobiles, using a notch filter will not resolve the situation. Instead mobile overload can be resolved by placing an attenuation in the receive path, prior to the first preamp, effectively reducing the sensitivity of the receive system.

5.18 System Performance Action Plan

The following is a list of performance action plans or activities that need to be completed on a daily, weekly and monthly basis.

1. Daily check lists
2. Weekly plan
3. Monthly plan
4. New site integration
5. New service integration
6. Quarterly drive tests

5.18.1 TIC lists

The following are brief TIC lists that can be used by performance engineers and their managers as quick reminders of activities that need to be done.

Performance engineer

Daily TIC list (9 A.M.)
1. Examine exception report.
2. Review datafill conflict reports.
3. Review status operations for outages and other performance issues.
4. Review customer trouble reports for area.

Weekly TIC list (Monday 9 A.M.)
1. Examine exception reports (are there trends?).
2. Examine key factors and performance reports.
3. Generate plan for week.

Monthly (first Friday of every month)
1. Examine key factors and performance reports.
2. Identify the 10 worst-performing sectors in region by each key metric.
3. Generate a plan for what needs to be done for each sector and cluster.
4. Identify what was completed versus last month's plan

Performance manager

Daily TIC list (9 A.M.)
1. Examine exception report (system and regional level).
2. Review status operations for outages and other performance issues.
3. Review customer trouble report summary.

Weekly TIC List (Monday 12 P.M.)
1. Examine exception reports (are there trends?).
2. Examine key factors and performance reports.
3. Review regional plans for action and correct if necessary.

Monthly (first Friday of every month)
1. Examine and communicate key factors and performance reports, regional and system.
2. Identify progress against key metrics for system and each region.
3. Identify the 10 worst-performing sectors in system by each key metric.
4. Generate a plan for what needs to be done for each sector.
5. Identify what was completed versus last month's plan.

5.18.2 Weekly reports

The following is the basic material that should be included in the weekly plan of action from system performance engineers to their managers.

Weekly report material	Completed
10 worst-performing sectors identified by key factor category	
Planned activity by day (general)	
Requests for assistance	
Key factor summary of area versus monthly and quarterly goals	
New cell activations	

The following is the weekly report material that needs to be provided to the director of engineering by the system performance manager.

Weekly report material	By area	System
10 worst-performing sectors identified by key factor category		
Planned activity by day (general)		
Requests for assistance		
Key factor summary of area versus monthly and quarterly goals		
New cell activations		
Comments about planned deviations and how to correct them		

5.18.3 Monthly plan format

The following is a brief description regarding the format of the monthly plan of action that needs to be generated for each of the separate areas within system performance. The plan is meant to indicate what the general focus will be for the region over the next month. The plan is meant as a general guide from which major system problems can be identified and the department as a whole will be aware of what their counterparts are doing. Effectively the engineers for system performance with their manager show what their plan is for helping improve the performance of the network over the next few weeks. It is suggested that each area is handled separately so as to maximize the work effort by everyone else. For example, Monday afternoon and Tuesday morning could be set aside for conducting these presentations.

The informal presentation takes a maximum of 2 h the first time and 1 h from then on. A rotation program can be done for the various performance engineers. The presentation should be given to the director of engineering.

System performance engineer. The topics to include involve the following items.

1. Identification of problem areas by
    a. Customer complaints
    b. Statistics (key factors and performance indicators)
    c. FER, BER, and SQE; C/I; E_b/N_o
    d. Coverage < −86 dBm
2. What the suspected problems are (they cannot all be coverage) in rank order
3. What actions are to be taken (they cannot all be new cell or retune)
4. When they expect to have the problems resolved
5. The schedule of new sites and the status of datafill efforts
6. Results of last month's plans versus accomplishments
7. Performance relative to key factors and performance for each area

The following is the material that should be presented by the manager of system performance on a monthly basis to the engineering and operations management. The material is similar to the information that the individual system performance engineers generate with the exception that all the areas are incorporated into this plan.

Manager of system performance

1. Identification of problem areas by
    a. Customer complaints
    b. Statistics (key factors and performance indicators)
    c. FER, BER, and SQE; C/I; E_b/N_o (E_b/I_o)
    d. Coverage < −86 dBm
2. What the suspected problems are and rank order them (*focus only on top 5 problems*)
3. What actions are to be taken
4. When they expect to have the problems resolved
5. The schedule of new sites and the status of datafill efforts
6. Results of last month's plans versus accomplishments
7. Performance relative to key factors and metrics for each area

Key performance indicator	Goal	Area	System
Blocking %			
Dropped lost call % (LOS)			
Access failure %			
Handoff failure %			
Usage minutes/lost call			
Radio utilization rate			
% area > −85 dBm (RSSI design threshold)			
% area >XX FER/BER/(C/I)/SQE			
% area >XX kbit/s			
Sites on-air			
No. of trouble tickets/no. of subs			
No. of trouble tickets			
Trouble ticket response time			

Chapter

6

Circuit Switch Performance Guidelines

Once a system is operating in a live environment, its performance will need to be constantly measured and optimized. In addition, as service problems arise, troubleshooting will have to be conducted using various techniques as they pertain to a wireless network. This chapter discusses the major network performance metrics, ways to monitor and optimize these metrics, and procedures to troubleshoot call delivery problems.

6.1 Network Performance Measurement and Optimization

6.1.1 Switch CPU loading

All switching equipment contains processors, whether the node architecture is of a distributed or hierarchy design, including the main processor [or central processing unit (CPU)] and the subordinate or regional processors it controls. When determining the processor loads of a switch with a hierarchy-based design structure, you should take into account all processors and their specific functions in the delivery of a mobile telephone call. You need to determine which processors have the highest traffic levels, and thus, which are the most susceptible to reaching an upper threshold and creating possible problems as the traffic in the system increases. A processor load study for a switch architecture based upon a CPU should measure this processor's traffic load rather than that of any of the secondary processors the CPU controls. The secondary processor loads should also be reviewed but not as frequently.

When the switch or node is in a stand-alone configuration, i.e., not in an application environment, there is a baseline (initial) load present on each of its processors. This load consists of the basic administration and maintenance processes the CPU and its subordinate processors perform. This load value is a good indication of how well the switch vendor has designed its product. A typical baseline value may be in the 5 to 7 percent range. A 10 percent baseline load would be rather high. Obviously a higher baseline load means that less processor capacity is available to the application environment. Consult your vendor for this design information when performing your processor load analysis. Many other factors can add to the baseline load and overall load of the system switch processors, e.g., a new switch operating system and new system features or functions of the node. The switch vendor should provide the percent increase or decrease in the processor load for each of these network changes. Furthermore, the network engineering department working in conjunction with the vendor's support personnel should verify these changes after they have been implemented to assure that they meet the design specifications previously defined.

A switch vendor should also be able to supply you with the maximum operating limit of the processor. When reviewing this value, take into consideration that a processor cannot operate at 100 percent capacity for a number of reasons. First, as a processor reaches its maximum operating level it begins to shed tasks assigned to it for processing. These tasks are prioritized with regard to the operation of the switch. For instance, the administration and operational tasks of a switch may be given a greater priority than the actual call-processing tasks. This design is sometimes used by the manufacturer to assure that the switch does not encounter a catastrophic failure whereby all communications and control of the node are lost. Thus, the input and output (I/O) communication functions of the switch are assigned the highest priority. The operation of the I/O functions are necessary so that the switch operators can troubleshoot and correct any system error or reduce the processor load, whichever caused the disruption. Further priority assignments may place the call-processing tasks at a higher priority than the tasks used in the collection of the system operating statistics, since the processing of mobile calls is more important than the collection of operational data for a short period of time. Once the loss of operational data is noticed it is assumed that corrective action would take place immediately without the loss of system revenue. The loss of statistical data from the switch is an indication that a system could be reaching its upper capacity, especially during the system busy hour.

Another reason that a processor cannot operate at 100 percent capacity is due to the fluctuations in the traffic load on the switch and thus in the processor itself. Given a sampling rate and a time in which to sample the processor capacity, the final load value is an average of all the sample measurements. Obviously the greater the sample rate, the more accurate the average value of the processor load. (If a switch allows you to set the sample rate for the processor load measurement, be careful not to set this value too high because it may affect the call-processing functionality of the switch itself.) You cannot detect all the high traffic instances (spikes) that occur during the sampling period if you use the average value of the processor load measurements. Although an average processor load value might read 80 percent, there may very well have been a spike of 95 percent. It is with these concepts in mind that the operational loads of a processor are set to a value less than 100 percent of its full capacity. For some switches this value may be 75 percent, while for others it is 95 percent.

The actual assignment of priority levels to the various tasks in the switch and the level in which the processor load shedding begins is dependent upon the switch vendor and the design the vendor chose for its product. While these switch design concepts are critical to the operation of your system, the environment in which the switch operates is equally as important to the performance of the processor and the time it will take to reach its maximum load. Table 6.1 is an example switch processing task prioritization.

The methods used to measure the processor load of a switch vary from one manufacturer to another. For some vendors this procedure involves simply issuing a command to the switch for a specified time and then collecting the printed data from the operations department. Other switch vendors, however, may require a collection program to be set up specifying the sample rate and

TABLE 6.1 Example Switch Processing Task Prioritization

Switch function	Priority
I/O communication functions	Highest level
Call-processing functions	Secondary level
System billing functions	Third level
System statistics collection functions	Fourth level

TABLE 6.2 Switch Processor Load Data for Market "x" (1995)

Month	Processor load, %
March	58.2
April	60.1
May	63.0
June	65.3
July	64.2
August	64.8
September	66.2
October	67.1
November	68.0

Vendor-specified switch baseline processor load: 10.0%
Vendor-specified maximum switch processor load: 72.0%

the starting and ending times of the sampling period. Once the method of measurement is established, the CPU load data of the switch can be collected, tracked, and analyzed. It is recommended that this load be measured during the system busy hour (typically between 4:00 and 5:00 P.M.) on a daily basis. This data can then be averaged for the 10 highest traffic days of the month for graphing and monitoring purposes. See Table 6.2 and Fig. 6.1 for an example of this process.

Upon reviewing the graph in Fig. 6.1, it is evident that the maximum processor load for switch A will be reached by the system in mid- to late January

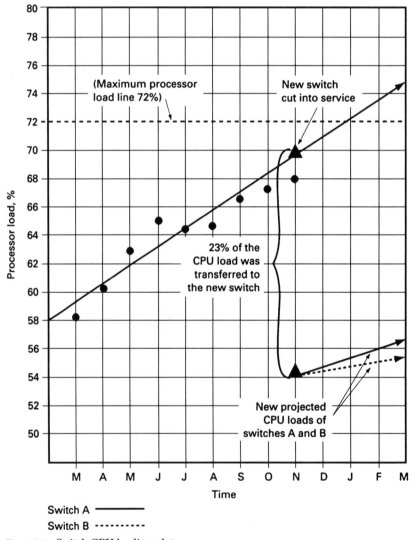

Figure 6.1 Switch CPU loading plot.

1996. Before this limit is reached, plans for a new switch (switch B) should begin to relieve this loading in a 6- to 10-month time frame. Either a new switch must be added to the network or various system parameters will require adjustment.

Using the guidelines previously discussed, if switch B is cut into service at its scheduled target date, 2 months before the projected maximum switch load is reached (November 1995), then an off-loading of switch A is experienced. This off-loading is a direct result of the transfer of system traffic to the new switch. Prior to the actual cut date the network and radio frequency (RF) engineers estimate the amount of traffic to off-load. Once switch B has been in service the actual off-loading value should be determined and the amount of variation in the design checked for accuracy against the initial projection. In Fig. 6.1, switch A is off-loaded 23 percent (69.8 to 54.0 percent CPU load decrease) by switch B.

In addition to the off-loading aspect of this project the CPU loading trends of switches A and B need to be determined and the projection process continued for a multiswitch network. The projection of CPU loads for switches A and B are shown in Fig. 6.1.

As a temporary solution to the CPU loading problem a number of different system parameters can be adjusted in the switch. For instance, if a system is using the mobile registration feature whereby all system mobiles signal to the network their location in regularly scheduled periods based upon a specified algorithm (this process is further described later in this chapter), then a set of parameters exists for use as a temporary solution.

By increasing the registration interval the total number of mobile registrations (location signals) being sent to the switch will decrease over a given time period. This, in turn, reduces the number of registrations the switch has to process, further reducing the switch processor load. However, this change comes with the network effect of reducing (slightly) the number of incoming calls to the mobiles since the accuracy of the location function (mobile registrations) has been reduced. This is not as critical as it first appears since the majority of calls in a cellular system are typically the mobile-to-land type (approximately 80 percent), but it still has a negative effect upon the network.

Again, this type of parameter adjustment is meant to be temporary until a final solution is prepared. Other parameter changes may include turning off the directed retry feature during the system busy hour in an attempt to reduce the amount of redirected calls at this critical load period. Some vendors' switches perform maintenance functions continually over a 24-h period while the switch is processing calls. These functions may be halted during the system busy hour, again as a temporary solution to relieve the processor load on the particular switch in trouble.

The purpose behind monitoring the performance of the processor is to prevent a critical situation by utilizing one of the temporary solutions mentioned. For this reason the CPU load of all the network switches should be monitored on a regular basis and after any new switch software loading or hardware upgrades have been completed in the network. This process is meant to keep the engineering staff informed of the CPU load and to determine the effect of these network changes on the switch's performance. It is also a good idea to monitor the CPU load after any major changes in the switch's database translations or if a new system feature is being introduced in the network.

6.1.2 Switch call-processing efficiency

This value is the call volume to CPU loading ratio.

$$\frac{\text{Switch call volume}}{\text{Switch CPU load}} = \text{switch-processing efficiency} \qquad (6.1)$$

This metric can be used to monitor the call-processing efficiency of the switch as well as to indicate if a possible problem exists in the system. For example, a handoff border may need optimizing or a registration interval may need adjusting.

The load of a processor is typically based upon the number of calls processed per second by the switch (the processor load traffic). However, for a mobile call, many factors affect the processor load on a per-call basis, for instance, if the call was from a roaming mobile or a home subscriber, if the subscriber used a special feature, or if there was a handoff involved at any time during the call. All these factors contribute to the processor load for any given call. Therefore, any new feature introduced by your marketing department and utilized by your subscriber base will have an effect on the system. It may increase the processor load of the switch, even though the average number of mobile calls processed during the system busy hour would remain about the same. It is for this reason that the load of the processors should be measured and graphed along with the number of calls processed per second by the switch during this same time period. Comparing these two metrics by graphing them together is an excellent way to observe the call-processing efficiency of a network switch. We now present an example.

Example 6.1: Switch Processing Efficiency A system has two network switches, A and B, processing 15 calls per second at processor loads of 50 and 60 percent, respectively, during the system busy hour. Obviously switch B is the least efficient switch of the two. If the same system features are present on both switches and the traffic patterns, number of cells, number of trunks to the public switched telephone network (PSTN), etc., are about the same, then this difference may indicate a problem with switch B. The performance of switch B should be analyzed in more depth with possible optimization of its cells and global system parameters being conducted. For instance, if there existed a high-traffic area on switch B where the RF parameter settings of the cell sites were in error, then this could affect the switch's efficiency along with other system variables such as voice channel blocking on the network facilities.

The data for this plot should be taken from the daily system busy hour for both the processor load measurement and the number of calls processed per second measurement and not the monthly average data. Figure 6.2 is an example plot of this data to reference.

6.1.3 Switch/node downtime (service outage)

Obviously, this is an important metric that needs to be monitored and tracked on a daily, monthly, and yearly basis to determine the rate, duration, and cause of the occurrence and what procedure and personnel were in place to correct the problem. If a system has multiple switching offices, then compare the switch outage reports against the various mobile switching centers to determine where improvements in personnel and operating procedures can be made. Also, determine if the outage is due to errors in the operating procedures of the switch or if it is a quality issue with the switch vendor's software or hardware. This metric can also be used to track trends of declining or improving vendor support. Perhaps there is a scheduling problem with the number of network activities taking place in the system at a given time. Trying to perform too many complex and original projects in the network simultaneously or in a short time frame could be the cause of many system service outages. Consult your operational or network management groups to address this performance issue in more detail. The example service outage report in Fig. 6.3 may assist in this effort.

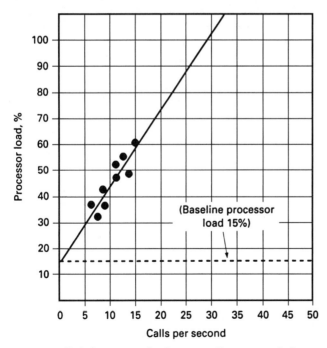

Figure 6.2 Switch processor load versus calls per second plot.

```
System outage report for market X

Report date  _____

Report number  _____

Switch operator  _____

Outage date  _____

Outage start time  _____

Outage end time  _____

Description of outage  _____

Switch data available  _____

Personnel notification  _____
```

Figure 6.3 Service outage example report.

6.1.4 Switch service circuit loading

The amount of service circuits ordered and installed in your switch at the time it was put into service was determined by the vendor's network engineers. The quantities of senders, receivers, conference cards, etc., are dependent upon the estimated traffic load expected to be processed by the switch. Most switches have a set amount of these circuits installed as part of an initial switch configuration. However, as the system and the switch begin to carry more traffic, the loading on the processors as well as the service circuits increases. There was an actual case where a live switch stopped processing calls and was off the air because some of the service circuits were overloaded!

There is no need for such a situation if circuits are monitored on a regular basis and proper planning and ordering of new circuits is completed well in advance of an overload condition. A typical switch may include 200 multifrequency (MF) sender channels, 80 MF receiver channels, 80 dual-tone multifrequency (DTMF) receiver channels, and approximately 100 conference circuits. Note, however, that the use of these circuits is related to the usage of system features and other services offered by your company. Any changes to the system's service may affect the loading on the service circuits. Again, after any major system change or new service offering, quickly monitor your system performance statistics for changes in performance and traffic load increases.

6.1.5 Switch/node total erlangs and calls volume

The total erlangs and calls carried by the switch during the system busy hour should be monitored to determine the switch's percent traffic distribution in the system and to determine each nodes call-processing efficiency. This metric is also used to trend the loading of the individual nodes in the system. It is important to determine where the next critical loading limit will be encountered and when to begin plans to relieve this load. For example, if a network of four switches all carry the same percent of the system traffic, then as the call volume increases eventually all the switches will have to be off-loaded at the same point in time. This would be too large a task. It would be better to stagger the system load on the switches so that the times each will need to be off-loaded will be staggered. This spreads the planning and work needed for increasing system capacity over a larger time frame and, thus, makes it easier to manage.

6.1.6 Switch/node alarms

A report of the switch's alarms can be used to evaluate the performance of an individual switch or an entire network. It also serves as a good indicator of whether the node and corresponding interconnected network are operating in a clean and efficient manner. This data will contain the alarms that (although not necessarily critical) indicate minor problems with the switch database, cell sites, trunks, etc., or these alarms may be due to data that were loaded in the switch but are scheduled for activation at a later time.

For example, if a cell site is scheduled to be brought into service in a few days and the data has been loaded early in preparation for this cut, then the switch may give some minor alarms that these devices are out of service. This is not a problem as long as the site *does indeed* get cut into service as scheduled! Otherwise this data could remain and cause unnecessary alarms and CPU loading. The same holds true for new data links and other network facilities that usually have their data loaded prior to being brought into service. Also, take note of any invalid or old mobile code ranges, devices out of service, problems with trunk groups, and other possible causes of network alarms.

6.1.7 Switch memory settings and utilization

Some switch manufacturers have the ability to dynamically set the amount of memory available for each function in the switch. While this is a great method to utilize the node's memory to the best extent possible, it also requires monitoring of these memory assignments quite closely. If not enough memory is allocated, this will reduce or even halt the particular function altogether. Such memory limitations may even cause a switch outage! To prevent this from occurring, monitor the memory utilization in your system switches and audit them on a regular basis to assure they have a consistent setting in each node.

6.1.8 Switch timing source accuracy

Timing of the network nodes is critical for proper operation of the system. It is recommended that you perform audits of the network timing sources on a regular basis (at least once a month) to assure they adhere to a stratum 2 level of accuracy. Table 6.3 provides a listing of the specifications for the various stratum levels.

6.1.9 Auxiliary node performances

Monitor the performances of the network auxiliary nodes with the same level of detail that you monitor the major switch nodes but on a less frequent basis. They are an important part of your network, and their performance must be reviewed. Some auxiliary node examples are

1. System digital access cross connects (DACs), multiplexers, and other dynamic transmission equipment used in the voice network
2. Signal transfer points (STPs) and other signaling system 7 (SS7) data network equipment
3. Voice mail systems
4. Network monitoring systems used to access and query the network switches and nodes (including network data routers and multiplexers)

6.1.10 Node performance summary

The following is a summary of the major metrics used to measure the performance of a network node.

1. Measure the load on critical node processors.
2. Measure the processing efficiency of the node.
3. Track and monitor the outages of the node.
4. Measure the load on the switch service circuits.
5. Measure and report on the total erlangs and calls processed by the node during the system busy hour.
6. Monitor the node alarms.
7. Track and monitor the node's memory utilization and capacity.
8. Check the node timing sources for a stratum 2 level of accuracy.

TABLE 6.3 Stratum Levels and Associated Timing Accuracy

Stratum level	Timing accuracy
1	$\pm 1.0 \times 10^{-11}$
2	$\pm 1.6 \times 10^{-8}$
3	$\pm 4.6 \times 10^{-6}$
4	$\pm 32.0 \times 10^{-6}$

6.2 Network Link Performance Measurement and Optimization

6.2.1 Network link performance

When monitoring a network for data link performance, the key items of interest are the links that interconnect the system switches and other major nodes. Cell site data links that do not utilize SS7-type data links, although important, will not be considered at this time.

Critical metrics used in the measurement of a link's operating performance will be applied to the SS7-type signaling protocol since this is currently the most widely used industry format for data transmissions used in providing call delivery between switches of differing manufacturers. However, the concepts of these basic metrics can be applied to other transport-type protocols such as X.25.

6.2.2 Link traffic loading

SS7 data links are typically designed to operate at a maximum 40 percent (0.4 erlangs) utilization. Though the link can operate at a much higher level, should another link in the link set experience trouble, then the possibility of congestion exists.

The traffic levels on an SS7 data link can be measured by using either the reporting system of the switch or separate monitoring equipment such as a protocol analyzer patched into the network in a nonintrusive (passive) manner. Some switches require (as in the case of the processor load data previously discussed) a collection program to be written to collect the data necessary to calculate the traffic on a given data link in erlangs. This method is used to determine the number of messages transmitted and received on a per-second time interval. To come up with the number of erlangs of traffic the link is carrying, determine the number of bits per message and divide by the data rate of the link (56 kbits).

Here are two equations for measuring link performance:

No. of message signaling units (MSUs) transmitted + no. of MSUs retransmitted + no. of MSUs received = no. of messages carried per second

(Messages per second) (bytes per message) (8 bits per byte)
$$= \text{link traffic carried in 56,000 erlangs}$$

This same data can be acquired in a much quicker and easier fashion from a network analyzer if it is set up properly. However, not all systems have the ability to patch into each network link and collect this data. Under these circumstances it is better to have the switch provide this data and have the network engineers process it along with the other remaining system performance statistics.

If a data link is approaching a 40 percent utilization load on a regular basis, then plan to add another link to the link set. If a data link is showing a utilization rate of 90 percent or above and is causing service problems, first collect as much system data as possible and then reset the link and monitor the traffic levels once more. The reset may clear the problem. If it doesn't, then a possible rerouting of the link may be necessary. Consult your network manager before performing any such actions to receive the proper authority and consultation for conducting this work.

Similar to a node processor an SS7 data link will shed (not transmit) lower-priority messages in the event of a link problem for the sake of maintaining the transmission and throughput of the higher-order maintenance messages. There are four levels of priority assigned to the SS7 messages for maintaining data link performance. The lowest level (level 0) is assigned to normal SS7 messages, while the highest level (level 4) is assigned to link messages [link status signaling units (LSSUs)] used for maintaining and administering the link during conditions of congestion and errors.

6.2.3 Link retransmissions

Another metric used to measure the performance of a data link is its percent of message retransmissions. A high number of retransmissions indicates a problem in the data network that requires further investigation. A typical retransmission level for a normal data link might be in the range of 0.1 to 0.4 percent. To calculate the percent retransmissions use the following equation:

$$\frac{\text{Number of MSU octets retransmitted}}{\text{Total number of MSU octets transmitted}} \times 100 = \text{retransmission \%}$$

6.2.4 Link errors

The maintenance and administration of an SS7 data link will include the monitoring of the number of message signaling units received in error. This is a function of the SS7 protocol and should be monitored for evaluation of the link's performance. Again, this data can be collected by one of two methods: by a protocol analyzer or by the statistical collection functions of the switch itself. This metric should be recorded over 1-h intervals for a typical operational day (approximately 7:00 A.M. to 8:00 P.M.). As a guide, 400 signaling units in error over an hour period would represent a nominal link performance.

The actual physical data link itself should be tested for an acceptable bit error rate prior to placing any data link into service. The American National Standards Institute (ANSI) specification is stated as follows:

>99.5% error-free seconds [<432 error-seconds/day at a DS0 rate (64 kbit/s)]

This basically means that when a data link facility is ordered from the local exchange carrier (LEC) or interexchange carrier (IXC), it must be of a 64-kbit/s rate and operate 99.5 percent of the time without errors for a trial period of at least 24 h before it is acceptable to be placed in service to carry live SS7 message traffic. Consult the ANSI Signaling System 7, MTP T1.111.2 (section 3) specification for more information about this test requirement.

6.2.5 Link changeovers

If the number of errors on a link exceeds the specified maximum limit, then the link may be taken out of service and the traffic destined for that link will be rerouted to another link in the corresponding link set. Once the problem has been resolved, the traffic will be rerouted back to the corrected data link and the routing of data messages will resume its original configuration. This change in the routing of the message traffic and status of the links in the link set is termed a *changeover*. Obviously a large number of such events is a problem in the network and requires further investigation. As a guide, four changeovers in an hour period would be an acceptable level of performance.

6.2.6 Link active time

A data link may experience brief outages during its normal operation in the network. However, excessive downtimes for a data link means that the message traffic assigned for routing over this facility has been reassigned to another link while it attempts to resolve the problem. Initially this does not appear to be a problem since there is sufficient capacity on the other links in the link set. However, this type of network management will lead to further troubles if left unattended. In this particular situation the other links must carry more traffic than their normal intended design levels. Under these circumstances the capacity of the entire link set has been reduced. Should another link experience a problem, then the entire set (containing two links) may need to be taken out of service. Even on a temporary basis this type of message rerouting will begin to cascade and get out of control. As a basic guideline, link performance problems should be resolved within a regular operational week. This metric should also be used to determine if a link's outage performance is steadily increasing, decreasing, or remaining constant.

6.2.7 Link performance summary

The following is a summary of some basic metrics and guideline values for measuring the performance of an SS7 data link, link set, and corresponding network facilities. There are many other parameters and methods used to conduct monitoring and performance evaluations of this type.

1. Provide the basic link and link set definition for identification purposes.

2. Measure the actual traffic on the individual links for a specified time period (typically the system busy hour) by either collecting the number of MSUs received and transmitted and using the equations provided or use a protocol analyzer for a direct reading of this data. The traffic level for the link set needs to be determined as well.

3. Measure the number of retransmissions on a data link. As a guideline an SS7 data link with a retransmission rate between 0.1 and 0.4 percent would represent a nominal performance.

4. Measure the number of MSUs received in error. If this value exceeds a given threshold (e.g., 400) for a specified time interval (e.g., 1 h), then more in-depth monitoring and analysis of the link needs to be conducted. When commissioning a new data link or testing an existing facility, check the bit error rate as specified by ANSI for a 24-h time period.

5. Measure the number of link changeovers in 1 h intervals. As a guideline, four changeovers in an hour period would represent a nominal performance.

6. Determine the overall service outage times of the link in terms of hour intervals. As a guideline, any link service outages should be resolved in one operational week.

6.3 Network Routing Performance Monitoring and Management

These metrics apply to both the voice and data networks of a wireless system for measuring the routing efficiency and accuracy and for detecting possible problems in the network.

6.3.1 Routing efficiency (voice and data)

Located in the switch is a subset of the database whose specific function is to analyze the dialed digits of the mobile subscribers when they place a call and to route these calls based upon the results of the analysis. This database subset is typically called the number translations of the system. Other names exist such as translations and B# analysis. In spite of their many names these databases are basically designed to perform the same function, analyze dialed digits and route a call based upon the results of the analysis. This process applies to the voice network; however, a similar database is developed and operated to perform the routing of SS7 data packets within the network between nodes much the same way mobile calls get routed. The routing efficiency is, therefore, a measure of how accurate the call and data packet routing tables operate. First, let's discuss the routing efficiency of the mobile calls in a system.

The translations tables that are responsible for the analysis of the mobile's dialed digits and the routing of the calls to their proper destination (the PSTN or to another cell site for a mobile call termination) are developed by the switch vendor, the software engineers within the network engineering department, the marketing department, and the legal department. This last group is included since it is responsible for the company's tariff compliance which determines where and how calls can be routed outside the network to other telecommunication-type companies. The definition of the initial dialing plan (mobile dialing patterns) and any subsequent changes must be approved by the directors of both the marketing and network engineering departments.

The actual implementation of the design, future alterations, and the maintenance of these tables is the sole responsibility of the software group. Any changes to these tables will immediately affect the call delivery service of the system mobile subscribers. So this is obviously an important design and operational aspect of the system. In order to maintain these important tables, procedures should be put in place by management to assure the accuracy of this database, to prevent unauthorized personnel from gaining access to these tables, and to provide a recent backup of the data in case a switch disaster occurs. With all this said, the purpose of this section is to provide a few basic methods to use in the measurement of the system's routing efficiency and management of the switch translation tables.

Every switch has the ability to provide call delivery statistics to some degree. Although the forms of the data vary from switch to switch, there is a certain amount of data that is basic to each manufacturer. For instance, the utilization of recorded announcements is a common statistic that should be readily obtained for review and analysis. The usage on such devices (actual magazines in the switch) will be of great benefit in measuring the routing accuracy of the translations and to detect changes in the calling patterns of your mobile subscriber.

Every recorded announcement in the switch has a number associated with its message. This number should be included at the end of the recorded message along with some type of node identification. This will help immensely in the troubleshooting process described in Sec. 6.3.2. Some typical recorded announcements are given in the following table.

Message number	Call category	Recording
01	No page acknowledge	The mobile you have called does not answer. Please try your call later—message number A01.
02	Facility problem	We're sorry your call cannot be completed at this time. Please try your call again—message number A02.
03	Unauthorized for use	Your phone is not authorized for use at this time. For further assistance please call 611 from your phone and reference message number B03.

With the utilization statistics for these recordings you can determine if you are routing calls to the correct facility. For example, if there is an abnormally high number of message 02 recordings taking place, then either there is a problem with one of the system trunk groups (voice facilities) or there is blocking occurring on some of these facilities. It would be a good idea to have the software and traffic engineers review this problem and determine if it is a capacity issue with the network facilities or whether it is a routing problem that has entered the translations.

A large number of message 03 recordings may indicate a problem with the translations in regard to the node to mobile number range assignment and routing. If a new mobile code is loaded into a multiple-switch system and the routing it not set up properly, subscribers assigned to these numbers will not receive service. Their mobiles will not be validated, and thus, they will receive recording 03. By checking the routing for this mobile range, the problem could be resolved completely or a step in the troubleshooting process reduced.

An example call delivery statistics report is shown in Fig. 6.4. by reviewing the report an engineer could look for trends and any large changes in the percentages of the system call mix and other call categories. Some useful categories for monitoring and improving the routing tables are the total calls completed percentage and the intermachine trunk (IMT) unavailable percentage. The total calls completed in a system will vary from month to month, but large percentage changes should not occur without some logical explanation. If no obvious network changes have taken place recently, then review the past month's translational changes for possible errors. It is a good idea to have a process established to archive recent translation and switch software load changes as a means to accomplish this review. Many software engineers store past changes on their computers at work. However, a more formal and defined method should be used that makes this data accessible to other members in the department as a backup in the event the originator is unable to respond to a problem.

As previously mentioned, the total calls completed is useful to know as is the IMT unavailable category. Typically these facilities are designed with a 0.001 percent grade of service (GOS). A value higher than this would indicate a blocking level on the system facilities higher than the design specified. Should this occur, have the software engineers review the network routing tables for errors. If no significant errors are found, then consult with the network engineers to determine if the system IMTs need expanding due to an unexpected growth in the system traffic.

In addition to the mobile call routing tables the SS7 network routing tables need to be reviewed as well. Misrouting of SS7 data messages can lead to link congestion, message looping, and possible system service outages. These tables should be maintained much like the routing tables used in mobile call delivery. An overall review of the SS7 routing tables is recommended once a quarter. This type of review is especially important in a network with a large amount of SS7 network growth. As new links are added or reconfigured the routing becomes more complex and chances for an error increase.

One method of checking the routing accuracy of the SS7 data network is to monitor the SS7 data links for occurrences of the various destination point codes (DPCs) of the network. The SS7 data messages routed over each individual system data link are defined by the SS7 routing tables. By viewing the distribution of SS7 messages (sorted by the DPCs) transmitted over a particular data link an engineer can determine if the link is carrying message traffic that another network link could route more efficiently. As a network grows many times, new links are available that provide a more direct route for the transmission of SS7 messages.

This method is also useful in detecting network messages with unknown DPCs. Messages of this type should be tracked for their source of origin and, if invalid, blocked before they are routed in the network.

These are just a few examples of how to utilize the data available in a network to optimize the routing of mobile calls in the system and to check the routing of SS7 messages in the data network. Other data useful in monitoring system mobile call delivery are available from your information systems department. Since this department is responsible for the processing of the system billing records and producing the customers' bills, it will have access to a large quantity of mobile call delivery data. Consult the manager of this department for a listing of reports and data available to the company.

It is recommended that a system software design review and audit take place on a regularly scheduled basis. A reasonable time frame to assure accuracy for this aspect of the network design is once a month for the mobile call delivery routing tables and, as previously stated, once a quarter for the SS7 routing tables.

6.3.2 Network routing performance summary

The following is a summary of steps used in measuring and maintaining the routing performance of the system translation tables.

1. Obtain a copy of the translation tables for reference.

2. Obtain statistics on the utilization of the system's recorded announcements.

3. Obtain statistics on the network call delivery.

4. Monitor the utilization of the recorded announcements and the call delivery category percentages and note any unusual changes in the values.

```
System call mix:
    Percent of total system calls (M-L):      80.0
    Percent of total system calls (L-M):      15.0
    Percent of total system calls (M-M):       5.0
System calls by category:
```

Category	Percent of total system calls
Completed calls	78.00
Unacknowledged mobiles	09.45
Invalid mobiles	00.05
Invalid ESN	00.10
RF channel unavailable	00.23
Land trunk unavailable	00.06
IMT unavailable	00.05

Figure 6.4 System call delivery statistics report.

5. Correlate large changes in the data to possible errors in the routing tables or the current network design.

6. Develop formal procedures to track and correct any errors that may develop in the routing tables. Also, keep a current copy of the tables in a secure place at all times and limit access to the tables to experienced authorized personnel.

6.4 Network Software Performance

Every node in the system has an operating system. This software provides the ability to operate the system nodes for call delivery, billing, maintenance and administration of the equipment, etc. The actual performance of the operating system itself is very often overlooked in a network performance evaluation. The following are some metrics to use to complete a basic evaluation. The following are some metrics to use to complete a basic evaluation.

The initial software load and subsequent loads of a switch typically require additional switch hardware to provide more memory capacity. The amount of memory needed for a particular software load is dependent upon the type of operating system design utilized by the vendor, the number of system features supported by the load and active in the system, etc. For some switch vendors this is not a critical issue; for others new software loads require a large increase in the amount of memory required. This can sometimes be expensive, so be aware of this issue when preparing the system's budgetary input to the company.

Another critical issue to be aware of when reviewing a switch's operating system is the amount of processor load increase that a new software release will introduce. Sometimes a new load will increase the CPU load by as much as 5 percent. This value depends upon which system features are used and the kind of system the switch or node is operating within. A new software release can also increase or decrease the call delivery completion rate and the call delivery times (actual time it takes for the call to be completed). These metrics can be measured in a simple manner by performing call testing in the network using a standard call test plan and known drive test routes. The results should be compared to an already established baseline test (a call test conducted during normal or stable system times) for observing any abnormalities and determining if the new load caused any degradation in the service in the system.

6.5 Network Performance (General Data)

The following is a list of general system performance and design data that can be used as a measure of the system's operation performance and capacity. This data will also be helpful when you conduct specific performance and growth studies of the network switches.

1. Cell site data
   - Number of cell sites currently in service on a switch-by-switch basis
   - Number of cell sites projected to be cut in service by the end of the year

2. Subscriber data: Number of subscribers that are
   - Currently assigned in the switch
   - To be assigned by the end of the year
   - Assigned with the voice mail feature
   - Assigned with other network features (traffic information service, etc.)

3. System and switch traffic data
   - Average system busy hour usage (erlangs) (10 high day average per month)
   - Estimated network maximum traffic level
   - Average switch busy hour usage on a switch-by-switch basis
   - Estimated switch maximum traffic level
   - The usage per subscriber (erlangs)
   - Percentage of subscribers registered and active in the system during the busy hour
   - Number of registration attempts on the system during the system busy hour
   - Registration interval assignments on a switch-by-switch basis
   - Number of call attempts in the system during the system busy hour
   - Number of calls completed in the system during the system busy hour
   - Number of blocked calls in the system during the system busy hour
   - Number of dropped calls in the system during the system busy hour
   - Number of intrasystem handoffs during the system busy hour
   - Number of intersystem handoffs during the system busy hour
   - Average call holding time (s)
   - Average call setup time (s)
   - Switch processor load on a switch-by-switch basis
   - Switch efficiency percentages on a switch-by-switch basis
   - Total number of voice channels in service on a switch-by-switch basis (both RF and land)

6.6 Network Call Delivery Troubleshooting

Resolving call delivery problems in a cellular environment can be a lengthy process. It takes time to review the customer's complaint, collect necessary data, begin working with the customer (if possible), and conduct call testing if necessary. In this section some basic troubleshooting procedures will be given as an aid to this task as well as some examples of actual resolved call delivery problems.

6.6.1 Network call delivery troubleshooting procedures (initial steps)

When presented with a network call delivery problem do not immediately rush off and begin delving into a large, time-consuming call testing and analysis project. Take a few preliminary steps that may, in the end, save you a tremendous amount of time and aggravation. First, determine who is reporting the problem. Are they following the previously established trouble reporting and resolution process? It has always been challenging to a manager of a group of systems engineers to guide them through their normal work load while they are receiving calls outside set procedures from other departments in the company about mobile call delivery problems. We've seen three or four engineers get separate calls from another department to resolve a call delivery problem for a single customer's complaint. All these calls were received at about the same time. No managers were notified, and so each engineer began working on the problem separately. They stopped their current scheduled work to start troubleshooting the problem. After some time, it came to light that they were all working on the same problem but none of them knew what the others were doing. A few phone calls from another department stopped the work of an entire group in the engineering department! So we stress again, follow agreed-upon procedures to resolve customers' call delivery complaints. Tell your engineers that if they receive a call regarding troubleshooting outside their normally scheduled projects to redirect the call to their manager.

Another great assistance when troubleshooting a problem in the system is to have the person reporting the trouble provide a written description of the problem along with a date and name.

Next, determine if the call is legitimate. In other words does this type of call actually exist in the network? A number of times we have heard of a customer complaint about using a feature in the network only to find out that it was not available. Or if it was available its functionality was removed from the system for one reason or another. To stop this type of confusion from occurring it is recommended that a manual be developed and maintained of all the calls supported in the system for both the home subscribers and the various types of roaming subscribers. This listing will be a great reference when performing call delivery troubleshooting and in conducting network hardware and software acceptance call testing.

Finally, find out what activity is currently taking place in the system for that day or week.

1. Is there a new software load being implemented?
2. Have there been some recent changes in the translation tables?
3. Was there a change in the network configuration?
4. Was there a system outage?
5. What activities have the operations department been conducting?
6. Were there any recent changes to the validation systems in the network?
7. Did a previous system that your company had a roaming agreement with cancel its contract? Consult your accounting and finance department on

this and request that they notify the engineering department when any such changes do take place.

6.6.2 Troubleshooting procedures (first level)

Note, before starting the troubleshooting process, that it is recommended to have the customer care department quantify the problem to determine the magnitude of the customer impact. This will assist in scheduling the troubleshooting work among the other scheduled engineering projects. Obviously if the problem is critical and there is a large service outage, this troubleshooting will take precedence over other scheduled engineering work.

There are a number of levels to troubleshooting call delivery problems in a cellular network. The first level begins with the customer service department. At this level the customer trouble tickets are generated and basic problem resolution is conducted. The most common problems are an incorrect subscriber profile with an invalid mobile phone number to electronic serial number (ESN) assignment in the switch, an improperly activated mobile code (code management problem), a customer assumes he has features that were not assigned to him, and a customer does not know how to use her cellular service and is dialing improperly.

Many of these problems can be resolved by noting what call treatment was received by the customer. A *call treatment* is the classification of the various types of call terminations. For instance, a call completed to the intended party would be classified as a successful call termination. A call that is sent to a busy recording because the called party was on another phone call would be classified as a busy call treatment. The call treatments are a very useful tool in troubleshooting call delivery problems. Each call treatment should include a switch code at the end of the recording that designates which switch in the network provided the call treatment and, thus, where the call was terminated. This is very important data.

Table 6.4 lists some basic troubleshooting tips. The message number at the end of all recorded announcements is the switch description followed by the recorded announcement number; e.g., message number A01 represents switch A in the network, and 01 represents message 01 within switch A.

6.6.3 Troubleshooting procedures (second level)

If the call delivery problem cannot be resolved at the first (customer service) level, then the engineering department will have to get involved and begin the second level of troubleshooting. This level involves the initial steps previously mentioned as well as call testing (see Sec. 6.6.4). The first task after performing these steps is to verify that the mobile is indeed registering in the system (provided that the system is using autonomous registration). Most systems in service today use this function to locate the mobile within the system. All mobiles must first signal the system to be validated for placing and receiving

TABLE 6.4 System Call Treatments and Possible Causes

Call treatment	Causes
Recoding 01: The mobile customer you have called does not answer. Please try to call later—message number A01.	No page response. No mobile answer. Mobile is outside of the service area. Mobile is turned off, and the phone is not transferred. Mobile is not registering. Mobile may have an incorrect station class mark. Possible system problem.
Recording 02: The mobile you have called has been temporarily disconnected—message number A02.	The customer is not active in the system (voluntary cancellation). The customer is not active in the system (involuntary cancellation).
Recording 03: We're sorry your call cannot be completed as dialed. Please check the number and dial again—message number A03.	The customer has dialed improperly.
Recording 04: You are not allowed calls to this number. For more assistance please call our customer service number at 611 and reference message number A04.	The customer has calling restrictions on their profile such as local dialing only or international calling blocked, or the called number may be blocked by the switch (e.g., 900s, 976s, 0+ dialing, etc.).
Recording 05: We're sorry your call cannot be completed at this time. Please try to call later—message number A05.	All circuits are busy. System problem.
Recording 06: Your mobile phone is not authorized for service at this time. For additional assistance please call customer service at 611 and reference message number A06.	Customer's profile has an ESN mismatch. Customer's ESN is on the deny list. Roamer's mobile phone number is not active in the switch. There is no roaming agreement with this cellular company.

calls. If a mobile is not registering, then check the profile of the subscriber in the switch. If these data are correct, then check the mobile and RF environment to assure these areas are set up properly. For more information on the mobile registration process refer to Electronic Industries Association (EIA)/Telecommunications Industry Association (TIA) specification 553.

If the mobile is registering, you can begin call testing to try to uncover the call delivery problem. Again, most of the time it is a switch database error or a problem with the validation of the mobile. The latter is becoming more predominant due to the increased use of more complex and sophisticated fraud-prevention systems. It is a necessity to have a listing and description of the systems in use in your network when performing call delivery troubleshooting.

If the problem is further embedded in the system, then more extensive call testing and tracing must be completed. A trace on the call will reveal a great deal of information such as the number dialed by the mobile unit, the results of the translations of the dialed digits by the switch, the route the call was sent to, and the final treatment of the call. You may want to have the network software engineers and your local vendor support team get involved at this point to try to resolve this problem quickly while all these resources are available.

After a problem has been solved, a log of the problem and its correction should be kept. This will serve as a means to track the work completed and as reference material for future troubleshooting and to teach company personnel how to resolve problems on their own.

6.6.4 Troubleshooting call testing procedures

The first step in conducting test calls for troubleshooting is to set up the test mobile. This task requires a mobile phone that is easily programmed, a well-charged battery and wall charger, a test phone number, and the ESN of the mobile. The next step is to build a subscriber profile for the test phone in the switch database. This will include activating the number in the database, assigning this number to the test mobile ESN, assigning system features to the test mobile, and assigning the category of service to the test mobile (examples are local service only, local and toll service, and local plus toll plus international dialing service).

Note that it is a very good idea to obtain a copy of the switch subscriber commands along with an explanation of each field for use as a reference while conducting this call testing. It is also helpful to have an example of a previous test mobile profile that has already been set up in the switch to use as a template. The collection of useful switch commands and example subscriber profiles should be carried over to other types of applications. For instance, printouts from past call traces would serve as excellent reference material when performing this type of work at a later date.

Next collect the customer trouble tickets from the customer service department. These reports will contain the data needed to test and possibly resolve the call delivery problem in question. A standard customer trouble ticket will have the customer's mobile phone number, the ESN, the category of service assignment, a listing of the features assigned to the customer, a description of the call delivery problem including the type of treatment received when attempting the call, and where this problem occurred. With this data you can access the magnitude of the problem and begin the troubleshooting process.

Another task you need to perform before call testing can begin is to check with your accounting and finance department for any requirements it may have regarding this use of the system's service. Actual call testing may take a few hours, or it may take months. The billing for the airtime of these test mobiles will still be produced and given to the accounting and finance department for processing. If the department is not informed of the call testing taking place, it may consider these calls fraudulent and take the necessary action to block service to these mobiles. Therefore, set up a formal procedure to specify the test numbers to be used in call testing, the duration of the call testing, the purpose of the call testing, and the name of the person responsible for this testing. This information will assist the accounting department in distinguishing between the system use by actual paying customers and that generated by call testing from the engineering and operation departments (see Fig. 6.5).

It is recommended that you have terminal access to each network switch all in one location to perform call delivery troubleshooting. This access is necessary so that you can change the subscriber profiles of the test mobiles, check on the status of the mobiles' activity in the switch, and perform call traces. The location of the call testing must also have good coverage by the system. In a system with multiple switches it is a good idea to build test cells programmed to different control channels and make them available for this type of work.

In this test configuration the test mobile can be programmed to scan for only one of the test cell's control channels. This allows call testing to take place on all network switches at any given time at one switch location. Having these test cells and test mobiles available is important not only for conducting call delivery troubleshooting but also for performing call testing in regards to accepting new switch software loads and bringing new network switches into service.

Another suggested requirement for performing network call testing is to have test call log sheets produced beforehand to assist in recording the completed test calls and their results. This data will be valuable in the overall analysis of the call delivery problem (see Fig. 6.6). Once the profile of the test mobile is built in the switch database and the other departments within the company have been

```
To: Accounting Finance Dept.

From:  _____
        Network Engineering Dept.


Date:  _____


RE: Request for test mobile activation


Mobile test:  _____

Mobile ESN/ESNs:  _____

Purpose of call test:  _____

Start date of test:  _____

End date of test:  _____


Additional comments:
```

Figure 6.5 Example accounting and finance form used in performing network call testing.

```
System switch: _____

Tester: _____

Date: _____

Test mobile: _____

Test mobile ESNs: _____

Type of test call: _____

Start time: _____

End time: _____

Duration of call: _____

Call treatment expected: _____

Call treatment received: _____


Additional comments:
```

Figure 6.6 Example test call log form.

given proper notification, call testing can begin. The next step is to check the test phone to make sure it operates properly in the system. Check to see if the phone registers, can make outgoing calls and receive incoming calls, and can activate and deactivate the features it has been assigned. If there is a problem with performing these functions, then either the phone is not set up properly or there is a problem with the system itself and further data must be collected.

If the test mobile works properly, then call testing can begin. If the problem is a customer complaint, then review the trouble ticket from this customer and try to mimic the exact call by having the test mobile register or obtain service from the same network switch that the customer claims to have a problem. Then try to make the same type of call the customer is having problems with. Take note of the type of call treatment you received when making the call. Did the call get completed? Did you receive the same treatment that the customer described in the trouble report? Did you receive a switch recorded announcement? All these data are valuable when troubleshooting a system call delivery problem.

Call testing summary steps

1. Develop a testing lab complete with test cells and terminals assigned to every switch in the network for use in troubleshooting and completing call testing of network hardware and software.
2. Develop a test call log sheet.

3. Obtain the customer trouble tickets from the customer service department for determining the type of problem calls, the mobile number of the subscriber, the features assigned in the customer profile, and the quantity and magnitude of the customer complaints.

4. Obtain a mobile phone and an unassigned number in the switch for call testing. If possible, choose a mobile phone that is easy to program and a test number that is in the same 100s block that the customer's number is located. [A 100s block is a group of phone numbers clustered by the last three digits (0-99) of a 10K range, e.g., 201-555-1200 to 201-555-1299 would represent a 100s block of phone numbers.]

5. Set up the subscriber profile of the test mobile in the switch database. Specify the type of service the mobile is to have and the features needed for testing.

6. Notify the other departments in the company of this call testing by using standardized forms and formal agreed-upon procedures.

7. Program the mobile phone for operating on the designated network switch. Check the phone for proper operation.

8. Obtain, for reference, a listing of all the recordings on the network switches and a listing of the mobile codes stored in the system. The list of mobile codes should contain the corresponding node's designations where these codes are actually located. Obtain a list of all the service categories and their definitions as they are used in your system. Finally, a list of relevant and frequently used switch commands along with a set of example switch printouts would serve well as quick reference material, further expediting the troubleshooting process.

9. Begin call testing by trying to mimic the type of mobile call in question. Record your findings on the test call log sheets.

6.6.5 Network call delivery troubleshooting summary

The following list is a summary of the troubleshooting steps discussed in this chapter.

1. First-level steps (customer service)
   - Determine if the customer service department has attempted to resolve the problem.
   - Determine if the customer trouble ticket is completed.
   - Identify the magnitude of the problem by quantifying the problem.
   - Make the customer's profile available for use in further troubleshooting.

2. Network engineering call delivery troubleshooting initial steps
   - Don't rush.
   - Determine who is reporting the problem.
   - Determine if other company personnel are working on the problem.

- Make sure proper problem resolution procedures are being followed.
- Obtain a typewritten description of the problem.
- Determine if this is a legitimate call.
- Find out what other activities are occurring in the network.

3. Network engineering second-level steps
   - Complete the initial steps for network engineering.
   - Conduct call testing and call tracing.
   - Log and record the results of the problem resolution.

These are just some initial ideas to assist in correcting call delivery problems that may arise in your network. Obviously, these steps can be expanded and augmented with your own data and experiences to provide more accurate and effective troubleshooting procedures.

6.7 Network Call Delivery Troubleshooting Examples

Example 6.2: Network Timing Problem This problem actually occurred between two multiswitch cellular systems interconnected with direct voice and data links (see Fig. 6.7). It was given to me in the network engineering department after the customer service technicians could not resolve the complaint. The problem was stated as such:

> There was an intermittent call delivery problem occurring for system A customers roaming in system B. During selected times, calls to these customers would fail and receive an out-of-car recording as their call treatment. The problem was recorded as taking place during the system busy hour.

Intermittent trouble often signifies a load-related problem in the network. Somewhere a process was being overloaded or a timer was being expired. This was most likely the problem here since the customer complaints all indicated that the problem occurred during the system busy hour. We worked with system B personnel to set up a test number and began call testing. After some time, we were able to duplicate the problem. Upon review of the call trace in system B, it was noted that the call timed out while waiting for the mobile acknowledgment to be returned.

Upon further discussion with personnel from system B, it was brought to our attention that there had been a recent change in their network. System B personnel had *cut* a new switch in their network (switch 2) and had changed the routing of system A's calls in the redesign. Instead of routing the calls directly to each of their switches as they were previously doing, they were now using one of their switches as a tandem (switch 1). Thus, a call to a system A mobile roaming in system B would first get routed to switch 1 for processing. The mobile would be paged, and if no acknowledgment was received, then the call was redirected to switch 2. This additional processing caused the total duration of the call set up to exceed the maximum limit specified in system A's network and so the call failed and was routed to an out-of-car recorded announcement.

Once we knew what the problem was, the solution was easy: set up the original routing. The routing was corrected, and call testing determined that the problem was solved. In summary, make sure first-level troubleshooting is completed, acquire all necessary data and recent network changes, conduct call testing, perform a call trace to get more detailed information about the call problem, and finally test and record the solution.

We have seen a similar problem where a number of mobiles were not being properly validated due to an error in the network SS7 routing tables. The problem occurred for subscribers stored in the systems home location register (HLR), but only during the system's peak busy hour. The subscribers experiencing the problem were receiving an invalid mobile recording from the switch until the SS7 routing error was fixed and the timing was no longer an issue.

Example 6.3: System Code Management Problem Errors in mobile code management are the leading cause in call delivery problems. This is due to the high subscriber growth, subscriber churn (turnover rate), the demand for more mobile codes, and the constant shuffling of these numbers within the network to accommodate this growth. Under these conditions the chance for errors to occur is tremendous. Most cellular companies have a person or group dedicated to the task of managing the mobile codes in their networks. Many names have been given to this team including the directory inventory group and the telephone number inventory group. The code management function and personnel are discussed further in Chap. 9.

This group must monitor the usage of each mobile code (10,000s block of numbers provided by the LEC), order new codes based upon the remaining amount of unassigned phone numbers available, and the expected subscriber growth provided by the marketing department, and keep track of where to store these codes in the cellular network. This is an important and time-consuming task to say the least.

When a customer first purchases a mobile phone, he is assigned a telephone number along with a chosen category of cellular service and any other system

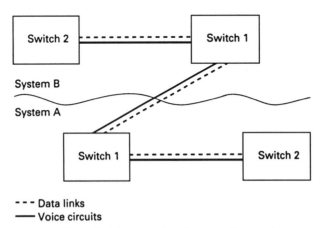

Figure 6.7 Example 6.2 system A and system B network configuration.

features that might prove useful. All this information is important, but the mobile number assignment is the most critical from a customer and system standpoint for the following reasons. When a mobile code (usually a 10,000 block of numbers) is loaded in the system, it is assigned a location in one of the network nodes as well as a serving central office from the LEC. If an error should occur in any part of this assignment, then these numbers cannot receive service. It is highly unlikely that all 10,000 will be assigned to customers, but you can see the effect of an error in this part of the network. Therefore, management of these numbers is vital to prevent call delivery problems and to assure these resources are utilized in the most efficient manner possible.

Suppose a number of mobile codes (some partial and some whole) are moved from a switch to an HLR in a multimode system and an error occurs in the switch database. The error was made when the location of one of the partial codes was incorrect. The code is now rendered useless, and all the mobile subscribers assigned to any numbers in that range will no longer be validated by the system and will be denied service. Typically these type of changes occur in the evening with only partial (random) call testing being completed to assure accuracy in the database. If the problem goes unnoticed, the customer service department will begin to receive calls from customers complaining about their service disruption. After a while, a manager in this department will notice the trend and call the network engineering department to resolve the problem quickly. A quick review of the previous night's activity and database changes will reveal the error and the problem can be resolved. This type of problem is especially prevalent in call delivery scenarios involving mobiles roaming in other markets where the difficulty of managing codes is further increased between two companies.

Example 6.4: Network SS7 Problem Since many of the cellular systems in North America use SS7 to communicate between each other for performing mobile call delivery, any errors that take place in this data network tend to have an immediate and large impact on the customer base. With the development and use of new network protocols, the need for trained and experienced network personnel, and the frequent network changes (both in hardware and software), the chances of an error occurring in the SS7 network is great. Although this type of network has exceptional error detection and recovery, functionality errors still occur in the network and cause call delivery problems.

Another actual call delivery problem took place when an SS7 routing error occurred in a signaling point and caused messages to begin looping around the network. The message looping caused the constant retransmission of the messages between two STPs. This, in turn, caused the higher-level link management functions to begin to intervene and start transmitting messages across the interconnecting C links in an effort to correct this problem. This caused a halt in the transmission of the lower-priority messages (MSUs) used for call delivery, which, in turn, stopped service to mobiles from other systems roaming in this network. (See Fig. 6.8 for more details.)

When a protocol analyzer was placed on the link set and its performance statistics viewed, the problem became clear. The traffic levels on the C-type links within the link set were in the high 90s (almost at 100 percent congestion). This was causing

422　Chapter Six

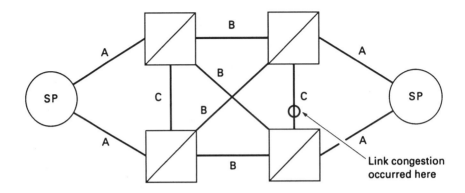

Figure 6.8　Example 6.4 SS7 network.

the messages used for call delivery to get blocked, and thus calls to mobiles utilizing these facilities were stopped. To correct this problem a sample data file was collected of the SS7 messaging on the links for later analysis and then the link set was reset.

The reset cleared the problem temporarily (for 1 to 2 days), and upon analysis of the SS7 message data captured the looping problem was discovered and fixed. This type of problem could have been avoided if the SS7 data network performance statistics were reviewed more closely and on a more frequent basis.

These are just a few examples of some call delivery problems that were resolved within a wireless network. It would be very productive to assign the engineering staff the task of documenting resolved network problems (not just call delivery problems) for use in teaching engineers how to troubleshoot and fix wireless network problems. This effort would also assist the engineering staff in their understanding of the network's overall operation.

Chapter 7

Billing and Charging in a Wireless Network

7.1 Basic Billing Process

This chapter briefly covers the various elements that comprise a billing and charging system. Whole companies have been established for the sole purpose of handling the billing and charging aspects of wireless mobility systems.

7.1.1 Billing process description

A billing system is comprised of a series of independent applications whose final output is the production of a customer invoice. The major components of a billing system are as follows:

1. *Call detail record generation, collection, and storage application.* This is used to record the details of the call. A basic call detail record (CDR) contains, at the very minimum, the start time of the call, the total call duration, the originating number, and the terminating number. The CDRs are stored until the time an invoice needs to be produced or the final billing stage.

2. *Guiding application.* This process matches the calls to the various customer-calling plans. This application uses the start and end numbers and the duration and time of the call to determine what the charge should be based upon the calling plans of the customer account.

3. *Rating application.* This application applies the rate for the individual guided calls. The rating gives the call a value to be charged at the time of billing but does not include any discounts, promotions, or taxes.

4. *Billing application.* This application is performed once a month or during defined cycles. This process collects all the rated calls that have been stored

over the past 30 days and adds any discounts or promotions that are associated with a customer's account. In addition, taxes and credits are also applied at this point in the billing process.

5. *Invoicing application.* Once the billing process is complete, a file is created that includes all the customer's invoice information. This file is then sent to a print house to be converted into paper invoices. These invoices are then sent to the customer. If a company has implemented an electronic statement system, then customers can choose to have their invoices in diskette, tape, or even e-mail format.

From a wireless carrier's perspective, as customers use the mobile service the billing process, in turn, builds and collects the CDRs using charging software operating on the mobile switch. These CDRs are then transferred from the mobile switch to a separate billing server (a computer configured to support billing process software) or directly to a billing company. Once the CDRs are collected, they are then rated. Rating a call involves examining the associated CDR to determine if the call is, for example, a local call that falls within a local-calling-are plan, a toll call, or an 800 number.

Further, the time the call was made is also used in rating the call. If the mobile call is made during the day, it takes place during peak network time and will be charged at a higher rate than if the call were made during the evening or off-peak times. Once each call is rated, it is then stored until the time an invoice is processed and sent to the customer. In the case where the mobile customer is a roamer, the CDRs associated with this subscriber will be sent to a third-party company (referred to as a *clearinghouse*) for processing and billing other mobile carriers whose customers have roamed on this wireless system. Figure 7.1 is a flowchart of this process.

As it applies to the wireless carrier network the CDR itself is a formatted collection of information about a chargeable event (mobile call, activation or deactivation of call forwarding, etc.) occurring on the mobile switch. The CDR itself must contain the originating number of the calling party, the terminating phone number, and the start and end times necessary for billing the call.

CDRs associated with a roaming mobile customer are formatted differently than normal home subscriber records. In order to support the exchange of revenue between different wireless carriers that have common roaming agreements a standard record format called a cellular intercarrier billing exchange roamer (CIBER) is used. This is a proprietary protocol for the exchange of roaming billing information (voice as well as data) among wireless telecommunication companies as well as other billing vendors and data clearinghouses. Figure 7.2 is a flowchart of this process.

7.1.2 Billing cycles and billing data filtering

In order to better manage the billing of an entire customer base the processing of CDRs and the subsequent generation of customer bills are completed in scheduled periods. These periods are called billing cycles and take place every

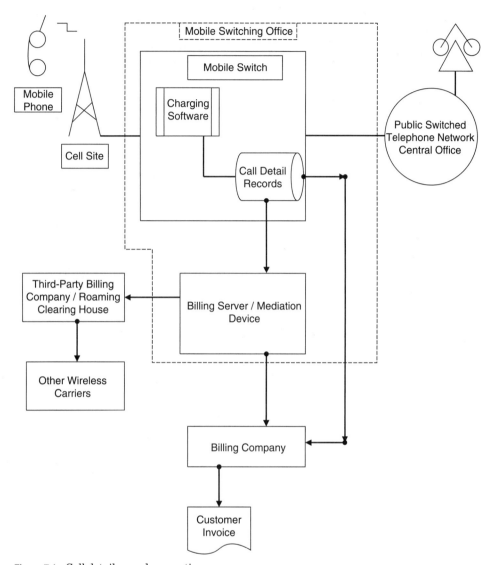

Figure 7.1 Call detail record generation.

month. The number of billing cycles required would depend upon the amount of mobile customers the carrier serves and the volume of CDRs generated on average. For instance, if the carrier has 80,000 customers and the billing company can process and send out 20,000 bills per week, then there will be four billing cycles. Although this is a simple example, you can see the advantage of breaking this process up into parts and billing the customers in cycles. One advantage is the ability to implement quality checkpoints and monitor any billing errors that might have occurred during one of the cycles. If there is a

426 Chapter Seven

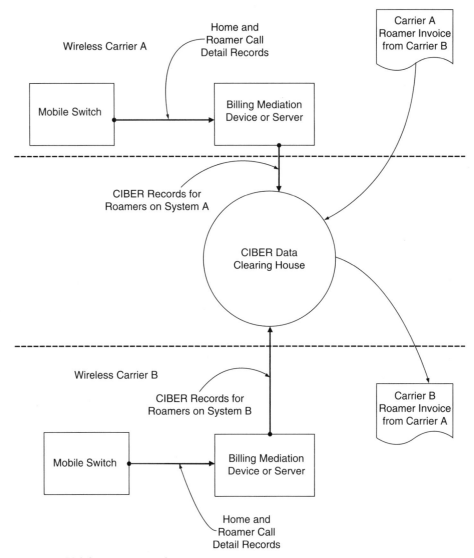

Figure 7.2 Mobile roamer record processing.

problem with the billing of a particular service, then it can be detected in one billing cycle and corrected before the remainder of the customers are affected.

A billing cycle is further defined by taking a subset of the carrier's total customer base and assigning it to a specific cycle. For example, if a carrier has 80,000 customers and 20,000 customers are billed in the first billing cycle, these customers can be identified and processed by their mobile telephone numbers. In this example if 30,000 telephone numbers are used to serve these customers, the range may be 732-528-0000 to 732-528-2999. The billing application software can then filter all the billing records whose telephone numbers fall within this range and process, rate, and bill these customers. Take note that the electronic serial number is another field containing information that can be used to filter CDRs for specific applications.

The mobile telephone number populated in the CDR can also be used for other types of applications where the identification of an individual customer or subset of customers needs to be completed. For example, if the customer care department needs to address a billing question from an individual customer, then its support system will need to be able to access the billing database and use the mobile's phone number to collect all the billing information about this customer for a specific period of time. Other needs exist in the carrier's organization where information about specific customers is necessary and the use of the mobile subscriber's phone number in the CDR for filtering the billing database can be used.

7.1.3 Call detail record description

Information contained in a CDR varies depending upon the events of the mobile customer. A call or other activity by a mobile customer will trigger the mobile switch to begin generating a corresponding CDR. However, if a mobile experiences a handoff during the call, then this event will be recorded in the corresponding CDR. A call that does not hand off during its duration will have a blank in this field. Take note that not all mobile activity will generate a CDR. A mobile registration or other authentication will not generate a switch CDR or populate one of its fields.

Variation of the CDR structure can also exist depending upon the type of technology deployed. A time-division, multiple-access (TDMA) system will have a different CDR structure than a global system for mobile communications (GSM). Differences in the type of switch manufacturer are another contributor to differing CDR structures. Regardless of the variations that may exist amongst CDR formats some common information is consistent in all record types.

The following is a list of common field types:

- User identity (i.e., mobile telephone number)
- Network identity (system identification number)
- Terminal or mobile unit identity (electronic serial number)
- Calling party telephone number (mobile or land)
- Called party telephone number (mobile or land)
- Start time or time of initiation of trigger event
- End time
- Call duration
- Call record type (home subscriber, roaming subscriber, handoff, etc.)
- Serving cell site

- Allocated trunk group
- Call feature (call forwarding, call waiting, three-party conference)

7.1.4 Call detail record generation and collection

As previously mentioned the generation of a CDR is triggered by an event taking place on the carrier's mobile switch. If the call is initiated by a mobile customer (also known as the *calling party*), then the switch initiates the generation of a CDR. If, for example, a call is completed and lasts 2 min, then the total billable time will be recorded in the CDR. Once this CDR is completed, it is closed and cannot be further altered by the switch. One can think of the CDR as a small file that contains a bulk of data about the activities of mobile customers recorded in discrete events. For example, if the call that the mobile customer originated encounters a handoff during the call, then this activity or event will be recorded in the CDR.

CDRs can also be generated in the event of

- A call termination to the mobile subscriber
- A feature activation or deactivation such as call forwarding, call waiting, or three-party conference
- A call that tandems (is routed in and out of the switch and thus is not terminated) through the mobile switch and is routed to the voice mail system so that the calling party can leave the mobile subscriber a message

After a period of time, 15 min for example, the charging software will write the CDRs for this given period into a file on the switch disk drive. This file is called a *circular file* since blocks of CDRs are continually written to it until the end of the file is reached. Once this point is reached, the software begins to overwrite the CDRs in the beginning of the file. Once these "aged" (e.g., oldest blocks) of CDR data are overwritten, the data are lost and cannot be recovered. This process typically takes 24 h depending upon the size of the blocks, the rate at which CDRs are generated and stored, and the size of the CDRs.

What keeps the switch from losing billing data is the use of another process that continually transmits the blocks of CDRs to an external billing server or directly to the billing company at specified times, every 4 h for example. In this process as CDRs are generated and written to the circular file as subfiles or blocks on the switch disk, the same software is also transmitting the prior stored data to an interfaced billing server or company. The parameters defining the file sizes, rates of storage, and transmission times are established by the information technology department and the engineering and operations departments in conjunction with the switch vendor. It is important that these processes be communicated to all parties involved especially the operations department since it is this department's personnel who are responsible for the overall operations and maintenance of the mobile switch. Figure 7.3 is a functional diagram of this process.

7.1.5 Call detail record back-office processing

Because of the volume of CDRs that a wireless network can generate, there are a number of design considerations that need to be taken into account when setting up the billing systems. First, if a billing server or mediation device is interfaced to the mobile switch, then CDRs can be stored more reliably than if they were left on the switch circular file.

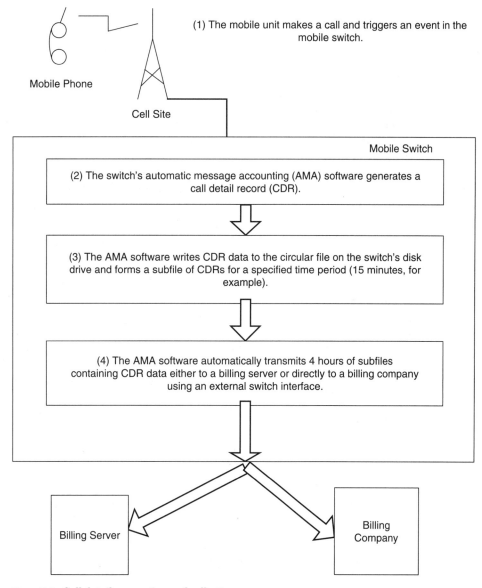

Figure 7.3 Call detail generation and collection.

Once the CDRs have been transmitted to the billing mediation device, they can be stored, manipulated, copied, filtered, or deleted depending upon the business needs of the carrier. Another important aspect of having the CDRs stored on an external device is security and reliability. Once the CDRs are stored on the billing mediation device, the switch is no longer responsible for maintaining this data, which it is not designed to perform. It is supposed to switch phone calls, not process CDR data for any extended period of time.

Copying of CDRs may be necessary if more than one application requires this data. For instance, the billing company may need a complete set of CDRs from the switch to bill mobile customers while the fraud department may need the same data to detect fraudulent activity in the network. To accomplish this, the billing mediation device will need to have the capability to replicate each CDR and transmit it to each application separately.

Filtering of CDRs may also be necessary depending upon the needs of the various back-office business applications. As mentioned earlier if roaming mobile CDRs are generated and stored on the billing mediation device, then it will be necessary to filter all these records and store them in a separate file for eventual transmission to the roaming clearinghouse for processing. These records should not be sent to the billing company for processing since it will charge the carrier for the filtering and deleting of these records. This extra processing plus the added cost of transmitting these records can be a significant expense. Only those CDRs that are absolutely required should be sent to each business process (e.g., billing and rating, roaming, fraud, financial reporting) as opposed to sending all the CDRs to each of these end applications. This is why it is important to determine what records are generated for each mobile event and define exactly what records are needed by each of these business applications.

Defining which CDRs are needed for each back-office application is the responsibility of the business operations department or possibly the information technology department. A simple process to identify critical CDRs is as follows:

1. Identify and document all the wireless services offered to the carrier customers. (This information can be obtained from the marketing department.)

2. Identify and document all the possible events on the mobile switch that can generate a CDR (e.g., a mobile call or a feature activation or deactivation)

3. Develop a test plan with each event detailed for a given set of mobile test numbers.

4. Conduct each test and determine the CDRs that are generated for each event using the mobile switch CDR verification feature. (By using this feature and filtering on the test mobile's phone number, a user can actually view the contents of all the CDRs that get generated for that specific event.)

5. Review each CDR and determine which records are needed to rate the call, are needed by the fraud system, and are needed by the finance, information technology, and engineering departments to generate their reports.

Although this process sounds simple, completing it can take weeks or months. However, without performing this type of analysis, it is very difficult to know when a call record is valid or invalid. As previously mentioned, there may be many CDRs that are generated but are not useful in the end business applications. These will need to be identified and filtered (written to a separate file and deleted from the valid CDR file). Examples 7.1 and 7.2 describe nonbillable activities.

Example 7.1 A mobile subscriber misdials a number, and the switch records this event and routes the call to a recorded announcement informing the customer of the error. This call, although completed, was not what the customer intended to do and thus the call should not be billed. The CDR associated with this call needs to be filtered and reviewed by the information technology and operations departments and should not be further transmitted and processed by the downstream back-office billing systems.

Example 7.2 A mobile customer dials 611 and reaches the carrier's customer care department. In this call scenario the carrier may not want to charge the customer for air time (i.e., use of the network) while it is the concerns addressing service. Again the CDRs associated with these types of calls can be filtered and further reviewed by the information technology or operations department. These records are sometimes referred to as edit reports.

7.2 Call Test Plan for Billing Verification

Testing of the wireless system for verification of proper CDR generation is the responsibility of the information technology and operations departments' personnel. In the initial deployment of a wireless system a complete call test plan needs to be executed to determine if all the wireless services are functioning properly and to validate the billing of each of these services. Additionally, if any changes are made to the network, then call testing needs to be conducted to assure that the proper CDRs are being generated and the billing process has not been affected.

Some examples of network changes requiring CDR generation and billing verification include

- New software loads and patches in the mobile switch
- Changes to the mobile switch translations
- Addition of cell sites and trunks to the network
- Certain hardware changes to the mobile switch (e.g., memory, interfaces, or disk drives

The following is an example call test plan that can be used for basic billing verification:

1. Mobile-to-land test calls
    - Mobile to 911—emergency call

- Mobile to 611—carrier customer care
- Mobile to 411—information
- Mobile to 0—operator
- Mobile to 7-digit local call
- Mobile to 10-digit local call
- Mobile to 11-digit national long-distance call
- Mobile to 011 + international call

2. Mobile-to-mobile test calls
   - Mobile-to-mobile 7-digit call
   - Mobile-to-mobile 10-digit call

3. Land-to-mobile test calls
   - Local landline to mobile 7-digit call
   - Local landline to mobile 10-digit call
   - National long distance to mobile 10-digit call

4. Mobile feature tests
   - Mobile call-forwarding unconditional (activation or deactivation)
   - Mobile call-forwarding conditional
   - Mobile call-waiting (answer or no answer)
   - Mobile three-party conference

5. Roaming mobile test call
   - Roaming mobile call origination
   - Roaming mobile call termination
   - Roaming mobile feature activation or deactivation

6. Short message service
   - Mobile receives short message service
   - Mobile originates short message service (if applicable)

7.3 Billing Verification Methods

Upon completion of the testing the next step is analyzing the data. Beyond the obvious observations of determining if a CDR was indeed generated for every valid call or mobile activity, a review and comparison can also be made of other vital billing information. For instance, call start and end times and consequently the call duration can be checked at the point the call was logged and in the actual CDR. It can also be checked after the call has been rated and printed to an invoice. By comparing the logged time, the times populated in the CDR, and the call time and duration on the invoice, variations can be determined and, if the variations are large enough, a plan can be developed to investigate these discrepancies.

In regards to call times and durations, be aware that some switches and billing systems round up to the nearest minute for every completed call. This can lead to a significant source of discrepancy in total usage time customers experience versus what they are billed. Some carriers advertise billing to 1-s accuracy. In reality all billing processes and systems will have some errors. The goal is to keep these to a minimum by constantly monitoring and improving these processes. Billing can be unforgiving and thankless work. Although the general impression is that once these systems are in place they are foolproof, this is hardly the case.

7.4 Nonbillable Billing Events

Additional checks and comparisons to the testing data include a check of the nonbillable calls in the network, for instance, 800 number calls (the landline portion), 611 calls to customer care, 911 calls (done in accordance with the local emergency provider!), and calls to recorded announcements. Checks also need to be made for calls that originate on new cell sites that were recently commissioned. Check for calls that tandem to another mobile switch (if in a multiple-switch system) and for calls that are forwarded to another number, and make sure the number of transfers is limited. Finally, a check of any new services that are introduced by the marketing department is important and must be conducted prior to the launch of the new service. Further, the engineering and operations departments should already be on the marketing department's distribution list for new products and services and should receive this update every month in addition to being part of the planning processes for these new services. Additional billing verification topics will be discussed in Chap. 8.

Chapter

8

Revenue Assurance in a Wireless Network

8.1 Revenue Assurance Basics

Revenue assurance has aspects that involve a variety of organizations within a wireless company. Typically the revenue assurance department, or team, is structured within the finance department. It is here that the programs and plans to improve revenue assurance are developed and executed. The basic areas to focus upon for improving revenue assurance are the following processes:

1. *Fulfillment process.* Begins by collection of the customer's data at the point of sale. This data includes the customer's name, personal data (e.g., address, employer), and type of service desired (e.g., rate plan, service plan). Typically a system is used to collect this data using a graphical interface provided to the salesperson at the point of sale. Once these data are collected, they are loaded into a database. These data are then made available to other business systems (provisioning systems, billing systems, customer care systems, etc.). Take note that at this point in the fulfillment process the customer's mobile phone number is assigned along with other vital network data.

2. *Provisioning process.* Involves the formatting of customer data and assigned network parameters (e.g., mobile phone number, electronic serial number) for downloading into the various required network elements (e.g., mobile switch, voice mail system, short message system). The provisioning system must interface to each system's unique user interface for the purposes of adding, deleting, and changing customer data on a day-to-day basis.

3. *CDR generation and mediation process.* As described here includes the recording of mobile activities on the mobile switch and the subsequent transmission of this data to an external system (billing server or mediation device) for storage, additional processing, and eventual transmission to the end business application.

4. *Rating process.* As described here is the process by which CDR data, representing individual calls, are assigned a particular value to be charged at the time the calls are billed.

5. *Billing process.* The final step which results in an invoice being sent to the customer.

In addition to these processes a revenue assurance plan will also include a comparison and analysis of the system usage and performance data acquired from the various company sources. This task is basically a check and comparison of the system usage and other performance data obtained from the call detail records (CDRs) and the statistical data obtained from the mobile switch.

The focus of the following discussion is on the areas where the engineering and operations departments' personnel become involved to support the activities and programs proposed by the revenue assurance department. The areas where engineering and operations support are most needed are the provisioning processes, the CDR generation and mediation processes, and verification of network traffic and usage data as they apply to various business applications.

At this point a clarification of terms is needed. The term billing can be used very loosely to identify many processes and systems deployed in a wireless company. Strictly speaking, however, *charging* is defined as the process of collecting data to allow for the monitoring of resource usage, accounting, and/or billing, while *billing* is defined as a process that is used to generate the final customer invoice.

Since the network engineering and operations departments' personnel are responsible for the design, implementation, and operation of the network equipment necessary to support the wireless carrier's service, it is necessary that they also support the provisioning of subscriber databases resident on these systems. Further, since the same personnel are responsible for the mobile switch and other network elements that may produce data necessary for the generation of a final bill, it is required that they be involved in the testing and verification of these processes. Once these critical business processes are addressed, the next step is to begin verifying system usage data and performing analysis using other types of system metrics and data.

8.2 Analysis and Reconciliation of Customer Databases within a Wireless System

8.2.1 Wireless network customer database types and description

The actual number of customer or subscriber databases required in a wireless network will depend upon the technology used, the design of the network, and the services to be offered. In a typical wireless system there is a mobile switch containing a subscriber database called a home location register (HLR). This database may also be located in a separate node external to the switch and connected using SS7 data links; however, in either configuration the database must exist and be maintained for the mobile service to be supported.

In addition to the HLR subscriber database there is a customer database located within the voice mail system and the short message system. Each of these subscriber databases will be structured differently, and the interface used to input and output customer data will be different as well. It is the job of the information technology department to maintain these interfaces and to assure that every type of change to these databases is accurate and complete. In reality after a system has been operating for a period of time, there will be discrepancies that will arise. A wireless system also has customer databases within the billing system, the customer care system, and the fulfillment system that need to be taken into consideration. Developing a plan to reconcile all these databases is where the information technology, engineering, and operations departments need to work together to identify and, where possible, correct any errors that exist.

Here is a list of example customer databases in a typical wireless network:

- Home location register (located on the mobile switch or as a stand-alone system)
- Voice mail system
- Short message system
- Billing system
- Provisioning system
- Fulfillment system
- Fraud system

8.2.2 Customer database provisioning

Since the process to provision customer data within the network is the responsibility of the information technology department, it will not be discussed in further detail with the exception of a few comments. Since it is the responsibility of the engineering and operations departments to maintain the network equipment, it is also their responsibility to maintain the software of each of these network elements. Thus, any time new software is scheduled to be loaded into a network element, it must first be reviewed by the engineering and operations departments for possible changes to the user-interface commands and functionality. Any changes to these commands or interfaces need to be communicated to the information technology personnel for them to, in turn, change their system interfaces. Failure to notify the information technology department of these man-machine interface (MMI) changes can result in transactions (changes, additions, or deletions to customer accounts) failing.

If there are problems with the provisioning process, it will affect customers' service and ultimately their satisfaction. In addition, if customer accounts that should have been deleted from the system are allowed continued service, fraudulent activity can result and the carrier will lose revenue. Figure 8.1 is a functional diagram of this process.

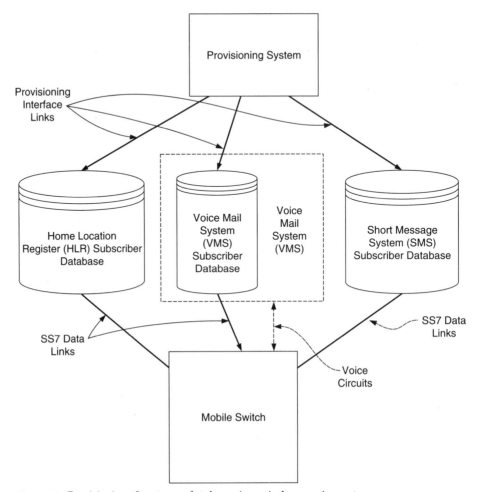

Figure 8.1 Provisioning of customer databases in a wireless carrier system.

8.2.3 Determining the total number of subscribers in a system customer database

The initial basic step in verifying the accuracy of the customer databases within a wireless system is to obtain and compare the total subscriber counts. Be aware that there may be many ways to obtain this information. In some systems this number is available to the user by entering a simple command on the system. In other systems it requires the intervention of the vendor or manufacturer. Determining what the total customer or subscriber count means is important but may not be easy. Take into consideration the number of commands that some systems require to add complete customer information to the database, sometimes referred to as "building the customer profile." In order for a subscriber to obtain service on the wireless network they must have certain critical information present in the HLR such as a phone number, an electronic serial number, and a voice mail routing number. Completing all this information in a database may take as many as five or more user commands and may involve many different files within the system. One can see where the complexity begins in such a task especially in a system where there may be millions of customers.

Further a count of the total number of subscribers in a network database may be affected by the state (active, inactive, suspended) of the customers' accounts. These states apply mainly to the HLR and the billing databases. Other databases such as voice mail for example will not have these states.

The comparison of system subscriber databases is not a straightforward matter. There are exceptions that need to be taken into consideration for the analysis to be accurate. Take the issue of test numbers and other numbers needed by the wireless system to support service. If a test account and test number are loaded into the HLR database but not loaded into the billing system, then there will be a discrepancy in the total valid customer count. Therefore, if test numbers and accounts are needed by the operations department to test the system, they will need to be identified and communicated to the information technology department and the finance department so they can be included in the business processes as exception accounts.

In addition to test accounts, there are instances where restricted numbers are assigned to a valid customer or test phone in the network. These restricted numbers, assigned by the engineering department, are typically used as routing numbers or internal test numbers. In the IS-41 specification there are roamer routing numbers that are used to deliver calls to home subscribers roaming in another wireless networks. Restricted numbers need to be communicated to the information technology, finance, and fulfillment departments so that they will not be assigned or used inappropriately.

Some examples of restricted phone numbers within a wireless network are

- IS-41 roaming routing numbers
- Voice mail routing numbers (e.g., busy forward)
- Internal network address numbers
- Public switched telephone network (PSTN) test numbers

In determining the total number of subscriber accounts in each of the system customer databases note the following concerns:

1. You need to determine if the total customer count for each system is a tabulated number that is incremented and decremented as accounts are added and deleted and what the means are by which this total count is established. Some systems have a utility that will check each entry (for instance, each phone number assignment possible) and provide a count of the numbers that are currently assigned on the system.

2. The customer accounts in the voice mail system and the short message system subscriber databases do not have states associated them. Either the

account is present (customer is active and valid) or not present (customer is inactive and not valid).

3. In the HLR and the billing system databases there are various states that an account can be assigned depending upon the customer's current credit status. These states will make it more difficult to perform a total subscriber comparison between system databases. However, as a simple guideline, a system of 500,000 subscribers may have at one time 2 percent of the total subscriber base in a suspended state while the number of inactive subscribers may be approximately 5 percent. These values can be reviewed with members of the customer care and information technology departments and the percentages validated.

4. All the test numbers and test accounts must be identified and eliminated from the total subscriber count to make the final number more accurate.

8.2.4 Postpaid and prepaid service issues

The previous discussion assumed that the wireless system was a postpaid system as opposed to a prepaid system. In a postpaid system all customers have completed a service agreement with the wireless carrier to pay on a monthly basis *after* their service has been provided. In a prepaid wireless service arrangement the customers must pay for their service *prior* to it being made available to them. This is an important point since it can have a large effect on how customers are counted and tracked in a wireless system. Some carriers choose to assign all prepaid customers to a complete 10,000 number block (NPA-NXX-0000 to 9999), while others break this assignment down into smaller-number blocks. This mixing of prepaid and postpaid customers makes it more difficult when trying to reconcile system databases. Taking the time to develop a plan that all departments will understand and accept will reduce the problems and conflicts associated with this service distinction in the future.

8.3 Revenue Assurance Billing Verification

8.3.1 CDR volume benchmarks

As a wireless system begins commercial operation, the volume of CDRs generated will rise as the system traffic increases. In accordance the number of CDRs that are edit-rejected will similarly increase. One method to ensure that the systems and processes are functioning properly is to implement checks at various points in the billing system. These checkpoints can be easily deployed. For instance, by monitoring the amount of CDRs generated and then transmitted from the switch to the billing mediation device, a check can be made to assure that this volume does not change dramatically in the day-to-day operations of the system. If a large variation does appear, then further investigation of this deviation is necessary. This type of change could be due to a new software load in the mobile switch or a problem with the billing mediation device.

Another checkpoint can be implemented at the billing mediation device by monitoring the number of edit-rejected CDRs that are produced on a daily basis. Again a large variation from the daily average (established as a benchmark value) warrants further investigation. An increase in these types of CDRs could signal a problem with the network. In this case the problem could be the implementation of a new wireless service or new cell site that has not yet been implemented in the billing system. This situation can occur when there is a lack of communication among the departments within the company. The marketing department may propose a new service that the engineering department implements without involving the information technology or billing department. This problem needs to be addressed quickly since the company is losing revenue while the problem exists.

An increase in the number of edit-reject CDRs could also signal a problem with the customer's understanding of the wireless service. If, for instance, a new service is introduced where the customer must dial a unique dial pattern that she is unfamiliar with, there may be numerous misdials before the call is successfully completed. These misdialed calls show up as edit-rejected CDRs and will increase the daily volume for a period of time until the customers become familiar with the service.

Another example where an increase in the edit-rejected CDR volume signals a problem in the wireless network is when a new trunk group is added to the network but does not get loaded into the billing system. If this occurs, calls that are routed over this trunk group will not get billed and again the company loses revenue.

At this point one can see the value of benchmarking and monitoring the CDRs generated and transmitted throughout the billing system. A way to remedy these situations and prevent them occurring again in the future is to put in place processes that are well defined and well documented and that every department is required to follow. These processes will help the personnel in each department understand the impact their work has upon the company.

CDR benchmarking and monitoring summary

1. Apply a checkpoint where the CDRs are transmitted from the switch to the billing mediation device. Compare the total number of CDRs and files transmitted by the switch to the number received by the billing server.
2. Monitor the number of edit-reject CDRs on a regular basis. Again benchmark the daily volumes to establish a running average.
3. Review the edit-rejected CDRs for problems in the billing system, cell sites and trunks that have not been added to the billing system, problems with customers misdialing as a result of not understanding their wireless service, etc.
4. Develop processes to ensure company information about new services; network changes and additions, and billing system alterations are communicated throughout the various departments. From the engineering and

operations departments' perspective some processes to develop include a new cell site turn-up process, new trunk group add process, and translation changes review and approval.

8.3.2 Verification of bill contents

In Section 8.3.1, we discussed methods to determine if the basic billing record information was being generated and transported to the proper systems within the wireless network. The next step is to check the accuracy of the bill itself. The actual testing of the bill contents needs to be coordinated with the information technology and billing departments to ensure that the results coincide with the billing designs implemented. Review and analysis of the call times (start, stop, and duration) is one important basic check. However, there are many other verification tests necessary to ensure the billing process is accurate.

Basic call testing using a stopwatch, a log sheet, and a test plan will suffice as an initial test to check call times. Additional testing of nonbillable calls needs to be conducted to ensure customers are not being charged for services that are not specified in the service agreement. Again, a detailed service description can be obtained from the marketing department and used as a guideline to develop a test plan and complete this type of verification testing.

Further, testing of roaming services needs to be conducted with the assistance of another network where a roaming agreement has been established. This type of testing requires a test mobile number and account from the partner network. Basic call testing using the roamer test mobile can be conducted and the final invoice compared to the test log sheets for analysis. Again, discussions with the marketing department need to take place in order to determine the applicable roaming rates and costs associated with this type of service. In addition to verify usage and airtime charges, features will need to be tested as well. It is not uncommon to have differences in the features offered to home customers versus those offered to roaming customers. Typically the home customers have a more extensive feature list than roaming subscribers.

Finally, a test that is not obvious is identification of a dropped mobile call (failed calls due to a loss of radio signal). Some wireless carriers have programs in place to identify dropped calls and then credit the customer when this occurs. Take note that dropped calls are inherent in a wireless network in that radio coverage is planned for the most populated areas and will eventually end as the system border is reached. In this case established mobile calls that drive outside of the agreed-upon coverage area will need to be identified and distinguished from a dropped call that occurs within the network. A method to separate border dropped calls from internal network dropped calls is use of the cell identification number (ID). All cells in the wireless network are numbered, and mobile calls that take place have the cell ID populated in the CDR.

This cell information can be used to filter valid dropped calls (calls dropped on the border of a wireless network) from invalid dropped calls (calls dropped within the wireless system).

In summary, verification of the mobile bill content includes

1. Testing of valid mobile services (mobile-to-land call, land-to-mobile call, mobile-to-mobile call, call forwarding, call waiting, etc.)
2. Testing of the mobile call times (start time, end time, and duration)
3. Testing of nonbillable mobile services (911 calls, customer care calls, etc.)
4. Testing of roamer services using a valid wireless roaming partner's network
5. Testing of dropped calls (border and internal network) and associated credit applications

8.4 Comparison, Analysis, and Verification of System Usage and Performance Data

Wireless carriers generate their revenue from the mobile customers' usage of the network referred to as *airtime*. This usage is measured and metered using CDR data from the mobile switch. However, usage data can also be derived from the performance statistics collected and made available from the mobile switch as well. A comparison and analysis of system data from these two sources as well as other analysis is presented in Sec. 8.4.1.

8.4.1 CDR and switch statistical call data comparison

A valuable comparison for a revenue assurance program to include is the total system usage for a given period of time, 1 month for example, obtained from both the CDR data and the mobile switch statistical data. The total amount of usage from the CDR data is readily available from either the information technology department or from the billing company. These data are typically compiled on a regular basis as CDR data are processed. At the end of the month, system reports are generated and made available to the various company departments. Determining the total system usage data from the statistics of the mobile switch is more complicated.

In a typical wireless network the mobile switch has both cell site and trunk groups interconnected to allow mobile calls to take place within the wireless system and to external networks. During normal operation the mobile switch will collect usage data from both the trunk groups and the cell sites. Usage data from the cell site RF channels will be presented on an individual basis and then aggregated to represent the usage of the entire cell. In a similar manner a trunk group usage is presented on an individual trunk basis and then aggregated to represent the whole trunk group. The key difference of the usage data obtained from the switch is that it is based upon peg counters.

A *peg counter* is a method of measuring activity on the switch or system as discrete events. For instance, if a mobile customer calls a landline customer, the switch will assign a trunk (voice circuit) to this call and interconnect the two networks. When this event occurs, the switch will increment a counter associated with this device, or *peg* the event. Then for a given period of time, 30 min, for example, the total number of peg counts can be multiplied times the average call hold time, and a total usage value can be obtained for this resource. This usage value will be provided as total erlangs. Mathematically this is presented as

(Total peg counts for a 30-min period) (average call hold time)
$$= \text{total device usage in erlangs}$$

This total usage in erlangs can be converted to minutes by simply dividing by 60. By adding all the usage data for every trunk in a trunk group the total usage for that facility can be obtained. In a similar fashion usage data for every cell site can be obtained. The engineering and operations departments then download this data from the switch and process it to arrive at daily, weekly, and monthly usage totals for all the cell sites and trunk groups. This data can now be used to arrive at the total system usage for a daily, weekly, or monthly period; however, some rules must be followed to make this value accurate.

For all land-to-mobile (L-M) and mobile-to-land (M-L) calls the call usage can be obtained from measuring the outgoing and incoming network trunk groups. Mobile-to-mobile (M-M) calls will not utilize a trunk to the PSTN but will register usage on two or more cell sites. To arrive at a total system usage value it is better to take the total usage from the incoming and outgoing trunk facilities and add the estimated M-M call usage using the call distribution derived from the CDR data. The call distribution of a wireless network is the percentage of M-L, L-M, and M-M calls. For example, in a typical mobile network the distribution might be 50 percent M-L calls, 45 percent L-M calls, and 5 percent M-M calls. By taking the M-M call percent from the most recent call distribution and adding it to the total M-L and L-M usage total, a final system usage total can be obtained.

For example, if the total M-L call usage for a week is 100,000 min and the total L-M call usage for the same week is 75,000 min, then by adding the two plus an additional 5 percent (175,000 × 0.05 = 8750 min), you arrive at a total system usage for the week of 183,750 min. This is just an estimate of the usage, but it can be compared to the total system usage for the same period taken from the CDR data. Further refinements to this example will be necessary for calls that are routed to the voice mail system or to another mobile switch (tandem calls) in a multiple-switch network.

8.4.2 System usage data applications

As mentioned earlier, system usage data can be leveraged by many applications in a wireless organization. Here are a few examples.

One of the greatest operational costs for a wireless carrier is the interconnection facility costs (telephone circuits from the local exchange carrier, etc.). The usage of these circuits and their performance is of particular interest to engineering and operations departments' personnel as well as to the financial and legal departments' personnel.

From an engineering and operational perspective the performance of the network interconnection facilities (trunks) is critical to the operation of the network. Any outages or circuit downtime results in a loss of service to the mobile customers. These service outages are measured and monitored by the operations staff and the data presented to the engineering department as input for designing additional redundancy in the network. These data are also provided to the legal department and used in contract negotiations with the landline carrier. The legal staff compares the outage data with the performance criteria stated in the private line or dedicated interconnection agreement. If the facilities have experienced outages greater than those specified in the contract, then the landline carrier is responsible for a penalty payment to the wireless carrier. The role of the revenue assurance team is to make sure this process is documented and completed on, at least, a quarterly basis since significant revenue can be realized.

Further, the switched interconnect agreement between the wireless carrier and the landline carrier will include a provision for mutual compensation to take place between the two companies. In simple terms, this agreement states that traffic originating from a network and terminating on another carrier's network will be charged to the originating company, and vice versa. In order to fulfill this agreement each company must measure the traffic that originates and terminates in its own network, filter this traffic by carrier, and then charge each carrier accordingly. This task normally requires the use of the CDR data and then separation of the traffic of each carrier (both terminating and originating). Determining the usage of each individual carrier can be accomplished by filtering upon the subscriber phone numbers.

As calls are completed in a network, their corresponding CDRs are generated including the originating and terminating phone numbers. As an example to further clarify this concept take two carriers, wireless carrier A and landline carrier B. If a landline customer (carrier B) calls a home mobile active on the wireless carrier A network and the call is completed, a corresponding CDR will be generated (on the mobile switch) that contains the phone number of the landline subscriber populated in the originating, or *calling,* party field and the phone number of the mobile customer populated in the terminating, or *called,* party field. By filtering the CDR data and collecting all the records that have the wireless carrier's (carrier A) phone numbers in the terminating field and landline (Carrier B) phone numbers in the originating field and then adding all the usage times for every call for a given time period (typically 1 month), the total usage that can be charged to the landline company can be obtained. This type of CDR data processing must be developed and completed within the information technology department with the assistance of the engineering and legal departments. Figure 8.2 provides a description of this process.

446 Chapter Eight

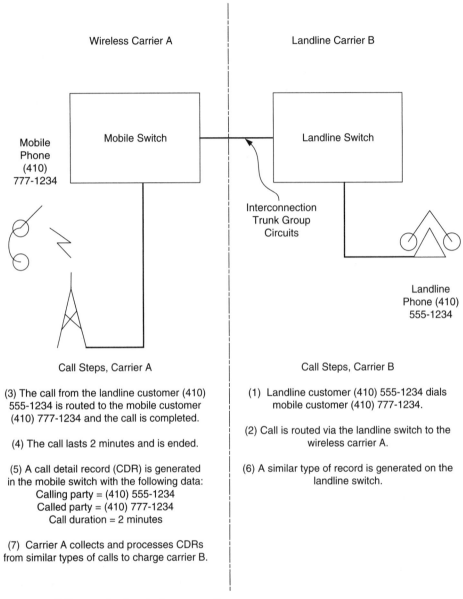

Figure 8.2 Call example of mutual compensation.

Take note that a similar situation will exist on the landline (carrier A) network where a record is generated on the landline switch that contains the same data populated in the calling party, called party, and call duration fields. Further, the landline carrier will total up all the usage that originates from the wireless carrier A network and terminates on its system for charging back to the wireless company. This type of business arrangement is referred to as *mutual,* or *reciprocal,* compensation.

Another method used to estimate the traffic per carrier is to use the trunk usage data. By summarizing both the incoming and outgoing trunk usage for a month and then applying a call-mix distribution to this value, a good approximation to the total traffic or usage data for a given landline carrier can be obtained. The success of this method will depend greatly upon the configuration and design of the network. If only one local exchange carrier (LEC) is used for delivering local calls, then the model is simple enough to apply.

For example, if 100,000 min of usage or traffic is delivered on the outgoing trunks to the LEC for a given 1-month period, an estimate of the total traffic delivered to this carrier can be calculated by multiplying by the percentage of calls that are local calls. This percentage can be obtained from the call-mix distribution of the wireless system. To define this concept further, a simple call mix will include the following call types: local calls (LEC), long-distance calls (IXC), operator calls (0+), and information or 411 calls. By processing the CDR data, the total number of each of these call types over a monthly period can be determined. Then by dividing each call type total by the total calls occurring on the system for the same monthly period, a percent of call type distribution that is specific to that particular wireless system can be obtained.

This call-mix distribution has many applications to other types of analysis or business forecasting. For instance, it can be used to project the increase in traffic to the local landline carrier (LEC) and thus provide an estimate of the increased costs of this service for the carrier's yearly operating budget. A similar forecast can also be made for long-distance service costs, and directory assistance service costs using this call-mix distribution and current system traffic data.

8.5 Summary

Because of the large overlap in disciplines, a well-supported and executed revenue assurance program must include knowledgeable personnel from the finance, engineering, operations, marketing, and information technology departments at a minimum. This discussion has been an introduction to this topic mainly from an engineering and operations personnel perspective. The sharing of data between departments can be very beneficial to a company since similar business and system models will be developed to fulfill specific departmental needs but could be easily applied to other areas in the organization. The information technology department, because of its involvement and interaction with the billing company or personnel will have the ability to construct a multitude of reports based upon the processing of CDR data. The finance department will have industry models used in the completion of the company's capital and operational budgets. The engineering and operations departments will have specific design and operational data about the actual system. The marketing department will have information about current and future services that are valuable in the development of an initial revenue assurance plan. Even the most basic checks and monitoring can yield impressive results to a carrier in operation.

Chapter 9

System Documentation and Reports

This chapter will discuss the various levels of documentation and system reports that need to be generated by the technical community within a wireless mobile system. The reports presented are recommendations and are structured in a generic fashion for guidance in establishing the various types of reporting which is required for any operating system. The reports can easily be parsed down or added to based on the particular structure of your company, the platforms used for service delivery, and, of course, the services themselves.

Reports referenced throughout this chapter relate to both written and verbal reporting. The objective with performance reports is to facilitate the proper management of the network as a whole which will result in improved performance and utilization of the capital infrastructure. The reports, if structured properly, can and will be used to prevent and correct system-related problems. They should also address requirements for designing the network in support of sales initiatives and help facilitate the use of add-on selling by helping inform the sales staff of their client's usage patterns.

Regardless of the type or function of the reports a hierarchical approach to system report generation is strongly recommended due to the information requirements needed by different levels of management in an organization. Specifically a report that is used directly by a network engineer should be different from the report issued to a vice president on the same topic. While the hierarchical approach for report generation and dissemination seems straightforward, there are many reports that go directly up the chain of command in a raw format.

It is suggested that each report include a description of how it was produced. The description should include the following:

1. All the data contained in its report and the sources of the data

2. All the equations used in the calculations and the field in which they are applied

3. A listing of the software used in the processing of the data

It is also suggested that each report be checked for accuracy by running a set of sample values through the processing algorithm. A manual run of the same data using a calculator or slide rule should be performed and compared to the values obtained by using the processing algorithm. Although this seems like a rather basic issue, reports that are never checked can contain misinformation that can cause numerous errors.

There are some basic concepts to report generation and dissemination that you need to address for every report you generate. It is important to obtain the answers to the follow seven questions:

- What is the report's purpose?
- Who will generate the report on an ongoing basis?
- Who will act on the information that is in the report?
- Who will receive the report?
- Is the report needed?
- What format should the report be in, i.e., electronic or paper?
- How will the report be processed?

It is strongly recommended, if reporting processes have not been fully implemented, that you identify all the reports currently being generated in the various departments within engineering alone. A simple review of all the reports that are being generated will most likely result in several of the reports being identified as no longer being needed. The review of the reports will also point to duplications in efforts within the organization itself. The duplication of various reports can be eliminated, and resources can be better utilized to focus on other topics of direct interest to the company.

The reports themselves should each have a document control number and be stored in a central location. The central location recommended for depositing all the engineering documents is the engineering library which should contain all the meeting and project notes and all system reports. To facilitate access to the information these documents should also be resident on the engineering local area network (LAN). The engineering library should consist of both electronic and physical media since both forms are utilized in the technical community. For example, not all vendors present material on a CD-ROM, and trade magazines are still predominantly physical media at this time.

If an engineering library does not exist in your company, it is strongly suggested that you develop one. The departmental administrator or a separate assistant should be placed in charge of the engineering library. This individual should be responsible for overseeing the signing out and return of manuals, reports, etc., and have the responsibility of updating the library and discarding outdated material. Examples of materials to have in the engineering library include industry specifications, current vendor manuals and reports, and, of course, copies of request for proposal (RFP) responses.

It is just as important to ensure that there is sufficient information in the engineering library as to prevent unwanted material from being stored there. The physical media location should not be a technical dumping ground for industry magazines, software manuals, and other items collected by the engineering staff.

The same issue goes for the electronic media in that the file structure used should be organized in a logical fashion to facilitate retrieval of information. Only a few key personnel should have electronic storage privileges (permission to add files to the engineering library); everyone else should have read-only privileges.

9.1 Reports

Fundamentally, company personnel need to have information on every element in the system and relevant performance data. There must be enough performance data to satisfy personnel that the services offered are at or exceed the desired design goals. The performance data are also meant to help personnel dimension the network elements by providing information on how the network is handling the various types of services with their relevant traffic loads.

The general categories of reports which need to be generated and reviewed for potential action are listed here for quick reference. The list can be added to, as you decide what you need to respond to on a daily, weekly, monthly, quarterly, and yearly basis.

- Bouncing congestion hour traffic report (node and service)
- RF network performance report
- Packet switch performance report
- Circuit switch/node performance report
- Telephone number inventory report
- IP number inventory report
- Facility usage report
- Facilities interconnect report (data)
- System traffic forecast report
- Network configuration report
- System growth status report
- Exception report (morning, weekly)
- Customer care report
- Project status reports (currently and pending)

- System software report
- Upper management report
- Company meeting report
- Network briefings

9.2 Objectives

The creation and dissemination of reports is meant to foster action toward improving the network's operation, capital utilization, and minimization of operating expenses. However, a report cannot elicit action unless there is some underlying goal from which to measure the report's information against. For instance, if the objective is to offer packet services to the customers, then the throughput and percentage of area values are not relevant, only the ability to offer the service. Obviously this is a gross simplification, but in order to generate reports a series of objectives need to be formulated.

Therefore, the design objectives set forth ultimately define the quality level expected for an operating system. The design objectives need to not only factor in the financial constraints normally placed upon an operator but also sales and marketing's desires for improving the performance and marketability of the wireless system.

The design objectives are derived through a combination of downward- and upward-directed activities. However, the objectives defined by corporate will ultimately determine which specific design objectives are sought after in the local market. The following is a brief listing of some of the major technical design objectives, which, of course, can be expanded upon as needed.

Radio engineering (design and performance)

Lost calls	<1%
Blocking levels (erlang B)	1%>, <2%
Radio utilization	80%
Bit error rate (BER)/frame error rate (FER)	<1%
Handoff successes	99%
System access success	99%
Registration successes	99%
Insufficient reverse signal strength indicator (RSSI) access	<1%
Usage/lost calls	50 min
C/I/SQE (AMPS/TDMA/iDEN)	22 dB
Customer disconnects (poor system)	<1%
Trouble ticket response	< 24 h
% system > −85 dBm	95%
% system > 100 kbit/s	40%

Erlangs per subscriber	TBD
Erlangs/radio [or base transceiver station (BTS)]	TBD
Mbit/s per subscriber	TBD
Mbit/s per (or BTS)	TBD

Network engineering (design and performance)

Total subscribers	TBD
Total voice-only subscriber	TBD
Total data-only subscribers	TBD
CPU processor occupancy	<80%
Subscriber limit	<90%
PSTN blocking	< 1% (erlang C or Poisson)
Radio base station (RBS) blocking	< 1% erlang B (conflict)
Intermachine trunk blocking	< 0.2% (Poisson or erlang C)
PSTN circuit utilization	80%
Voice distribution	
M-M	10%
M-L	50%
L-M	40%
Packet circuit utilization	80%
Packet call distribution	
M-M	1%
M-L	90%
L-M	9%
BTS-to-BSC link utilization	80%
Intermachine trunk utilization	80%
Active subscriber per design	TBD
Voice mail	
Subscribers	TBD
Disk storage	TBD
Voice mail/fax	TBD
Route blocking	0.2% (erlang C or Poisson)
Voice dispatch (iDEN)	
Subscribers	TBD
Route blocking	0.2% (erlang C or Poisson)
Microwave facility reliability	99.995%
Fiber facility reliability	99.999%
PSTN facility reliability	99.999%
Telephone number inventory utilization	80%

Public IP number inventory utilization	80%

Operations

Circuit switch reliability	100%
Packet switch reliability	100%
Fiber facility reliability	99.999%
Microwave facility reliability	99.995%
PSTN facility reliability	99.999%
Digital cross connect (DXX) reliability	100%
Cell site reliability	99.9%
Radios out of service	<2%
Customer service trouble tickets	
Acknowledge to customer service < 2 h	100%
Trouble ticket technical resolution	
<4 h	50%
<8 h	75%
<24 h	95%
Cell site maintenance schedule	TBD
Call completion rate	75%

Implementation

	2Q2002	4Q2002	1Q2003	2Q2003	etc.
Properties identified	TBD				
RF properties accepted	TBD				
Construction begun	TBD				
Construction completed	TBD				
Cells accepted by engineering options	TBD				

Punch list resolution	<10 days

Obviously the list of design objectives can and should be expounded upon. The specific nodes or RF technology–related issues can be included, e.g., a PN or channel reuse plan or DCCH blocking or CPU loading for the AAA. However, the material included in this section should provide some guidance on where to begin crafting the design objectives that all activites for the technical community should revolve around.

9.3 Bouncing Congestion Hour Traffic Report (Node and Service)

The bouncing congestion hour (BCH) traffic report identifies the system and individual node busy hours, which in turn allows network or local congestion problems to be identified. The BCH report should be generated on a biweekly and monthly basis. With this interval the report provides design-oriented data and not operational-related data, the latter of which is captured in the exception report. The BCH report should include weekend data and weekday traffic information for a 24-h period. This level of information should be disseminated only to the network and RF engineering for the network on a regular basis.

The system-level report should be presented in both tabular and graphic form for ease of understanding. The information should be broken down to represent individual days of the month and be compared to the previous years' data for trending information, when the system is more mature, or on a monthly basis for initial system or service inception. The bouncing busy hour data for the system level should be the highest traffic volume period for any given day, over a 1-h period, and may be different for voice and Internet protocol (IP) data traffic.

Tables 9.1 to 9.4 show examples of BCH report formats. Naturally as with all reports, the specific infrastructure of the network and the services offered need to be included in the decision as to what needs to be monitored. For instance, if a best effort for IP traffic is all that is offered, then monitoring the congestion level for the Internet will not be as important as ensuring that all attach requests for a packet session are made.

The biweekly report can also be used for a sales support tool which is where the individual customer's traffic is monitored and compared to what was signed up for. Based on the usage levels coming from the system the information should be discussed with the customer to ensure that the service chosen with sales support was correct. In addition this can be a chance to identify a configuration problem where the customer may be using the service inefficiently and incurring a higher cost than necessary. For the BTS BCH report (Table 9.5) the number of expected radios or TCHs could also be added for completion.

The data in Tables 9.1 to 9.5 are straightforward and require little explanation. However, it is not recommended that any averaging method be utilized for these particular reports, e.g., weekly. Instead the individual bouncing congestion hour data for each element need to be reported.

As mentioned previously the reports should only be disseminated to the network and RF engineering groups. The same report should, however, be distributed to the technical director for the network and RF engineering on a monthly basis. This report should not be distributed to higher levels of management since it is really meant for the working level to utilize.

TABLE 9.1 Circuit Switch BCH Report

Switch no.	Time	% CPU load	Port no.	BCH traffic	Blocking %	Utilization

TABLE 9.2 Packet Platforms BCH Report

Node	Time	Port no.	Service	% CPU load	CIR, Mbit/s	Mbit/s avg.	Mbit/s peak	Bandwidth available	Link utilization, %	Congestion, s	No. of retransmissions	OOS, min
ATM												
Router (IP)												
Frame relay												

Note: CIR = committed information rate, OOS = out of service.

TABLE 9.3 BSC BCH Report

BSC ID	Time	% CPU load	Port no.	CIR, Mbit/s	Erlang/ Mbit/s avg.	Erlang/ Mbit/s peak	Bandwidth available	Link utilization, %	Congestion, s	OOS, min

The customer BCH report should be disseminated to the sales and marketing departments on a monthly basis. Obviously more detail can be added to the BCH report like DCCH loading as well as DAP (iDEN) related loading issues. Also in reviewing Tables 9.1 to 9.3, the inclusion of the CPU load is listed and is obviously not a port-related event.

An example of where a BCH report would be utilized is in the redistribution of RF channels. The BCH report would be used for RF channel redistribution based on changing traffic patterns in the network. One might *pull,* or remove, channels from an otherwise underutilized sector during one season only to have to add channels back into the same sector during the next season.

If the traffic engineer would have reviewed the bouncing busy hour report for the past year, the trend in traffic would have been noticed and the physical removal of RF channels would not have taken place. The correct alternative would have been to leave the physical channels in their current place, remove them from service, and secure more radios for the vendor. The key concept to always keep in mind is that RF channel reallocation is a costly endeavor to the company in terms of dollars and worker time from engineering and operations.

9.4 RF Network Performance Report

The RF performance report is meant to communicate to various engineering groups the RF network's current level of system performance. It should contain some tabular and graphical data and be generated on a weekly basis.

The RF performance report should contain as a minimum the lost call, blocking, handoff, and attempt failure statistics for voice calls and dropped packet sessions for IP. The report should be distributed to the RF performance engineering group as a whole on a weekly basis. On a biweekly basis it should also be distributed to the director of engineering.

The report should utilize a weekly average for the system. If the system is subdivided into individual areas of responsibility, the report should also have the individual regions displayed on the same chart. Please keep in mind that the design objective used on this chart needs to be the same as the yearly goals and objectives set forth for the technical community.

The radio system performance report will be quite lengthy, especially when different technology platforms and service offerings are factored in. For instance, if the system has IS-136 and AMPS in the 800-MHz band, besides cellular data packet data (CDPD), and then deploys GSM/GPRS in the PCS band, it is not practical to display all the performance criteria on one page.

TABLE 9.4 Microwave Point-to-Point Platform BCH Report

PtP no.	Channel no.	Transmission freq.	Receiver freq.	Bandwidth, MHz	Link size	Link utilization	Link peak utilization (24 h)	Type of link multiplexed/ home run	Mbit/s of usage	Protected/ unprotected	Outage daily, s	Outage weekly (rolling 7 days), min	BTS sites

TABLE 9.5 BTS BBH Report

	BTS (cell site)				Voice			Data (IP)		
Cell ID	Sector	No. of radios	No. of voice TCH	No. of data TCH	BCH erlangs	% blocking	Radio utilization, %	BCH, Mbit/s	Congestion, s	No. or % retransmissions	Radio utilization, %

Therefore, it will be necessary to break each technology or band into its own separate report. However, a cross-reference report will need to be created since one service may have more performance problems then another due to some unique system configuration issues, e.g., reuse or combining. If at all possible, it is advisable to keep the RF performance report to one legal size page.

You should also try to delineate the voice and data traffic from each other. The reason for this lies in how the system treats a voice session versus a packet session. For instance, GPRS can utilize a dynamic TCH assignment method, which is different than the fixed TCH assignment method that is pursued for initial system deployments.

In addition CDMA systems could and will at some time in the system's life cycle have IS-95 and CDMA2000 equipment. The inclusion of IS-95 equipment, whether base station or subscriber, will have an impact on the Walsh code availability, thereby posing the possibility of reducing the potential voice and/or data throughput capabilities with that sector or site.

Lastly the iDEN dispatch traffic needs to be dimensioned accordingly to allow for voice traffic to take place. Since iDEN utilizes a common set of neighbors for both interconnect and dispatch but relies on ILA and DLA boundaries, respectively, it presents unique DCCH and TCH requirements.

The list of technology-specific issues can and does continue approaching a dissertation level of documentation. But regardless of what or how the technology platforms are deployed and utilized in your network, there is one common theme: You must craft the performance report to support the company's technical goals for RF performance while at the same time reducing the mind-boggling volume of information.

What we have done in the past is craft a first-tier report and then a second-tier report to address the volume of information. The first tier has the more basic metrics which are included in the key metrics and performance report, while the second tier has all the requisite details from which proper troubleshooting can take place. You can see that maybe a third tier could include specific call processing and AMA investigations for individual radio or subscriber investigations or problem identification.

Therefore, in Table 9.6, we have put together some, but not all, of the information that will be needed for an RF performance engineer. The list is generic and with little effort you should be able to expand or reduce the list.

9.5 Packet Switch Performance Report

Tables 9.7 and 9.8 relate to the packet and circuit switching aspects of a wireless system. The components that comprise the network elements will differ somewhat by operator because of the services offered. But the following should provide sufficient detail from which to begin monitoring and refining the network. The description of reports assumes that there is an ATM switch along with a router but that other services are handled by another service provider.

TABLE 9.6 RF Performance Metrics

Cell/sector	Cell type	BSC	Technology	Colocated tech. different	Antenna common/ different	MSC	BSC	LA/ILA	RA/DLA
Usage erlangs	kbit/s downlink	kbit/s uplink	No. of radio channels	No. of voice TCH	No. of data TCH-DL	No. of data TCH-UL	% blocking total	% blocking voice	% blocking data
No. of registrations	% registration failures	No. of access attempts	% access failures	No. of pages	% paging failures	No. of location updates	% location update failures		
LC% voice	LC% IP	Min./LC	Mbit/s per LC						
No. of HO sources	% HO failures sources	No. of hand-ins (target)	% hand-in failures (target)	No. of neighbors	No. of soft HOs	Hand downs	Hard HOs	No. of HOs (access + hand-in)	
BER/FER/SQE UL > x%	BER/FER/SQE DL > x%	% packet sessions, kbit/s, reduced	% packet retransmissions						

TABLE 9.7 ATM Switch Report

System name:		Date:
ATM switch no.:		IP address:
Metric	Goal	Present value
Type (edge/core)		
No. of ATM ports		
No. of IP ports		
No. of FR ports		
No. of CES ports		
CPU utilization %		
No. of PNNI		
No. of PVC		
No. of SVC		
% on-net traffic		
Cell error rate		
Severely errored cell block rate		
Cell loss ratio		
Cell misinsertion rate		
Cell transfer delay		
Mean cell transfer delay		
Cell delay variation (CDV)		
% AAL1		
% AAL2		
% AAL5		

Note: FR = frame relay, CES = circuit emulation service, PNNI = private network-to-network interface, PVC = private virtual connection, SVC = switched virtual connection, AAL = ATM adaptive layer.

9.6 Circuit Switch/Node Performance Report

The circuit switch/node metrics report is intended to assist in the monitoring and dimensioning of the switches in the network. This report should be generated once a week for each switch or node in the system and be provided to the network engineering department for review. This report should include, as a minimum, the following data fields:

Performance data	Recommended sample times
Central processor (CPU) load/utilization value	System busy hour %
Secondary processors (SPs) load/utilization value	System busy hour %
Port capacity assigned	Weekly average
Port capacity available	Actual

Subscriber capacity assigned	Weekly average
Subscriber capacity available	Weekly average
Memory capacity assigned	Weekly average
Memory capacity available	Weekly average
Switch/node I/O capacity assigned	Weekly assignments
Switch/node I/O capacity available	Actual
Service circuits load/utilization values (for senders, receivers, tone generators, etc.)	System busy hour %
Switch/node outages (number and duration of occurrences)	Daily recordings

As previously mentioned, this report should be generated once a week as part of the normal maintenance and operation of the network switches and nodes. However, if a problem is encountered or if any metric listed in the table approaches an operational limit, then this report or the metrics of interest should be collected and reviewed on a daily basis. This will provide better accuracy when monitoring the switch or node of interest. The proposed circuit switch/node metrics report is shown in Table 9.9.

TABLE 9.8 Router Performance Report

System name:	Date:	
Router no.	Function:	IP address:

Metric	Goal	Present value
Type (edge/core)		
No. of T1/E1 ports		
No. of OC3/STM1 ports		
CPU utilization %		
No. of PVC		
No. of SVC		
Peak, Mbit/s		
Avg., Mbit/s		
Cell latency		
Max. no. simultaneous connections		
Connection utilization %		
Packet retransmission %		
On-net traffic %		
Inbound traffic %		

TABLE 9.9 Circuit Switch/Node Metrics Report

System name:	System X
Date:	
Report week:	
Node:	Switch 1
CPU load/processor occupancy:	47%
Secondary processor 1 load:	29%
Secondary processor 2 load:	20%
Secondary processor 3 load:	23%
Node port capacity:	985/1200 matrix ports
Node subscriber capacity:	45,000/70,000 subscriber records
Node memory capacity:	10 M/35 M
Node I/O capacity:	12/21 I/O ports
Service circuits loading:	
Sender circuits:	30%
Receiver circuits (MF/DTMF):	43%
Conference circuits:	23%
Node outage data:	No outage for this report period
Start time of outage:	
End time of outage:	
Duration of outage:	
Reason for outage:	

9.7 Telephone Number Inventory Report

The use of the telephone number inventory report is vital for monitoring the telephone numbers usage, e.g., how many actual numbers are assigned to the customer base at a given time. The report also is used for predicting, projecting, and trending the directory inventory growth. Through knowing the telephone number growth patterns an engineer can order additional codes for future use from the local exchange carrier (LEC) or the governing body for code administration.

This report needs to be generated on a monthly basis and should indicate as a minimum the complete breakdown of all the directory numbers used in the network. The report also needs to include the actual network nodes where the numbers are stored. The telephone number inventory report should also indicate the status of future codes soon to be introduced in the network and when they are expected to be released. Additionally the report needs to include the actual central offices, where the codes are served out of to ensure proper planning and expedite troubleshooting.

The telephone number inventory report should be distributed to the network engineer responsible for the inventory tracking and forecasting, the network engineering manager, and the director of engineering. An example of the telephone number inventory report is shown in Table 9.10.

9.8 IP Number Inventory Report

The use of the IP addresses and their related subnets is essential for any wireless mobile system, especially one offering mobile IP. The IP addresses are also one of the most prevalent forms of identification for all elements within the wireless system and are a vital service supplied to the customers, whether static or dynamic host configuration protocol for mobile IP. There are both public and private addresses which need to be tracked. The IP number inventory report is vital for monitoring the IP address usage, i.e., how many there are and how they are distributed at a given time.

The report also is used for predicting, projecting, and trending the IP inventory depletion and subsequent growth. Through knowing the IP address growth patterns network engineering can order additional public addresses when needed or reserve the correct number of private addresses for future use.

This report needs to be generated on a monthly basis and should indicate as a minimum the complete breakdown of all the IP addresses used in the network. The IP number inventory report should also indicate the status of future addresses soon to be introduced in the network and when they are expected to be released. Additionally the report needs to include the actual elements or nodes, where the codes are served out of to ensure proper planning and expediting troubleshooting.

The IP number inventory report should be distributed to the network engineer responsible for the inventory tracking and forecasting, the network engineering manager and the directory of engineering. Its format is crude, but the general concept of reporting can be achieved by defining which elements in the network require IP codes, those which are private and public, and, of course, their distribution or assignment method. An example of the IP number inventory report is shown in Table 9.11.

The reason that mobile IP is separated from the network addresses several issues, DHCP and the lease issues with them. You should also address in the IP report what PDSN or routing area the IP addresses may be associated with, depending on how you established the network configuration.

9.9 Facility Usage Report

The facility usage report's purpose is to track the various interconnect facility usage levels in the network. The report should be issued on a monthly basis and used for many issues associated with improving the network's facility performance. It should be used to determine the best locations for the point of presence (POP) locations in the network. The selection of the POP locations, if

TABLE 9.10 Telephone Number Inventory Report

System name:
Weekending:
Date:

NPA	NXX	XXXX	Available	Central office	Resident switch	Route name	Tandem	Tested	No. released	No. active	Utilization rate	Comments
914	365	7000–7999	8/15/95	ORB	WDB1/003	BBEN9	ZERK	1000	1000	500	50%	
201	968	2000–2999	8/26/95	PARM	ERU07/011	CPG2/4	NWK	1000	500	200	20%	

System Documentation and Reports   467

TABLE 9.11   IP Number Inventory Report

System:

Date:

	Node	IP block assigned	Subnet range	% utilized	IP address	Subnet	Element	Service
Network								
Public addresses								
Private addresses								
Mobile IP								
Public addresses								
Private addresses								

done properly, will minimize the network infrastructure cost for delivering calls.

The report needs to also be used for verifying that the interconnect bills received for operating the network are valid. The interconnect bills should be reconciled against the facility usage report to ensure that there is nothing out of the ordinary being reported or billed. The reconciliation of the facility usage bill is normally the responsibility of a revenue assurance department.

The facility usage report should be distributed to the network facilities engineer responsible for facility usage dimensioning, the network manager, and the director of engineering. The director of engineering should forward a sanitized version of the facility usage report to upper management with the interconnect bills. An example of what a facility usage report should look like is shown in Table 9.12.

9.10   Facilities Interconnect Report (Data)

The facilities interconnect report (data) is intended to display the current configuration and performance of the network data links for detecting and resolving data problems and to plan for the dimensioning of these facilities as the network traffic grows. These are not the switch-to–cell site links but rather the switch-to-switch or switch-to–network node links responsible for call processing and database inquiries.

One of the most prevalent types of interswitch and internode interconnection protocols used in the cellular industry is the ANSI standard System Signaling 7 (SS7). We will assume these types of data links for this report. Other such transmission protocol types will use similar metrics to measure data link performance like those used for IP, frame relay, and ATM.

Thus, for evaluating link performance this report should include, as a minimum, the following data fields:

TABLE 9.12 Sample Facility Usage Report

System name:

Month:

Date:

Trunk no.	Usage	Band	% band usage	% variation previous month	Design %
NJ004-05	8545	1	25	4	25
		2	35	1	35
		3	30	5	25
		4	10	6	15

Basic link configuration data

1. Defined network point codes
2. Number of defined link sets in the network
3. Number of links defined in each link set

Performance data

1. Link traffic load (erlangs) (system busy hour data only)
2. Link set traffic load (erlangs) (system busy hour data only)
3. System traffic load (erlangs) (system busy hour data only)
4. Link active and inactivity times [total peak service hours (0700–2000)]
5. Link set change over count [total peak service hours (0700–2000)]
6. Link retransmission % [total peak service hours (0700–2000)]

This data can be collected by using either the network switches or separate monitoring equipment such as a protocol analyzer patched into the network links in a nonintrusive (passive) manner.

This report should be generated on a weekly basis with the responsibility of its review assigned to the data engineering group within the network engineering department. If, however, a problem is noticed on a link set or a link, then a more detailed report about this problem facility should be obtained to conduct more in-depth troubleshooting.

The actual link or link set performance threshold settings are based upon the ANSI standards manual to ensure accuracy and uniformity. A sample facilities interconnect report format is shown for reference in Table 9.13.

When monitoring the key facilities reports for packet data, the number of retransmissions along with the PCR as compared to the SCR needs to be reviewed and monitored for potential resizing, either up or down depending on

TABLE 9.13 Facilities Interconnect Report Example

Facilities Interconnect Report

System name: System X

Date: _____

Network point code assignments:
  Node name   255 - 1 - 1
  Node name   255 - 1 - 2

Network link definitions:
  Link set: 255 - 1 - 1   Links: SLC - 1
                                       SLC - 2
  Link set: 255 - 1 - 2   Links: SLC - 1

Link traffic data:
Link set 255 - 1 - 1	0.38 erlang	SLC - 1 0.20 erlang
		SLC - 2 0.18 erlang
Link set 255 - 1 - 2	0.15 erlang	SLC - 1 0.15 erlang

System data link traffic load: 0.43 erlang

Link service data:
Link set 225 - 1 - 1	Links: SLC - 1	No outages
	SLC - 2	0900–1000 hour, 02:00 min
	Link changeover count:	2
	Link routing error count:	0
	Link retransmission %:	SLC - 1 0.15%
		SLC - 2 0.01%
Link set 225 - 1 - 2	Links: SLC - 1	No outages
	Link changeover count:	0
	Link routing error count:	0
	Link retransmission %:	SLC - 1 0.01%

---

the usage. It is also recommended that the facility report be broken down into subcategories, e.g., on-net, off-net.

9.11 System Circuit Switch Traffic Forecast Report

The system circuit switch traffic forecast report is a plot of the total system usage monthly average for an entire year. The actual data is traffic for the 10 highest system busy hour days for the network. This report is used to predict yearly trends in the system traffic which will show the season changes to the network traffic patterns. A yearly traffic trend is shown in Figure 9.1. This traffic trend plot can easily be changed to include IP traffic, when the mobile IP traffic forecasting becomes more relevant for the system.

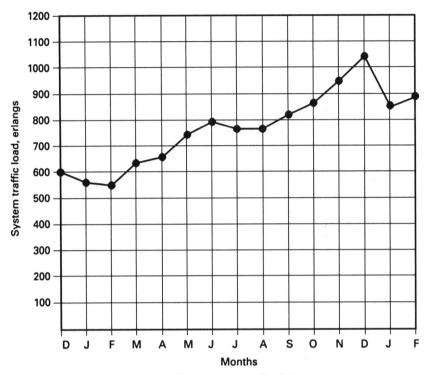

Figure 9.1 Example yearly system circuit switch traffic plot.

9.12 Network Configuration Report

The network configuration report is a collection of diagrams showing the network configuration as it exists. The network configuration report needs to be updated upon every major network change, e.g., new switch, node additions, or new central office exchanges. The network configuration report should be distributed to all network engineering and operations department personnel.

The required tables and diagrams for the network configuration are:

1. Voice interconnect diagrams
   - Concentration node to PSTN LEC central offices and IXC tandems
   - Internal system voice facilities for call delivery
2. Data network diagram
   - IP links
   - Frame relay links
   - ATM links
   - SS7 data links
3. Concentration node–to–base station assignments
4. Point-to-point microwave links

5. BSC-to-BTS assignment
6. Auxiliary system interconnection diagrams
   - Servers

A sample of a network configuration report for a CDMA2000 system is shown in Figure 9.2. In the figure the link names are not included, but depending on the granularity, afforded their specific IDs can and should be included.

9.13 System Growth Status Report

The system growth status report includes the various key elements associated with the system growth plan put out on a quarterly basis and helps ensure that the design requirements put forth in the growth plan are being met and still remain valid. It should be generated on a monthly basis and folded into the quarterly system growth studies that take place. The simple intention is to report on the status of the critical system growth indicators and make sure that the system is performing as expected for the predicted growth. The system growth status report, shown in Table 9.14, is a simple tabular report.

9.14 Exception Report

The exception report is one of the most invaluable tools available for daily troubleshooting. It should be generated by operations and engineering combined and contain the basic information regarding the network from a maintenance point of view. The report should be distributed to the network manager, RF performance manager, RF performance engineers, and operations managers as a minimum. The directors for both engineering and operations would also benefit from seeing the report.

The one primary problem that will occur is the coordination efforts. It might be best to have one group (preferably operations) be directly responsible for the generation of the report and a key contact in the other group provide tactical data.

The system exception report helps identify the current level of maintenance issues in the network at any given time. This report should help improve the troubleshooting response for engineering. The format of the report is shown in Table 9.15 for reference.

The information in this report is time sensitive and its dissemination is critical to the parties that need it. It is strongly suggested that if there is an e-mail system used by the system operator that this report be stored in a central location for quick reference.

The following is a brief description of the items included in the system exception report.

Regarding: This is the date the data and information contained in the report are relevant to. The reason for naming the report information date as

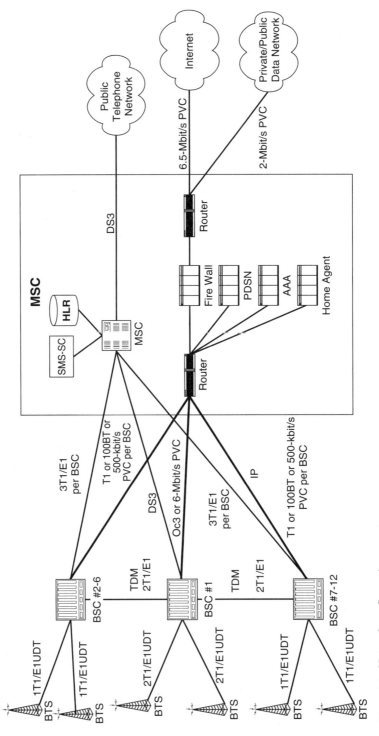

Figure 9.2 Network configuration report.

TABLE 9.14 System Growth Report

	Current	Design	Variance
Total erlangs			
Total Mbit/s			
Peak erlangs transported			
Average BBH erlangs transported			
Peak Mbit/s transported			
Average BBH Mbit/s transported			
Average erlangs/customer			
Average Mbit/s per customer			
Average erlangs/BTS			
Average Mbit/s per BTS			
% LC			
% blocked calls			
% attempt failures			
% handoff failures			
% mobile IP sessions dropped			
% radios OOS			
Backhaul system availability			
Switching network availability			
Packet network availability			
Overall system availability			
Number of circuit switches			
No. of BSCs			
No. of radios			
No. of XCVR/TCI/CE			
Total base stations			
No. of macro base stations			
No. of micro base stations			
No. of pico base stations			
No. of in-building/tunnel sites			
No. of 802.11 nodes			
No. of PtP radio links			

Note: XCVR = receiver, TCI = transcoder, CE = channel element.

TABLE 9.15 System Exception Report

Regarding: 6/14/02 (24 h)			
Total calls completed	XXXXXX		
Voice usage (erlangs)	XXXXXX		
IP usage (Mbit/s)	XXXXXX		
System Performance			
		Objective	6/14/02
Blocking		<2%	XXX
Access failure		<1%	XXX
Dropped calls		<2%	XXX
Mobile IP sessions dropped		<2%	XXX
Handoff failures		<1%	XXX
Radios OOS			XXX
Cell site outages			XXX
T1 outages			XXX
Planned outages: (description)			
Unplanned outages: (description)			
New cells added:			
New radio and network elements added:			

regarding is to avoid the inevitable confusion of listing the date the report is generated on and the date the report is meant to reflect information about.

Total calls completed: This indicates for the 24-h period regarding the report (0000 to 2400) all the calls that were completed on the system.

Voice usage: This indicates for the 24-h period regarding the report (0000 to 2400) the total voice traffic usage, in erlangs, for the system.

IP usage: This indicates for the 24-h period regarding the report (0000 to 2400) the total usage, in Mbit/s, for packet, IP, and traffic.

System performance: This is where the statistics for the system are displayed and compared to the system design objective. This period for this is again 24 h (0000 to 2400).

Cell site outages: This reports how many cell sites, sectors, were out of service during the course of the last 24 h. The time for the outage being counted is less than 1 min. If a radio is out of service for a site, it is not considered a cell site outage; only when the entire sector or even the cell is out of service is this included. The terminology needs to include both planned and unplanned outages because this is the time the sites are not available to the customer base.

T1 outages: This reports the approximate amount of time T1s are out of service for the 24-h period. For example, if one T1 is out for 1 min and another T1 is out for 15 min, then the total T1 outages is 16 min for the 24-h period. The terminology needs to include planned and unplanned outages.

Planned outages: This is a narrative section that describes for the previous 24 h what the planned outages were. For example, if a BSC is upgraded and this was planned, this is indicated here.

Radios OOS: This is the raw count of how many radios are out of service during the last 24 h. This terminology does not include resets that take place, only the physical radios that are not in service.

New cells added: This is where new cells that have been added into the network are included. The name of the cell site plus its cell ID should be included here.

New radio and network elements added: This is where the indication of any new elements to the overall wireless network are made, including all fixed plant items.

9.14.1 Weekly and regional exception report

The weekly and regional exception report is really geared toward the RF performance team or engineer. The objective is to help formulate a weekly battle plan following the seven-step process defined in Chap. 7.

The weekly exception report is a systemwide view, while the regional exception report is done obviously on a regional basis. The reason for the two different approaches is to help the RF performance manager ensure that the department resources are being properly allocated and directed. Both the weekly and regional exception reports should be produced on a daily basis and then on a weekly basis averaging the data over the bouncing busy hour (BBH) timer period.

The material in the weekly and regional exception report is similar in nature, e.g., the format, to that of the morning exception report. The key distinction between the reports is that the regional exception report pertains to a very specific geographic area within the network. The geographic area within the network is the area a particular performance engineer is responsible for. Therefore, there could be from 2 or more regional exception reports generated on a daily basis and distributed to the people indicated on the distribution list. The weekly and regional exception report's structure is shown in Table 9.16.

The exceptions part of the regional RF exception report is where exceptions to the goals for the key metrics are displayed. The variance used to cull out the worst-performing sectors will change over time and needs to be modified as the system performance improves. It is suggested that the three key metrics be used initially since these impact the most on the customer. The suggested

476   Chapter Nine

TABLE 9.16 Weekly and Regional Exception Report

Regarding: 7/24/02 (24 h)

Total calls completed	XXXXXX			
Voice usage (erlangs)	XXXXXX			
IP usage (Mbit/s)	XXXXXX			

	System Performance	
	Objective	7/24/02
Blocking	<2%	XXX
Access failure	<1%	XXX
Dropped calls	<2%	XXX
Mobile IP sessions dropped	<2%	XXX
Handoff failures	<1%	XXX
Radios OOS		XXX
Cell site outages		XXX
T1 outages		XXX

Exceptions (i.e., variance violations):

Cell/sector	Blocking	Lost calls	Handoff failures	Access failure	Mobile IP sessions dropped

starting point is to have all the sectors that exhibit greater than 10 percent of the system normal to be flagged as an exception for each of the categories.

9.15 Customer Care Report

The customer care report is one of the key metrics for receiving information about the quality of the network. It should be integrated with the trouble reporting and resolution system used by customer care and performance engineering. The frequency of the customer care report should be biweekly, and the report should be distributed to all the managers in engineering and operations. The level of detail and information content in the report is shown in Table 9.17.

9.16 System Status Bulletin Board

The system status bulletin board is an essential element in the continued process of conveying critical system information in a uniform and public format. It is a central location where network performance data should be displayed. The bulletin board should be displayed on a centralized wall in the engineering department. The centralized location will enable other company personnel, besides engineers, to view the data and possibly comment on it. The

TABLE 9.17 Sample Customer Care Report

System name:

Date:

	Region 1	Region 2	Region 3	System
Network complaints	11	12	17	40
No. of lost calls	3	2	5	10
No. of interference complaints	5	9	5	19
No. that did not get onto system	3	1	7	11
No. of mobile IP complaints	1	5	2	0
No. of trouble reports issued	4	6	5	15
No. of trouble reports closed	3	6	5	14
No. of outstanding trouble reports	1	0	0	1

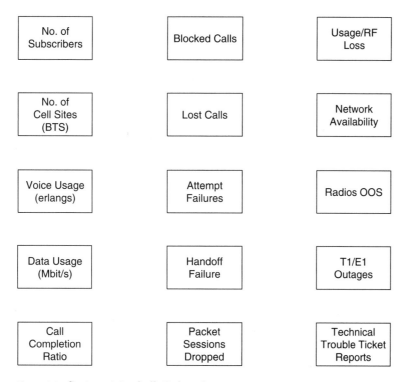

Figure 9.3 System status bulletin board.

TABLE 9.18 Sample Project Status Report

System name:

Date:

Dept.	Project name	Priority	Originator	Lead dept./person	Due date	Project plan	Capital funding	Comments
RF Engineering	Alpha	1	Marketing	RF/Smith	7/30/02	5/01/02	5/17/02	On-target
Network	Tree	2	Engineering	Network/Gervelis	7/30/02	5/15/02	5/24/02	On-target

bulletin board if used correctly will foster improved motivation by having the group's hard work displayed. The suggested types of reports that should be included in the system status bulleting board are shown in Figure 9.3.

9.17 Project Status Report (Current and Pending)

The project status report tracks all the major projects currently under way or proposed by engineering for the coming year. It should have a rolling 1-year projection as the best possible position to report on. The report helps identify major network activities that are to or will take place. It should match the organization's goals and objective and needs to be generated and issued on a biweekly basis to all managers for engineering, operations, and implementation. This will ensure that everyone in the company's technical community is fully aware of all the projects engineering is currently or will be involved with. The format for the report is shown in Table 9.18.

It is imperative that when the project is formed, the project plan include an executive summary, objective, project prime, review dates, milestones, workforce loading, budget impact, impact to other departments, pass/fail criteria, and a method of procedure. When the project is completed, the postimplementation review and report needs to be generated by the project leader.

The executive summary for the project plan and the postanalysis can easily be used for conveying the objective and final outcome of the project to upper management. The general recommended format is illustrated here.

Project plan report format

1.0 Executive summary (one page)
    1.1 Objective
    1.2 Expected Time
    1.3 Workforce and infrastructure requirements
    1.4 Projected cost
    1.5 Positive network impact
    1.6 Negative network impact
2.0 Project description
    2.1 Project leader
    2.2 Project team
    2.3 Project milestones
    2.4 Related documents
3.0 Design criteria
    3.1 Basic design criteria
    3.2 Project review dates
    3.3 Method of procedure
    3.4 Hardware changes
    3.5 Software changes
    3.6 Pass/fail criteria

3.7 Preimplementation test plan
3.8 Postimplementation test plan
4.0 Resources
  4.1 Workforce projections
    Weekly
    Monthly
  4.2 Infrastructure requirements
  4.3 Interdepartment resources required
  4.4 External department impact
    Marketing
    Customer care
    MIS
    Operations
    Implementation
    Vendor
5.0 Budget
  5.1 Total project cost
  5.2 Capital budget impact
  5.3 Expense budget impact
  5.4 Comparison of project costs to budget, planned and actual

9.18 System Software Report

The system software report is another essential report. It identifies the current software configuration for the network for all the managers and relevant engineers in engineering and operations. The report should only be one page in length and be issued on a biweekly or monthly basis. A very rudimentary system software report is shown in Table 9.19. Naturally if there is more than one vendor or type of device located in the network, then the list should be expanded upon.

9.19 Upper Management Report

The upper management report is an essential element in a wireless mobile system's operation. The use of a series of quick and concise information to upper management will satisfy their need for critical knowledge on the performance of the network. The amount and types of reports that can be sent to the upper management of any company range from voluminous to sparse. It is recommended that the following information be produced and disseminated to all the management in the technical area of the company on a monthly basis.

The information contains data extracted from most of the reports recommended to be generated in this chapter. The difference in the reports lies mainly in the format and amount of information content delivered. A proposed upper management report is shown in Table 9.20.

TABLE 9.19 System Software Report

System X

Date:

	Current	Tested	Next load and expected date
Class 5 switch			
Class 5 switch CPU			
Class 5 switch matrix			
Class 5 switch database			
Router			
ATM switch			
Voice mail			
DXX			
PPP microwave			
BSC			
BTS			
AAA			
Servers			
STP			

Note: DXX = digital cross connect, STP = signaling transfer point.

9.20 Company Meetings

Meetings are essential in the process of report generation. The types and frequency of the actual meetings is indicative of the focus placed on designing and improving the network. The critical point is that there needs to be an effective balance between the number of meetings held and the need to convey information in a resource-limited environment.

It is recommended that you review the current meeting levels and see if you are either meeting too frequently or not enough. A suggested meeting structure is proposed for various levels in the organization.

Director to director—once a week

Director to manager—once a week

Manager to manager—once a week (you need to talk with your counterparts on a more frequent basis)

Engineering and marketing—once a month

Engineering department meetings—once a month

Engineering and customer care—once a month

TABLE 9.20 Upper Management Report

System name:		Date:		
	Type	Goal	Measured	Action needed
		Customers		
No. of subscribers				
No. of mobile IP subscribers				
No. of roaming subscribers				
No. of 801.11 LAN/WANs				
		System Usage Measurements		
Total erlangs				
Total Mbit/s				
Peak erlangs transported				
Average BBH erlangs transported				
Peak Mbit/s transported				
Average BBH Mbit/s transported				
Average erlangs/customer				
Average Mbit/s/customer				
Average erlangs/BTS				
Average Mbit/s per BTS				
		Network Performance		
Radio				
% LC				
% blocked calls				
% attempt failures				
% handoff failures				
% mobile IP sessions dropped				
% radios OOS				
Network				
Circuit switch CPU load/utilization rate				
Subscriber capacity				
Switch/node port capacity				
Memory capacity				
HLR subscriber capacity				
PDSN CPU load/utilization rate				
AAA subscriber capacity				
Microwave radio PPP availability				

TABLE 9.20 Upper Management Report (*Continued*)

System Availability
RF system availability
Backhaul system availability
Switching network availability
Packet network availability
Overall system availability
Planned outage time (min)
Unplanned outage time (s)
<div align="center">Configuration Information</div>
No. of MSCs
No. of circuit switches
No. of BSCs
No. of radios
No. of XCVR/TCI/CE
Total base stations
No. of macro base stations
No. of micro base stations
No. of pico base stations
No. of in-building/tunnel sites
No. of 802.11 nodes
No. of MSCs
No. of PPP radio links
ATM switches
Servers
No. of E-mail servers
No. of E-commerce servers
No. of Web hosting servers
Routers
DAC 3/1/0
DAC 1/0
Facilities
No. of T1/E1
No. of DS3/E3
No. of PSTN trunks (T1/DS3)
No. of IP trunks (T1/DS3)
No. of frame relay trunks
No. of ATM trunks

Engineering design reviews—once a month

Project approval meeting—biweekly

Network growth plan—every 3 months

Except for the engineering design reviews, project approval, and network growth plan the meetings should not last more than 1 h. It is recommended that the weekly status meeting between the groups takes place at the beginning of the week. Prior to each meeting, an agenda should be provided so participants can stay focused.

Regarding company meetings it is imperative that the marketing department plan initial meetings with the engineering department prior to the onset of any new product or service. The objective is to assess the feasibility of the product or service offering. These meetings can take place more frequently than once a month.

The main and auxiliary system capabilities should be discussed with marketing as well to ensure that these system capacities can support the projected growth. The marketing department needs to provide for any new product and services it plans to launch to the customer base.

It is important that the following key elements, or rules, be followed to ensure that company meetings are successful. A meeting is successful if it meets its stated purpose, for example, a decision is reached or the required follow-up action is defined and delegated.

Premeeting steps
1. Plan what the meeting is about and its desired outcome.
2. Identify who should attend the meeting.
3. Develop the meeting agenda and distribute it in advance of the meeting.
4. Ensure that there is sufficient room and audiovideo equipment for the meeting.

Meeting steps
1. Introduce everyone and clarify everyone's role in the meeting.
2. Review the agenda.
3. Cover each agenda item one at a time.
4. Allow for sufficient feedback on each topic and keep people focused.
5. Close all discussions on the topic at hand before moving on to the next item on the agenda.
6. Summarize all decisions and agree upon action items.
7. Draft the agenda for the next meeting and agree upon a time.
8. Close the meeting by thanking everyone for attending.

Postmeeting steps
1. Write and distribute meeting minutes promptly.
2. File the agenda, minutes, and other key documents in the engineering library and LAN.
3. Follow up on all open items to ensure closure.

9.21 Network Briefings

Network briefings are important to report the status of the network to the rest of the company on a regular basis. Specifically the network briefing is a once-a-month open dialog between the technical departments and the remaining company departments. The network briefing enables a large volume of information to be conveyed, which enables a uniform dissemination of information to take place.

It is recommended that the network briefings take on a format that is uniform and also timely for the general audience that will attend. The meeting should be scheduled to last no longer than 2 h at the most. The choice of using an overhead presentation or a more informal approach is dependent upon your company's culture.

However, the format for the network briefing is as follows:

1. *Introduction.* The various speakers and the agenda to be used for the meeting are introduced.
2. *Engineering activities.* Last month's activities are discussed and are usually a follow-up to the previous month's comments. In addition, the current month's activities are discussed with expected outcomes. In general the talk is high level and not detail oriented.
3. *Construction.* A representative from implementation and real estate discusses the current build program for the company. The topics covered will be the recent completion of sites for the network. In addition to the sites built, a list of the sites currently under construction are also discussed.
4. *Planned network and technical projects.* The facilitator of the meeting then goes over the projects and major events that are planned to take place between this meeting and the next with respect to the network.
5. *Discussion.* Members of the audience ask questions regarding any network issues they have.
6. *Closing.* The meeting is closed and the next meeting date and location is mentioned.

9.22 Reporting Frequency

Most performance reports are historical in nature, meaning they are reports of what has happened and not about what is going to happen. Through proper selection of the reports and metrics, utilized inferences with regard to future activities can be extracted.

In the list in Table 9.21 there are several types of reports to be categorized and defined. There are basically four levels (1 to 4) of reports expected to be utilized in any wireless mobile system. Level 1 reports are utilized by the engineers or technicians in the daily performance of their jobs. Level 2 is used by managers for each of the departments, utilizing a reduced set of data, to determine the performance of the key areas within their functional responsibility.

TABLE 9.21 Performance Report Frequency

Report	Frequency
Exception report (morning, weekly)	Daily or weekly
Bouncing congestion hour traffic report (node and service)	Weekly
RF network performance report	Weekly
Packet switch performance report	Weekly
Circuit switch/node performance report	Weekly
Telephone number inventory report	Monthly
IP number inventory report	Monthly
Facility usage report	Monthly
Facilities interconnect report (data)	Monthly
System traffic forecast report	Monthly
Customer care report	Weekly
Project status report (current and pending)	Biweekly
System software report	Monthly
Upper management report	Monthly
Network briefings	Monthly
Growth plan	Quarterly or every 6 months

Level 3 reports are utilized for reporting to senior management and across department functions and include information on the key performance metrics and how the company is performing against these goals. Level 4 reports are those that are reported upward to, say, the board so that higher-level personnel are assured that the network is operating properly.

Chapter 10

Network and RF Planning

The network and radio frequency (RF) growth planning for any network is exceptionally important because it defines the how, what, when, and why of the company's technical direction. The direction put forth in this plan is used to define most projects that need to be completed by the technical community in a specified time frame. This chapter discusses how to put together a growth plan that incorporates both the network and RF aspects in the design process. These two aspects are linked in various points in the growth planning, and it is necessary to incorporate both to ensure a proper design is completed.

10.1 Planning Process Flow

The recommended planning process flow is shown in Fig. 10.1. Note that for the design to have any relevance, input is needed from various departments within the company. Without critical marketing, sales, customer care, and, of course, legal and regulatory guidance, the technical design will in all likelihood not support the services the company is seeking to sell.

Therefore, each of the functional disciplines within the company need to be involved with the overall high-level design review. There would, of course, be several layers within each functional discipline involving a preliminary design review and a critical design review. Regardless of the amount of sublayers within the preliminary design review, there would also be a final design review meeting where the work of all the different functional groups is combined into a unified package which has both network and RF components. Additionally the network components should also include the transmission planning aspects and not just the infrastructure requirements.

It is important to note that while operations and implementation are not included in the design process by name, they are involved with the system planning. Although the engineering department is responsible for the generation of the technical design, it needs to coordinate its information with operations

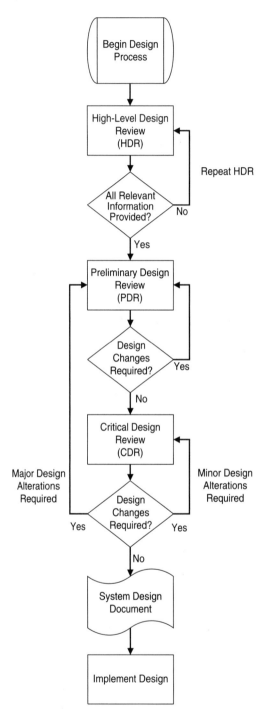

Figure 10.1 Planning process flow.

and implementation to ensure that the planned design is possible and that specific network requirements from operations are not excluded by mistake. If your organization has the performance group incorporated into the operations division or department, then it will need to be directly involved with the RF planning aspects.

10.2 Methodology

Establishment of the methodology to be utilized for the formulation of the network and RF growth plan is essential in the beginning stages of the report to ensure that the proper baseline assumptions are used and to prevent labor-intensive reworks. Although more subjective at times the methodology that is followed is essential in establishing a good design which correlates with the business plans of the wireless company. The overall design process involves a good look at what services need to be supported, where they will be supported, and, of course, how they will be supported. The venerable what, where, when, and how of any decision process will determine the methodology for the design.

What. Defines what you are trying to accomplish with the design

Where. Clarifies the issue of where the service will be introduced

When. Defines a time frame to follow

How. Clarifies the concept of how this will be realized

The methodology chosen determines the fundamental direction of the design itself. The methodology obviously should not be left to the purview of the technical services group, but senior management from marketing, sales, customer service, new technology, operations, implementation, network engineering, and, of course, the RF engineering group need to have direct involvement.

For a successful design to take place, reviews and interim steps in the process, to obtain projection information from other departments which are critical to the growth assumptions, will need to take place. It is strongly advised that the three-step process of high-level design review be followed by a preliminary design review and then the critical design review.

The general flow of the design process is important because it needs to work regardless of the system's age, technology used, and organization. Often problems arise in the design due to short cuts being taken due to time constraints, lack of technical discipline by management, or both.

Some of the methodology issues that need to be identified at the beginning of the study are

1. Time frames for the study to be based on
2. Subscriber demographic information, i.e., geographic locations where subscribers are clustered
3. Services offered, both future and legacy issues

4. Subscriber usage projection (current and forecasted by month and/or quarter) by service type
5. Design criteria (technology-specific issues)
6. Baseline system numbers for building on the growth study
7. Cell site construction expectations (ideal and with land-use entitlement issues factored in)
8. Maximum and minimum off-loading for cell sites
9. Utilization factors for central processing units (CPUs) and network equipment including transmission facilities
10. Resiliency requirements for transmission, network and RF (i.e., an availability of 99.XXX percent)
11. Regulatory and lawful intercept requirements
12. New technology deployment and time frames
13. Staffing requirements
14. Budget constraints
15. Date design and budget is due

At each stage of the design process the department heads need to sign off on the design process phases, where relevant. For example, in the high-level design review phase, the head of marketing needs to sign off on the objectives, after review by peers, indicating that this is what the technical services group should implement. For the preliminary design review, the technical services department head signs off on the design phase, while the critical design review is signed off by corporate.

The time frames for the report should be established prior to the report generation. They will define what the baseline, foundation, and how much of a future look the report will present. The baseline time frame that the data used for generating the report is critical since the wrong baseline dates will alter the outcome of the report.

The amount of time the report projection will take into account (1, 2, 5, or even 10 years) is also critical for the analysis since it will have a dramatic effect on the final outcome. In addition to the projection time frame it is important to establish the granularity of the reporting period.

The particular marketing plans also need to be factored into the report itself. The marketing department's plans are the leading element in any network and RF growth study. The basic input parameters provided by the marketing department are listed next.

1. The projected subscriber growth for the system over the time frame for the study
2. The projected millierlangs per subscriber expected at discrete time intervals for the study

3. The increase or dilution for the subscriber usage over the time frame for the study
4. Types of subscriber equipment used in the network and percent distribution of customer premise/purchased equipment (CPE) projections (e.g., portable or mobile units in use and their percent distribution)
5. Special promotion plans over the time frame of the study like local calling or free weekend use.
6. The projected amount of mobile data users over the time frame of the study
7. The projected kbit/s per subscriber expected at discrete time intervals for the study
8. Top-10 customer complaint areas in the network requiring coverage improvements

There are a multitude of other items needed from the marketing department for determining network and RF growth. However, if you obtain the information on the basic eight topics listed above from the marketing department, it will be enough to adequately start the network and RF growth study.

The establishment of the design criteria used for the report is another key criterion when putting together the plan. It is recommended that the design criteria used for the study be signed off by the director of engineering to ensure that nothing is missed. The design criteria include the following as a minimum.

1. Marketing input
2. RF spectrum available for the growth plan
3. Type of grade of service (GOS) table to be used for plan, i.e., erlang B 2 percent GOS
4. Minimum and maximum off-loading factors for new cells (voice and data)
5. Coverage requirements (could be different for voice and data)
6. Identification of coverage sites
7. Busy hour (BH) peak voice traffic, 10-day high average for month
8. BH peak mobile data traffic, 10-day high average for month (or best guess)
9. Infrastructure equipment constraints
10. Resiliency for radio, fixed network, and transmission
11. Facility concentration, e.g., remote hubs
12. Interconnection call and data delivery requirements
13. Digital and analog radio growth
14. 911 phase 2 implementation
15. Communications Assistance for Law Enforcement Act (CALEA)/lawful intercept requirements

16. New technology and service considerations
17. Cell site configurations used for new cells
18. Baseline system numbers

This list should be used as the foundation for establishing the design criteria for the RF and network growth plan.

The baseline system numbers should be listed in the design criteria for the growth plan. They define the time period for basing the growth projection from. For example, a decision to use June data instead of July data might have a significant impact on the plan as would using peak versus average values for traffic. It is also important to note that voice and mobile IP will in all likelihood have different BHs.

The construction aspects of the system will also need to be factored into the report. These pertain to the proposed new cell sites, or network equipment, that are, or will shortly be, under construction. In addition, a realistic time frame for construction needs to be stated in the report, for example, although 100 sites are needed in the next 3 months only 20 will actually be available for operation.

New technology deployments, for example, converting from an IS-136 system to a CDMA2000 system or simply implementing general packet radio services (GPRS) into a global system for mobile communications (GSM), will also need to be factored into the plan. Other technology deployment issues that need to be factored in might be the cell site infrastructure equipment where a new vendor's equipment is planned to be deployed. Also involved with technology deployments could be the introduction of a new switch that has improved processor capacity for key trigger points which change the requirements for future growth.

The budget constraints imposed on any operator must be folded into the network and RF growth plan. Failure to incorporate budget constraints into the plan will only make the plan unrealistic and squander everyone's time needlessly. Often if the plan calls for more sites than budgeted for, as is the case most of the time, then it is imperative to help establish a ranking order for the sites to establish which are truly needed versus which would be nice to have.

The last part of the preparation for the report is finding out when the report itself is due, the level of detail required, and who is assigned to work on the actual report.

10.3 Traffic Tables

You will need several tables to create the network and RF growth plan. The most important traffic table is the erlang B table, an example of which is shown in Table 10.1. The use of an erlang B table is extremely valuable for putting together any projected equipment requirements by defining the amount of individual channels, or time slots, depending on the technology

TABLE 10.1 Erlang B Table

Channels	\multicolumn{6}{c}{Blocking probability}					
	P01	P012	P015	P02	P03	P05
1	0.0101	0.0121	0.0152	0.0204	0.0309	0.0525
2	0.153	0.168	0.19	0.223	0.282	0.381
3	0.455	0.489	0.535	0.602	0.715	0.899
4	0.869	0.922	0.992	1.09	1.26	1.52
5	1.36	1.43	1.52	1.66	1.88	2.22
6	1.91	2	2.11	2.28	2.54	2.96
7	2.5	2.6	2.74	2.94	3.25	3.74
8	3.13	3.25	3.4	3.63	3.99	4.54
9	3.78	3.92	4.09	4.34	4.75	5.37
10	4.46	4.61	4.81	5.08	5.53	6.22
11	5.16	5.32	5.54	5.84	6.33	7.08
12	5.88	6.05	6.29	6.61	7.14	7.95
13	6.61	6.8	7.05	7.4	7.97	8.83
14	7.35	7.56	7.82	8.2	8.8	9.73
15	8.11	8.33	8.61	9.01	9.65	10.6
16	8.88	9.11	9.41	9.83	10.5	11.5
17	9.65	9.98	10.2	10.7	11.4	12.5
18	10.4	10.7	11	11.5	12.2	13.4
19	11.2	11.5	11.8	12.3	13.1	14.3
20	12	12.3	12.7	13.2	14	15.2

used. It is important, however, to note that the erlang B model is to be used for circuit switched, i.e., voice, and not packet-based, Internet protocol (IP), services.

To facilitate inclusion of an erlang B table in a typical growth plan for any wireless network, the equation is given below. The erlang B equation lends itself to easier insertion into a spreadsheet through use of a lookup table, which is a standard function in all spreadsheet programs commercially available.

$$\text{Grade of service} = \sum_{n=0}^{n} \frac{E^n}{n!} \quad (10.1)$$

where E = erlang traffic and n = number of trunks (voice channels) in group.

If the traffic is data traffic, how it is projected and accounted for in the estimation of network elements is dependent upon whether it is contention based or not. If the data traffic is contention based, i.e., subject to radio blocking, one could make the argument that the use of erlang C would be appropriate; this is because the IP traffic or other similar traffic can be placed in a queue. However, the inclusion of erlang C into spreadsheets can be problematic due to its complexity.

Additionally for the contention-based traffic the issue of overbooking, depending on the service offering, needs to be accounted for in the process. We will assume that the traffic is symmetrical, i.e., it has the same downlink and uplink. Therefore, a simple example of the overbooking would be to offer 56 kbit/s as the mobile IP for, say, 100 subscribers that could be simultaneously active for that sector. Using the linear approach this could equate to 5.6 Mbit/s being required for the facility. However, if the service offering is 20:1 and everyone is requiring bandwidth at the same instant, then this becomes a radio network requirement of 208 kbit/s. Obviously the issue of overbooking and active subscribers needs to be defined, besides the service offering requirement.

If the traffic is not contention based the issue is what format the data will be encapsulated in. You have to determine if the traffic will be transported as IP or encapsulated in another format, like frame relay or asynchronous transfer mode (ATM), or have a dedicated time-division multiplexing (TDM) bandwidth, i.e., a T1/E1.

A simple example of TDM usage for IP traffic would be where you have 280 kbit/s of traffic and consider it the peak load for any time. TDM, however, is dedicated; once it is assigned for this effort, that is all it is allowed to do. This is rather inefficient, and it is always difficult to calculate the amount of bandwidth for a TDM circuit.

Using the E1/T1 DS0 as the fundamental transport bandwidth size of 64 kbit/s, the 280 kbit/s requires a total of 4.375 DS0s, but this is really 5 DS0s out of a possible 24/30 usable time slots. The reason we mention this is that this could lead to sharing facilities, reducing backhaul requirements through use of drop and insert equipment, or use of an ATM platform.

10.4 System Expansion

The decision regarding the system expansion sometimes produces an endless do loop because of the interaction between the various decisions. This situation is why a due date is needed from which to make a valued decision. The system expansion decision is usually made at the conclusion of the high-level design review, but could also take place, depending on the magnitude of the issue, at the preliminary design review.

A common decision that all wireless operators have to make regards the inclusion of other wireless technology platforms, e.g., possible bifurcation of the system. The bifurcating or even pentfurcating of the system could create a system which has advanced mobile phone system (AMPS), IS-136, CDPD, and GSM/GPRS services with 802.11 associated with the GPRS offering. This can be further complicated by the possibility of cellular and PCS bands being used in the same market by the wireless operator. Additionally the inclusion of trimode subscriber units, soon to be quad-mode or even technology agile units, complicates the design since in the percentage of the traffic load each service gets, either through growth or through traffic steering, needs to be determined.

In addition, CDMA systems, if including cellular, have AMPS, CDMA, CDPD, IS-95, CMDA2000, and possible 802.11 offerings, which can also include the PCS band as well. GSM by itself, which in the United States occupies the PCS band, has GPRS and 802.11 to contend with. While the integrated dispatch enhanced network (iDEN), which as of this writing occupies the specialized mobile radio (SMR) band, may only have to contend with the inclusion of 802.11.

The procedures associated with a two and a half generation (2.56)/third generation (3G) system design are similar to that followed for a 2G, or even 1G, wireless system. There is amazing similarities for implementing 2.5/3G into an existing system as was the case when 2G was introduced into cellular systems. The one major difference in many of the design decisions is the inclusion of the packet data serving node (PDSN) to support wireless data mobility.

There are fundamentally three variants to system designs for a wireless communication system.

- Existing system expansion (no new access platforms)
- New system design (700-MHz or PCS C block)
- Introduction of new technology platform to existing system

All the components which comprise the path the radio signal takes as well as how the individual base stations are integrated into a larger system need to be factored into the system design. Additionally the method of voice and data service treatment needs to be included.

The specific procedures which need to be followed, of course, vary depending on the market and individual technology platform being installed and the type of legacy system which is in place, if any. It is important to restate that you need to know what your objective is from the outset of the design process and that objective needs to be linked to the business and marketing plans for the company. Following the direction of design discovery, i.e., we will build it and they will come, has had some very negative consequences in the wireless industry to date.

10.4.1 New wireless system procedure

This section provides a brief list of the general design procedures which need to be performed whether the design is for a new or existing 2.5G or 3G system. The design process for a new 2.5G or 3G system is basically the same as that followed for a new 2G or even 1G wireless system. However, the subscriber usage needs to include in both voice and packet data usage.

1. Obtain the marketing plan and objectives.
2. Establish the system coverage area.
3. Establish the system on-air projections.
4. Establish technology platform decisions.

5. Determine the maximum radius per cell (link budget).
6. Establish environmental corrections.
7. Determine the desired signal level.
8. Establish the maximum number of cells to cover the area.
9. Generate the coverage propagation plot for the system.
10. Determine the subscriber usage.
11. Determine the usage per square kilometer (voice and packet).
12. Determine the maximum number of cells for capacity.
13. Determine if the system is capacity or coverage driven.
14. Establish the total number of cells required for coverage and capacity.
15. Generate a coverage plot incorporating coverage and capacity cell sites (if different).
16. Reevaluate the results and make assumption corrections.
17. Determine the revised (if applicable) number of cells required for coverage and capacity.
18. Check the number of sites against the budget objective; if the number of sites is excessive, reevaluate the design.
19. Using the known database of the site's overlay on the system design, check matches or close matches ($<0.2R$; $R =$ radius of design site).
20. Adjust the system design using site-specific parameters from known database matches.
21. Generate propagation and usage plots for the system design.
22. Evaluate the design objective with time frame and budgetary constraints and readjust if necessary.
23. Issue search rings.

The process is primarily radio-access driven, but the PSDN needs to be accounted for as well. However, the limiting case for wireless mobility is typically associated with the radio environment.

10.4.2 2.5G or 3G migration design procedure

The process for introducing a 2.5G or 3G platform into an existing wireless system needs to account for the impact the reallocation of the spectrum will have on the legacy system. Also the design needs to address the new platforms and modifications to the existing platforms which are needed to facilitate the introduction of the new system. The following is a brief listing of the main issues which need to be addressed when integrating a new platform into an existing system.

1. Obtain the marketing plan.
2. Establish the technology platform introduction timetable.
3. Determine new technology implementation tradeoffs.
4. Determine new technology implementation methodology (footprint and 1:1 or 1:N).
5. Identify coverage problem areas.
6. Determine the maximum radius per cell (link budget for each technology platform).
7. Establish environmental corrections.
8. Determine the desired signal level (for each technology platform).
9. Establish the maximum number of cells to cover areas.
10. Generate the coverage propagation plot for the system and areas showing before and after coverage.
11. Determine the subscriber usage [existing, and new (packet and voice)].
12. Determine the subscriber usage by platform type.
13. Allocate the percent system usage to each cell.
14. Adjust the cell maximum capacity by the spectrum reallocation method (if applicable).
15. Determine the maximum number of cells for capacity (technology dependent).
16. Establish which cells need capacity relief.
17. Determine new cells needed for capacity relief.
18. Follow steps 14 to 23 in the list in Sec 10.4.1.

This list and the one in Sec. 10.4.1 can be easily crafted into a checklist to be followed by the design team. Obviously these lists are generic and need to be tailored to specific situations. However, these should provide sufficient guidance to organize the process for a successful design that will meet the customer and business objectives for the company.

10.5 Traffic Projections

The process and methodology for conducting system traffic engineering, i.e., determining the amount of physical and logical resources that need to be in place at different points and nodes within the network to support the current and future traffic, are rather more straightforward since you have existing information from which to make decisions upon. For future forecasts the level of uncertainty grows exponentially the farther the forecast or planning takes you into the future. However, many elements in the network require long lead times, ranging from 3 weeks to over 1 year, to implement. Obviously the goal of traffic engineering is to design the network and its subcomponents to not only meet the design criteria, which should be driven by both technical, marketing, and sales, but also to achieve them in a cost-effective manner. It is not uncommon to have conflicting objectives within a design, e.g., wanting to ensure that customers have the highest quality of service and grade of service (QOS/GOS) for both voice and packet data but having a limited amount of capital from which to achieve this goal. Therefore, it is important to define objectives at the outset of the design process, and then to have some interim decision points where the design process can be reviewed and altered, if required, by increasing the capital budget, revisiting the forecast input, or altering the QOS/GOS expectations.

Since there will be different variants to circuit-and packet-switched services offered, the variations are vast. However, there are some commonalities which can be drawn upon. There are several methods which can be used for calculating the required or estimated traffic for the network. There is no simple approach because of the level of perturbations that exist between technology platforms and service offerings by each wireless operator. However, we can provide a guide from which you can, when applying engineering skills, craft the traffic projections to meet your individual market requirements. There are several key points within the network where the traffic engineering calculations need to be applied.

- Base transceiver station (BTS) to subscriber terminal
- BTS to base station controller (BSC)
- BSC to PDSN
- BSC to mobile switching center (MSC)
- MSC to public switched telephone network (PSTN)
- PDSN to private and public data network

There are several situations and an unknown level of perturbations which can occur in the estimation of traffic for a system. In an ideal world the traffic forecast would be projected by integrating the marketing plan with the business plan and coupled that with the products which should be integral to both the marketing and business plan. However, reality is much harsher and usually very little information is obtainable by the technical team from which to dimension a network.

Packet data traffic has been low, but it appears that it will become more predominant. One method, of course, to increase usage is to reduce the usage fee which would encourage more loading.

10.6 Radio Voice Traffic Projections

There are several methods for estimating the amount of voice and packet traffic an operator can pursue with regards to planning its network. The air interface, which represents the scarcest resource in the network, is dimensioned with the highest blocking probability. Typically, network designers dimension the air interface according to a 2 percent blocking probability (erlang B). Voice metrics are well understood in terms of usage and forecasting using the millierlang/subscriber approach which when coupled with an erlang B table nets the amount of radios or traffic channels (TCHs) required to support the traffic load for that particular sector or cell. However, when dealing with packet data forecasting for a system, without historical usage patterns for the demographic segment associated with the market, you need to take on a more artistic approach to forecasting.

Determining voice traffic projections for a wireless mobile system is a reasonably well documented process. One of the biggest factors in determining the voice usage for any growth plan is obtaining both good baseline data, e.g., what the current capacity is, and fairly accurate projections about voice traffic growth which includes usage patterns. Because of the forecasting timing differences the reliance on historical data with linear trending is the most accurate method for the initial forecast and dimensioning. The linear trending method works well with AMPS, IS-136, CDMA, GSM, or iDEN regardless of some of the specific technical issues.

For packet data forecasting the key point is that without specific applications that the system is trying to address, the estimation of traffic is rather dubious since it bases several assumptions upon each other. For example, streaming video may be the desired application, but the real service is more e-mail and outlook synchronization. Therefore, for packet data usage forecasting there are two approaches to traffic calculation: forecast and discovery.

The forecast approach involves a detailed analysis of existing data traffic and working with the marketing and subscriber sales force a take rate and estimated bandwidth for each subscriber can be achieved. This then is distributed across the regions, or appropriate BTSs, to arrive at the appropriate packet data forecasted volume, which then can be equated into the requisite radio and network elements necessary to support the estimated load.

The other approach is to discovery approach where, using the whole system or just the core of the network, one or more packet channels are deployed, which can be individual TCHs or 1XRTT-capable channels. There are two ways to implement the discovery approach. The first involves a method similar to forecasting, except there is no actual usage data to draw upon. Therefore, you will need to determine the number of subscriber units that are or will be packet data capable and then multiply that number by a (kbit/s)/subscriber value and assume that each will be operational during busy hour initially and weight the traffic volume over the BTSs involved. The second way, which is used more widely, is to deploy packet-data-capable network elements in the system making the packet service available. Normally this approach is implemented when traffic data forecasting is not available or was never attempted. For GPRS systems a few TCHs are typically allocated for packet data on the initial channel which has the broadcast control channel (BCCH). For a CDMA system 1XRTT-capable channels and channel elements are deployed enabling packet services to be available for those cells or sectors. The objective with the GPRS or CDMA discovery approach is to enable the system to be packet capable and then measure the actual usage once services are being used by subscribers and not system testers.

Obviously the forecasting method for voice and especially packet is the desired approach to traffic projections. But reality often deals a different hand, and the discovery approach is more or less the method that is used initially until real usage patterns have been established. The usage part of the forecasting method is probably the easiest to implement by offering the service free. However, achieving a revenue level that will pay for the infrastructure and the recurring expenses is the true challenge with forecasting.

From our experience it is rare that you ever have all the material needed for forecasting, either in content or quality. The reason for this is that most of the projection information needs to be supplied by another division within your organization. The issue is not the division's competence; the technical community needs marketing forecasts about 3 or 6 months prior to when the division is required to forecast that expectation.

10.6.1 IS-136

IS-136 is a 2G system which utilizes digital modulation format and occupies 30 kHz of spectrum for each physical radio channel. Each of the radio channels are divided into six time slots of which two are used per call, e.g., three subscribers per physical radio. The digital control channel (DCCH) in IS-136 occupies two of the six time slots, and, therefore, if a physical radio also has a DCCH assigned to it, only two subscribers can use the physical radio for communication purposes.

Typically the relationship is that for every digital radio there are three potential traffic channels. For a radio that has a DCCH the capacity is limited to two traffic channels, while for a radio that does not have a DCCH assigned to it, a total of three traffic channels are available.

Table 10.2 will help you in assigning the amount of traffic capability the various digital channels are capable of when mixed in an analog environment. The same configuration of 19 physical radio channels was used for all the examples and a simple manipulation to your particular situation can be made by using Table 10.2.

For PCS bands, since there are no analog channels needing to be displaced, with the exception of current microwave users, the channel capacity can be extracted directly from the use of an erlang B calculation. Please keep in mind that the radio channel which has a DCCH has only two traffic channels available when calculating the capacity for the cell or sector.

For example, if your usage for a sector was estimated to be 11.1 erlangs over the study period for digital usage alone, then utilizing an erlang B table at 2 percent GOS the amount of DTCs required to support the traffic load would be 18. Since there are 3 DTCs per radio carrier, this equates to 6 carriers having a total amount of 18 TCHs. However, since one DCCH is needed, this example

TABLE 10.2 IS-136 Traffic

No. of analog	Channels			Erlangs, 2% erlang B		
	No. of	Digital		Analog	Digital	Total*
		DTC	DCCH			
19	0			12.3	0	12.3
18	1	3	0	11.5	0.602	14
	1	2	1		0.223	13.2
17	2	6	0	10.7	2.28	15.8
	2	5	1		1.66	16.6
16	3	9	0	9.83	4.34	17.5
	3	8	1		3.63	16.6
15	4	12	0	9.01	6.61	19.3
	4	11	1		5.84	18.4
14	5	15	0	8.2	9.01	19.3
	5	14	1		8.2	20.2

*Assumes trunking efficiency.

would require the use of a seventh radio carrier, having overdimensioned by 2 DTCs, or you can accept more blocking and have a GOS of 3 percent erlang B by using 6 radio carriers.

10.6.2 CDMA

Traffic forecasting for CDMA does not lend itself to a spreadsheet method for determining the dimensioning of the various access-related elements. Instead, the use of a computer modeling program is warranted due to the cell breathing, variable assignment of channel elements and Walsh codes.

However, when performing a traffic forecast for a CDMA system the design engineer needs to determine prior to the initiation of the study the amount of cells or area which will have CDMA2000 capability and of those whether 1X-EV, 1X-DO, or 1X-DV will be deployed. The issues of how many CDMA carriers and frequency assignments (FAs) will be available and whether packet services will be assigned to a dedicated carrier or commingle with the voice allocations will need to be determined. Lastly you will also need to know the amount of subscribers which have legacy, IS-95 only, handsets so you can effectively derate the Walsh code availability when migrating from an IS-95 to CDMA2000 configuration.

The capacity for a CDMA cell site is driven by several issues. The first and most obvious issue is that traffic modeling for a CDMA cell site involves how many channel cards the cell site is configured with. For an IS-95 system there are a total of 55 possible TCHs, available for use at a CDMA cell site out of a total of 64 Walsh codes. However, for a CDMA2000 1XRTT system there are a total of 128 Walsh codes which are allocated according to the subscriber unit type and the data rate allowed or available for that cell or sector. It is important to note that regardless of the amount of Walsh codes available, unless the channel cards are installed that can support the traffic load and service type the full potential is not realizable.

A second issue that fits into the traffic calculations for the site involves system noise. There is a simple relationship between system noise and the capacity of the cell site. Typically the load of the cell site design is somewhere in the vicinity of 40 to 50 percent of the pole capacity, with a maximum of 75 percent.

The third major issue in determining the capacity for a CDMA cell is the soft handoff factor. Since CDMA relies on soft handoffs as part of the fundamental design for the network, this must also be factored into the usable capacity at the site. The reason for factoring soft handoffs into the capacity is that if 33 percent of the calls are in a soft handoff mode, then this will require more channel elements to be installed at the neighboring cell sites to keep the capacity at the desired levels.

The effective number of traffic channels for a CDMA carrier is the number needed to handle the expected traffic load. However, since soft handoffs are an integral part of CDMA, they need to also be included in the calculation for capacity. For each traffic channel that is assigned for the site a corresponding piece of hardware is needed at the cell site also. Additionally the shorter Walsh codes also are a resource that needs to be accounted for with packet data usage. However, whether the service being handled by the BTS is voice or data, it will require the allocation of traffic channels. The actual number of traffic channels for a cell site is determined using the following equation:

Actual traffic channels = effective traffic channels + soft handoff channels

The maximum capacity for a CDMA cell site should be 75 percent of the pole. However, typical traffic loads result in a maximum of 20 TCHs for IS-95 and approximately 40 for CDMA2000.

The BTS controls the interface between the CDMA2000 network and the subscriber unit. Therefore, when a new voice or packet session is initiated, the BTS must decide how to best assign the subscriber unit to meet the services being delivered. The BTS in the decision process not only examines the service requested but also must consider the radio configuration and the subscriber type and, of course, whether the service requested is voice or packet. The resources the BTS has to draw upon can be both physically and logically limited depending on the particular situation involved.

The following is a brief summary of some of the physical and logical resources the BTS must allocate when assigning resources to a subscriber.

Fundamental channel (FCH)—number of physical resources available

FCH forward power—power already allocated and that which is available

Walsh codes required and those available

Total FCHs used in that sector

The physical resources the BTS draws upon also involves the management of the channel elements which are required for both voice and packet data services.

Integral to the resource assignment scheme is the Walsh code management. For 1XRTT (whether 1XEV, 1XDO, or 1XDV) there are a total of 128 Walsh codes to draw upon as compared to IS-95 which has 64 Walsh codes.

For CDMA2000-1X the voice and data distribution is handled by parameters which are set by the operator which involve

Data resources [percentage of available resources which includes FCHs and supplemental channels (SCHs)]

FCH resources (percentage of data resources)

Voice resources (percentage of total available resources)

Topic	Percentage	Resources
Total resources	100	64
Voice resources	70	44
Data resources	30	20
FCH resources	40	8

Obviously the allocation of data and FCH resources directly controls the amount of simultaneous data users on a particular sector or cell site.

The channel element (CE) dimensioning obviously will be based on the requirements for both voice and data services. The total number of channel elements required will be the summation of both the FCH and SCH elements defined for voice and data services.

The new channel element that is being offered by all the major vendors is compatible with the existing IS-95 system and can be directly substituted for an existing channel element. However, full replacement of all CEs is not a practical solution based on logical deployment options used where IS-95 legacy systems should be left in place.

For simplification a CE is required for

Each voice cell

Each leg of the soft handoff

Each overhead channel

Each data call

The dimension of the CEs is done in increments of 32 or 64 for CDMA2000-1X capable CEs. Since CEs typically come in 32/64 cards leading to the issue that even though only 20 CEs are required, a 32-CE card must be acquired. If 33 CEs are needed, a choice needs to be made whether to underequip or obtain another CE card which will bring the count to 64 when only 33 are needed.

The CDMA2000-1X full capacity is derived based on a fixed environment and availability of 128 Walsh codes. Obviously the percentages shown depend on the mix of voice and data traffic within the system as well as mutual interference.

A rule of thumb to follow is that for sites requiring less then 40 CEs, a 32-element card should be used, but for sites requiring more than 40 CEs, a 64-CE card should be used. This rule of thumb is based on the pooled CEs allowing for better trunking efficiency.

Obviously packet data services have different implications when introduced into a radio environment as compared to a fixed network environment. More specifically for the radio link, how packet data are handled is dependent on whether the radio link is sharing its resources with voice services or is data-only. The data-only possibility is available with CDMA2000-1XEV if data services are only permitted on the new channel. However, regardless of this issue when involved with the wireless link, data services are still a best effort. In addition signaling traffic has a higher priority than voice, but voice services have a higher priority then packet data.

Regarding the packet data resource dimension, it is important to remember that the packet session is considered *active* when data are being transferred. During this process a dedicated FCH for traffic signaling and power control exists between the mobile and the network. In addition the high-speed SCH can be utilized for large data transfers. An important issue that needs to be considered in the allocation of system resources is that while the session is active, channel elements as well as Walsh codes are consumed by the subscriber and system independent of whether data are actually being transferred.

The packet session alternatively is considered *dormant* when there is no data being transferred, but a point-to-point protocol (PPP) link is maintained between the packet server data network (PSDN) and the subscriber. It is important to also note that no system resources are consumed relative to channel elements or Walsh codes while the packet session is considered dormant.

Example 10.1 If you have forecasted that a total of 10 erlangs voice usage is required to be supported along with 4 erlangs of soft handoff traffic, this gives rise to the need to support an aggregate of 14 erlangs of circuit-switched voice traffic. Additionally there is a need to provide three sessions of 60 kbit/s of packet data during the busy hour for that sector. The packet throughput will be symmetrical to simplify the example. The number of TCHs required for voice is 17 at a 2 percent GOS using erlang B prior to the inclusion of soft handoff traffic. Therefore, a total of 21 TCHs are required to support the expected voice traffic load.

Looking at the packet data load, 60 kbit/s of packet data has to be supported, and this is handled by a short Walsh code of 8, equating to 3 CEs. Therefore, a total of 24 CEs are needed to support both packet and voice services for that sector. The CEs listed do not include the additional CEs needed for the various control and signaling channels to support a CDMA voice or packet session.

It is important to note that when commingling of voice and packet data on a CDMA channel the before-mentioned short Walsh code 8 reduced the available Walsh codes for voice traffic by 37.5 percent, assuming a total of 128 Walsh codes. This should not be an issue since the pole as well as the available CEs will be the real limiting case.

Also depending on the percentage of IS-95 only subscribers that exist within the network, you should derate the available Walsh codes to 64, form 128, if the subscriber base consists of more than 25 percent IS-95-only subscriber units. Obviously each system is different and the degrading, if any, should be made after review of the existing Walsh codes utilization, which is available as a typical metric.

For reference, Tables 10.3 and 10.4 show the relationship between the radio configuration (RC), the spreading rate (SR), Walsh codes, and packet data rates.

10.6.3 GSM/GPRS

Because all GSM traffic is circuit switched, network dimensioning is a relatively straightforward process, once the traffic demand per cell is specified. The process largely involves determining the amount of traffic to be carried in the busy hour and dimensioning the network according to an erlang B table using a 2 percent GOS as the reference or dimensioning objective.

TABLE 10.3 Walsh Codes versus Data Rates

	RC	256	128	64	32	16	8	4
SR1	1	NA	NA	9.6	NA	NA	NA	NA
	2	NA	NA	14.4	NA	NA	NA	NA
	3	NA	NA	9.6	19.2	38.4	76.8	153.6
	4	NA	9.6	19.2	38.4	76.8	153.6	307.2
	5	NA	NA	14.4	28.8	57.6	115.2	230.4
SR3	6	NA	9.6	19.2	38.4	76.8	153.6	307.2
	7	9.6	19.2	38.4	76.8	153.6	307.2	614.4
	8	NA	14.4	28.8	57.6	115.2	230.4	460.8
	9	14.4	28.8	57.6	115.2	230.4	460.8	1036.8

Notes: Data rates in kbit/s; NA = not applicable.

TABLE 10.4 Packet Data Rates

		Forward	
RC	SR	Data rates	Characteristics (R)
1	1	1200, 2400, 4800, 9600	1/2
2	1	1800, 3600, 7200, 14,400	1/2
3	1	1500, 2700, 4800, 9600, 38,400, 76,800, 153,600	1/4
4	1	1500, 2700, 4800, 9600, 38,400, 76,800, 153,600, 307,200	1/2
5	1	1800, 3600, 7200, 14,400, 28,800, 57,600, 115,200, 230,400	1/4
6	3	1500, 2700, 4800, 9600, 38,400, 76,800, 153,600, 307,200	1/6
7	3	1500, 2700, 4800, 9600, 38,400, 76,800, 153,600, 307,200, 614,400	1/3
8	3	1800, 3600, 7200, 14,400, 28,800, 57,600, 115,200, 230,400, 460,800	1/4 (20 ms) 1/3 (5 ms)
9	3	1800, 3600, 7200, 14,400, 28,800, 57,600, 115,200, 230,400, 460,800, 1,036,800	1/2 (20 ms) 1/3 (5 ms)
		Reverse	
1	1	1200, 2400, 4800, 9600	1/3
2	1	1800, 3600, 7200, 14,400	1/2
3	1	1200, 1350, 1500, 2400, 2700, 4800, 9600, 19,200 38,400, 76,800, 153,600, 307,200	1/4 1/2 for 307,200
4	1	1800, 3600, 7200, 14,400, 28,800, 57,600, 115,200, 230,400	1/4
5	3	1200, 1350, 1500, 2400, 2700, 4800, 9600, 19,200, 38,400, 76,800, 153,600, 307,200, 614,400	1/4 1/2 for 307,200 and 614,400
6	3	1800, 3600, 7200, 14,400, 28,800, 57,600, 115,200, 230,400, 460,800, 1,036,800	1/4 1/2 for 1,036,800

For a GSM channel, referred to as a transmitter (TRX), there are a total of eight time slots, or traffic channels (TCHs), each capable of carrying an individual voice conversation. However, when dimensioning a GSM cell or sector, it is important to remember that there are other channels that need to be allocated that are related to signaling, namely, the BCCH, common control channel (CCCH) and stand-alone dedicated control channel (SDCCH).

The allocation or dimension of the BCCH is such that it is always associated with the first TRX of the site or sector, and one is required per sector. The BCCH occupies time slot 0 of the first TRX leaving a total of seven TCHs for handling voice traffic. The CCCH shares time slot 0 with the BCCH and, except for provisioning, the inclusion of a CCCH is taken care of with the BCCH allocation. The SDCCH allocations vary depending on the amount of circuit-switched traffic expected for the site.

To determine the amount of SDCCHs required, use the following method.

$$x = \text{number of call attempts per second} + \text{number of location updates per second}$$

If $x = 1$ attempt per second $+$ 2 location updates per second, then

$$x = \frac{3 \text{ activities}}{s} = 3 \times 3600 = \frac{10{,}800 \text{ activities}}{h}$$

$$y = \text{average SDCCH occupancy} = 4 \text{ s} = \frac{0.001111 \text{ h}}{\text{activity}}$$

$$\text{SDCCH traffic} = x \times y = 10{,}800 \times 0.00111 = 12 \text{ erlangs}$$

$$\text{SDCCH erlang} = 12 \text{ erlangs} = 15 \text{ channels (10\% GOS erlang B)}$$

$$\text{SDCCH time slots} = \frac{\text{SDCCH}}{8} = 1.875 \approx 2$$

Therefore, two SDCCH time slots are needed.

To determine the number of radios required, use the following method.

$$\text{Number of radios} = \text{number of TCHs} + \frac{\text{SDCCH}}{8} \text{ time slots} + \frac{\text{BCCH}}{\text{CCCH}}$$

Therefore,

$$\frac{\text{BCCH}}{\text{CCCH}} = 1 \text{ time slot}$$

$$\frac{\text{SDCCH}}{8} = 2 \text{ time slots}$$

$$\text{TCHs (10 erlangs)} = 17 \text{ TCHs at 2\% GOS erlang B}$$

Therefore, $17 + 1 + 2 = 20$ time slots are needed. Since there are 8 time slots per GSM carrier, $20/8 = 3$ GSM carriers are needed.

Here is another approach when the number of GSM carriers is known. A cell with 7 TCHs (BCCH, CCCH, and SDCCH/4 sharing time slot 0) can accommodate approximately 2.9 erlangs. For a 2-TRX cell, with 14 TCHs (time slot 0 on one carrier used for BCCH and CCCH and time slot 1 used for SDCCH/8), the cell can accommodate approximately 8.2 erlangs. For a 3-TRX cell with 22 TCHs (one time slot allocated for SDCCH/8), the cell can accommodate approximately 14.9 erlangs.

For GRPS the channel that is used for transfer of actual user data over the air interface is called the packet data traffic channel (PDTCH). All PDTCHs are unidirectional—either uplink or downlink. This corresponds to the asymmetric capabilities of GPRS. One PDTCH occupies a time slot, and a given subscriber unit with multislot capabilities may use multiple PDTCHs at any given instant. Furthermore, a given subscriber may use a different number of PDTCHs in the downlink versus the uplink. In fact, a subscriber could be assigned a number PDTCHs in one direction and zero PDTCHs in the other.

If a subscriber is assigned a PDTCH in the uplink, the subscriber must still listen to the corresponding time slot in the downlink, even if that time slot has not been assigned to the subscriber as a downlink PDTCH. Specifically, the subscriber must listen for any PACCH transmissions in the downlink. The reason is the bidirectional nature of the PACCH, which in the downlink is used to carry signaling from the network to the subscriber, such as acknowledgments.

This, of course, leads to some important issues related to dimensioning the radio elements for a GSM/GPRS system. There are many different methods in which GPRS can be allocated into each TRX, sector, site and system. None is better than the other; the choice is driven primarily by each carrier and the method that it chooses to implement. For the examples which follow it will be assumed that GPRS (packet-switched) traffic will share the radio resources with the voice network.

Voice and GPRS each have their own allocation within each TRX (8 time slots), known as *territories*. It is important to note that the circuit-switch load always takes priority over the packet-switch load. The circuit- and packet-switch allocation of the time slots within the TRX is dependent upon the number of carriers, TRXs, per sector and per site, and the load needed. For example, in a 1 + 1 + 1 site configuration (in other words for any sector which only carries one TRX), the circuit and packet switch allocation may be as indicated in Fig. 10.2. Note that the single initial TRX that has the BCCH allocation does not frequency hop.

In Fig. 10.2, time slots 6 and 7 are allocated as dedicated GPRS capacity. This means that voice can use all the time slots, except for time slots 6 and 7. This is normally accomplished by the allocation of a percentage to each TRX, depending upon the number of TRXs. A dedicated GPRS capacity is normally smaller than or equal to a default GPRS capacity.

When estimating the dedicated GPRS capacity the need to determine the amount of packet-carrying capacity needs to be established. As mentioned previously the two methods for estimating packet capacity are the forecasted and discovery methods.

Keeping in mind the GPRS coding scheme (CS) 1–4 capabilities (see Table 10.5), when allocating two TCHs, the GPRS capacity can range from 18.12 to 42.8 kbit/s. The use of the dedicated GPRS channels is equivalent to a committed information rate for that sector. Obviously the subscriber usage and overbooking factors need to come to bear when you estimate the packet traffic load. Table 10.6 is an example of calculating the dedicated GPRS capacity.

The percentage figures in Table 10.6 are calculated based upon the total TCHs in a sector and under the assumption that a 1-TRX sector has a combined con-

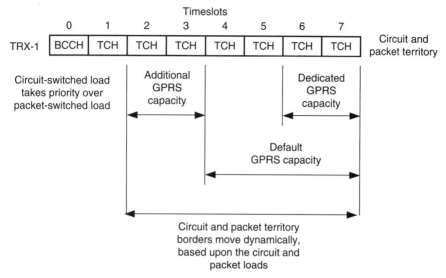

Figure 10.2 One TRX GSM/GPRS allocation scheme.

TABLE 10.5 Coding Scheme

Coding scheme	Data rate (kbit/s) per TCH
1	9.06
2	13.4
3	15.6
4	21.4

TABLE 10.6 Dedicated GPRS Example

No. of TRXs	No. of dedicated GPRS TCHs	% TCHs dedicated
1	2	30
2	2	15
3	2	10
4	2	7

figuration with one time slot for signaling and the sectors with more than one TRX have noncombined configuration with two or more time slots for signaling.

You can also allow additional time slots to be used for GPRS only when the circuit load is low and the packet load is high, which would be a default GPRS capacity, in which the time slots for packet switch would grow from time slots 6 and 7 to time slots 4 to 7. Again, this is normally accomplished by the allocation of a percentage to each TRX, depending upon the number of TRXs. A

TABLE 10.7 Default GPRS Example

No. of TRXs	No. of default GPRS TCHs	% default GRPS TCHs	No. of dedicated GPRS TCHs	Max. no. of GPRS (CS1)	Max. GPRS kbit/s (CS1)	Max GPRS kbit/s (CS4)
1	3	45	2	5	45.3	107
2	3	22	2	5	45.3	107
3	3	14	2	5	45.3	107
4	3	10	2	5	45.3	107

default GPRS capacity is normally larger or equal to a dedicated GPRS capacity as illustrated in Table 10.7. The percentages in Table 10.7 are calculated as for Table 10.6.

A carrier may finally decide that its GPRS capacity is so high that it outweighs the demand for voice, or at some time later, it may choose to add a second TRX to that sector, but not in time, in which an additional GPRS capacity would be allocated for time slots 2 and 3. In this case, the time slots for packet switching would grow from time slots 4 to 7 to time slots 2 to 7. However, the use of allocating any additional time slots is very rarely seen in one-TRX sectors; nevertheless, it is there for when and if the need arises. Please keep in mind that circuit switching always takes priority over packet switching, and as the need for voice increases in a TRX, and the GPRS needs decrease, the two time slots that GPRS has left to fall back on, for sure, are 6 and 7.

In the event that the carrier has, for example, a 2 + 2 + 2 site configuration, or in other words, for any sector that carries two or more TRXs, the circuit- and packet-switch allocation may be as indicated in Fig. 10.3. TRX-1 time slots would be for voice only, and TRX-2 time slots would be allocated as follows:

$$\text{Dedicated GPRS capacity} = \text{time slots 6 and 7}$$

$$\text{Default GPRS capacity} = \text{time slots 4 to 7}$$

$$\text{Additional GPRS capacity} = \text{time slots 2 to 7}$$

Note that the second, third, and so forth TRX can frequency hop. However, it is recommended that packet services be allocated to the TRX which does not frequency hop as in this example.

Using an example where you have forecasted that a total of 10-erlangs voice usage is required to be supported along with the need to provide for that sector 23 kbit/s of packet data during the busy hour. The packet throughput will be symmetrical to simplify the example. Therefore, the number of TCHs required for voice is 17, using the calculations done at the beginning of this section. A total of 2 SDCCHs are needed along with the BCCH equating to a 20-TCH requirement to support circuit-switched usage. However, to support 23 kbit/s of packet data you need to determine which CS to use; for this example a CS-1 will be used since it is worst case. Therefore, a total of 3 TCHs are

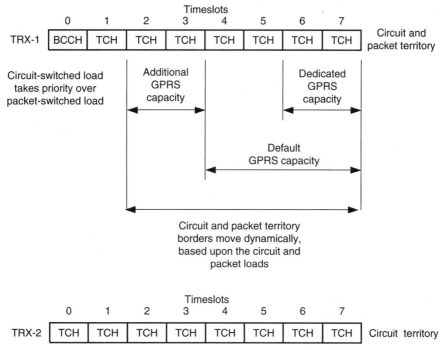

Figure 10.3 TRX GSM/GPRS allocation scheme.

required for packet data during the busy hour. A total of 23 TCHs are needed, and this falls nicely into three GSM channels which have 24 TCHs in total available for allocation.

10.7 Radio Data Traffic Projection

The inclusion of mobile data traffic in a wireless system can be a very complex issue. The mobile data traffic volume is directly dependent upon the type and quality of services that will be offered and how they will be transported. The traffic estimation process involves not only the radio link but also the other fixed facilities which comprise the network. Additionally, the inclusion of 802.11 (WiFi) services, which utilize the unlicensed bands, commonly referred to as license exempt, complicate the situation for estimating wireless data usage in a *local area network* (LAN) environment.

Mobile data traffic is a contention-based service, and, therefore, the issue of blocking, i.e., denied access due to radio facilities, will continue to prove to be difficult. The difficulty lies in the nature of mobile data in that it is packet based and not circuit switched, meaning that you can and should overbook the service.

Therefore, the forecasting of the mobile IP traffic would be much more simplified if the usage patterns were understood better. But in light of this issue the forecast, or growth, could be extrapolated from the business plan or simplified marketing plans, which would specify a specific growth level desired. The equation to determine total traffic for an existing system is

Total traffic = existing traffic + new traffic expected

The new traffic expected could be a simple multiplication of the existing traffic load. For instance, if the traffic is 25 Mbit/s and the plan is to increase the traffic by 25 percent over the next year, then the 1-year traffic forecast would be altered by increasing the current traffic load at each node by that amount and determining the requisite amount of logical and physical elements needed in addition to any load sharing that might be achievable.

The ultimate question that the designer must answer is, How do you plan on supporting the traffic, both data and voice, with prescribed services? Since there are numerous types of services available for both circuit-switched as well as packet-switched system, some generalizations need to be made in order to have a chance at arriving at some conclusions necessary for input into the design phase. Therefore, the symbols defined in Table 10.8 will be used to help define the different classifications of transport services required.

For a new or existing mobile IP system, the hardest part is knowing where to begin. However, one of the key parameters you need to obtain from marketing or sales is the penetration rate (take rate for each of the services types offered). This can be achieved via several methods: You can use a general approach where a standard percentage is used for, say, packet services. Or you could determine the amount of packet data subscribers from the number of handsets expected to be purchased for resale in the market.

Voice services are included in Table 10.8, because if the wireless system offers packet services, it will also offer voice, and the spectrum and other radio resources will be in competition for those resources. If the wireless system does not offer voice services, as is the case with some PCS C block providers, then the issue of voice and possibly switched data can be removed from the mix.

Regardless, the first step in any traffic study is to determine the population density for a given market. In the case of an existing system the population density and primary penetration rates are already built into the system due to

TABLE 10.8 Circuit and Packet Data

Symbol	Service type	Transport method
S	Voice	Circuit switch
SM	Short message	Packet
SD	Switched data	Circuit switch
MMM	Medium multimedia	Packet
HMM	High multimedia	Packet
HIMM	High interactive multimedia	Packet

TABLE 10.9 Cell Area Equations

Cell type	Cell area	Sector area (3-sector cell)
Circular	πR^2	$\pi R^2/3$
Hexagonal	$2.598 R^2$	$2.598 R^2/3$

known loading issues. However, especially for new services, like packet data, the process of determining the population density for a given area followed by the multiplication of this by the penetration rate will greatly help in the determination of the expected traffic load from which to design the system.

Population density. A measure of the quantity of people that could possibly utilize the service for a given geographic area. When determining population density, it is important to note that for the same geographic area there could be different population densities. For example an area could have 100,000 pedestrians per square kilometer, but only a vehicle density of 3000 per square kilometer.

Penetration rate. This is a measure of the percentage of people in the population density that will utilize a particular service. For example, out of the 100,000 possible users only 5 percent may want a particular service; therefore, the possible usage may only be experienced by 5000 people for that service offering. An important issue is that each service offering will likely have a different penetration and it is very possible that based on the amount of services offered that the total penetration rate could exceed 100 percent because of the various service offerings.

Cell site area. The geographic area that a cell site or its sector will cover is determined either by computer simulation or for a rough estimate by a two-dimensional approach. The formula for determining the area for a cell is shown in Table 10.9. The radius for the cell site is determined from the link budget and is dependent upon numerous issues. However, the use of a standard cell radius for a given morphology is recommended to be used for the initial design phase. The cell coverage area for a later phase in the design process can be determined through use of computer simulations which should factor in the cell breathing issues that are evident in CDMA systems or dynamic allocation of TCHs for GSM/GPRS.

QOS. Quality of service is a term that has many meanings. For this discussion QOS is a description of the bear channel's capability for delivering a particular grade of service. The GOS is typically defined as a blocking criteria, and for circuit-switched data it is defined by erlang B, erlang C, or Poisson equations. For packet data, the relationship for QOS/GOS is blocking, erlang C, and delay.

With the introduction of packet data the traffic modeling for packet-switched data involves the interaction of the following items:

Number of packet bursts per packet session

Size of packets

Arrival time of packet burst within a packet session

Arrival times for different packet sessions

The packet usage is relatively an unknown area for wireless mobile systems on a mass market basis as of this writing since the killer application appears to be wireless mobility, i.e., transport ubiquity.

The issue of where, when, and how much you dimension a system for packet data will always be a debate between the marketing and technical teams. However, in light of the fact that packet data usage is in its infancy, there is little guidance from which to go forth and design the network from. However, ITU-R M.1390 has some guidelines for dimensioning. Tables 10.10 to 10.15 are extracted from that specification. The values in the tables should be used as a guide to establish packet loading for dimensioning when market-specific data are not available for numerous reasons. It is important to note that of the services listed in Tables 10.10 to 10.15, only MMM and HMM are asymmetrical; the rest are symmetrical service offerings.

Note that the penetration rates for the services are the same and come to more than 100 percent for any location. It is also important to note that the numbers indicate that 73 percent of the system usage is expected to be voice oriented and that 10.8 percent is for SM; 3.51 percent is for SD, MMM, and HMM; while 6.75 percent is for HIMM. Tables 10.12 to 10.14 will help provide additional insight into possible traffic dimensioning requirements.

The next step is to determine the traffic forecast by user by service type. The method for achieving this value is determined by the equation for each of the service types and locations defined (building, pedestrian, vehicular). (See Tables 10.12 to 10.14.)

$$\text{Traffic per user} = \text{BHCA} \times \text{call duration} \times \text{activity factor}$$
$$= 0.9 \times 180 \times 0.5 = 81 \text{ call seconds}$$

during the system busy hour for downlink or uplink voice service for a building environment. The number of circuits required for circuit-switched voice,

TABLE 10.10 Net User Bit Rate

Service type	Downlink, kbit/s	Uplink, kbit/s
S	16	16
SM	14	14
SD	64	64
MMM	384	64
HMM	2000	128
HIMM	128	128

TABLE 10.11 Penetration Rates

Service type	Penetration rates, %		
	Building	Pedestrian	Vehicular
S	73	73	73
SM	40	40	40
SD	13	13	13
MMM	15	15	15
HMM	15	15	15
HIMM	25	25	25

TABLE 10.12 Busy Hour Call Attempts

Service type	Building	Pedestrian	Vehicular
S	0.9	0.8	0.4
SM	0.06	0.03	0.02
SD	0.2	0.2	0.02
MMM	0.5	0.4	0.008
HMM	0.15	0.06	0.008
HIMM	0.1	0.05	0.008

switched data, and HIMM services is determined via erlang B, while the remaining packet data services are determined via erlang C.

The next step is to define the next set of variables which need to be established to help dimension the rest of the packet network. For symmetrical services the dimensioning is relatively straightforward. However, for asymmetrical services a few more details are required which are used for the selection and performance of the PDSN.

$$\text{Transmission time (s)} = \text{NPCPS} \times \text{NPPPC} \times \text{NBPP} \times \frac{8 \text{ bits per byte}}{1024 \text{ kbit/s}}$$

$$\text{Total session time} = \text{packet transmission} + [\,(\text{PCIT} \times (\text{NPCPS} - 1))\,] + [\text{PIT} \times (\text{NPPPC} - 1)]$$

$$\text{Activity factor} = \frac{\text{packet transmission time}}{\text{total session time}}$$

10.8 RF System Growth

The cell site growth projects in a network are an ongoing process of refinements and adjustments based on a multitude of variables, most of which are not under

TABLE 10.13 Call and Session Duration

Service type	Call duration		
	Building	Pedestrian	Vehicular
S	180	120	120
SM	3	3	3
SD	156	156	156
MMM	3000	3000	3000
HMM	3000	3000	3000
HIMM	120	120	120

TABLE 10.14 Activity Factor

Service type	Downlink	Uplink
S	0.5	0.5
SM	1	1
SD	1	1
MMM	0.015	0.00285
HMM	0.015	0.00285
HIMM	1	1

TABLE 10.15 MMM/HMM Packet Data

Type	Description	Downlink	Uplink
NPCPS	No. of packet calls per session	5	5
NPPPC	No. of packets per packet call	25	25
NBPP	No. of bytes per packet	480	90
PCIT	Packet call interarrival time	120	120
PIT	Packet interarrival time	0.01	0.01

the control of the engineering department. However, the cell site growth analysis can be used to help direct the limited resources of the company.

The final output of the cell site growth section of the plan is to identify the number of cell sites required for the network and their required on-air dates. In addition to the number of new cells any radio equipment expansion is also included.

The suggested process to use for putting together the cell site and radio growth requirements discussed here assumes that there is not an automated cell building program in place for your network. The automated cell building program involves decision support systems which distribute traffic loads according to cell site parameters input into the modeling system.

Most wireless operators have prediction modeling tools that have the ability to spread the traffic load, whether it is voice or data. Depending on the sophistication of the software tool the ability to factor in data traffic or even dispatch traffic may be problematic at best. In these instances it is suggested that the data or dispatch traffic be treated as a circuit-switch load for ease of computing.

However, the method shown in Secs. 10.8.1 to 10.8.3 uses an AMPS system as the underlying example. It is the intention here that if you can follow the methodology used for AMPS it can be easily extrapolated for other wireless technologies. In addition the inclusion of multiple-technology platforms within a wireless system further necessitates the need to understand the approach. Lastly through understanding the approach verification of the software tools, or rather reliance after verification, is appropriate.

10.8.1 Coverage requirements

The first step in the RF system growth process is to determine all the coverage requirements needed for the network. They will be rank-ordered in the final process to ensure that all the proper input parameters are taken into account. The RF coverage identification process is as follows:

1. Coverage requirements identified by
   a. Marketing and sales
   b. System performance
   c. Operations
   d. Customer care
   e. RF engineering
2. Generate a propagation plot of the system, or subregions, that reflects the current system
3. Generate a propagation plot of the system, or subregions, that reflects the current system and known future cell sites that are under construction
4. Utilizing physical field measurements, generate a plot of the system, or subregions
5. Compare the plots from parts 2 and 3 with areas identified in part 1 for correlation
6. Compare the field measurement plot against parts 1 and 2 for correlation
7. Using the design criteria used by RF engineering determine how many sites will be needed to satisfy the design goals.
8. Using the list of cell sites identified in step 7, rank-order them according to the point system methodology.

The point system involves five key parameters ranked on a scale of 1 to 5 based on severity. Each of the five categories receives a value which is then

TABLE 10.16 Ranking Example

Cell	Coverage	Traffic potential	Customer care	Marketing	System performance	Total
101	3	1	2	4	2	48
102	2	4	3	1	4	96

multiplied by each field to arrive at a ranking. The ranking methodology uses the following key fields:

1. Coverage
2. Traffic [erlang or (kbit/s)] potential
3. Customer care problems
4. Marketing and sales needs
5. System performance requirements

Table 10.16 will help clarify the ranking methodology. This table shows that cell 102 has a higher weight than 101 in the identification of prioritization. The ranking methodology should be applied to all the potential cell sites in a network. Establishing a ranking system addresses the budgetary issues of having a method in place to determine where to cut or add sites to the build program.

10.8.2 Capacity cell sites required

The next step in the RF system growth plan is to determine the sites required for capacity relief. This involves determining which sites or sectors are at their current capacity and require some level of relief. The capacity relief can come from a variety of options:

1. Radio additions
2. Parameter adjustments
3. Antenna system alterations
4. New cells

There are more capacity relief methods available for redistributing the traffic loads of cell sites. For example, in CDMA you should limit the amount of soft and softer handoffs since they rob system capacity. Additionally the issue of subscriber type and legacy issues will always play a factor in the growth planning part.

The first step in identifying the capacity cell sites required for a network is to use a spreadsheet to determine where the problems are anticipated. The recommended spreadsheet format is shown in Table 10.17 with supporting equations listed below.

TABLE 10.17 Growth Projection

System: XYZ

Date:

RF growth

	Baseline			Quarter of interest (4Q2002)							
							Growth projection				
Cell site	Erlangs (A)	Radio carriers (B)	TCH (C)	Erlangs projected (D)	Off-load factor (E)	Acquired erlangs (F)	Total erlangs (G)	New TCH (H)	Total TCHs needed (I)	New radio carriers (J)	Total radio carriers (K)
1A	5.4	17	17		(0.4)	0					
2A	4.0	12	12								
101	0	0	0		0.6	3.1	3.1	12	12	12	12

The format shown is then carried out for each of the quarters involved in the initial study. The following is meant to help explain the individual columns in Table 10.17.

A. This is the current erlangs carried by site or sector. The baseline erlang value is based on the data collected in the design criteria. For example, if the baseline is the average of the 10 busiest days in June, the erlang value for the site's contribution is then placed here.

B. This is the number of physical radio carriers employed at the site.

C. This is the number of physical channels currently available at the site or sector at baseline.

D. This section is the projected erlangs for the site or sector based on the forecasted traffic load using the design criteria. The value that is used for the lookup is arrived at using the following equation:

Projected system erlangs
$$= \text{(millierlang per subscriber)} \text{(total system subscribers for quarter)}$$

$$\text{Cell site system capacity} = \frac{\text{erlangs for sector}}{\text{total system erlangs}}$$

This part uses the data from the previous quarter. The value will not be 100 percent correct, but it is more than sufficient to use for planning purposes.

$$\text{Projected erlangs} = \text{(system erlangs projected)}\text{(cell site system capacity)}$$

E. The off-load factor is used for shedding traffic to another cell site. The methodology used for arriving at how to shed traffic is discussed later. However, if it is determined that you can shed 40 percent of the site's projected traffic to another cell site, then a value of 0.4 is entered.

F. This is the acquired erlangs the site receives from adjacent cell sites as a result of adjustments made to them. The acquired traffic portion for the entire system should equal the total amount of traffic offloaded (column E). The values arrived at in this section pertain to how the traffic is distributed in the off-loading cell site. For example, cell 5 may off-load cell 3 by 40 percent, but the traffic will not all go onto one sector, usually. Therefore, you must determine how this is distributed around the site. Continuing, cell 5 will have cell 3's traffic distributed with sector 1 getting 70 percent and sector 2 getting 30 percent. We would strongly suggest that you keep meticulous records about how much each site off-loads to the other and the distribution percentages used.

G. The total erlangs is the result of off-loading and acquired traffic to the site. The formula to use is

$$\text{Total erlangs} = \text{(cell site projected erlangs)}\text{(off-load value)} + \text{acquired traffic}$$

The total erlang value is then used with the same erlang lookup table used for the RF channels and individual carriers needed.

H. New TCH is simply the number of new individual channels required. In the case of AMPS the obvious issue is that this is the same as the number of carriers, excluding the control channel.

I. Total TCHs needed is simply the number of new individual channels required added to the existing TCHs:

$$\text{Total TCH needed} = \text{existing TCHs} + \text{new TCHs}$$

J. New radio carriers refers to the number of actual individual radio carriers required. For AMPS the value is a 1:1 with the TCHs but for, say, IS-136 it can be 1:2 or 1:3 depending on the number of physical radios deployed at the site.

K. Total radio carriers is simply the number of new radio carriers required added to the existing radio carriers operational at the site or sector:

$$\text{Total radio carriers} = \text{existing radio carriers} + \text{new radio carriers}$$

Obviously when looking at Table 10.17, a recursive feedback loop process is really implemented. It involves additional off-loading and acquired erlang values until the proper design criteria is met.

Regarding the acquired traffic for a new site it is suggested that a minimum value be added to the overall value for every sector added. The added erlang value will be used to ensure that a minimum of one radio channel is put in place for every new cell. However, doing this will inflate the overall system erlang value beyond the projection for that quarter only.

Lastly based on the projection, the amount of radio spectrum, for the usage defined, may not be adequate. Obviously the issue then is how to handle the situation.

10.8.3 RF traffic off-loading

The off-loading of traffic, either voice or data, to another cell site can be accomplished through an elaborate computer simulation method or educated engineering guesses. The elaborate computer modeling capability is the ultimate desired method since it eliminates guesswork involved in arriving at the off-loading values. However, the computer model method is only as good as the model itself. If you are using the computer model method the off-loading and acquired values should be readily extracted from the algorithm output and input into the spreadsheet either manually or automated.

If you are not fortunate enough to have a sophisticated computer modeling method, then the following method is recommended. This method has been used repeatedly with a high level of success. A four-cell subsystem is shown in Fig. 10.4. The situation depicted will not require the network manager to address any coverage-related issues on the site. It can be solved by the introduction of a new cell site for the sole purpose of providing traffic relief to adjacent cell sites.

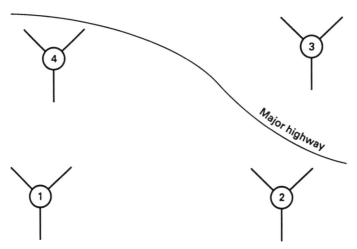

Figure 10.4 Sample four-cell system.

Traffic projections were done using the spreadsheet method, and it was found that sites 3 and 4 required some form of traffic shedding. It was determined that the only available method of ensuring enough traffic shedding for both sites was to introduce cell site 5, shown in Fig. 10.5. The placement of cell site 5 in the network can ensure that it will off-load several of the adjacent cell sites by the percentages listed in the figure. Off-loading values were determined through an interactive process involving the following data.

1. Current coverage of zone by the existing cell sites
2. Determination of voice and control channel dominant server areas, by cell and sector
3. Evaluation of the cell site's configurations in the zone
4. Incorporation of cell 5 into data provided for parts 1, 2, and 3
5. Comparison of propagation plots to actual field measurement data
6. Establishment of off-loading and acquired percentages

The off-loaded traffic is then assigned to the sectors of cell 5 as shown in Fig. 10.6. The values represented in Figs. 10.5 and 10.6 are then input into the tables of the growth plan to ensure the desired effect.

10.9 Fixed Network

The fixed network portion of the planning phase obviously includes both circuit-switched and packet-switched services. The inclusion of packet-switched services is evident in a wireless system that just processes voice since the requisite signaling required for call delivery is in fact packet based. Therefore, the

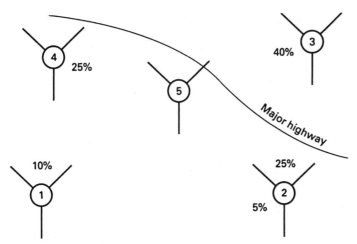

Figure 10.5 Offload percentages.

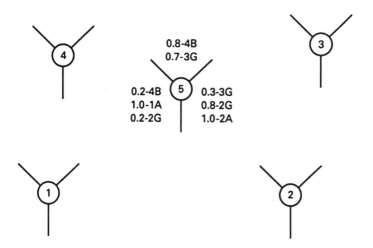

Figure 10.6 Offload.

introduction of packet data services for wireless mobility involves the introduction of the PDSN and the associated planning that needs to accompany it.

For planning purposes the integration of the RF design has always been, will continue to be, an integral part of the fixed network design. The fixed network design can proceed in parallel with the RF design process since many of the same marketing inputs serve both interests. However, the final configuration associated with the number of BTSs, their location, and the BSC and other node concentration considerations are dependent upon the RF group.

The introduction of mobile data has brought to focus many issues associated with packet data. But this issue is not new to fixed network design since many of the enablers for offering voice services has included the need for data services.

Therefore, the introduction of the mobile IP has expanded on the use of data services, primarily IP related, for the fixed network design process. Although you might argue that this is an oversimplification, including different protocols [frame, IP, ATM, TDM, synchronous digital hierarchy (SDH), synchronous optical network (SONET), etc.] has been a staple of the network engineer.

There are certain commonalities that are applicable for both the voice and data services which involve utilization rates. The facility utilization goal for the network should be 70 percent of capacity for the line rate over the time period desired. The time period that should be used is a 9-month sliding window which needs to be briefly revisited on a monthly performance report and then during a quarterly design review, following the design review guideline process.

The facility will need to be expanded once it is understood that 70 percent of the utilization level will be exhausted within 9 months with continued growth showing exhaustion (100 percent utilization rate) within 18 months. The reason for the large lead time lies in the implementation cycle that is characteristic of the fixed network environment. If your market significantly deviates from the suggested lead time recommendation you should make the requisite changes.

For all the platforms, you should use the following general guidelines for the fixed network:

Processor occupancy: 70 percent

Packet data platforms: SCR 25 percent, PCR 90 percent

Port capacity: design for 70 percent port utilization

10.10 Circuit Switch Growth

The circuit switch growth study will be broken down into three major areas: processor capacity study, port capacity study, and mobile subscriber database capacity study.

10.10.1 Switch processor capacity study

There are many methods and models used to determine the current processor capacity of a switch or node in a particular network and to predict when an upper threshold will be reached due to increased traffic loads. In this section we will discuss only a simple study for use in managing you network switches and nodes. More complex and accurate studies can be formulated by working with you equipment vendors and obtaining their assistance in measuring your unique system loads.

The steps for conducting a switch processor load study are as follows:

1. Assemble and list all design assumptions of the study.
2. Measure and collect all processor load data and then plot this data versus time on a graph.

3. Project this plot using either regression analysis or actual subscriber forecast data from the marketing department. By using regression analysis on the data collected, the best-fit straight-line approximation for the data can be obtained. The plot of this line would represent the projected load of the processor for the future months, assuming uniform traffic growth. However, for a more accurate projection use the system subscriber forecast developed by the marketing department.

4. Obtain the switch vendor's suggested processor load limit with an agreed-upon variance and plot this on the processor load graph. Note when the load of the processor reaches the specified upper threshold.

5. Present all data and analysis to your upper management during the quarterly design review. Develop a plan for expanding the network or for adjusting the proper switch and system parameters to relieve this processor load in the time frame provided by study. The point where the processor load plot intersects the maximum limit threshold mark is the time period in which planned relief of this switch should take place by either expanding the network (adding more switching capacity) or performing system parameter changes (reduce registration intervals, etc.).

Example 10.2: Switch Processor Capacity Study The actual measured data from the switch were collected during the system busy hour for the 10 highest traffic days of the month for a period of 6 months (October 1994 to March 1995). The processor load was projected out to the end of the year using both the regression analysis method and the subscriber forecast developed by the marketing department. Using the regression analysis method the traffic growth of the system is assumed constant and is based upon the previous 6 months' processor load data. Using the data supplied by the marketing department the processor load can be related to the expected subscriber growth patterns in the following manner. Take the current traffic load on the system at 1000 erlangs. Then take the current total number of subscribers in a system at 200,000.

Find the expected amount of traffic each subscriber contributes to your network by dividing the total system traffic (1000 erlangs) by the total number of subscribers in your network (200,000). This value equals 0.005 erlangs per subscriber. It would be fair to assume that subscribers added to the system in the future would have similar calling patterns for your region and, thus, would contribute about the came amount of traffic to the network.

Now, take the subscriber growth data from the yearly marketing forecast. Add the number of subscribers expected to be added to the system every month to the current total subscriber count and then multiply this value by 0.005 erlangs per subscriber. This gives you the traffic increase on the system on a monthly basis. By assuming the system traffic to be directly proportional to the processor load, you can now obtain the projected processor load on a monthly basis as well.

The vendor recommended maximum processor load is 75 percent with a 25 percent overload factor. The baseline load (no call traffic present) for the switch processor as specified by the vendor is 15 percent.

Measure and collect all processer load data and plot the results on a graph (see Table 10.18 and Fig. 10.7).

Then project the processor load plot, and specify the maximum processor load limit and mark this load level on the processor load plot. Report the time period the switch will reach its projected maximum processor limit. According to the plot in Fig. 10.7, the switch for system X will reach a maximum processor load in early October 1995.

10.10.2 Switch port capacity study

The switch port capacity study is meant to take the current number of voice circuits assigned in the switch matrix and the total number of projected ports required over a given time period and determine if a maximum capacity will be exceeded. Every switch has a limited number of ports available for

TABLE 10.18 System Data for Market X for Processor Load Study, March 1995

Current subscriber count	200,000
Current system load (busy hour erlangs)	1000
Current erlangs per subscriber	0.005
Current processor load (monthly average)	60%

Marketing subscriber growth forecast for the 1995 year

Assume a 3 percent subscriber growth rate for the months of April to September. Then increase the rate to 8 percent owing to the expected sales from the holiday season marketing promotions.

Month	Sys. Subs.	Month	Sys. Subs.	Month	Sys. Subs.	Month	Sys. Subs.
Jan.	188,000	Apr.	206,000	Jul.	224,000	Oct.	252,000
Feb.	194,000	May	212,000	Aug.	230,000	Nov.	268,000
Mar.	200,000	Jun.	218,000	Sep.	236,000	Dec.	284,000

Projected system traffic levels (erlangs) and proportional processor loads (%)

Month	Sys. Tr. (erl)	Proc. Load (%)	Month	Sys. Tr. (erl)	Proc. Load (%)
Apr.	1030	61.8	Sep.	1180	73.8
May	1060	63.6	Oct.	1260	78.6
Jun.	1090	65.4	Nov.	1340	83.4
Jul.	1120	67.2	Dec.	1420	88.2
Aug.	1150	69.0			

Actual measured processor load data

Year	Month	Proc. Load (%)	Year	Month	Proc. Load (%)
1994	Oct.	56.2	1995	Jan.	56.4
1994	Nov.	58.8	1995	Feb.	57.5
1994	Dec.	59.0	1995	Mar.	60.0

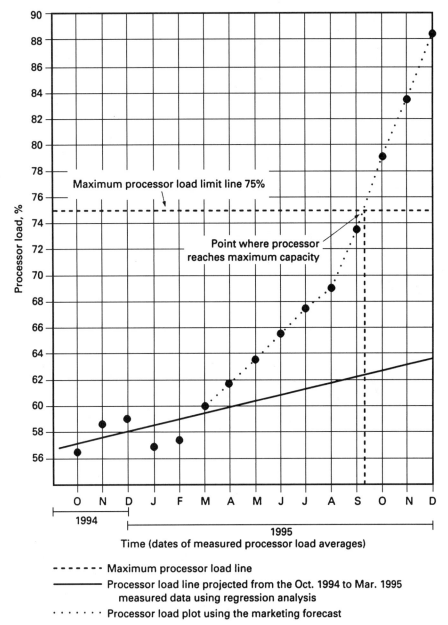

Figure 10.7 Switch processor load plot for market X.

assignment. This value can be obtained from your switch vendor and used to conduct your own port capacity study. The actual ports assigned to a switch can be broken down into two different types, RF voice circuits and standard landline interface circuits. Each type of circuit requires a different piece of switch hardware or magazine for interfacing to either a cell site or a landline facility. The maximum number of actual switch ports available for assignment will differ for both types, so make certain to study each for possible capacity limitations.

Once the switch port capacity limits are obtained, the next step is to determine the current port assignments for the switch. These data can be collected by printing out the database of the switch pertaining to the actual assignments of both landline and RF circuits. Tabulate the total number of both port types.

The next step requires input from the RF and network engineering groups. The RF engineering group will need to provide the projected number of coverage cell sites to be added to the system over the specified time period. The network engineering group will need to provide the projected capacity cell sites and the number of landline circuits to be added over the same time interval. These inputs determine the projected port growth for the switch. The actual data should be tabulated in such a manner that the projected port growth is shown in monthly increments. See Table 10.19 for an example system data report and see Fig. 10.8 for an example port capacity plot.

Example 10.3: Projected Switch Port Capacity Expansion For the remainder of the 1995 year engineering expects to add 15 coverage cell sites and 8 capacity cell sites. The RF and network engineering groups estimated that one T1 span (24 circuits) would be all that was required per each cell site. In addition to these ports the network group estimated that there would be a total of 240 landline ports added to the system during the remainder of the 1995 year. These data, given in Fig. 10.8, are tabulated monthly.

A review of the system data and the switch port capacity plot in Fig. 10.8 indicates that if the current trend in RF port additions continues, then a new switch will be required to provide additional capacity some 8 months past November 1995, sometime in July of 1996.

10.10.3 Switch subscriber capacity study

The switch subscriber growth study will mainly apply to smaller systems since networks of larger magnitudes have begun to implement home location register (HLRs) as a central database storing subscriber records for the system. However, this basic study can be applied to HLRs as well.

As in the port capacity study just discussed, the switch has a limitation on the number of subscriber records that can be stored at a given time. Again, this value will be available from your switching vendor along with any assistance in conducting this and other studies for managing your network.

Network and RF Planning

TABLE 10.19 System Data for Market X for Switch Port Study, March 1995

Current number of RF switch ports available	1500
Current number of switch ports (RF) assigned	900
Maximum switch port capacity (RF)	2000
Current number of land switch ports available	2000
Current number of switch ports (land) assigned	1500
Maximum switch port capacity (land)	10,000

Month	RF Ports/Channels Additional/Total*	Land Ports/Circuits Additional/Total†
Apr.	3 cells = 72 ports/972	48/1548
May	4 cells = 96 "/1068	72/1620
Jun.	3 cells = 72 "/1140	48/1668
Jul.	3 cells = 72 "/1212	24/1692
Aug.	4 cells = 96 "/1308	—/1692
Sep.	2 cells = 48 "/1356	48/1740
Oct.	2 cells = 48 "/1404	—/1740
Nov.	2 cells = 48 "/1452	—/1740
Dec.	/1452	—/1740

*Total RF ports added: 552.
†Total land ports added: 240.

Once the switch subscriber capacity limit is known the following data are required. A copy of the subscriber forecast from the marketing department which will give the subscriber growth on a monthly basis. Also required are any special case numbers for use in network call delivery such as dynamic routing numbers (DRNs) (numbers assigned in the switch for completing calls to your subscribers roaming in other systems). Assemble this data in tabular form and analyze them to determine the subscriber capacity needs of the system. See Table 10.20.

Based upon the data in Table 10.20, three switches will be needed to provide the necessary subscriber capacity to store 284,000 records plus the minimal, but additive, DRNs. With a conservative subscriber growth of 3 percent a month, another switch or subscriber database node would have to be added to the network by February of 1996. If your network is interfaced to another system with a set of established data links, then it is possible to store your system's numbers in one of the other system's nodes for temporary relief of this subscriber capacity limitation. This sort of arrangement would be possible if the two markets are on good business terms.

In summary the processor, port, and subscriber capacities of the network switches and other system nodes should be monitored weekly by the network engineering department. The collected data should be analyzed and presented

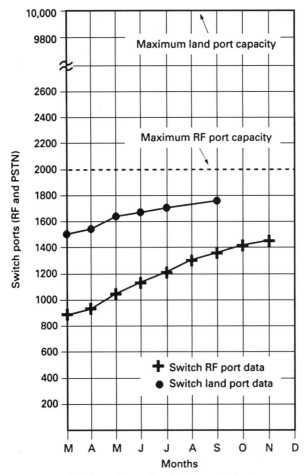

Figure 10.8 Switch port capacity plot for market X.

TABLE 10.20 System Data for Market X for Subscriber Capacity Study, March 1995

Maximum switch subscriber capacity	100,000
Number of assigned DRNs per switch	300
Current subscriber count	200,000
Projected subscriber count end of year 1995	284,000

at the engineering department's quarterly design review for use in evaluating the performance of each switch and the system as a whole. In addition, this data should be used as input to the company's overall budget for planning the expansion of the network necessary to accommodate the projected growth of the system.

10.11 Network Interconnect Growth Study (Voice)

The network interconnection growth study (voice) is the collection of current transmission facilities [T-spans, DS3s, plain old telephone service (POTS) lines, microwave links, etc., either purchased from the local exchange carrier (LEC) or privately owned by the company itself] and the assessment or analysis as to whether these facilities need to be expanded or decommissioned. These facilities are the required medium to interconnect the mobile switches to the PSTN and each individual cell site for providing cellular service to area mobile subscribers. For ease of accounting and management the term facilities previously mentioned should only include the voice circuits if possible. The facilities used for the network data links should be broken out into another study. See Sec. 10.12 for more details.

To conduct this study first collect the current system interconnect data. These data include all the voice circuits or trunk groups interconnecting the network switches. Note that an individual voice circuit is sometimes referred to as a *trunk*. Thus, a group of trunks is called a trunk group. Trunk groups that interconnect network switches are sometimes referred to as *intermachine trunks* (IMTs). Also considered as interconnect data are the trunk groups to the PSTN. This will include trunks to the LEC and the interexchange carrier (IXC) as well. The next major category of facilities are the cell site trunks. The number of trunks assigned to an individual cell site will vary depending on the number of voice channels it can support and the actual channels that are assigned by the RF engineering group. This value can range anywhere from one to three T-spans.

You will need to include the projected number of coverage and capacity cell sites and their associated voice channel count to be added to the system over the specified time period from the RF Engineering Group. In addition, the projected number of PSTN and IMT trunks to be added and decommissioned in the network from the Network Engineering Group needs to be included. These figures will depend upon the current traffic loads in the network as well as future projected traffic growth and system configuration changes (adding new switches, nodes, etc.).

Finally, include any POTS lines purchased by the company as part of the operation of the network. Put these data into tabular form and include a column specifying the number of facilities to be purchased or expanded on a monthly basis and the number to be decommissioned due to low usage or a change in the network configuration. See Table 10.21 for an example layout of this data.

Next, compare the monthly facilities totals to the current switch port capacities. If there is a shortage in the number of ports needed in any one month then either order more ports for the switch or, if necessary, order a new switch for expanding the network. In a multiple switch environment it is necessary to move (reassign) cell sites to other switches for purposes of balancing the port assignments. And processor load across the network.

This study, once completed, will provide an excellent input to the switch port study discussed in Sec. 10.10.2.

TABLE 10.21   System Data for Market X for the Network Interconnection (Voice) Growth Study, March 1995

Current number of land (PSTN) switch ports available	2000
Current number of PSTN (LEC) trunks	1000 (approx. 42 T-spans)
Current number of PSTN (IMT) trunks	500 (approx. 21 T-spans)
Current number of cell or RF switch ports available	1500
Current number of cell site trunks	900 (40 T-spans)
Current number of POTS lines	20

The following example shows an RF and network cell site projection.

Example 10.4   RF and network engineering cell site projections show that a total of 15 coverage sites and 8 capacity sites will be added to the system over the next 9 months according to the schedule given in Table 10.22. Each site will be brought into initial service with one T-span. Network engineering's system landline (PSTN) facility projections show a total of 240 trunks will be added to the system according to the schedule in Table 10.23.

As the table attests, there is sufficient capacity in the system to accommodate the growth in the amount of network facilities growth from a switch port standpoint. Other system considerations must be taken into account when analyzing the interconnect facilities. For instance, is there enough external hardware to support this network growth? DACs and patch panels may have to be ordered as well. The data presented in this study need to be discussed with your finance and revenue assurance departments for reconciliation of the bills associated with these facilities. The traffic engineer working on the system should be working closely with the finance department to find ways to route the network traffic over the least-expensive facilities. Perhaps by consolidating portions of the network traffic into larger volumes a better rate can be obtained from the LEC. These and other design and cost issues should be discussed among these departments. As a final note, it is important for the operations personnel to perform their own audit of the network interconnect facilities. They are a great source to rely upon since they work with this equipment on a daily basis.

10.12   Network Interconnect Growth Study (Data)

The network interconnect growth study (data) is similar to the voice circuit study with the exception that this review pertains only to the facilities used to transmit network-related data for use in call delivery, subscriber validation, etc. The actual data for this study should be obtained from the network engineering group and should include such data as the current number SS7 data links active in the system and the number of links expected to be added in the system due to network growth and reconfiguration. We suggest that these facilities be kept separate since they are usually ordered and tracked in a different manner than the voice circuits.

TABLE 10.22 System Cell Site Projection

Month	Coverage sites	Capacity sites
Apr.	2	1
May	2	2
June	1	2
July	2	1
Aug.	3	1
Sep.	1	1
Oct.	2	
Nov.	2	
Dec.		

TABLE 10.23 System Network Interconnect Facilities Analysis

Facility and ports	Apr.	May	June	July	Aug.	Sep.	Oct.	Nov.	Dec.
RF ports required	972	1068	1140	1212	1308	1356	1404	1452	1452
RF ports available	1500	1500	1500	1500	1500	1500	1500	1500	1500
Total ports removed: 528	432	360	288	192	144	96	48	48	
Land ports required	1548	1620	1668	1692	1692	1740	1740	1740	1740
Land ports available	2000	2000	2000	2000	2000	2000	2000	2000	2000
Total ports removed: 452	380	332	308	308	260	260	260	260	

When actually conducting this study, review all the planned projects for the network to determine if new data links will be required, the number of links and hardware needed, and the time frame for delivery of these items. All this formation will be necessary to complete the data interconnect study.

The following is a list of the parameters that are valuable in the design and operation of a cellular system. These data should be updated on a regular basis as input to the quarterly network design review and should be compared to the previous quarter's data to detect large changes in any specific parameter and to determine if the modeling and prediction efforts by the network group are improving or remaining consistent.

- Size of initial subscriber base
- Projected growth of subscriber base over a 2-year period
- Estimated usage per subscriber (millierlangs per subscriber)
- Estimated calls per subscriber
- Estimated calls per second

- Estimated average call holding time (data taken from other known systems in the area, or a typical industry value used)
- Estimated switch initial traffic (erlangs)
- Projected switch traffic growth over a 2-year period
- Estimated switch calls processed per second
- Projected switch calls processed per second over a 2-year period
- Estimated number of cell sites in service at the time of the system cut
- Projected cell site growth over a 2-year period
- Estimated number of PSTN trunks in service at the time of the system cut
- Projected number of PSTN trunks required over a 2-year period
- Estimated number of IMT trunks in service at the time of the system cut
- Estimated number of data links (SS7) in service at the time of the system cut
- Projected number of data links (SS7) required over a 2-year period
- Auxiliary systems in service at the time of system cut:
  Voice mail system (initial and projected required capacity)
  Switch manager
  Validation systems (precall or postcall)
  Fraud systems
  Billing systems
  Network management system

10.13 PDSN

The introduction of packet data services not only requires the focus on the radio environment but also on all the supporting elements which comprise the wireless system. Therefore, the following major elements need to be factored into the design of a mobile IP system.

Mobile switching center (MSC)

Base station controller (BSC)

Base station transceiver (BTS)

Packet data service node (PDSN)

Several nodes or elements directly associated with radio elements are listed here. The reason the BSC and BTS are listed in the fixed network design requirements is because connectivity needs to be established between the BTS and the BSC, whether it is via landline services or a microwave link. The BSCs are listed not only because they route packet and voice traffic, which requires a certain link dimensioning, but also because the BSCs can be local or remote to the MSC depending on the ultimate network configuration deployed.

The fixed network design includes not only element dimensioning but also dimensioning the links which connect the various nodes or elements in order to establish a wireless system. Some of the connectivity requirements involve the elements listed next. It is important to note that all elements require some level of connectivity whether it is from the DXX to the MSC or between the voice mail platform and the switch. However, the list which follows involves elements which usually require an external group to interface with while the internal nodes are more controllable.

1. Link between BTS and BSC (usually leased line)
2. Link between BSC and MSC (if remote)
3. BTS/BSC concentration method
4. Connectivity of the PDSN
    a. Router (for internal)
    b. Router (for external)
    c. AAA
    d. Home agent (HA)
    e. SCS and other servers
5. Interconnection to the public and/or private data networks

Note as a general practice that the routers used for the packet data network applications should not be utilized for other functions, like company LAN work.

In reviewing the various connectivity requirements, the PDSN needs to have connectivity with the following major nodes:

1. Radio network
2. Either a public or private data network or both.
3. AAA server
4. DHCP server
5. Service creation platform which contains the configuration, policy, profile, subprovisioning, and monitoring capability

The PDSN is usually connected to the packet data network via a OC3/STM1 or 100baseT connection. The choice of which bandwidth to utilize is determined not only by the proximity of the PDSN to the BSC but also by traffic requirements. In summary the PDSN design is based on many factors including the following basic issues:

1. Number of BSC locations
2. Access type supported (simple IP, mobile IP, etc.)
3. Connectivity between nodes
4. Network performance requirements

10.14 IP Addressing

The issue of IP addressing is important to a mobile IP system design. The inclusion of the IP address scheme in the planning phase of the wireless system is essential to prevent untold problems from occurring. Therefore, the introduction of simple IP and mobile IP with and without a virtual private network (VPN) requires the use of multiple IP addresses for successful transport of the packet services envisioned to be offered. It is, therefore, imperative that the IP addresses used for the network be approached from the initial design phase to ensure a uniform growth that is logical and easy to maintain over the life cycle of the system.

Not only does the introduction of packet data require an IP address scheme for the mobile portion of the system, each of the new platforms introduced needs to have its own IP address or range of IP addresses. Some of the platforms requiring IP addresses are

PDSN

Foreign agent (FA)

HA

Routers

Some of these new devices require the use of private addresses as well as some public addresses. However, since the range of permutations for IP address schemes is so vast and requires a specific look at how the existing network is set up and because of future desires, a generic discussion on IP address schemes is provided.

The IPv4 format is shown in Tables 10.24 and 10.25. IPv6 or IPng is the next generation and allows for QOS functionality to be incorporated into the IP offering. However, this discussion will focus on IPv4 since it is the protocol used today and has legacy transparency of IPv6.

Every device which wants to communicate using IP needs to have an IP address associated with it. The addresses used for IP communication have the following general format.

Network number	Host number
Network prefix	Host number

There are, of course, public and private IP addresses. The public IP addresses enable devices to communicate using the Internet, while private addresses are used for communication in a LAN/WAN Intranet environment. The mobile IP system will utilize both public and private addresses. However, the bulk of the IP addresses will be private in nature and depending on the service offering will be dynamically allocated or static in nature.

You also need to determine how to include 802.11 in order to facilitate mobile IP and corporate LAN/WAN through the wireless network. Tables

10.24 and 10.25 represent the range of public and private IP addresses that can be used. The private addresses will not be recognized on the public Internet system and that is why they are used. Also it is necessary to reuse private addresses within sections of a network. Since the packet system is segregated based on the PDSN, each PDSN can be assigned the same range of IP addresses. Additionally, based on the port involved with the PDSN, the system can be segregated into localized nodes. This allows for reuse of private IP addresses ensuring a large supply of a seemingly limited resource.

To facilitate the IP addressing, the use of a subnet further helps refine the addressing by extending the effective range of the IP address itself. The IP address and its subnet directly affect the number of subnets which can exist and from those subnets the amount of hosts which can also be assigned to that subnet. It is important to note that the IP addresses assigned to a particular subnet include not only the host IP addresses but also the network and broadcast address, as shown in Table 10.26. For example, the 255.255.255.252 subnet which has two hosts requires a total of four IP addresses to be allocated to the subnet, two for the hosts and one for the network and the other for the broadcast address. Obviously as the amount of hosts increases with a valid subnet range, the more efficient the use of IP addresses becomes. For instance, the

TABLE 10.24 Public IP Address

Network address class	Range
A (/8 prefix)	1.xxx.xxx.xxx to 126.xxx.xxx.xxx
B (/16 prefix)	128.0.xxx.xxx to 191.255.xxx.xxx
C (/24 prefix)	192.0.0.xxx to 223.255.255.xxx

TABLE 10.25 Private IP Address

Private network address	Range
10/8 prefix	10.0.0.0 to 10.255.255.255
172.16/16 prefix	172.16.0.0 to 172.31.255.255
192.168/16 prefix	192.168.0.0 to 192.168.255.255

TABLE 10.26 Subnets

Mask	Effective subnets	Effective hosts
255.255.255.192	2	62
255.255.255.224	6	30
255.255.255.240	14	14
255.255.255.248	30	6
255.255.255.252	62	2

255.255.255.192 subnet allows for 62 hosts and utilizes a total of 64 IP addresses. Why not use the 255.255.255.255.192 subnet for everything? This would not be efficient either, so an IP address plan needs to be worked out in advance since it is extremely difficult to change once the system has been implemented.

The following rules apply when developing the IP plan for the system; the same rules are used for any LAN or ISP that is designed. There are four basic questions which help define the requirements:

1. How many subnets are needed presently?
2. How many are needed in the future?
3. What is the number of hosts on the largest subnet presently?
4. What will be the number of hosts on the largest subnet in the future?

You might be wondering why the use of multiple hosts should be factored into the design phase. The reason is that it is possible to have several terminals for a fixed application using a single subscriber unit or fixed unit.

Therefore using these methods an IP plan can be formulated for the wireless company's packet data platforms. It is important to note that the IP plan should not only factor into the design the end customers' needs but also the wireless operators' needs. Specifically the mobile IP operators' needs will involve IP addresses for the following platforms as a minimum. The platforms requiring IP addresses are constantly growing as more and more functionality for the devices is done through signaling network management protocol (SNMP).

- Base stations
- Radio elements
- Microwave point to point
- Subscriber units
- Routers
- ATM switches
- Workstations
- Servers (AAA, HA, FA, PDSN)

The list can and will grow when you tally up all the devices within the network both from a hardware and network management aspect. Many of the devices listed require multiple IP addresses in order to ensure their ability to provide connectivity from points A to B. It is extremely important that the plan follow a logical method. Some wireless network equipment may also require an IP plan that incorporates the entire system and not just pieces.

A suggested methodology is to

1. List out all the major components that used or that could be used in the network over a 5- to 10-year period.

2. Determine the maximum amount of these devices that could be added to the system over 5 to 10 years.
3. Determine the maximum number of packet data users per BSC.
4. Determine the maximum number of packet data users per PDSN.
5. Determine the maximum number of mobile IP users with and without VPN.
6. Determine the maximum number of simple IP users with and without VPN.

The reason for the focus on the amount of simple and mobile IP users lies in the fact that these devices will have the greatest demand for IP addresses due to their sheer volume in the network.

Naturally your particular requirements will be different. However, the concept presented here has been beneficial and should prove useful. If more information is sought on IP address schemes an excellent source for information is available on the Web at www.cisco.com. Regardless of the particular IP address requirements for the wireless network the IP address scheme needs to be identified and included in the wireless planning process.

10.15 Head Count Requirements

For this section the head count for the technical services' group needs to be defined on a monthly basis. The current head count as of the second quarter of 2001 will be the base line. Included with the head count forecast will be planned head counts for the years 2001, 2002, and 2003 (see Table 10.27).

When planning the resource requirement budget, owners should adhere to the head count phasing provided in the targeting exercise, unless revisions are communicated and approved. They should keep the following in mind:

- Look closely at head count ramp-up especially early in the year. Compare the first quarter head count to the latest actual head count to determine whether desired hiring levels are actually achievable.
- Assume head count turnover occurs throughout the year and budget accordingly for hiring expenses.

An organization chart will also be necessary as part of the headcount forecast. The organization chart will need to include all the technical services identified by name, title, and function. If the position is vacant, then use vacant for the name in the organization chart.

10.16 Budgeting

The budgeting plan needs to account for all capital and expense items associated with the network and RF growth plan and identify the total cash requirements needed for the growth plan period. Therefore, for any successful and meaningful growth plan the identification of the capital and operating expenses need to be defined.

TABLE 10.27 Head Count

Headcount	Current	4Q/XX	1Q/XX	2Q/XX	3Q/XX	4Q/XX
Permanent						
RF						
Network						
Outside plant operations						
Inside plant operations						
Implementation						
Miscellaneous						
Outsourced						
RF						
Network						
Outside plant operations						
Inside plant operations						
Implementation						
Miscellaneous						

The actual format for the budgeting issues associated with the plan should correlate with the format used for the technical budget submitted for the time frame this report covers. The budgeting aspects needed to be included are all new capital and expense requirements. The new capital and expense requirements need to be compared against the current budget already submitted and variances identified.

The capital and operating expenses for the budget process will typically be asked for on a monthly basis. The operating expenses should be broken down into monthly increments. However, the capital spending is much harder to define on a monthly basis. Therefore, it is best to list the capital requirements on a quarterly basis.

For example, referenced platform types are identified next. It is important to note that the list of major capital items includes the cost of the fixed network equipment (FNE), including hardware and initial software load, installation, and all associated construction costs.

1. Cell site
2. Towers, TI
3. Switching centers (MSC)
4. Class 5 switches
5. DXX
6. Packet data platforms (PDSN)

7. Voice mail platforms
8. Microwave systems
9. SS7
10. Anciliary switching and packet platforms
11. CALEA and E911 equipment
12. NOC

Operating expenses can and should include all the expenses needed to keep the system running, including the ability to treat the various voice and data services. The treatment usually is associated with the backhaul or interconnection leased line costs. The operating expenses also include software maintenance agreements, even though the software loads may include bug fixes. Other important topics for the operating expenses include ac/dc power, cell and MSC maintenance, landscaping, and HVAC maintenance.

Typically the finance department has a set of guidelines and codes that it would like to have the capital and operating expenses presented in. The inclusion of the finance department format is to eliminate the eventual reformatting of the budget material to match their requirements. This is not to say that you take the finance department's format and blindly use it. However, it is worth including the finance department in the initial and final design process to expedite the effort.

10.17 Final Report

The final report for the network and RF growth plan is the part of the project where all the efforts put forth to date are combined into a uniform document. There are many methods and formats that can be used for putting together a growth plan report. When crafting the final network and RF growth report, it is exceptionally important to remember who your target audience is. The report itself will be used by both upper management and the engineering department to conduct the actual planned network growth.

Therefore, the output of the design effort will be a report that will need to follow a uniform structure which captures the salient issues as well as facilitates the design review process. The outline shown is a recommended format to follow. However, your particular system requirements will in all likelihood require some variations to the proposed structure.

1. Executive summary
2. Introduction
3. Subscriber forecast
4. Expansion or migration plan
5. RF system

6. Circuit switch
7. PDSN
8. Interconnection
9. IP scheme
10. Implementation plan
11. Head count requirements
12. Budget

Regardless of the actual format used, the process is important to follow and essential for success.

10.18 Presentation

When presenting the material to upper management and your fellow engineers, it might be beneficial to craft two versions of the presentation: one for the technical departments as a whole and the other for upper management to view. Both presentations should involve a combination of visual aids and handouts. The visual aids should consist of several free-standing charts depicting the current and future network configuration. Additional visual aids include the use of overhead slides describing the key attributes of the plan. The handouts distributed should reflect the exact same information that is shown in the overhead projections.

The engineering presentation should take about 2 to 4 h to present, depending on whether it is a quarterly update or a yearly plan. The material presented should be of sufficient detail to ensure that all the departments within the technical organization understand the general implications of the information in the report.

It is recommended that a member of each department, usually a member on the growth plan, provide a summary of his or her group's projects planned over the report period. The discussion of the various subplans also needs to include projected start and end dates for each of the topics discussed.

The upper management report is very similar in nature to the presentation given to the technical groups of the company. The primary difference in the presentations is the time frame and emphasis of material. The upper management presentation shouldn't last more than 1 h. Visual aids should be used. It is recommended that the technical aspects take on a very high level approach and not focus on details, unless you have been instructed to do so. The presentation should include a few charts showing the growth trends of the network, critical triggers for the network, new cell sites expected, and cash requirements. The props used for the discussion with technical department should also be presented as background material.

Chapter 11

Organization and Training

The structure of your company's technical department and the quality of its personnel and its continuing education program will directly influence the success or failure of your company.

With this in mind, this chapter addresses two very important organizational aspects: organizational structure with specific responsibility delineated and training requirements for engineering departments.

11.1 Technical Organization's Structure

Many wireless organizations begin centralized at their infancy and at a later date move toward being more decentralized, followed by becoming centralized again. The centralization/decentralization pendulum tends to swing with each management change. The reason is that management attempts to correct an underlying problem by changing the organizational structure.

It can be argued that if an organization has a plan from the beginning that addresses its vision and matches its business plan, then the need for dramatic organizational changes would be minimized. The organization's problems would then largely result from external influences, e.g., the market condition, and not from internal pressures resulting from its structure.

The technical organization can be centralized, distributed, or a blend of these two.

The centralized approach has the advantage of potentially achieving economies of scale by eliminating redundant functions in each of the markets or areas. An example of a centralized function is new technology research for the company. It does not have to be accomplished with every department or division within a company. Having one group lead the effort will ensure uniformity, accountability, and the probability that the direction picked is coordinated between the various departments in the company.

The centralized approach, however, has the disadvantage of being defocused on market requirements. This can come about, for example, if the organization does not have any local knowledge, such as switch port assignments or services actually needed to be offered, about the technical configuration for the network.

The decentralized approach has the advantage of being more market sensitive and flexible than the centralized approach. An example of this flexibility to the market environment would involve the continuous configuration of the network based on voice and data traffic patterns. The idea of the decentralized approach is that the decisions that will affect the market are brought as close to the customer as possible.

The disadvantage with the decentralized approach is that redundant work is performed. The decentralized approach lends itself to localized procedures that foster inefficiencies and a lack of knowledge transport which often leads to the problem being repeated in another market when some simple communication could have conveyed how it might have been avoided. The decentralized approach also does not lend itself to any engineering practice procedures which is essential in the rapidly changing world of wireless communications.

The blended organizational approach takes advantage of both the centralized and decentralized forms. The structure should be such that centralization occurs at corporate headquarters while locality to the markets is done through a distributed method.

Regardless of which approach is taken toward the organizational structure, the span of control (direct reports) for each level of management should not go beyond seven or eight for any organization due to difficulties in managing such an organization for a prolonged period of time.

The technical organization comprising engineering, construction (implementation), and operations should be structured based on functional requirements. These need to be the driving force behind an organization's structure, not current personnel's views.

11.2 Technical Organization's Departments

Before delving further into the issue of centralized, decentralized, or blended organization structures the various technical departments need to be defined. The proposed list of technical departments is evident in any wireless mobile system; it just may be labeled differently. When looking over the list of functions, the names are not as important as the roles. Therefore, when matching your organization with this configuration keep in mind that there is no one organization which fits all situations due to skills and other legacy issues.

However, the key roles listed in the proposed technical organization involve a multitude of disciplines, which comprise a well-rounded organization. The disciplines needed for the organization are

1. Technical directors
2. New technology
3. Budgetary

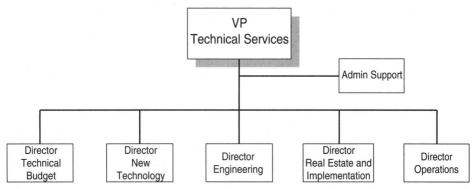

Figure 11.1 VP technical services organization.

The key roles listed in the proposed technical organization are budgetary, new technology, engineering, real estate and implementation, and operations directorates. (See Fig. 11.1.) The general functions of each of the directorates are

1. *Budget director.* Responsible for the capital and expense tracking, variance reporting, forecasting, and purchase order handling for the entire technical organization.

2. *Engineering director.* Responsible for the design of the network and the technical performance aspects.

3. *Real estate and implementation director.* Responsible for the acquisition, leases, civil work, and construction of the various projects put forth by engineering.

4. *Operations director.* Responsible for ensuring that the equipment installed in the network is maintained and operating at its peak performance. Also responsible for the operation of the network operations center (NOC) which can consist of a singular location or multiple NOCs.

5. *New technology director.* Responsible for ensuring that new technology is constantly being pursued that will meet the engineering, operations, implementation, and marketing requirements.

11.3 Engineering Organization

Regardless of whether it is part of a centralized or decentralized hierarchy, the engineering organization needs to ensure that the network is operating at the desired performance criteria. It needs to take the leadership role in defining the overall direction of the technical community. The engineering organization and responsibilities should be driven by market requirements.

The engineering department is responsible for the planning and design of the communication system from an RF and network perspective. Figure 11.2 depicts an organizational structure that can be used to support either an individual

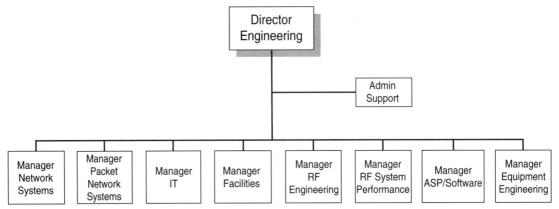

Figure 11.2 Engineering department structure.

market or a region which includes several markets. It is appropriate for worldwide, regional, country, or city levels.

Many variants to the proposed engineering organizational structure are possible and do currently exist. However, the structure is basically the same no matter how many systems or subsystems the organization needs to be responsible for. When looking at a possible variant to this proposed organization, the individual responsibilities of the groups need to be properly defined.

The *network systems* group is responsible for the architectural engineering of the network growth as well as for evaluating new network designs and performing network troubleshooting. It plans the switch dimensioning and module growth and determines when a new switch is needed for the network with its proposed location. The group is responsible for the voice network dimensioning and all value-added systems that are adjuncts to the voice network, e.g., voice mail. This group is also directly responsible for recorded announcement assignments associated with any network. Additionally the group is responsible for all product development design aspects for the voice-related services for the network that are derived by marketing.

The *packet data network* portion of the engineering group is responsible for all data packet and protocol issues associated with the data network. The group is responsible for the dimensioning, layout, and functionality of the PDSN and other packet-related services, e.g., CDPD. Some of the platforms involved include the routers, AAA, and ATM switches besides the STPs in support of the CCS/SS7. Additionally, some of the protocol issues for the data network include IS-41, integrated services (digital network) user part (ISUP), advanced intelligence network (AIN), and SS7. This group plays an important and pivotal role for dealing with out-of-market interfaces for the network. Additionally the group is responsible for all product development design aspects for the packet-related services for the network that are derived by marketing.

The *IT group* is responsible for the LAN/WAN configurations that are utilized throughout the corporation. This group is traditionally located within its own organization, but with the convergence of IP, ATM, and telephony, associated platforms need to be put under engineering control.

The *facilities management* group is responsible for all aspects associated with facilities and utilities management. It is responsible for ensuring the trunk designs for the network are adequate and properly dimensioned for growth. This group also orders the actual facilities for the network and ensures they arrive at the predetermined time for the project. Further, this group is responsible for dimensioning the DACs in the network and interfacing all the issues associated with the IXEs.

The *RF engineering* group is responsible for the macrocell and microcell planning and design efforts. The RF engineering group is also responsible for the RF planning and design efforts including the PPP links. Additional responsibilities include frequency management of the network, intersystem coordination, and regulatory (FCC) and aeronautical compliance (FAA) for new sites. There can also be multiple mini-project teams within a wireless mobile system. This group is also responsible for designing into the existing network multiple RF technology platforms. However, as the systems mature the need for a strong capital build program will not be as prevalent as in the past.

The *system performance* group is responsible for the overall performance of the existing RF network of the system. This group is responsible for ensuring that all the new cell sites in the network perform at a specified quality control standard. The group must also interface with customer care for problem ticket resolution.

The *ASP/software engineering* group is responsible for all the software loads put into the network and their impact. The group is responsible for leading any software first office application (FOA) for the switches and cell sites and for overseeing all major feature introductions into the network. The software engineering group also produces all the data translation information that is used for the circuit switches, base stations, and SLAs. In addition, this group is responsible for validation of the billing system whenever there is a software load change for the cell sites or switch and for system software audits to ensure that the software and translations are consistent across multiple nodes in the network and that all unwanted data are removed. The group is responsible for all the application-specific software utilized in the system and for customer-specific requirements, like 802.11 connectivity. It ensures uniformity of applications for ease of expansion, troubleshooting, and cost savings.

Equipment engineering is responsible for the physical layout and equipment requirements for the MSC, remote nodes, cell sites, ancillary platforms, and customer premise equipment (if required). Its role is to ensure configuration management in the network.

It is strongly recommended that the engineering organization not be broken down into a more decentralized function than that proposed here. Specifically the current trend is to operate in localized teams for an organization. The major disadvantage with trying to operate an engineering department through use of localized teams is that there could be a loss of configuration management and economies of scale.

It is also strongly recommended that the RF engineering and system performance groups operate under different managers because of the constant demand for resources that occurs with the build program of any network. It is essential for the short- and long-term health of the network that these two groups remain independent of each other so there is a check and balance system in the network design. This is imperative because of the large amount of capital and expense dollars expended on a yearly basis.

Figure 11.3 shows the next layer down within the engineering directorate. Naturally some of the functions may not be necessary depending on the services offered, e.g., mobile IP if only voice services are offered.

11.4 Operations

The operations directorate is responsible for managing the network operations center (NOC), maintaining the RF, IP, ATM, and circuit switch components as well as the potential customer equipment used to support corporate LAN/WAN applications. The difference with mobile IP is that if it is pushed to support LAN applications for 802.11, it may require the installation of remote equipment which needs support and occasional upgrades, sometimes requiring site visits.

The operations department shown in Fig. 11.4 is meant to provide a general guide as to how the organization can and should be structured with the advent of mobile IP services. It is important to note that if you are only providing a pipe, then the customer application section is not really relevant. However, if product and service differentiation is the desired goal, you should form a group to support the customer's LAN/WAN needs.

The NOC manager and the supporting staff will be responsible for monitoring the network components and dispatching the requisite staff to respond. The RF technician manager will be responsible for the operation and maintenance of not only the cell sites (BTS), but the customer terminals (802.11 nodes) and many microwave backbone systems.

The IP/ATM technology manager will be responsible for the operation and maintenance and implementation of engineering work orders for those platforms. The class 5 (MSC) manager will be responsible for the operation and maintenance and implementation of engineering work orders for those platforms.

The customer application manager will be responsible for upgrading the customer terminals for corporate LAN/WANs at the customer facilities, if applicable, and providing technical support to the customers for application and connectivity problems. The next layer down in the operations organization is shown in Fig. 11.5.

11.5 Real Estate and Implementation

The real estate and implementation directors are responsible for securing, constructing, and commissioning the designs put forth by engineering. (See Fig. 11.6.) The groups in the directorate are site acquisition, lease management, implementation manager, and bid manager.

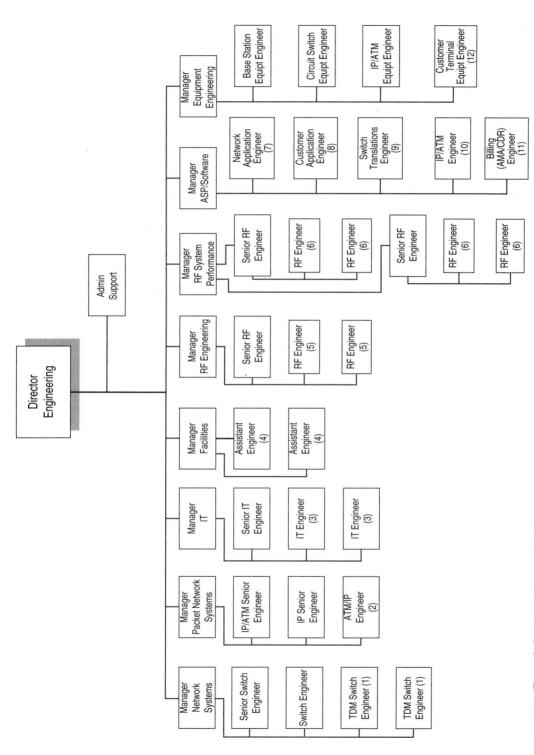

Figure 11.3 Expanded engineering organization.

549

550 Chapter Eleven

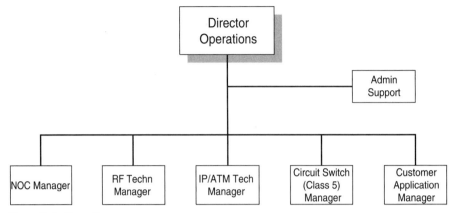

Figure 11.4 Operations.

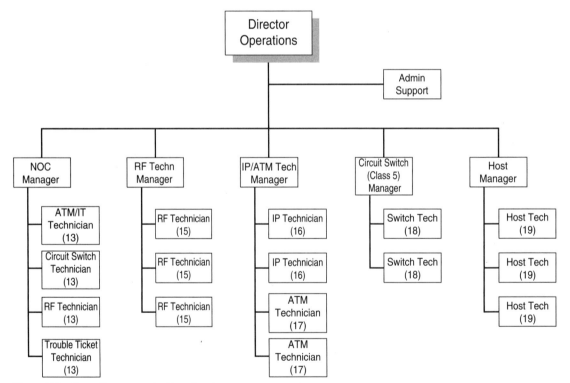

Figure 11.5 Operations organization structure.

The site-acquisition manager is responsible for obtaining the cell site and switch and remote node concentration locations as well as securing access for installations for inbuilding applications. The lease manager is responsible for managing the multitude of leases that will exist as the system grows. The implementation manager is responsible for physically constructing the various

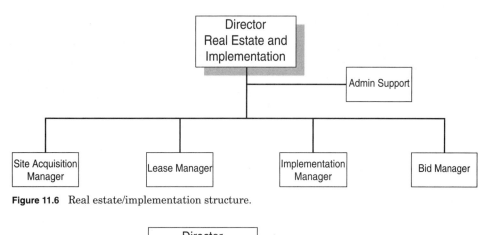

Figure 11.6 Real estate/implementation structure.

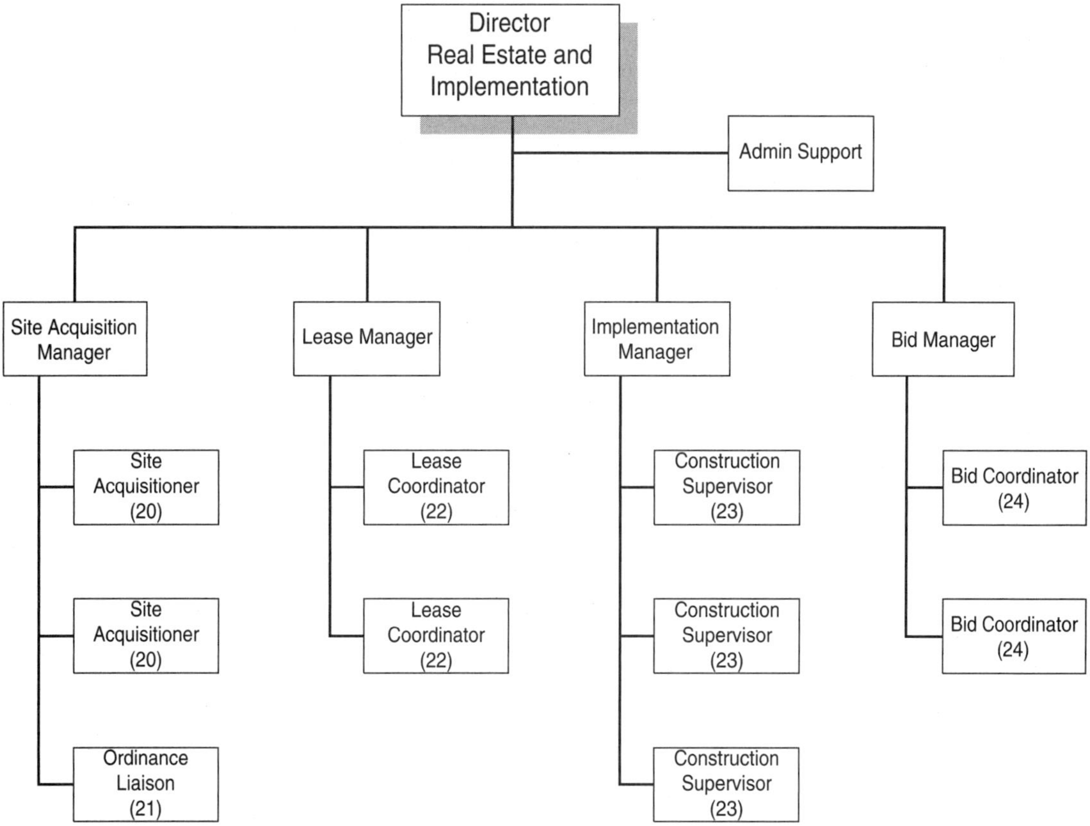

Figure 11.7 Real estate and implementation expanded organization.

facilities and commissioning them for conformance to engineering guidelines. The bid manager's job is to issue, manage, and award the numerous bids that will be issued to subcontractors and vendors.

The next layer down in the real estate and implementation directorate is shown in Fig. 11.7.

11.6 New Technology and Budget Directorates

Figure 11.8 is an example of the directorates for new technology and budget control for a wireless mobile system. The budgetary roles involve tracking, variance resolution and forecasting the capital and expense budgets for each of the various groups under this organization. The budgetary group in the organization plays a vital role in helping secure the necessary funding for various projects.

The budget staff listed in the chart is responsible for tracking, variance reporting, forecasting, and purchase order handling for the entire technical organization.

The technical support group's role is to establish, track, and report on the individual network's performance. Its role is to be a central clearinghouse for technical performance requirements leading to a standard set of criteria for all to use. This group should also ensure a best practice approach is accomplished between all the markets allowing for good ideas and procedures to be shared.

11.7 Head Count Drivers

Head count drivers, or rather ratios, should be used to dimension any organization, especially the technical community within a wireless mobility system. The drivers will need to be reviewed on a regular basis with the advent of new technology, better management systems, service changes, and, of course, the market conditions which dictate cash flow for the organization.

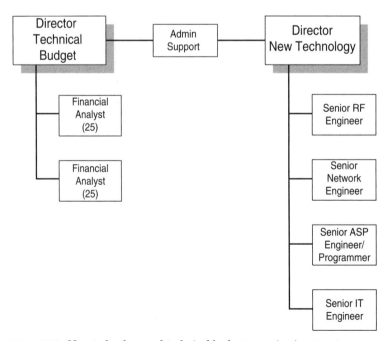

Figure 11.8 New technology and technical budget organization structure.

Ordinarily the head count drivers determine how large an organization should be as a function of success, since the other direction is not a desired alternative. The goal is to determine when and where the hiring needs to take place. Obviously with every new hire additional burdens are placed on the organization, on administration, and, of course, on space and support material the new hire will need for effectively doing the job he was hired to perform.

The personnel drivers for technical head count involve the following major items:

Number of customers

Traffic volume

Services offered

Number of base stations

Number of corporate LAN/WANs

This list is not all-inclusive but does serve as a good starting point from which to determine your staffing requirements. Revenue, of course, is a major driver for the wireless mobile organization as a whole, but the technical organization is typically a cost center and not a revenue center. Since the technical organization is primarily a cost center, the drivers shown above are cost oriented.

Now we will look at the particular head count drivers. Referring back to the various organization structures shown in Figs. 11.3, 11.5, and 11.7, the drivers listed in Table 11.1 can be used. Obviously the drivers are dependent upon the five basic items listed previously but the example that follows will help structure the formulation of the drivers and the subsequent organization structure.

The headcount drivers can be used to dimension the initial and future organization. The head count data then can be entered into the operating expense part of the budget. Of course, some of the head count, associated with design and implementation, should be allocated to the capital budget.

11.8 Hiring

There are several methods that are used for securing personnel for an organization. The paths for hiring can take several directions, from relying on the entire staffing from internal sources to that of almost complete outsourcing. But no matter which method or variant to the staffing method is chosen, hiring will need to take place in order to ensure that the finite resources are properly allocated.

The selection of skilled staff is indeed an arduous one since there is little time available to properly select the candidate and an improperly selected candidate can cause undue stress, which he or she was initially intended to relieve.

TABLE 11.1 Technical Services Head Count Drivers

Driver no.	Title	Dimensioning (per)
	Engineering	
1	Circuit switch engineer	500 ports
2	ATM/IP engineer	500 VPI/VCIs
3	IT engineer	75 internal company workstations
4	Assistant engineer	500 leased lines
5	RF engineer	25 new base stations
6	RF performance engineer	100 base stations
7	Network application engineer	200,000 subscribers
8	Customer application engineer	500 LAN/WAN (802.11)
9	Switch translations engineer	150,000 subscribers
10	IP/ATM engineer	500 VPI/VCIs
11	Billing (AMA/CDR) engineer	250,000 subscribers
12	Customer terminal engineer	1000 LAN/WAN (802.11)
	Operations	
14	NOC technician	1.25 per shift
15	RF technician	25 base stations (macro/micro) and 50 picosites
16	IP technician	200 customer terminals (802.11) and 40 internal company workstations
17	ATM technician	10 BSCs
18	Switch technician	200 ports
19	Host technician	100 host terminals
	Real estate and implementation	
20	Site acquisitioner	50 base stations builds per year
21	Ordinance liaison	Per market
22	Lease coordinator	250 base stations and 500 host terminals
24	Construction supervisor	25 base station builds per year
25	Bid coordinator	Per market

Note: VPI = virtual path identifier, VCI = virtual channel identifier.

Therefore, in most operations hiring qualified individuals is essential but should be approached with caution. We have hired both successfully and unsuccessfully regardless of the level of preemployment interviewing and screening done. The simple process we have used is as follows:

1. Establish a written job function.
2. Realistically assess the skills obtainable with the salary and benefits offered.
3. Understand the current labor market.

4. Solicit the position.
   - Colleagues
   - Paper
   - Headhunter (resume cycler): fee can be negotiated and should be based on the candidate's staying for 12 months (otherwise no payment will be made)
5. Establish a preinterview screening process (two people reviewing resumes as a minimum).
6. Narrow the list of possible candidates and schedule a time for interviews (4-h per interview).
7. Conduct brief phone interview and establish time for next-level interviews.
8. Establish interview screening method.
   - Three or four interviewers
   - Define interview grading matrix
   - Have 10 predefined questions per interview
   - Give each interviewer 1 h to interview
   - Review grading matrix results (that of interviewers as well as next-level management)
9. Check references (if not previously done).
10. Bring candidate back for second round of interviews and have them meet the individuals they would work with.
11. If the candidate meets all requirements and seems to be a fit, make an offer.
12. Establish orientation training and set up a location to sit and provide the new-hire with associated equipment needed to perform job function.

Obviously there are other steps involved and they need to be coupled with human resources, but the process is meant to minimize incorrect hires.

Depending on the urgency and duration of the staffing needs, the use of outsourcing may have been required or desired. Outsourced personnel can also be transitioned to permanent hires since the positions are filled utilizing a normal interview process. This approach is the most economical method for obtaining a working staff right away, but know that some organization rifts may occur if management is not involved when these new hires are brought on board permanently.

11.9 Smart Outsourcing

Smart outsourcing involves utilizing outplacement firms, commonly referred to as consultants, for the purpose of providing the necessary workforce to meet the short-term staffing goals of the company.

We refer to utilizing outsourcing as *smart outsourcing* only if there is a plan of how to manage this resource. Often the outsourcing is loosely managed or not managed at all. Without a proper outsourcing plan to manage the temporary or unique skilled resources, you are effectively managing by the discovery method, i.e., you do not know what you will really get.

The key to smart outsourcing is to utilize selected consulting and outplacement firms and establish a long-term relationship with them. Smart outsourcing should be part of your overall organization plan every year, depending on the level of outsourcing you want to utilize. The amount of outsourcing is largely driven by constraints either in building up a permanent staff or limitations in hiring the required personnel in the time at hand. Also, smart outsourcing can be used to keep the existing staff focused on ensuring that revenue, i.e., the customers and network, is being optimized instead of, for example, chasing new technology issues or generating RFPs.

Smart outsourcing should not be used as a replacement for hiring entry-level staff. If this is the case then in all likelihood the use of outsourcing is misapplied. However, outsourcing can be used to fill the entry-level positions for a transitional period, which needs to have a closure date, while permanent staff is sought.

Outsourced technical resources can augment the existing technical community within a wireless operation in the following areas:

Training

Design reviews

Special projects

Request for proposal (RFP) generation and analysis

Staffing shortages

Mentoring program

Smart outsourcing should be part of your organization plan, but it needs to be managed to ensure the resource is best used to benefit the organization both from a short-term and long-term aspect.

11.10 Training

Technical training for any organization is a difficult task since there is never enough time to ensure proper training of personnel. This section lists proposed general topics that should be utilized by the various engineering departments to ensure that there is a minimum level of formal training for all the groups.

It is strongly recommended that every new employee and existing employee be assigned a mentor for guidance. With the demand for immediate results for any issue the use of a mentor program as an augmentation to formal training will expedite the training of personnel. The selection of who is qualified to mentor individuals needs to be taken with extreme caution to ensure that bad design practices are not propagated to new members of the organization.

The training proposed for the engineering organization is divided in three sections, management, RF, and network. The rationale behind not selecting individual training programs for each of the individual organizations listed above is multifaceted. The first rationale is that there needs to be a certain degree of cross training between the major groups in engineering. The second rationale is that every organization has variances and individual market requirements.

The training that takes place for all the members of a technical organization should involve around 80 h of training per year. The 2-week training commitment is necessary for any technical organization to remain current and ensure its personnel are adequately trained. It is strongly suggested that you employ a training deconflict system to ensure that everyone receives training in the year but that personnel outages caused by training are not excessive.

The proposed training deconflict schedule should list all the training dates expected to take place for the employee and the department as a whole for the entire year. It is recommended that the initial deconflict schedule list only the months during which training is to take place, not exact dates. It is also important to list the personnel vacation plans, conferences, and major project milestone dates along with the training dates.

The following is a brief listing of suggested technical material that needs to be taken by various major subgroups within engineering. The list does not include specific courses required by your company like orientation and various diversity classes. It also does not define the number of classroom hours needed for each topic or the repeat interval for technical material. It is important to remember that certain courses should be required to be retaken at regular intervals to ensure a minimum level of competency is maintained at all times.

Management

Basic wireless communications	Fraud management
Network fundamentals	Statistics theory
Telephony call processing	Traffic theory
Switch architecture	Operations and maintenance
Project management	Disaster recovery
Interview techniques	SONET/SDH
Presentation techniques	Tariffs
General management	Budget training
Digital radio design	Interconnect
CDMA/GSM/GPRS/iDEN/IS-136	E-commerce/web design
Network design	VoIP
RF design	Real estate acquisition
IPv4 and IPv6	Packet switch design (IP and ATM)
SS7	

Network engineering

Network fundamentals	Traffic theory
Basic wireless communications	Statistics theory
Cellular call processing	Network services

Switch architecture
Project management
Interview techniques
Presentation techniques
Interconnect
SS7
ATM network design
Network design
RF design
DACS
PCM and adaptive PCM (ADPCM)
Voice mail
Translations switch
SONET/SDH
Call processing algorithms
Equipment grounding

Network architecture
ISDN (BRI/PRI)
LAN/WAN topology
Wireless PBX (WPBX)
Disaster recovery
Numbering plans
E-commerce/web design
AIN
Fiber-optics
Network maintenance
Switch maintenance
MFJ/descent decree
Tariffs
Data and voice transport
Performance troubleshooting

RF engineering

Basic wireless communications
CDMA/GSM/GPRS/iDEN/IS-136
Project management
Interview techniques
Presentation techniques
WPBX
Network design
Network architecture
Traffic theory
Statistics theory
Base site installation
Base site maintenance
Switch architecture

Digital radio design
Microwave system design
Presentation and speaking skills
Disaster recovery
Interconnect
Frequency planning
RF design
Performance troubleshooting
EMF compliance
Antenna theory
Grounding and lightning protection
Real estate acquisition

Operations

Basic wireless communications
LAN/WAN topology
Network maintenance
Base site installation
Base site maintenance

Network architecture
Fiber-optics
Switch maintenance
Grounding
Performance troubleshooting

Every employee, new and old, for a department needs to have a training program. The training program is more of a development program with the desired intention of helping to improve the skill sets of the individual. In addition to improving the skill sets for the individual the department as a whole will be improved through higher efficiency levels obtained with the new skill sets.

When crafting and authorizing any training program, it is essential that the subject matter is relevant to the employee's training program and job function. However, when a new employee arrives, it is imperative that he be given some guidance to help ease the transition period. Recommended items to include in establishing a new-hire training program are listed here. The details for each section will need to be tailored for the particular job function but the general gist of the program can be easily extracted here.

Objective. Define here what the stated objective is for the training program and its duration.

Description of job. Describe the actual job.

Reports. Describe the various reports needed to be generated by the individual and when they are due and to whom.

Reading material. Briefly describe the suggested reading material the person should have in her personal technical library. It is imperative in this section that the source locations for the various documents be spelled out.

Training program. Put together a training program for the individual. The mentor assigned to the person also needs to be identified.

Assigned project. List the particular projects assigned to the individual and his deliverables, with a time line.

Appendix A

Erlang B Grade of Service

Channels	1	1.5	2	3	5
1	0.01	0.012	0.0204	0.0309	0.0526
2	0.153	0.19	0.223	0.282	0.381
3	0.455	0.535	0.602	0.715	0.899
4	0.869	0.992	1.09	1.26	1.52
5	1.36	1.52	1.66	1.88	2.22
6	1.91	2.11	2.28	2.54	2.96
7	2.5	2.74	2.94	3.25	3.74
8	3.13	3.4	3.63	3.99	4.54
9	3.78	4.09	4.34	4.75	5.37
10	4.46	4.81	5.08	5.53	6.22
11	5.16	5.54	5.84	6.33	7.08
12	5.88	6.29	6.61	7.14	7.95
13	6.61	7.05	7.4	7.97	8.83
14	7.35	7.82	8.2	8.8	9.73
15	8.11	8.61	9.01	9.65	10.6
16	8.88	9.41	9.83	10.5	11.5
17	9.65	10.2	10.7	11.4	12.5
18	10.4	11	11.5	12.2	13.4
19	11.2	11.8	12.3	13.1	14.3
20	12	12.7	13.2	14	15.2
21	12.8	13.5	14	14.9	16.2
22	13.7	14.3	14.9	15.8	17.1
23	14.5	15.2	15.8	16.7	18.1

(Continued)

Channels	1	1.5	2	3	5
24	15.3	16	16.6	17.6	19
25	16.1	16.9	17.5	18.5	20
26	17	17.8	18.4	19.4	20.9
27	17.8	18.6	19.3	20.3	21.9
28	18.6	19.5	20.2	21.2	22.9
29	19.5	20.4	21	22.2	23.8
30	20.3	21.2	21.9	23.1	24.8
31	21.2	22.1	22.8	24	25.8
32	22	23	23.7	24.9	26.7
33	22.9	23.9	24.6	25.8	27.7
34	23.8	24.8	25.5	26.8	28.7
35	24.6	25.6	26.4	27.7	29.7
36	25.5	26.5	27.3	28.6	30.7
37	26.4	27.4	28.3	29.6	31.6
38	27.3	28.3	29.2	30.5	32.6
39	28.1	29.2	30.1	31.5	33.6
40	29	30.1	31	32.4	34.6
41	29.9	31	31.9	33.4	35.6
42	30.8	31.9	32.8	34.3	36.6
43	31.7	32.8	33.8	35.3	37.6
44	32.5	33.7	34.7	36.2	38.6
45	33.4	34.6	35.6	37.2	39.6
46	34.3	35.6	36.5	38.1	40.5
47	35.2	36.5	37.5	39.1	41.5
48	36.1	37.4	38.4	40	42.5
49	37	38.5	39.3	41	43.5
50	37.9	39.2	40.3	41.9	44.5
51	38.8	40.1	41.2	42.9	45.5
52	39.7	41	42.1	43.9	46.5
53	40.6	42	43.1	44.8	47.5
54	41.5	42.9	44	45.8	48.5
55	42.4	43.8	44.9	46.7	49.5
56	43.3	44.7	45.9	47.7	50.5
57	44.2	45.7	46.8	48.7	51.5
58	45.1	46.6	47.8	49.6	52.6
59	46	47.5	48.7	50.6	53.6
60	46.9	48.4	49.6	51.6	54.6

Channels	1	1.5	2	3	5
61	47.9	49.4	50.6	52.5	55.6
62	48.8	50.3	51.5	53.5	56.6
63	49.7	51.2	52.5	54.5	57.6
64	50.6	52.2	53.4	55.4	58.6
65	51.5	53.1	54.4	56.4	59.6
66	52.4	54	55.3	57.4	60.6
67	53.4	55	56.3	58.4	61.6
68	54.3	55.9	57.2	59.3	62.6
69	55.2	56.9	58.2	60.3	63.7
70	56.1	57.8	59.1	61.3	54.7

Appendix B

Erlang C Grade of Service

N	0.01	0.02	0.05
1	0.01	0.02	0.02
2	0.15	0.21	0.34
3	0.43	0.55	0.79
4	0.81	0.99	1.32
5	1.26	1.50	1.91
6	1.76	2.05	2.53
7	2.3	2.63	3.19
8	2.87	3.25	3.87
9	3.46	3.88	4.47
10	4.08	4.54	5.29
11	4.71	5.21	6.02
12	5.36	5.90	6.76
13	6.03	6.60	7.51
14	6.7	7.31	8.27
15	7.39	8.04	9.04
16	8.09	8.77	9.82
17	8.8	9.51	10.61
18	9.52	10.25	11.40
19	10.24	11.01	12.20
20	10.97	11.77	13.00
21	11.71	12.53	13.81
22	12.46	13.30	14.62
23	13.28	14.08	15.43

(Continued)

N	0.01	0.02	0.05
24	13.96	14.86	16.25
25	14.72	15.65	17.08
26	15.49	16.44	17.91
27	16.26	17.23	18.74
28	17.03	18.03	19.57
29	17.81	18.83	20.41
30	18.59	19.64	21.25
31	19.37	20.45	22.09
32	20.16	21.26	22.93
33	20.95	22.07	23.78
34	21.75	22.89	24.63
35	22.55	23.71	25.48
36	23.35	24.53	26.34
37	24.15	25.35	27.19
38	24.96	26.18	28.05
39	25.77	27.01	28.91
40	26.58	27.84	29.77

Appendix C

PCS 1900 GSM Channel Chart

A1	B1	C1	A2	B2	C2	A3	B3	C3
\multicolumn{9}{c}{A Block}								
512	513	514	515	516	517	518	519	520
521	522	523	524	525	526	527	528	529
530	531	532	533	534	535	536	537	538
539	540	541	542	543	544	545	546	547
548	549	550	551	552	553	554	555	556
557	558	559	560	561	562	563	564	565
566	567	568	569	570	571	572	573	574
575	576	577	578	579	580	581	582	583
584	585							
				B Block				
611	612	613	614	615	616	617	618	619
620	621	622	623	624	625	626	627	628
629	630	631	632	633	634	635	636	637
638	639	640	641	642	643	644	645	646
647	648	649	650	651	652	653	654	655
656	657	658	659	660	661	662	663	664
665	666	667	668	669	670	671	672	673
674	675	676	677	678	679	680	681	682
683	684	685						
				C Block				
736	737	738	739	740	741	742	743	744
745	746	747	748	749	750	751	752	753
754	755	756	757	758	759	760	761	762

(*Continued*)

A1	B1	C1	A2	B2	C2	A3	B3	C3
\multicolumn{9}{c}{C Block (*continued*)}									
763	764	765	766	767	768	769	770	771
772	773	774	775	776	777	778	779	780
781	782	783	784	785	786	787	788	789
790	791	792	793	794	795	796	797	798
799	800	801	802	803	804	805	806	807
808	809	810						
\multicolumn{9}{c}{D Block}									
586	587	588	589	590	591	592	593	594
595	596	597	598	599	600	601	602	603
604	605	606	607	608	609	610		
\multicolumn{9}{c}{E Block}									
686	687	688	689	690	691	692	693	694
695	696	697	698	699	700	701	702	703
704	705	706	707	708	709	710		
\multicolumn{9}{c}{F Block}									
711	712	713	714	715	716	717	718	719
720	721	722	723	724	725	726	727	728
729	730	731	732	733	734	735		

Bibliography

3GPP2 A.R0003, "Abis Interface Technical Report for cdma2000 Spread Spectrum Systems," Dec. 17, 1999.

3GPP2 A.S0001-A, "Access Network Interfaces Interoperability Specification," Nov. 30, 2000.

3GPP2 A.S0004, "Tandem Free Operation Specification," Nov. 8, 2000.

3GPP2 C.S0007-0, "Direct Spread Specification for Spread Spectrum Systems on ANSI-41 (DS-41) (Upper Layers Air Interface)," June 9, 2000.

3GPP2 C.S0008-0, "Multi-Carrier Specification for Spread Spectrum Systems on GSM MAP (MC-MAP) (Lower Layers Air Interface)," June 9, 2000.

3GPP2 C.S0009-0, "Speech Service Option Standard for Wideband Spread Spectrum Systems."

3GPP2 C.S0024, "cdma2000 High Rate Packet Data Air Interface Specification," Oct. 27, 2000.

3GPP2 C.S0025, "Markov Service Option (MSO) for cdma2000 Spread Spectrum Systems," Nov. 2000.

3GPP2 S.R0005-A, "Network Reference Model for cdma2000 Spread Spectrum Systems," Dec. 13, 1999.

3GPP2 S.R0021 Version 1.0, "Video Streaming Services—Stage 1," July 10, 2000.

3GPP2 S.R0022, "Video Conferencing Services—Stage 1," July 10, 2000.

3GPP2 S.R0023 Version 2.03, "High-Speed Data Enhancements for cdma2000 1x—Data Only," Dec. 5, 2000.

3GPP2 S.R0024 Version 1.0, "Wireless Local Loop," Sept. 22, 2000.

3GPP2 S.R0026, "High-Speed Data Enhancements for cdma2000 1x—Integrated Data and Voice," Oct. 17, 2000.

3GPP2 S.R0027, "Personal Mobility," Dec. 8, 2000.

3GPP2 S.R0032, "Enhanced Subscriber Authentication (ESA) and Enhanced Subscriber Privacy (ESP)," Dec. 6, 2000.

3PP2 P.S0001-A-1 Version 1.03, "Wireless IP Network Standard," Dec. 15, 2000.

Agilent, "Designing and Testing cdma2000 Base Stations," Application Note 1357.

Agilent, "Designing and Testing cdma2000 Mobile Stations," Application Note 1358.

Aidarous, S., and T. Plevyak, *Telecommunications Network Management into the 21st Century,* IEEE, 1994.

ANSI T1.111-1988, "Signaling System Number 7 (SS7)—Message Transfer Part (MTP)."

ANSI T1.115-1990, "Signaling System Number 7 (SS7)—Monitoring and Measurements for Networks."

ANSI T1.116-1990, "Signaling System Number 7 (SS7)—Operations, Maintenance and Administration Part (OMAP)."

The ARRL 1986 Handbook, 63d ed., American Radio Relay League, 1986.

The ARRL Antenna Handbook, 14th ed., American Radio Relay League, 1984.

Barron, Tim, "Wireless Links for PCS and Cellular Networks," *Cellular Integration,* Sept. 1995, pp. 20–23.

Bates, Regis J., *GPRS,* McGraw-Hill, 2001.

Bates, Regis J., and Donald W. Gregory, *Voice and Data Communications Handbook,* 4th ed., McGraw-Hill, 2001.

Bellcore, "Bell Communications Research Specification of Signaling System Number 7," TR-NWT-000246, Issue 2, Vol. 1, June 1991.

Bellcore, "Bell Communications Research Specification of Signaling System Number 7," TR-NWT-000246, Issue 2, Vol. 2, June 1991.

Bellcore, "Digital Network Synchronization Plan," GR-436-CORE.

Breed, Gary, "Receiver Basics, Part 1: Performance Parameters," *RF Design,* Feb. 1994, pp. 48–50.

Breed, Gary, "Receiver Basics, Part 2: Fundamental Receiver Architectures," *RF Design,* March 1994, pp. 84–89.

Brewster, R. L., *Telecommunications Technology,* Wiley, 1986.
Brodsky, Ira, "3G Business Model," *Wireless Review,* June 15, 1999, p. 42.
Brown, Bruce, "802.11a, 802.11b, and 802.11g Revisited," *ExTremeTech.Com,* Feb. 19, 2002.
Brown, Dan, Steve Joiner, Joel Rosson, and Toshiaki Satake, "MT-RJ: The Next Generation of Fiber-Based Networks," *Fiberoptic Product News,* Aug. 1998.
Carlson, A. Bruce, Paul B. Crilly, and Janet C. Rutledge, *Communication Systems,* 4th ed., McGraw-Hill, 2002.
Carmeli, Alon, "Gaining Momentum Using Two-Way Coax Plant for Data," *Communication Engineering & Design,* April 1997, p. 54.
Carr, J. J., *Practical Antenna Handbook,* TAB Books–McGraw-Hill, 1989.
Channing, Ian, "Full Speed GPRS from Motorola," *Mobile Communications International,* Oct. 1999, p. 6.
Chorafas, Dimitris N., *Telephony: Today and Tomorrow,* Prentice Hall, 1984.
Collins, Daniel, *Carrier Grade Voice over IP,* McGraw-Hill, 2001.
Daniels, Guy, "A Brief History of 3G," *Mobile Communications International,* Oct. 1999, p. 106.
DeGarmo, E. Paul, et al., *Engineering Economy,* 10th ed., Prentice Hall, 1997.
DeRose, James F., *The Wireless Data Handbook,* 4th ed., Wiley, 1999.
Dixon, Robert C., *Spread Spectrum Systems,* 3d ed., Wiley, 1994.
Engineering and Operations in the Bell System, 2d ed., AT&T Bell Laboratories, 1983.
Fink, Donald G., and H. Wayne Beaty (eds.), *Standard Handbook for Electrical Engineers,* 14th ed., McGraw-Hill, 2000.
Fink, Donald G., and Donald Christiansen (eds.), *Electronics Engineers Handbook,* 3d ed., McGraw-Hill, 1989.
Fulton, Fiona, "GPRS' Next Challenge: Billing," *Telecommunications,* Dec. 1999, pp. 65–66.
Goralski, Walter, *ADSL and DSL Technologies,* 2d ed., McGraw-Hill, 2002.
Gull, Dennis, "Spread-Spectrum: Fool's Gold?" *Wireless Review,* Jan. 1, 1999, p. 37.
Harte, Hoenig, McLaughlin, and Kikta, *CDMA IS-95 for Cellular and PCS,* McGraw-Hill, 1996.
Harter, Betsy, "Putting the C in TDMA?" *Wireless Review,* Jan. 2001, pp. 29–34.
Hartman, Dave, and Ted Rabenko, "Allocating Resources over a DOCSIS Network," *Communications Engineering & Design,* Dec. 1998, p. 76.
Held, Gil, *Voice & Data Interworking,* 2d ed., McGraw-Hill, 2000.
Holma, Harri, and Antti Toskala, *WCDMA for UMTS,* 2d ed., Wiley, 2002.
Hudec, Premysl, "Noise in CATV Networks," *Applied Microwave & Wireless,* Sept. 1997, p. 22.
IEEE Standard 184–1969, "IEEE Test Procedures for Frequency Modulated Mobile Communication Receivers."
Iler, David, "Telephony over Cable," *Communications Engineering & Design,* Dec. 1998, p. 22.
Jacobs, Jeffrey R., and Patrick C. Brown, "Life-Cycle Cost Savings in HFC Networks," *Fiberoptic Product News,* Aug. 1998, p. 55.
Jakes, William C. (ed.), *Microwave Mobile Communications,* Wiley, 1974.
Johnson, R. C., and H. Jasik, *Antenna Engineering Handbook,* 2d ed., McGraw-Hill, New York, 1984.
Kaufman, Milton, and Arthur H. Seidman (eds.), *Handbook of Electronics Calculations for Engineers and Technicians,* 2d ed., McGraw-Hill, 1988.
Keller, Gerald, Brian Warrack, and Henry Bartel, *Statistics for Management and Economics,* 3d ed., Duxbury Press, 1994.
Kesten, Gayle, "How to Construct an E-Commerce Site," *Varbusiness,* June 7, 1999, p. 79.
LaForge, Perry M., "cdmaOne Evolution to Third Generation: Rapid, Cost-Effective Introduction of Advanced Services," *CDMA World,* June 1999.
Lathi, B. P., *Modern Digital and Analog Communication Systems,* 3d ed., Oxford University Press, 1998.
Lee, W. C. Y., *Lee's Essentials of Wireless Communications,* McGraw-Hill, 2001.
Lee, W. C. Y., *Mobile Cellular Telecommunications,* 2d ed., McGraw-Hill, 1995.
Lindenburg, *Engineering Economics Analysis,* Professional Publications Inc., 1993.
Louis, P. J., *M-Commerce Crash Course,* McGraw-Hill, 2001.
Lynch, Dick, "Developing a Cellular/PCS National Seamless Network," *Cellular Integration,* Sept. 1995, pp. 24–26.
MacDonald, "The Cellular Concept," *Bell Systems Technical Journal,* 1979.
Masud, Sam, "Cable Operators Eye Telephony," *Telecommunications,* Nov. 1999, p. 35.
McClelland, Stephen, "Europe's Wireless Futures," *Microwave Journal,* Sept. 1999, pp. 78–107.
Molisch, Andreas F., *Wideband Wireless Digital Communications,* Prentice Hall, 2001.

Motorola Inc., "System Description Manual," 1987.
Mouly, Pautet, *The GSM System for Mobile Communications,* Mouly Pautet, 1992.
Muller, Nathan, *Desktop Encyclopedia of Telecommunications,* McGraw-Hill, 1998.
Muratore, Flavio, *UMTS Mobile Communications for the Future,* Wiley, 2000.
Newton, Harry, *Newton's Telcom Dictionary,* 14th ed., Flatiron Publishing, 1998.
O'Keefe, Sue, "TDD vs. FDD: The Next Hurdle," *Telecommunications,* Dec. 1999, p. 40.
Oba, Junichi, "W-CDMA Systems Provide Multimedia Opportunities," *Wireless System Design,* July 1998, p. 20.
Pawlan, Jeffrey, "A Tutorial on Intermodulation Distortion, Part 2: Practical Steps for Accurate Computer Simulation," *RF Design,* March 1996, pp. 74–86.
Prasad, Ramjee, Werner Mohr, and Walter Konhauser, *Third Generation Mobile Communication Systems,* Artech House, 2000.
Qualcomm, "An Overview of the Application of Code Division Multiple Access (CDMA) to Digital Cellular Systems and Personal Cellular Networks," Qualcomm, May 21, 1992.
Rappaport, Theodore S., *Wireless Communications,* 2d ed., Prentice Hall, 2002.
Reference Data for Radio Engineers, 6th ed., SAMS, 1983.
Rusch, Roger, "The Market and Proposed Systems for Satellite Communications," *Applied Microwave & Wireless,* Fall 1995, pp. 10–34.
Salter, Avril, "W-CDMA Trial & Error," *Wireless Review,* Nov. 1, 1999, p. 58.
Schwartz, Bennett, and Stein, *Communication Systems and Technologies,* IEEE, 1996.
Shank, Keith, "A Time to Converge," *Wireless Review,* Aug. 1, 1999, p. 26.
Smith, Clint, *Wireless Telecom FAQs,* McGraw-Hill, 2001.
Smith, Clint, *LMDS,* McGraw-Hill, 2000.
Smith, Clint, *Practical Cellular and PCS Design,* McGraw-Hill, 1998.
Smith, Clint, and Daniel Collins, *3G Wireless Networks,* McGraw-Hill, 2002.
Smith, Clint, and Curt Gervelis, *Cellular System Design and Optimization,* McGraw-Hill, 1996.
Stanton, Steve, "Testing W-CDMA Places New Demands on Designers," *Test/Measurement,* July 1998, p. 37.
Steele, R., and L. Hanzo (eds.), *Mobile Radio Communications,* 2d ed., Wiley, 1999.
Stimson, "Introduction to Airborne Radar," Hughes Aircraft Company, 1983.
TIA/EIA IS-2000-1, "Introduction to cdma2000 Standards for Spread Spectrum Systems," June 9, 2000.
TIA/EIA 553, April 1989.
TIA/EIA IS-127, "Enhanced Variable Rate Codec Speech Service Option 3 for Wideband Spread Spectrum Digital Systems," Sept. 1999.
TIA/EIA IS-718, "Minimum Performance Specification for the Enhanced Variable Rate Codec Speech Service Option 3 for Spread Spectrum Digital Systems," July 1996.
TIA/EIA IS-736-A, "Recommended Minimum Performance Standard for the High-Rate Speech Service Option 17 for Spread Spectrum Communication Systems," Sept. 6, 1999.
TIA/EIA IS-820, "Removable User Identity Module (R-UIM) for cdma2000 Spread Spectrum Systems," June 9, 2000.
TIA/EIA IS-2000-3, "Medium Access Control (MAC) Standard for cdma2000 Spread Spectrum Systems," Sept. 12, 2000.
TIA/EIA IS-2000-4, "Signaling Link Access Control (LAC) Specification for cdma2000 Spread Spectrum Systems," Aug. 12, 2000.
TIA/EIA IS-2000-6, "Analog Signaling Standard for cdma2000 Spread Spectrum Systems," June 9, 2000.
TIA/EIA-98-C, "Recommended Minimum Performance Standards for Dual-Mode Spread Spectrum Mobile Stations (Revision of TIA/EIA-98-B)," Nov. 1999.
TIA/EIA/IS-683-A, "Over-the-Air Service Provisioning of Mobile Stations in Spread Spectrum Systems," May 1998.
TIA/EIR IS-733-1, "High Rate Speech Service Option 17 for Wideband Spread Spectrum Communication Systems," Sept. 1999.
TIA/EIR IS-2000-2, "Physical Layer Standard for cdma2000 Spread Spectrum Systems," Sept. 12, 2000.
Watson, Robert, "Guidelines for Receiver Analysis," *Microwaves & RF,* Dec. 1986, pp. 113–122.
Webb, William, "CDMA from WLL," *Mobile Communications International,* Jan. 1999, p. 61.
Webb, William, *Introduction to Wireless Local Loop,* 2d ed., Artech House, 2000.
Webb, William, and Lajos Hanzo, *Modern Quadrature Amplitude Modulation,* IEEE, 1994.
Wesley, Clarence, "Wireless Gone Astray," *Telecommunications,* Nov. 1999, p. 41.

White and Duff, "Electromagnetic Interference and Compatibility," *Interference Control Technologies Inc.*, 1972.
Willenegger, Serge, "cdma2000 Physical Layer: An Overview," Qualcomm.
Williams, Taylor, *Electronic Filter Design Handbook,* 3d ed., McGraw-Hill, 1995.
Winch, Robert G., *Telecommunication Transmission Systems,* 2d ed., McGraw-Hill, 1998.
Wirbel, Loring, "LMDS, MMDS Race for Low-Cost Implementation," *Electronic Engineering Times,* Nov. 29, 1999, p. 87.
Witowsky, William, "VoP: Standards Remain Elusive," *Telecommunications,* May 1999, p. 53.
Yarbrough, Raymond B., *Electrical Engineering Reference Manual,* 5th ed., Professional Publications Inc., 1990.
Zyren, Jim, "802.11g Spec: Covering the Basics," *EE Design,* Feb. 2002.

Index

A through F frequency blocks in PCS, 7
A-band cellular operators, 4
   advanced mobile phone system (AMPS) and, 72–73, **78, 79**
   filters for, 36, **37**
   IS-136 and, 86, 92–93, **93–99**
above ground level (AGL) value, 204
above mean sea level (AMSL) value, 204
access failure metrics, 272, 277–278, **278**, 292–298, **295, 296**, 323–324, 371
   CDMA and, 323–324
   integrated dispatch enhanced network (iDEN) and, 312–315
access grant channel (AGCH), GSM, 148–150, 152
access point name (APN), GPRS, 175, 176
action plans for performance, 273, 388–391
activation (*see* cell sites; site activation)
active antennas, 33
advanced mobile phone system (AMPS), 3, 5, 21, 67, 69–79, **69**, 81, 269, 304
   A- and B-band operators in, 72–73, **78, 79**
   antenna interface frame (AIF) and, 71
   antennas and, 238–240, **238**
   base station controller (BSC) in, 70–71
   call setup scenarios in, 71, **72–74**
   carrier to interference ratio (C/I) in, 74
   CDMA and CDMA2000 in, 102, 103
   cell sites for, 69–71, **70, 71**
   channel band plan in, 77–79, **78, 79**
   design and, 217–219, **219**
   distance to radius (D/R) ratio in, 75, **76**
   expansion of system for, 494–497
   frequency reuse in, 74–77, **77**
   frequency spectrum for, 72–73, **75, 76**
   handoff in, 71–72
   mobile switching center (MSC) in, 70
   mobile telephone switching office (MTSO) in, 70
   performance in, 274–281, 315, 457
   public switched telephone network (PSTN) interface with, 70
   retuning frequency in, 346, 350–351
   voice traffic projections in, 499–500
air interface:
   general packet radio system (GPRS) and, 164, **164**, 180–181
   global system for mobile communication (GSM) and, 146–148
   IS-136 and, 85–86, **85**
alarm setting, for switches and nodes, 401
all servers busy (ASB) trouble:
   CDMA and, 321–323
   global system for mobile communication (GSM) and, 338–339

all servers busy (ASB) trouble (*Cont.*):
   IS-136 and, 308–309
amplifiers, 27, 63–64, 251–252
   tower–top (TTA), 63–64, **63**
amplitude modulation (AM), 29–32, 246
   receivers for, 40
   transmitter for, 27, **28**
antenna interface frame (AIF), AMPS and, 71
antennas, 26, 27, 32–34, 235–246
   above ground level (AGL) value, 204
   above mean sea level (AMSL) value, 204
   access failures and, 297
   advanced mobile phone system (AMPS), 238–240, **238**
   amplifiers for, 251–252
   base station, 235–236
   beam switching, beam steering, 65–66, **65**
   bidirectional antenna system for, 251
   cdma2000, 238–240, **238**
   change or alteration to, 245–246
   coverage area of, 33
   cross pole, 242–245, **244**
   design and, 235–246
   directional, 235
   distributed antenna system (DAS) for, 250
   diversity in, 236–237, **237**
   downtilting in, 383–385, **384, 385**
   effective radiated power (ERP) and, 56–58, 205
   electromagnetic field (EMF) compliance and, 212–216, **215**
   FCC regulation of, 245
   Federal Aviation Administration (FAA) guidelines for, 211
   figures of merit (FOMs), 34
   filters, 377–378
   gain in, 236
   general packet radio system (GRPS), 238–240, **238**
   global system for mobile communication (GSM), 238–240, **238**
   height of, 240
   horizontal separation in, 261–263, **262, 263**
   installation issues in, 237–240, **239, 240**
   intelligent, 64–65, **65**
   intermodulation distortion (IMD) and, 33
   IS-95, 238–240, **238**
   isolation in, 254–264
   leaky feeder amplification approach for, 251–252
   new site investigation and, 375–379, **376**
   omnidirectional, 235
   orientation in, 382–383

Pages shown in **boldface** have illustrations on them.

574 Index

antennas (*Cont.*):
  orthogonal transmit diversity (OTD) in, 237, 238
  power density chart for, 213–216, **213**
  radiation patterns and, 260
  search area request (SAR) and, 204
  selection of, 33
  signal to noise ratio (SNR) in, 65
  site qualification test (SQT) in, 205–208, **206–208**, 358–364
  slant separation in, 264, **264**
  source antenna correction (SAC) and, 263
  space transmit diversity (STD) in, 237, 238
  standing wave ratio (SWR) and, 34, 377
  tolerances, in installation, 241–242, **242, 243**
  types of, 34
  vertical separation and, 260–261, **261**
  voltage standing wave ratio (VSWR), 318
  wall mounting for, 241, **242**
  wireless LAN (WLAN)/in-building systems, 250, **250**
associated control channel (ACCH), iDEN, 185
asynchronous transfer mode (ATM), 3, 13, 494
AT&T, 2, 3
attach process, GPRS, 170–173, **171, 174**
attenuation, diffraction of radio waves, 55–56, **56, 57**
authentication:
  CDMA and CDMA2000 in, 103
  global system for mobile communication (GSM) and, 144, 152
authentication authorization accounting (AAA), CDMA and, 108–109, 324, 325
authentication center (AuC), GSM, 144
auxiliary node performance, 402

B-band cellular operators, 4
  advanced mobile phone system (AMPS) and, 72–73, **78, 79**
  filters for, 36, **37**
  IS-136 and, 86, 92–93, **93–99**
back office processing of CDRs, billing and, 429–431
balanced path, 59
band reject filters, 35–38, **36, 38**
banding approach, blocking, 299–300, **300**
bandpass filters, 35–38, **36, 38**
bandwidth, 1–3, 29
  CDMA and CDMA2000 in, 106, **107**
  local multipoint distribution system (LMDS) and, 9–11
  (*see also* frequency spectrum)
base site controller (BSC), CDMA and CDMA2000, 110
base station:
  antennas and, 235–236
  local multipoint distribution system (LMDS) and, 10–11
  site design checklist for, 265–267
base station controller (BCS):
  advanced mobile phone system (AMPS) and, 70–71
  CDMA and, 318, 321
  global system for mobile communication (GSM) and, 141–143, 145, 157–160, 338
  integrated dispatch enhanced network (iDEN) and, 183
  planning and, 534–535
  traffic projections and, 498
base station identification code (BSIC), GSM, 158
base station subsystem application part (BBSAP), GPRS, 168
base station system (BSS):
  general packet radio system (GPRS) and, 170, 172
  global system for mobile communication (GSM) and, 141, 144, 153
base station transceiver (BST), planning, 534–535
base transceiver station (BTS)
  CDMA and, 109–110, 318, 321

base transceiver station (BTS) (*Cont.*):
  global system for mobile communication (GSM) and, 141–143, 148, 157, 335–336
  traffic projections and, 498
basic rate interface (BRI), LMDS and, 12
basic trading areas (BTA), 7, 347
beam switching, beam steering antennas, 65–66, **65**
Bell Labs, 2, 3
Bell, Alexander, 2
benchmarks, 269–270
bidirectional antenna system, 251
billing and charging, 24, 423–433
  back office processing of CDRs in, 429–431
  billing applications for, 423–424
  call detail records (CDRs) in, 423, 424, **425**, 427–428
  call test plan for verification of, 431–432
  calling party in, 428
  cellular intercarrier billing exchange roamer (CIBER) format for, 424, **426**
  circular CDR files in, 428
  clearinghouse in, 424
  collection processes in, 428, **429**
  customer care report in, 476, **477**
  cycles for billing, 424–427
  data filtering for, 424–427
  guiding applications in, 423
  invoicing applications in, 424
  IP number inventory report in, 465, **466, 467**
  nonbillable events in, 433
  rating applications in, 423
  revenue assurance in wireless network and, 436
  switch subscriber capacity study in, 528–530
  telephone number inventory report in, 464–465
  verification methods for, 432–433, 440–443
  (*see also* revenue assurance in wireless network)
binary phase shift keying (BPSK), CDMA and CDMA2000 in, 107
bit error rate (BER), 90, **91**, 219, 272, 310, 371
  IS-136 and, 90, **91**, 310
  receivers and, 42, 47
blocking, 45, 272, 275–276, 298–303, **300**, 371
  receivers and, 45
blocking level, 233–234
Bluetooth, 17–18
  effective radiated power (ERP) in, 19
  wireless LAN (WLAN) and, 19, **20**
bore sites, 377
bouncing congestion hour (BCH) traffic report (node and service) in, 454–457, **455–459**
briefings, network briefings and, 485
broadband:
  cable systems and, 14
  local multipoint distribution system (LMDS) and, 8
broadcast control channel (BCCH):
  global system for mobile communication (GSM) and, 147, 158, 160, 178, 180, 337–339
  integrated dispatch enhanced network (iDEN) and, 184
  voice traffic projections in, 499–500, 506, 507
broadcast radio, 2
broadcasting, 246
BSC color code (BCC), GSM, 333–334
BSS application part (BSSAP), GSM, 143, **144**
BSS GPRS protocol (BSSGP), 166
BSS management application part (BSSMAP), GSM, 143
BSS operation and maintenance application part (BSSOMAP), 143
BSSAP+ protocol, GPRS and, 168
budgeting, 492, 539–541, 552

Index

bulletin board report, system status, 476–479, **477**
Bullington propagation model, 48
busy hour call attempts (BHCA), data traffic projections and, 514–515

cable systems, 13–14
call completion ratio, 272, 371
call delivery troubleshooting, 411–422
   call testing procedures in, 415–418, **416, 417**
   call treatment classification in, 413, **414**
   examples of, 419–422, **420, 422**
   first level procedures for, 413
   initial steps in, 411–413
   second level procedures for, 413
   summary of procedures for, 417–419
call detail records (CDRs), 424, **425**, 427–428
   back office processing of, 429–431
   billing and, 423
   call test plan for verification of, 431–432
   circular files in, 428
   collection using, 428, **429**
   fields for, information details in, 427–428
   generation and mediation process in, 435
   generation of, 428, **429**
   peg counters in, 444
   switch statistical call data comparison in, 443–444
   verification of billing and, 440–443
   volume benchmarks using, 440–442
call processing, 398–399, **399**
   call testing procedures in, 415–418, **416, 417**
   call treatment classification in, 413, **414**
   CDMA and CDMA2000 in, 123–124, **124**
   general packet radio system (GPRS) and, attach process in, 170–173, **171, 174**
   global system for mobile communication (GSM) and, 153, **154**
   integrated dispatch enhanced network (iDEN) and, 188–190, **189**
call termination:
   code division multiple access (CDMA) and CDMA2000, 125
   global system for mobile communication (GSM) and, 155–157, **156**
call test plan for verification, 431–432
call testing procedures, 415–418, **416, 417**
call treatment classification, 413, **414**
call volume, 400–401
called vs. calling party verification, 428, 444–447, **446**
capacity (Shannon–Hartley), 29
capacity of cell sites, 518–521
capacity studies, for switches, 524–526
capacity, switch subscriber capacity study in, 528–530
care of addressing (COA), CDMA and CDMA2000 in, 131
Carey propagation model, 48
carrier to interference ratio (C/I):
   advanced mobile phone system (AMPS) and, 74
   receivers and, 43
CDMA2000, 107
   (*see also* code division multiple access)
CDMAOne, 104–105
ceiling approach, blocking, 299
cell sites, 246, **247**
   activation and, performance issues in, 364–374, **368, 369**
   advanced mobile phone system (AMPS) and, 69, 70–71, **70, 71**
   bore sites, 377
   capacity of, 518–521
   CDMA and, 317, 323–324, 325, 328
   components of, 246, **247**
   coverage areas, 381

cell sites (*Cont.*):
   data traffic projections and, 513
   design issues in, 201, 380–381
   do not exceed line (DNEL) in, 348–349, **349**
   downtilting in, 383–385, **384, 385**
   drive map for, **208**
   drive testing in, 358–364
   electromagnetic field (EMF) compliance and, 212–216, **215**
   existing, performance evaluation in, 378–382
   Federal Aviation Administration (FAA) guidelines for, 211
   field test report for, 382, **382**
   filters, 377–378
   frequency planning in, 224–235, **227**
   frequency retuning in, 362, **363, 365, 366**
   general packet radio service (GPRS), 346
   global system for mobile communication (GSM) and, 330–331, **330–332**, 337–340
   improvement plans for, 381–382
   integrated dispatch enhanced network (iDEN) and, 185–186, 312, 314
   interference and, 380
   intermodulation in, 385–387
   IS-136 and, 306–307, 309
   $N = 12, N = 7, N = 4$ frequency assignment methods in, 227–235, **227, 228**
   new site investigation and, 374–379
   new site performance report for, 372, **373–374**
   orientation in, 382–383
   performance checklist for, 378–379, 410–411
   planning and zoning board approval in, 216–217
   post turn on test form for, 362, **363**
   power density chart for, 213–216, **213**
   retuning frequency in, 346–357, **348**
   search area request (SAR) in, 201–205, **203**
   site acceptance (SA) in, 208–209
   site acceptance form (SAF) in, 209, **210**
   site investigations in, 374–382
   site qualification test (SQT) in, 205–208, **206–208**, 358–364
   site rejection (SR) in, 209–211, **211**
   size of (macro-, micro-, pico-), 246, 249
   standing wave ratio (SWR) and, 377
   system growth and, 515–522, **519**
   types of, 246
cellular communication, 1, 3–5, **4**, 69, 83
   A- and B-band operators, 4, 36
   Cost231 propagation model, 50–55, **52**
   development and history of, 3–4
   free space path loss in, 49
   frequency spectrum for, 3, 5, 26
   handoffs in, 4
   improved mobile telephone system (IMTS) in, 4
   metropolitan and rural statistical areas (MSA/RSA) for, 4
   mobile switching centers (MSC) in, 4
   mobile telephone systems (MTS) in, 4
   public switched telephone system (PSTN) and, 4
   wireless application protocol (WAP), 17
   wireless local loop (WLL) and, 8
cellular data packet data (CDPD), 80, 190–194, **193**
   channel hopping in, 193
   expansion of system for, 494–497
   fixed end system (FES) in, 192
   frequency spectrum for, 192–193
   gaussian minimum shift keying (GMSK) in, 192
   handoff in, 194
   mobile data base station (MDBS) in, 191–192
   mobile data intermediate system (MDIS) in, 192
   mobile end system (MES) in, 191

576 Index

cellular intercarrier billing exchange roamer (CIBER) format, billing and, 424, **426**
central office, 21–24, **24**
central processing unit (CPU)
    call processing efficiency in, 398–399, **399**
    prioritizing tasks in, 394–398, **395**
    switch loading and performance optimization in, 393–398, **396**
centralized organizations, 543–544
challenge handshake authentication protocol (CHAP), 128, 131
changeovers, link, 405
channel band plan, AMPS, 77–79, **78, 79**
channel elements (CE), voice traffic projections and, 503–504
channel hopping, CDPD and, 193
channels, 1900 GSM channel chart, 567–568
charging gateway function (CGF), GPRS, 166
circuit emulation service (CES), voice over IP (VoIP) and, 20–21
circuit switch growth, 524–530
circuit switch performance guidelines, 393–422
    alarm setting, for switches and nodes, 401
    auxiliary node performance in, 402
    call delivery troubleshooting in, 411–422
    call processing efficiency in, 398–399, **399**
    downtime (service outage time) and, 399, **400**
    erlangs (total) and call volume in, 400–401
    general data on, 410–411
    link active time in, 405
    link changeovers in, 405
    link errors in, 404
    link performance measurement and optimization in, 403–406
    link retransmission rate in, 404
    link traffic loading in, 403–404
    memory settings and utilization, for switches, 401
    message signaling units (MSU) in, 403
    prioritizing tasks in, 394–398, **395**
    routing efficiency rating in, 406–409
    routing performance monitoring and management in, 406–410
    software performance and, 410
    spikes in traffic load and, 394–398, **395**
    switch CPU loading and, 393–398, **396**
    switch service circuit loading in, 400
    timing source accuracy in, for switches, 401–402, **402**
    (see also performance issues)
circuit switch/node performance report in, 462–463, **464**
circular CDR files, billing and, 428
class 5 switches, 24
classes of GPRS, 163
clearance, path, 60–62, **62**
clearinghouse, billing and, 424
cochannel interference, 248, 332–333, **333**
code division multiple access (CDMA) and CDMA2000, 67, 81, 101–140, **106, 108**, 269, 274
    access failure in, 323–324
    advanced mobile phone system (AMPS) and, 102, 103
    all servers busy (ASB) trouble in, 321–323
    antennas and, 238–240, **238**
    authentication, authorization, accounting (AAA) in, 108–109, 324, 325
    bandwidth in, 106, **107**
    base site controller (BSC) in, 110, 318, 321
    base transceiver site (BTS) in, 109–110, 318, 321
    benefits of, 102–103
    binary phase shift keying (BPSK) and, 107
    call and data processing in, 123–132
    care of addressing (COA) in, 131
    CDMA2000 and, 102, 104–107, **106, 108**, 315–328
    CDMAOne and, 104–105

code division multiple access (CDMA) and CDMA2000 (*Cont.*):
    cell overlap in, 322–323
    cell sites for, 317, 323–325, 328
    cellular communication and, 5
    challenge handshake authentication protocol (CHAP) in, 128, 131
    channel allocation in, 110–111, **111, 112**
    design and, 217, 219, **220**
    diversity and, 237
    dominant pilot problems in, 317
    dynamic host configuration protocol (DHCP) in, 128, 131
    encryption and authentication in, 103
    expansion of system for, 495–497
    forward and reverse links in, CDMA2000, 107
    forward channels in, 111–115, **113, 115**
    forward error rate (FER) in, 120
    frame error detection (FED) in, 120
    frame error rate (FER) in, 316
    frequency assignment in, 229–230
    frequency spectrum for, 110–111, **111, 112**
    fundamental channel (FCH) in, 110
    global positioning system (GPS) and, 318
    global system for mobile communication (GSM) and, 105
    handoffs in, 103, 110, 132–137, 316, 319–321, **319**, 326, 327–328
    hard handoffs in, 134
    home agent (HA) in, 109
    home location register (HLR) in, 109, 323, 325
    hybrid phase shift keying (HPSK) and, 107
    IS-95 and, 103–104, **103**, 106, **108**
    layer 2 tunneling protocol (L2TP) in, 128
    lost calls in, 316–319
    maintenance in, 318, 321, 324, 326
    microwave and, 101
    mobile assisted handoff (MAHO) in, 103
    mobile IP (3G) in, 129–132, **132–134**
    neighbor lists for, 327–328
    network interface card (NIC) settings for, 324
    1XDO and, 105
    1XDV and, 105
    1XEV and, 105
    1XRTT and, 105, 106, 119, 324, 326–328
    origination of mobile call in, **124**
    packet data rates for, 505, **506**
    packet data serving node (PDSN) in, 107–110, 127–132, 325–326
    packet data transport process flow in, 124–132, **126, 127**
    packet session access in, 324–326
    packet session throughput problems in, 326–328
    paging channel in, 111–113
    password authentication protocol (PAP) in, 128, 131
    personal communication services (PCS) and, 7, 102, 103
    pilot channel in, 111, 137–140, **138**
    pilot pollution and access failure in, 297, 317, 323
    pilot signal measurement message (PSMM) in, 317
    point to point protocol (PPP) and, 108, 126
    power control in, 120, 316–317
    power levels in, subscriber, 118, **118**
    pseudorandom number (PN) assignment in, 102, 137–140, **138–140**
    radio configuration (RC) in, 119–120, **120**, 122–123, **123**
    radio network using, 105–107
    registration area for, 323
    retuning frequency in, 346, 353–355
    reverse channels in, 116–119, **117, 118**
    round trip delay (RTD) in, 317, 323–324
    routers in, 109
    satellite and, 101

Index **577**

code division multiple access (CDMA) and CDMA2000 (*Cont.*):
 search windows in, 134–135, **135**
 simple IP in, 127–129, **128–130**
 soft handoffs in, 103, 133, 135–137, **137**
 spread spectrum technology and, 101
 spreading rate (SR) in, 119–120, **120**, 122–123, **123**
 supplemental channel (SCH) in, 110
 sync channel in, 111
 termination of mobile call in, **125**, 125
 three/3XRTT and, 106, 119, 121
 time division multiple access (TDMA) vs., 84, 103
 time division multiplexing (TDM) and, 110
 timeouts in, 328
 traffic load in, 317–318, 321, 325, 327
 virtual private network (VPN) in, 127–129, **130, 131**, 132, **134**
 visitor location register (VLR) in, 126–127, 323
 voice processing in, 325, 328
 voice traffic projections and, 499–505
 Walsh codes in, 109, 110, 122–123, **122**, 322, 325, 326, 328, 502, 505, **505**
 wideband CDMA (WCDMA) and, 105
 wireless application protocol (WAP), 17
coded digital voice color code (CDVCC), IS-136 and, 88
coding schemes, GPRS, 161–162, **162**, 344–346, **345**
collection processes, 428, **429**
 (*see also* billing and charging)
collinear antennas, 34
collision sense multiple access/collision avoidance (CSMA/CA), 19
colocation, 24, 246, 255–264
common control channel (CCCH):
 global system for mobile communication (GSM) and, 147, 160
 integrated dispatch enhanced network (iDEN) and, 184
 voice traffic projections and, 506, 507
company meetings and report generation, 481–484
competitive local exchange carriers (CLEC):
 interface with mobile system by, 23
 local multipoint distribution system (LMDS) and, 11
 voice over IP (VoIP) and, 20
complaints, 272, 371
configuration, network configuration report in, 470–471, **472**
congestion, bouncing congestion hour traffic report (node and service) in, 454–457, **455–459**
construction, 492
control channel:
 general packet radio system (GPRS) and, 164–165
 integrated dispatch enhanced network (iDEN) and, 184–186
 IS-136 and, 91–99
controlled vs. uncontrolled environments and EMF, 213–216
cordless telephone (CT-2), wireless local loop (WLL) and, 8
Cost231 propagation model, 48, 50–55, **52, 54**
coverage area, 248, 249, 381, 383–385, 517–518
 search area request (SAR) and, 204
 global system for mobile communication (GSM) and, 329–330
 local multipoint distribution system (LMDS) and, 11
critical design review (CDR) in, 199–201
cross connect switch (DXX), 24
cross mapping metrics, 272–273
cross pole antennas, 242–245, **244**
customer care report, 476, **477**
cycles, billing, 424–427

data processing, CDMA and CDMA2000 in, 123–132
data rates, CDMA, 505, **506**
data traffic projections, 511–515
databases, customer information:
 analysis and reconciliation of, 436–440

databases, customer information (*Cont.*):
 determining total subscriber numbers in, 438–440
 provisioning of, 437, **438**
DCS1800, 6
DCS1900, 81
dedicated control channel (DCCH)
 general packet radio system (GPRS) and, 165
 integrated dispatch enhanced network (iDEN) and, 185
degradation levels, receivers and, 43
demodulation, 26
departments in organizational structure, 544–545, **545**
desense, receivers, 46–47
design guidelines, 195–267, 491–492
 advanced mobile phone system (AMPS), 217–219, **219**
 antennas, 235–246
  (*see also* antennas)
 base station site checklist in, 265–267
 cell site design in, 201, 380–381
 code division multiple access (CDMA), 217
 colocation issues and, 246
 controlled vs. uncontrolled environments and EMF, 213–216
 critical design review (CDR) in, 199–201
 design guidelines for, 217–221, **218**
 electromagnetic field (EMF) compliance and, 212–216, **215**
 Federal Aviation Administration (FAA) guidelines for, 211
 frequency planning in, 224–235, **227**
 general packet radio services (GPRS) and, 221, **222**
 global system for mobile communication (GSM) and, 221, **222**
 high level design review (HDR) in, 199–201
 in-building systems, 247–254
 information needed for, 196–197
 integrated dispatch enhanced network (iDEN) and, 219, **221**
 IS-136, 219, **220**
 IS-95/CDMA2000 (1XRTT), 219, **220**
 isolation in, 254–264
 link budgets in, 221–224, **223, 225, 226**
 migration/legacy systems and, 197
 performance issues and, 198, 217–218
 personal communication systems (PCS) and, 221
 planning and zoning board approval in, 216–217
 power budget in, 212–216, **215**
 power density chart for, 213–216, **213**
 preliminary design review (PDR) in, 199–201
 process of, 197–201, **202**
 reradiator, 246–247
 RF and networking design guidelines in, 196
 search area request (SAR) in, 201–205, **203**
 site acceptance (SA) in, 208–209
 site acceptance form (SAF) in, 209, **210**
 site qualification test (SQT) in, 205–208, **206–208**, 358–364
 site rejection (SR) in, 209–211, **211**
 site type in, 246
 steps in, 196
 system radio channel expansion and, 233–235
 tunnel systems, 247–254, **252, 253**
differential QPSK (DQPSK), 31, **33**
diffraction of radio waves, 55–56, **57**
digital access cross connects (DAGs), auxiliary node performance in, 402
digital AMPS (D-AMPS), 5, 84–101, 219
 (*see also* IS-136)
digital color code (DCC), IS-136 and, 100, **101**
digital communication, 81, 83
digital control channel (DCCH), IS-136 and, 92–93, **93–99**
digital European cordless telecommunications (DECT), 8
digital mobile attenuation code (DMAC), IS-136 and, 307

578 Index

digital subscriber line (DSL), 298
digital traffic channel (DTC), IS-136 and, 86, **88**
digital verification color code (DVCC), IS-136 and, 91, 100–101, **102**, 305
direct transfer application part (DTAP), GSM, 143, 152
directional antennas, 33, 34, 235
dispatch application processor (DAP), iDEN, 182, 183, 187, 314
dispatch location areas (DLAs), iDEN, 189–190, **190, 192**, 313–314
dispatch only transmission, iDEN, 187–188, **187, 188**
distance to radius (D/R) ratio, 59, 75, **76**
distortion, receiver, 40
distributed antenna system (DAS), 250
distributed organizations, 543–544
diversity, 249
    antennas and, 236–237, **237**
diversity gain, 236
do not exceed line (DNEL), 348–349, **349**
documentation (*see* system documentation and reports)
domain name service (DNS), GPRS and, 176
dominant pilot problems, CDMA and, 317
downlink, 59
    global system for mobile communication (GSM) and, 146–147
    link budgets in, 221–224, **223, 225, 226**
downtilting, 383–385, **384, 385**
downtime (service outage time) and, 399, **400**
drive map, **208**
drive testing, 358–364, 371
dual tone multifrequency (DTMF), 148, 400
dynamic host configuration protocol (DHCP):
    CDMA and CDMA2000 in, 128, 131
    general packet radio system (GPRS) and, 176, 177
dynamic power control (DPC), IS-136 and, 307
dynamic range, receiver, 39, 41–43, **42, 44**
dynamic routing numbers (DRNs), 529

Earth curvature and path clearance, 61–62, **61**
eCommerce, LMDS and, 12
effective cell radius (ECR), 57
effective radiated power (ERP), 56–58, 205
    access failures and, 296–298
    global system for mobile communication (GSM) and, 337–339
    wireless LAN (WLAN) and, 19
efficiency, call processing, 398–399, **399**
efficiency, receiver, 39
800 numbers, 433
802.11 protocol, 10, 18–19
    (*see also* wireless LANs)
802.11a, 19
802.11b, 19
    (*see also* WiFi)
electromagnetic field (EMF) compliance, 212–216, **215**, 255
electromagnetic waves, 25–26, **26**
emergency services, 433
encryption, 80, 103
energy per bit to noise density ratio (Eb/No), receivers and, 43
engineering department organization, 545–548, **546, 549**
engineering library, 450–451
enhanced base transceiver (EBTS), iDEN, 182, 314
enhanced data rate for GSM evolution (EDGE), 67
enhanced specialized mobile radio (ESMR), 26, 246
    Cost231 propagation model, 50–55, **52**
    free space path loss in, 49
environmental attenuation, 55, **56**
equipment identity register (EIR), GSM, 145–146
equipment room, 21, 24
Erlang B, 234
    blocking and, 300–303

Erlang B (*Cont.*):
    grade of service, 493, 561–563
    traffic table, 493
Erlang C:
    blocking and, 300–303
    grade of service, 494, 565–566
    traffic table, 494
erlangs (total) and call volume in, 400–401
European Telecommunication Standards Institute (ETSI), 20
exception reports, 471–476, **474, 476**
expansion of system, 233–235, 494–497
    2.5 or 3G migration design procedure for, 496–497
    frequency allocation and, 233–235
    new wireless system procedure for, 495–496

facilities interconnect report (data), 467–468, **469**
facility usage report, 465–467, **468**
fade, 248
fast associated control channel (FACCH):
    global system for mobile communication (GSM) and, 148–150, 153, 155, 165
    IS-136 and, 86
fax Internet, 3
fax over IP (FaxIP), LMDS and, 12
Federal Aviation Administration (FAA) guidelines, antennas, 211
Federal Communications Commission (FCC), 7, 73, 245
figures of merit (FOMs), receiver, 34, 41–47
filtering, data for billing, 424–427
filters, 26, 34–38, **36, 38**, 377–378
    classification of (low–pass, high–pass, etc.), 35, **36**
    selection of, criteria for, 35, 38
    types of, 35
final assembly code (FAC), GSM, 145
final report, planning and, 541–542
first generation (1G) technology, 67
fixed end system (FES), CDPD and, 192
fixed network planning, 522–524
fixed wireless point to multipoint (FWPMP) system, 8
flowchart of performance, 281, **281**
forward channels, CDMA and CDMA2000 in, 111–115, **113, 115**
forward dedicated control channel (F DCCH), CDMA and CDMA2000 in, 115
forward error rate (FER), CDMA and CDMA2000 in, 120
forward fundamental channel (F FCH), CDMA and CDMA2000 in, 114
forward quick paging channel (F QPCH), CDMA and CDMA2000 in, 114–115
forward supplemental code channel (F SCH), CDMA and CDMA2000 in, 114
forward transmit diversity pilot channel (F TDPICH), CDMA and CDMA2000 in, 115
forward common control channel (F CCCH), CDMA and CDMA2000 in, 115
fractional T1/E1, LMDS and, 12
frame error detection (FED), CDMA and CDMA2000 in, 120
frame error rate (FER), 272, 371
    CDMA and, 316
    IS-136 and, 100, 310
frame relay, LMDS and, 12
free space and free space path loss, 48–49, 259
frequency correction channel (FCCH), GSM and, 147
frequency division duplexing (FDD), LMDS and, 13
frequency modulation (FM), 29–32, 246
    receivers for, 40
    transmitter for, 27, **28**
frequency range, receiver, 39
frequency reuse, 3
    advanced mobile phone system (AMPS) and, 74–77, **77**
    IS-136 and, 100–101

Index

frequency spectrum, 1, 224–235
  advanced mobile phone system (AMPS) and, 72–73, **75, 76**
  alteration of frequency plan in, form for, 230–233, **231**
  blocking level in, 233–234
  Bluetooth, 17–18
  CDMA and CDMA2000 in, 110–111, **111, 112**
  cellular communication, 3, 5, 26
  cellular digital packet data (CDPD) and, 192–193
  Cost231 propagation model, 50–55
  do not exceed line (DNEL) in, 348–349, **349**
  erlang B tables in, 234
  filters and, 35–38, **36**
  frequency planning in, 224–235, **227**
  general packet radio service (GPRS), 161, 342, 346
  global system for mobile communication (GSM) and, 140, 141, 146–147, **147**, 332–333, 337, 340, 342, 346
  industrial, scientific, medical (ISM) band, 17, 19
  instruction television fixed service (ITFS), 13–14
  integrated dispatch enhanced network (iDEN) and, 183–184, **184**, 312
  intermodulation in, 385–387
  IS-136 and, 306
  local multipoint distribution system (LMDS) and, 11
  method of procedure (MOP) in, 231–233
  multichannel multipoint distribution system (MMDS), 13–14
  multipoint distribution service (MDS), 13–14
  $N = 12, N = 7, N = 4$ frequency assignment methods in, 227–235, **227, 228**
  personal communication services (PCS) and, 6–7, **7, 83**
  retuning the network and, 346–357
  rules for assigning, 229–230
  specialized mobile radio (SMR), 3
  system radio channel expansion and, 233–235
  unlicensed national information infrastructure (UNII) band, 19
  wireless LAN (WLAN) and, 19
  wireless local loop (WLL) and, 8
  (*see also* bandwidth)
Fresnel zone, 61–62, **61**
fulfillment process, revenue assurance in wireless network and, 435
fundamental channel (FCH), CDMA and CDMA2000 in, 110

gain, diversity type, 236
gateway GPRS support node (GGSN), 166, 168–169, 176–178, 181
gateway MSC (GMSC), GSM, 145, 155
gaussian minimum shift keying (GMSK), CDPD and, 192
general packet radio service (GPRS), 161–181, **167**
  access point name (APN) in, 175, 176
  air interface for, 164, **164**, 180–181
  attach process in, 170–173, **171, 174**
  base station subsystem application part (BBSAP) in, 168
  base station system (BSS) in, 170, 172
  BSS GPRS protocol (BSSGP) in, 166
  BSSAP+ protocol in, 168
  cell sites for, 346
  charging gateway function (CGF) in, 166
  classes of, 163
  coding schemes in, 161–162, **162**, 344–346, **345**
  control channels in, 164–165, **164**
  dedicated control channel (DCCH) in, 165
  design and, 221, **222**
  dynamic host configuration protocol (DHCP) in, 176, 177
  expansion of system in, 494–497
  frequency spectrum for, 161, 342, 346
  gateway GPRS support node (GGSN) in, 166, 168–169, 176–178, 181
  global system for mobile communication (GSM) and, 161, 164

general packet radio service (GPRS) (*Cont.*):
  GPRS tunneling protocol (GTP) in, 168
  home location register (HLR) in, 167, 168, 172, 173
  international mobile subscriber identity (IMSI) in, 176
  IP addresses, URL, DNS in, 176
  logical link control (LLC) in, 175
  maintenance in, 342, 346
  mobile application part (MAP) in, 167
  mobile country code (MCC) in, 176
  mobile switching center (MSC) in, 166, 168, 173, 179
  modes in, 340–341, **341**
  network nodes in, 166–169, 181
  network service access point identifier (NSAPI) in, 175, 176
  packet access grant channel (PAGCH) in, 165
  packet associated control channel (PACCH) in, 165, 170, 172
  packet broadcast control channel (PBCCH) in, 165, 178
  packet common control channel (PCCCH) in, 164–165, 172
  packet control unit (PCU) in, 166
  packet data channel (PDCH) in, 164
  packet data protocol (PDP) context in, 166, 173–178, **175**
  packet data serving node (PDSN), 341
  packet data traffic channel (PDTCH) in, 165–166, 172
  packet notification channel (PNCH) in, 165
  packet paging channel (PPCH) in, 165
  packet random access channel (PRACH) in, 165
  packet session access in, 340–342, **341**
  packet session throughput and, 343–346
  packet temporary mobile station identity (P–TMSI) in, 169, 178
  packet timing control channel (PTCCH) in, 164, 165
  performance in, 329, 457, 460
  point to multipoint–multicast (PTM–M) notification in, 165
  quality of service (QoS) in, 175, 176, 177
  resource sharing in, 162–163
  routing areas (RAs) in, 166, 177–179, **179**, 345–346, **345**
  service access point identifier (SAPI) in, 175
  service provided by, 161–163
  serving GPRS support node (SGSN) in, 166–169, **167**, 172, 173, 176–179, 181
  short message service center (SMSC) in, 168
  speed of transmission/throughput in, 161, 163
  SS7 and, 167
  subnetwork dependent convergence protocol (SNDCP) in, 177
  subscriber identity module (SIM) in, 342
  temporary block flow (TBF) in, 169
  temporary flow identity (TFI) in, 169
  temporary logical link identity (TLLI) in, 169
  time slots in, 163, 165–166
  traffic calculation and network dimensioning in, 180
  traffic channels (TCH) in, 342
  traffic scenarios in, 169
  uplink state flag (USF) in, 169
  user devices in, 163
  visitor location register (VLR) in, 167, 168, 173, 179
  voice processing in, 341, 346
  voice traffic projections and, 499–500, 505–511, **509**
generations of mobile technologies, 67
global positioning system (GPS), CDMA and, 318, 321
global system for mobile communication (GSM), 5, 21, 67, 81, 83, 140–161, **141**, 269, 274, 304
  1900 GSM channel chart, 567–568
  access grant channel (AGCH) in, 148–150, 152
  air interface in, 146–148
  all servers busy (ASB) trouble in, 338–339
  antennas and, 238–240, **238**
  authentication center (AuC) in, 144
  authentication in, 152

580   Index

global system for mobile communication (GSM) (*Cont.*):
   base station controller (BSC) in, 141–143, 145, 157–160, 338
   base station identification code (BSIC) in, 158
   base station system (BSS) in, 141, 144, 153, 155
   base transceiver station (BTS) in, 141–143, 148, 157, 335–336
   broadcast control channel (BCCH) in, 147, 158, 160, 178, 180, 337–338, 339
   BSC color code (BCC) in, 333–334
   BSS application part (BSSAP) in, 143, **144**
   BSS management application part (BSSMAP) in, 143
   BSS operation and maintenance application part (BSSOMAP) in, 143
   call processing, mobile originated voice call in, 153, **154**
   call termination, mobile terminated voice call, 155–157, **156**
   CDMA and CDMA2000 in, 105
   cell sites for, 330–331, **330–332**, 337–340
   channel structure in, air interface, 148–150
   co channel interference in, 332–333, **333**
   common control channel (CCCH) in, 147, 160
   coverage area in, 329–330
   design and, 221, **222**
   development of, 141
   direct transfer application part (DTAP) in, 143, 152
   effective radiated power (ERP) in, 337–339
   equipment identity register (EIR) in, 145–146
   expansion of system for, 494–497
   fast associated control channel (FACCH) in, 148–150, 153, 155, 165
   frequency correction channel (FCCH) in, 147
   frequency spectrum for, 140, 141, 146–147, **147**, 332–333, 337, 340, 342, 346
   gateway MSC (GMSC) in, 145, 155
   general packet radio system (GPRS) and, 161, 164
   global system for mobile communication (GSM) and, 329–346
   GPRS attach using, 173, **174**
   handoffs in, 157–160, **159**, 331–332, 336–338, **337**
   home location register (HLR) in, 144, 155
   home public land mobile network (HPLMN) in, 150
   hopping methods in, 333–334
   hopping sequence number (HSN) in, 333–334
   international mobile identity (IMEI) in, 145–146
   international mobile subscriber identity (IMSI) in, 152
   interworking function (IWF) in, 146, 160
   location area (LA) in, 336, 338, 340
   location update in, 150–152, **151**
   lost calls in, 329–336
   maintenance in, 336, 338, 340, 346
   mobile application part (MAP) in, 152
   mobile assisted handoff (MAHO), 330, 332, **334, 335**, 336–338, **337**
   mobile station ISDN (MSISDN) in, 155
   mobile station roaming number (MSRN) in, 155
   mobile switching center (MSC) in, 143, 144, 145, 152, 153, 155, 157–160, 166
   neighbor lists for, 330–331, **330–332**
   network color code (NCC) in, 333–334
   notification channel (NCH) in, 148–150
   operation and maintenance link (OML) in, 143
   operations and support system (OSS) in, 141
   packet data serving node (PDSN), 341
   packet session access in, 340–342, **341**
   packet session throughput and, 343–346
   paging channel (PCH) in, 147, 155
   performance in, 315, 457
   personal communication services (PCS) and, 6
   power control in, 335–336
   pulse code modulation (PCM) in, 144

global system for mobile communication (GSM) (*Cont.*):
   random access channel (RACH) in, 148–150, 155
   retuning frequency in, 346, 352–353
   reverse signal strength indicator (RSSI), 330, 332, 336–339, **337**
   routing area (RA) in, 336, 338, 340
   short message service center (SMSC) in, 145
   signal strength in, insufficient (IS), 339–340
   signaling connection control part (SCCP) in, 143
   slow associated control channel (SACCH) in, 148–150, **150**, 158
   SS7 and, 143, 152
   stand-alone dedicated control channel (SDCCH) in, 148–153, 155–157, 160, 180
   subscriber identity module (SIM) in, 141–142
   switching system in, 141
   synchronization channel (SCH) in, 147
   temporary mobile subscriber identity (TMSI) in, 152
   time division multiple access (TDMA) in, 146
   time slots in, 147–149, **149**, 157–158
   traffic calculation methods in, 160–161
   traffic channel (TCH) in, 148–150, **150**, 153, 155, 157
   traffic load in, 334–335
   transcoding and rate adaptation unit (TRAU) in, 143–144
   type approval code (TAC) and final assembly code (FAC) in, 145
   up- and downlink for, 146–147
   visitor location register (VLR) in, 143, 145, 152, 155
   voice processing in, 341, 346
   voice traffic projections and, 499–500, 505–511, **509**
   wireless application protocol (WAP), 17
GPRS tunneling protocol (GTP), 168
grade of service (GoS), 301
   Erlang B, 493, 561–563
   Erlang C, 494, 565–566
   routing efficiency, 408–409
   traffic projections and, 498
grounding, 255
group call feature, iDEN, 182
growth of system, 515–522, **519**
   capacity cell sites and, 518–521
   coverage area and, 517–518
   network interconnect growth studies in, 531–534, **532, 533**
   traffic off-loading and, 521–522, **522, 523**
guard time, IS-136 and, 88
guiding applications, billing and, 423

H.323, voice over IP (VoIP) and, 20
handoffs:
   advanced mobile phone system (AMPS) and, 71–72
   CDMA and CDMA2000 in, 103, 110, 132–137, 316, 319–321, **319**, 326–328
   cellular communication and, 4
   cellular digital packet data (CDPD) and, 194
   failure of, 272, 307–308, 319–321, 336–338, **337**, 371
   global system for mobile communication (GSM) and, 157–160, **159**, 331–332, 336–338, **337**
   integrated dispatch enhanced network (iDEN) and, 186–187
   IS-136 and, 100, 307
   local multipoint distribution system (LMDS) and, 10
handover, 157
   (*see also* handoffs)
hard handoffs, CDMA and CDMA2000 in, 134
harmonics, intermodulation in, 385–387
Hata propagation model, 49–50, 48, **54**, 55
head count requirements, 539, **540**, 552–553, **554**
high-level design review (HDR), 199–201
high-pass filters, 35–38, **36, 38**
HiperLan/2, 20

Index 581

hiring practices, 553–555
history of data communication, 2–3
home agent (HA), CDMA and CDMA2000 in, 109
home location register (HLR), 436–437
   CDMA and CDMA2000 in, 109, 323, 325
   general packet radio system (GPRS) and, 167, 168, 172, 173
   global system for mobile communication (GSM) and, 144, 155
   integrated dispatch enhanced network (iDEN) and, 182
home public land mobile network (HPLMN), GSM and, 150
hopping, 333–334
   (*see also* channel hopping)
hopping sequence number (HSN), GSM, 333–334
horizontal separation, 261–263, **262, 263**
hub, LMDS and, 10–11
hybrid fiber/coax (HFC) networks, 14, **15**
hybrid phase shift keying (HPSK), CDMA and CDMA2000 in, 107

Ikegami propagation model (*see* Cost231 propagation model)
improved mobile telephone system (IMTS), 4
improvement plan, site, 381–382
IMT2000, 17, 68
in-building system design, 247–254
   (*see also* wireless LANs)
inband interference, 256–257, **256, 257**
industrial, scientific, medical (ISM) band, 17, 19
information requirements for performance/troubleshooting, 271–272
Institute of Electrical and Electronics Engineers (IEEE), 19
instruction television fixed service (ITFS), 13–14
integrated dispatch enhanced network (iDEN), 3, 5, 67, 181–190, **183**
   access failures in, 312–315
   associated control channel (ACCH) in, 185
   base station controller (BSC), 183
   broadcast control channel (BCCH) in, 184
   call processing in, 188–190, **189**
   cell selection, bandmap, and reselection in, 185–186
   cell sites for, 312, 314
   common control channel (CCCH) in, 184
   control channels in, 184–186
   dedicated control channel (DCCH) in, 185
   design and, 219, **221**
   dispatch application processor (DAP) in, 182, 183, 187, 314
   dispatch location areas (DLAs) in, 189–190, **190, 192**, 313–314
   dispatch only transmission in, 187–188, **187, 188**
   diversity and, 236
   enhanced base transceiver (EBTS) in, 182, 314
   expansion of system for, 495–497
   fast reconnect in, 186
   frequency spectrum for, 183–184, **184**, 312
   group call feature in, 182
   handoff in, 186–187
   home location register (HLR) in, 182
   interconnection location areas (ILAs) in, 190, **191, 192**, 312, 315
   international mobile subscriber identity (IMSI) in, 183
   lost calls in, 311–312
   maintenance in, 312
   metro packet switch (MSC) in, 182
   mobile assisted handoff (MAHO) in, 186–187
   operations and maintenance center (OMC) in, 182
   packet channel (PCH) in, 185
   performance in, 310–315, 460
   power control in, 186, 314
   primary control channel (PCCH) in, 184, 313–314
   provisioning checks for, 313
   random access channel (RACH) in, 184

integrated dispatch enhanced network (iDEN) (*Cont.*):
   random access protocol (RAP) and, 188
   receive signal strength indicator (RSSI) in, 185
   retuning frequency in, 346, 353
   service area (SA) in, 187
   short messaging service (SMS) and, 182
   speed of transmission in, 183–184
   temporary control channel (TCCH) in, 185
   time division multiple access (TDMA) in, 183
   traffic channel (TCH) in, 185, 314
   traffic load in, 312
   transcoder (XCDR) in, 182
   voice traffic projections in, 499–500
integrated systems digital network (ISDN), LMDS and, 12
intelligent antennas, 64–65, **65**
intercept equipment, 24
interconnect/telco room, 23
interconnection location areas (ILAs), iDEN, 190, **191, 192**, 312, 315
interconnection, network interconnect growth studies in, 531–534, **532, 533**
interexchange carriers (IXC), 23, 531
interference, 248, 256–257, **256, 257**, 380
intermachine trunks (IMTs), 531
intermediate frequency (IF), receivers and, 40
intermodulation, 255
intermodulation distortion (IMD), 33, 42, 47, 248, 385–387
international mobile identity (IMEI), GSM, 145–146
international mobile subscriber identity (IMSI):
   general packet radio system (GPRS) and, 176
   global system for mobile communication (GSM) and, 152
   integrated dispatch enhanced network (iDEN) and, 183
Internet access, 12, 14, 17
Internet protocol (IP), 1–2
   CDMA and CDMA2000 in, mobile IP (3G), 129–132, **132–134**
   CDMA and CDMA2000 in, simple, 127–129, **128–130**
   voice over (*see* voice over IP)
interworking function (IWF), GSM, 146, 160
invoicing applications, billing and, 424
IP addressing, 536–539
   general packet radio system (GPRS) and, 176
   IP number inventory report in, 465, **466, 467**
IP number inventory report, 465, **466, 467**
IP-in-IP tunnel, CDMA and CDMA2000 in, 129
IPv4, 536–539
IS-136, 3, 6, 67, 81, 84–101, **85**
   A- and B-band operators in, 86, 92–93, **93–99**
   advance mobile phone system, 91
   air interface signaling in, 85–86, **85**
   all servers busy (ASB) trouble in, 308–309
   bit error rate (BER) in, 90, **91**, 310
   call flow diagram for, **87**
   cell sites for, 306–307, 309
   coded digital voice color code (CDVCC) in, 88
   control channel in, 91–99
   design and, 219, **220**
   digital color code (DCC) in, 100, **101**
   digital control channel (DCCH) in, 92–93, **93–99**
   digital mobile attenuation code (DMAC) in, 307
   digital traffic channel (DTC) in, 86, **88**
   digital verification color code (DVCC) in, 100–101, **102**
   digital voice color code (DVCC) in, 91, 305
   expansion of system for, 494–497
   fast associated control channel (FACCH) in, 86
   frame error rate (FER) in, 100, 310
   frequency reuse in, 100–101
   frequency spectrum for, 306

582   Index

IS-136 (*Cont.*):
  full vs. half rate mobile in, 86
  guard time in, 88
  handoff failures in, 307–308
  handoffs in, 100
  lost calls in, 305–307
  maintenance issues for, 307
  mobile assisted handoff (MAHO) in, 100
  mobile switching center (MSC) in, 307–309
  modulation in, 100
  offset between transmit and receive in, 88–89, **89**
  performance in, 304–310, 457
  power control in, 307
  ramp time in, 88
  received signal strength level (RSSI) in, 100
  retuning frequency in, 346, 351–352
  reverse signal strength indicator (RSSI) in, 305
  shortened burst in, 91, **92**
  signal strength in, insufficient (IS), 309
  slow associated control channel (SACCH) in, 86
  speech coding in, 89–90, **90**
  static in, 309–310
  supplementary digital color code (SDCC) in, 100, **101**
  SYNC/synchronization in, 88
  time alignment in, 90–91
  time division multiple access (TDMA) in, 88–89, 100
  traffic load in, 306
  vector sum excited linear prediction (VSELP) in, 89
  vocoders for, 85–86
  voice channel structure for,86–88
  voice traffic projections and, 499–501, **501**
  wireless application protocol (WAP), 17
IS-36, 269
IS-66l, 6, 81
IS-95, 67, 81
  antennas and, 238–240, **238**
  CDMA and CDMA2000 in, 103–104, **103**, 106, **108**
  design and, 219, **220**
  wireless application protocol (WAP), 17
isolation, 254–264
  antenna radiation patterns and, 260
  calculating needed amount of, 257–258, **258**
  free space and free space path loss in, 259
  horizontal separation in, 261–263, **262, 263**
  interference and, 256–257, **256, 257**
  margin for, 261
  requirements for, 256–257, 259
  slant separation in, 264, **264**
  source antenna correction (SAC) and, 263
  vertical separation and, 260–261, **261**

Japan total access communication system (JTACS), 5

K factor, 61
Ki authentication algorithm, 144

"last mile" technologies, 2
layer 2 tunneling protocol (L2TP), 128
leaky feeder amplifier approach, 251–252
lease line, LMDS and, 12
legacy systems, 197, 496–497
line of sight (LOS), 248, 249
linear predictive coding (LPC), IS-136 and, 89
link active time, 405
link budget, 59–60, 221–224, **223, 225, 226**, 248, 249
link changeovers, 405
link errors, 404
link models, 48
link performance measurement and optimization, 403–406
  link active time in, 405
  link changeovers in, 405
  link errors in, 404
  link retransmission rate in, 404
  link status signaling units (LSSU) in, 404
  link traffic loading in, 403–404
  message signaling units (MSU) in, 403
link status signaling units (LSSU), 404
local area network (LAN), 247–248
  Bluetooth and, 17–18
  local multipoint distribution system (LMDS) and, 12
  wireless (*see* 802.11 protocol; wireless LAN)
local multipoint distribution services (LMDS), 8–13, **10**, 246, 269
  applications benefited by, 12
  architecture types used in, 13
  bandwidth of, 9–11
  competitive nature of, 11
  coverage area for, 11
  802.11 wireless LAN protocols and, 10
  frequency spectrum for, 11
  handoffs in, 10
  hub or base station for, 10–11
  microwave transmissions via, 10
  multiplexing in, 11
  physical transport layer functions in, 10
  quality of service (QoS) in, 11
  service layer functions in, 10
  voice over IP (VoIP) and, 21
location area (LA), GSM, 336, 338, 340
location update, GSM, 150–152, **151**
log periodic antenna, 34
log periodic dipole array (LPDA), 34
logical link control (LLC), GPRS and, 175
Longley-Rice propagation model, 48
loss, 248, 272, 278, **278**, 371
  Cost231 propagation model, 50–55
  effective radiated power (ERP) and, 56–58
  environmental attenuation in, 55, **56**
  free space path, 48–49
  link budget, 59–60
  link budgets and, 221–224
lost calls, 272, 275, **277**, 281–292, **285, 289–292**, 371
  CDMA and, 316–319
  global system for mobile communication (GSM) and, 329–336
  integrated dispatch enhanced network (iDEN) and, 311–312
  IS-136 and, 305–307
low-pass filters, 35–38, **36, 38**

macrocells, 246, 249
maintenance, CDMA and, 326, 328
man machine interface (MMI), in billing processes, 437
manager of system performance role, 390
maps, search area request (SAR) and, 204
Marconi, Guglielmo, 2
marketing plans, 490–491
measuring performance, 270, 272–273, 393–402
media access control (MAC), wireless LAN (WLAN) and, 20
media gateway control protocol (MGCP), VoIP, 20
memory settings and utilization, for switches, 401
message signaling units (MSU), 403
method of procedure (MOP), 231–233
metrics (*see* measuring performance)
metro packet switch (MSC), iDEN and, 182
metropolitan statistical areas (MSA), 4, 347
metropolitan trading areas (MTA), 7, 347
microcells, 246, 249

microwave, 2
  CDMA and CDMA2000 in, 101
  local multipoint distribution system (LMDS) and, 10
migration, 197, 496–497
mobile application part (MAP)
  general packet radio system (GPRS) and, 167
  global system for mobile communication (GSM) and, 152
mobile assisted handoff (MAHO):
  CDMA and CDMA2000 in, 103
  global system for mobile communication (GSM) and, 330, 332, **334, 335**, 336–338, **337**
  integrated dispatch enhanced network (iDEN) and, 186–187
  IS-136 and, 100
mobile country code (MCC), GPRS, 176
mobile data base station (MDBS), CDPD and, 191–192
mobile data intermediate system (MDIS), CDPD and, 192
mobile end system (MES), CDPD and, 191
mobile IP (3G), CDMA and CDMA2000 in, 129–132, **132–134**
mobile phone services, 2
mobile station ISDN (MSISDN), GSM and, 155
mobile station roaming number (MSRN), GSM and, 155
mobile switching center (MCS), 21
  advanced mobile phone system (AMPS) and, 70
  cellular communication and, 4
  general packet radio system (GPRS) and, 166, 168, 173, 179
  global system for mobile communication (GSM) and, 143, 144, 145, 152, 153, 155–160, 166
  IS-136 and, 307–309
  planning and, 534–535
  traffic projections and, 498
mobile telephone switching office (MTSO), 21, 70, 80
mobile telephone systems (MTS), 4, 74
modems, 2
modulation, 1, 26, 27, 29–32, **30**, 81, 83
  IS-136 and, 100
  local multipoint distribution system (LMDS) and, 13
monitoring performance, 270
monitoring systems, auxiliary node performance in, 402
Morse code, 2
Morse, Samuel, 2
multichannel multipoint distribution system (MMDS), 13–14, 269
multichannel sites, 246
multifrequency (MF) channels, 400
multiplexing, LMDS and, 11
multipoint distribution service (MDS), 13–14
multipoint distribution system (MMDS), 246
mutual compensation, called vs. calling parties, 446–447, **446**

narrowband, 3, 6
neighbor lists:
  CDMA and, 327–328
  global system for mobile communication (GSM) and, 330–331, **330–332**
$N = 12$, $N = 7$, $N = 4$ frequency assignment methods, 227–235, **227, 228**
network and RF planning, 487–542
  2.5 or 3G migration design procedure for, 496–497
  budgeting and, 492, 539–541
  capacity cell sites and, 518–521
  circuit switch growth and, 524–530
  construction and, 492
  coverage area and, 517–518
  data traffic projections in, 511–515
  design criteria in, 491–492
  final report in, 541–542
  fixed network and, 522–524

network and RF planning (*Cont.*):
  head count requirements in, 539, **540**
  IP addressing and, 536–539
  marketing plans and, 490–491
  methodology for, 489–492
  network interconnect growth studies in, 531–534, **532, 533**
  new technology deployment and, 492
  new wireless system procedure for, 495–496
  packet data serving nodes (PDSN) and, 534–535
  planning process flow in, 487–489, **488**
  presentation for, 542
  radio voice traffic projections in, 498–511
  switch capacity studies and, 524–526
  switch port capacity study in, 526–528, **527, 529**
  switch subscriber capacity study in, 528–530
  system expansion and, 494–497
  system growth and, 515–522, **519**
  time frames for, time reports for, 490
  traffic off-loading and, 521–522, **522, 523**
  traffic projections and, 497–498
  traffic tables in, 492–494
network briefings, 485
network color code (NCC), GSM and, 333–334
network configuration report, 470–471, **472**
network dimensioning, GPRS and, 180–181
network interface card (NIC), CDMA and, 324
network service access point identifier (NSAPI), GPRS and, 175, 176
new site investigation, 374–379
new technology deployment, 492, 552
new wireless system, expansion of system for, 495–496
1900 GSM channel chart, 567–568
911 numbers, 433
nodes, GPRS and, 166–169, **166**, 181
noise, 40, 43–45, 248
noise figure, receivers and, 43–45
nonbillable events, 433
Nordic mobile telephone (NMT), 5
North American digital cellular (NADC), 5, 6, 84
notch filters, 37
  (*see also* band reject filters)
notification channel (NCH), GSM and, 148–150

off-loading traffic, 521–522, **522, 523**
Okumura report, modeling of radio communication and, 49
omnidirectional antennas, 33, 34, 235
Omnipoint, 6, 81
1–db compression, receivers and, 45, **46**
1XDO, 67, 105
1XDV, 105
1XEV, 105
1XRTT, 67
  CDMA and CDMA2000 in, 105, 106, 119, 324, 326–328
  design and, 219
  voice traffic projections in, 499–500, 503
operation and maintenance link (OML), GSM and, 143
operations and maintenance center (OMC), iDEN and, 182
operations and support system (OSS), GSM and, 141
operations department organization, 548, **550**
operations room, 21, 23
optical fiber, 2, 14, **15**
optimizing performance, 393–402
organizational structure, 543–544
  centralized, 543–544
  departments in, 544–545, **545**
  distributed, 543–544
  engineering department organization in, 545–548, **546, 549**
  head count drivers in, 552–553, **554**

584 Index

organizational structure (*Cont.*):
  hiring practices in, 553–555
  new technology and budget department organization in, 552, **552**
  operations department organization in, 548, **550**
  outsourcing in, 555–556
  real estate and implementation department in, 548–551, **551**
  training in, 556–559
orientation, 382–383
orthogonal frequency division multiplexing (OFDM), wireless LAN (WLAN) and, 20
orthogonal transmit diversity (OTD), 237, 238
oscillation of RF waves, 25
out-of-band emission, 248
out-of-service reports, 272, 371
outsourcing, 555–556

packet access grant channel (PAGCH), GPRS and, 165
packet associated control channel (PACCH), GPRS and, 165, 170, 172, 508
packet broadcast control channel (PBCCH), GPRS and, 165, 178
packet channel (PCH), iDEN and, 185
packet common control channel (PCCCH), GPRS, 164–165, 172
packet control unit (PCU), GPRS and, 166
packet data, CDPD and, 190–194
packet data channel (PDCH), GPRS and, 164
packet data protocol (PDP) context, GPRS and, 166, 173–178, **175**
packet data rates, CDMA, 505, **506**
packet data serving node (PDSN):
  CDMA and CDMA2000 in, 107–110, 127–132, 325–326
  expansion of system for, 495–497
  general packet radio service (GPRS), 341
  global system for mobile communication (GSM) and, 341
  planning and, 534–535
packet data traffic channel (PDTCH):
  general packet radio system (GPRS) and, 165–166, 172
  voice traffic projections and, 507–508
packet notification channel (PNCH), 165
packet paging channel (PPCH), GPRS and, 165
packet random access channel (PRACH), GPRS and, 165
packet session access:
  CDMA and, 324–326
  global system for mobile communication (GSM) and, 340–342, **341**
packet session throughput:
  general packet radio service (GPRS), 343–346
  global system for mobile communication (GSM) and, 343–346
packet switch performance report, 460–462, **462, 463**
packet temporary mobile station identity (P–TMSI) in, 169, 178
packet timing control channel (PTCCH), GPRS and, 164, 165
paging, 246
  CDMA and CDMA2000 in, 111–113
  general packet radio system (GPRS) and, 165
  global system for mobile communication (GSM) and, 147, 155
paging channel (PCH), GSM and, 147, 155
passive antennas, 33
password authentication protocol (PAP), CDMA and CDMA2000 in, 128, 131
path clearance, 60–62, **62**
path loss, link budgets and, 221–224
PCS1900, 6
peg counters, 444
penetration rate, data traffic projections and, 513
performance issues, 269–391
  access failure in, 272, 277–278, **278**, 292–298, **285, 296**, 312–315, 323–324, 371
  action plans for, 273, 388–391
  advanced mobile phone system (AMPS) and, 315, 457

performance issues (*Cont.*):
  alarm setting, for switches and nodes, 401
  all servers busy (ASB) trouble in, 308–309, 321–323, 338–339
  analysis methodology in, 274–281
  auxiliary node performance in, 402
  benchmarks established for, 269–270
  bit error rate (BER), 272, 371
  blocking, 272, 275–276, **277**, 298–303, **300**, 371
  call completion ratio, 272, 371
  call delivery troubleshooting in, 411–422
  call processing efficiency in, 398–399, **399**
  cell site data on, 410–411
  circuit switch (*see* circuit switch performance guidelines)
  circuit switch/node performance report in, 462–463, **464**
  code division multiple access (CDMA), 315–328
  complaints, 272, 371
  cross mapping metrics in, 272–273
  data traffic projections in, 511–515
  design and, 198, 217–218, 380–381
  downtilting in, 383–385, **384, 385**
  downtime (service outage time) and, 399, **400**
  drive testing in, 358–364, 371
  dynamic routing numbers (DRNs), 529
  erlangs (total) and call volume in, 400–401
  exception report in, 471–476, **474, 476**
  existing cell site evaluation in, 378–382
  facilities interconnect report (data) in, 467–468, **469**
  facility usage report in, 465–467, **468**
  flowchart of, 281, **281**
  frame error rate (FER), 272, 371
  frequency planning and, 225–235
  general data on, 410–411
  general packet radio service (GPRS) and, 329, 457, 460
  global system for mobile communication (GSM), 315, 329–346, 457
  goals and objectives of, 269–270, 273
  handoff failure in, 272, 307–308, 319–321, 336–338, **337**, 371
  head count requirements in, 539, **540**
  information requirements for, 271–272
  integrated dispatch enhanced network (iDEN), 310–315, 460
  interference and, 380
  intermodulation in, 385–387
  IS-136, 304–310, 457
  key factors in, 271–274
  key indicators of, **391**
  link active time in, 405
  link changeovers in, 405
  link errors in, 404
  link performance measurement and optimization in, 403–406
  link retransmission rate in, 404
  link traffic loading in, 403–404
  loss, 272, 278, 371
  lost calls in, 272, 275, **277**, 281–292, **285, 289–292**, 305–307, 311–312, 316–319, 329–336, 371
  manager of system performance role in, 390
  measuring and optimizing, 270, 272–273, 393–402
  memory settings and utilization, for switches, 401
  message signaling units (MSU) in, 403
  monitoring in, 270
  monthly plan for, 389–391
  network briefings and, 485
  new site investigation and, 374–379
  optimizing, 393–402
  orientation, 382–383
  out of service reports, 272, 371
  packet switch performance report in, 460–462, **462, 463**
  personal communication services (PCS), 457
  prioritizing tasks in, 394–398, **395**

performance issues (*Cont.*):
   project status report (current and pending) in, **478**, 479–480
   provisioning checks for, 313
   quality of service (QoS) and, 281, 298
   radio voice traffic projections in, 498–511
   receivers and, 41
   reporting on, 273–281, 389, 457–460, **461**
   retunes in, 346–357
   reverse signal strength indicator (RSSI) in, 359–364
   routing efficiency rating in, 406–409
   routing performance monitoring and management in, 406–410
   signal quality estimate (SQE), 272, 371
   signal strength in, insufficient (IS), 309, 339–340
   site activation and, 364–374, **368, 369**
   site investigations in, 374–382
   site qualification test (SQT) in, 358–364
   software performance and, 410
   spikes in traffic load and, 394–398, **395**
   static in, 309–310
   statistics reports for, 275, **276**
   steps in evaluating, 270–271, 274–275
   subscriber data in, 411
   switch capacity studies and, 524–526
   switch CPU loading and, 393–398, **396**
   switch port capacity study in, 526–528, **527, 529**
   switch service circuit loading in, 400
   switch subscriber capacity study in, 528–530
   system and switch traffic data in, 411
   system circuit switch traffic forecast report in, 469, **470**
   system expansion and, 494–497
   system growth status report in, 471, **473**
   system performance engineer's role in, 390
   system software report in, 480, **481**
   system status bulletin board report in, 476–479, **477**
   system usage and, comparison, analysis, verification of, 443–447
   technology specific guidelines for, 304
   TIC lists for, 388–389
   time division multiple access (TDMA), 315
   time frame established in, 269–270
   time vs. money in, 271
   timing source accuracy in, for switches, 401–402, **402**
   traffic off-loading and, 521–522, **522, 523**
   traffic projections and, 497–498
   trouble reports, 272, 371
   upper management report in, 480, **482–483**
   usage, 272, 278, **278**, 371
   wall chart to display, 279–280, **280**
   (*see also* circuit switch performance guidelines)
personal communication networks (PCN), 6
personal communication services (PCS), 1, 6–7, 69, 81, 83, 304
   1900 GSM channel chart, 567–568
   A through F frequency blocks in, 7
   CDMA and CDMA2000 in, 102, 103
   Cost231 propagation model, 50–55, **52**
   design and, 221
   expansion of system for, 494–497
   free space path loss in, 49
   frequency spectrum for, 6–7, **7, 83**
   global system for mobile communication (GSM) and, 6
   metropolitan and basic trading areas (MTA/BTA) in, 7
   narrow- vs. wideband systems for, 6
   performance issues in, 457
   propagation models and, 48
   standards for, 6–7
   types of, 6–7
   wireless application protocol (WAP), 17
   wireless local loop (WLL) and, 8
   (*see also* cellular communication)

personal digital assistants (PDA), 17
personal digital cellular (PDC), 6
personnel:
   hiring practices for, 553–555
   smart outsourcing and, 555–556
   training, 556–559
phase modulation (PM), 29–32
   receivers for, 40
   transmitter for, 27, **28**
phase noise, receiver, 39
physical transport layer, local multipoint distribution system (LMDS) and functions, 10
Pi/4DQPSK, IS-136 and, 31, **33**, 100
picocells, 246, 249
pilot channel, CDMA and CDMA2000 in, 111, 137–140, **138, 141**
pilot pollution, 297, 317, 323
pilot signal measurement message (PSMM), CDMA and, 317
plain old telephone service (POTS), LMDS and, 12
planning (*see* network and RF planning)
planning and zoning board approval, 216–217
point to multipoint (PMP), 246
point to multipoint–multicast (PTM-M) notification, GPRS and, 165
point-to-point communication systems, 60–62
point-to-point protocol (PPP), CDMA and CDMA2000 in, 108, 126
Poisson ratio, blocking and, 300–303
population density, data traffic projections and, 513
port capacity, switches, 526–528, **527, 529**
post turn on (PTO), 201
postal, telegraph, telephone (PTT), 4
   (*see also* public switched telephone system)
postpaid services, 440
power budget, 212–216, **215**, 248–249
power control:
   CDMA and CDMA2000 in, 120, 316–317
   global system for mobile communication (GSM) and, 335–336
   integrated dispatch enhanced network (iDEN) and, 186, 314
   IS-136 and, 307
power density chart, 213–216, **213**
power levels, CDMA and CDMA2000 in, subscriber, 118, **118**
power room, 21, 23
preliminary design review (PDR), 199–201
prepaid services, 440
presentation, of final planning report, 542
primary control channel (PCCH), iDEN, 184, 313–314
primary rate interface (PRI), LMDS and, 12
prioritizing CPU tasks, 394–398, **395**
private branch exchange (PBX), 8, 19, 247
project status report (current and pending), **478**, 479–480
projections, traffic projections, 497–498
propagation, 26
propagation models, 48–55, **54**
propagation plots, search area request (SAR) and, 204
provisioning checks, iDEN and, 313
provisioning process, revenue assurance in wireless network and, 435
pseudonumber (PN) code, 224
   CDMA and CDMA2000 in, 102, 137–140, **138–140**
   retuning frequency in, 346
public switched telephone network (PSTN):
   advanced mobile phone system (AMPS) and, interface with, 70
   blocking in, 299
   cellular communication and, 4
   interface with mobile system in, 21
   network interconnect growth studies in, 531–534, **532, 533**
   traffic projections and, 498
   wireless local loop (WLL) and, 8
pulse code modulation (PCM), GSM and, 144

586 Index

quad–mode subscriber units, 494–495
quadrature amplitude modulation (QAM), 5, 31
quadrature phase shift keying (QPSK), 31, **32**, 107
quality of service (QoS), 281, 298
    data traffic projections and, 513
    general packet radio system (GPRS) and, 175–177
    local multipoint distribution system (LMDS) and, 11
    traffic projections and, 498
    voice over IP (VoIP) and, 20

radiation, electromagnetic field (EMF) compliance and, 212–216, **215**
radio, 2
radio configuration (RC), CDMA and CDMA2000 in, 119–120, **120**, 122–123, **123**
radio engineering, 25–66
radio frequency (RF) network performance report, 457–460, **461**
radio frequency (RF) waves, 25–26
radio spectrum (*see* frequency spectrum)
radio system, **198**
radio systems, 26–27, **27**
radio voice traffic projections in, 498–511
ramp time, IS-136 and, 88
random access channel (RACH):
    global system for mobile communication (GSM) and, 148–150, 155
    integrated dispatch enhanced network (iDEN) and, 184
random access protocol (RAP), iDEN and, 188
rating applications, billing and, 423
rating process, revenue assurance in wireless network and, 436
real estate and implementation department organization, 548–551, **551**
receive/reverse signal strength indicator (RSSI), 359–364
    global system for mobile communication (GSM) and, 330, 332, 336–339, **337**
    integrated dispatch enhanced network (iDEN) and, 185
    IS-136 and, 100, 305
receivers, 26, 39–40, **39**
    amplitude modulation (AM), 40
    bit error rate (BER) in, 42, 47
    carrier to interference ratio (C/I) in, 43
    degradation levels in, 43
    desense in, 46–47
    distortion, 40
    dynamic range of, 39, 41–43, **42, 44**
    efficiency of, 39
    energy per bit to noise density ratio (Eb/No) in, 43
    figures of merit for, 41–47
    frequency modulation (FM), 40
    frequency range, 39
    intermediate frequency (IF) and, 40
    intermodulation distortion (IMD) and, 42, 47
    isolation in, 254–264
    noise, noise figure and, 40, 43–45
    1–db compression in, 45, **46**
    performance criteria for, 39, 41–47
    phase modulation (PM), 40
    phase noise, 39
    selectivity of, 41
    sensitivity in, 40, 41, 43, 45
    signal to noise ratio (SNR) in, 42, 44, 47
    spurious free dynamic range (SFDR) in, 42
    third order intercept (IP3) in, 45–46
    tuning resolution in, 39
    tuning speed in, 39
reconciliation of customer databases, 436–440
reconnection, iDEN and, 186
registration area, CDMA and, 323

remote antenna driver/remote antenna signal processor (RAD/RASP), 14, **16**
repeater (*see* reradiators)
reports (*see* system documentation and reports)
reradiators, 246–247, 249
resource sharing, GPRS and, 162–163
retuning frequency, 346–357, 362, **363**
reusing frequency, 229
revenue assurance in wireless network, 435–447
    billing process and, 436
    call detail records (CDR) generation and mediation process in, 435
    called vs. calling party verification in, 444–447, **446**
    database analysis in, 436–440
    database provisioning in, 437, **438**
    determining total subscriber numbers for, 438–440
    fulfillment process in, 435
    mutual compensation in, called vs. calling parties, 446–447, **446**
    peg counters in, 444
    postpaid and prepaid services in, 440
    provisioning process in, 435
    rating process in, 436
    reconciliation of customer databases in, 436–440
    system usage and performance data in, comparison, analysis, verification of, 443–447
    verification of billing in, 440–443
    volume benchmarks using CDR in, 440–442
    (*see also* billing and charging)
reverse channels, CDMA and CDMA2000 in, 116–119, **117, 118**
reverse common control channel (R CCCH), 118
reverse dedicated control channel (R DCCH), 118
reverse enhanced access channel (REACH), 118
reverse pilot channel (RPICH), CDMA and CDMA2000 in, 118
reverse signal strength indicator (RSSI) (*see* receive/reverse signal strength indicator)
reverse supplemental channel (RSCH), CDMA and CDMA2000 in, 118
RF planning (*see* network and RF planning)
roaming:
    cellular intercarrier billing exchange roamer (CIBER) format for, 424, **426**
    global system for mobile communication (GSM) and, 155
round trip delay (RTD), CDMA and, 317, 323–324
routers, 24
    CDMA and CDMA2000 in, 109
    dynamic routing numbers (DRNs), 529
routing area (RA), GSM and GPRS, 166, 177–179, **178**, 336, 338, 340, 345–346, **345**
routing performance monitoring and management, 406–410
    grade of service and, 408–409
    routing efficiency rating in, 406–409
    summary of, 409–410
rural statistical area (RSA), 4, 347

satellite, CDMA and CDMA2000 in, 101
search area request (SAR), 201–205, **203**
search windows, CDMA and CDMA2000 in, 134–135, **135**
second generation (2G/2.5G) technology, 67, 80–84, 496–497
selectivity, receiver, 41
sensitivity, receiver, 40, 41, 43, 45
servers, 24
service access point identifier (SAPI), GPRS and, 175
service area (SA), iDEN and, 187
service circuit loading, 400
service layer, local multipoint distribution system (LMDS) and functions, 10
service level agreements (SLA), 21, 299

serving GPRS support node (SGSN), 166–169, 172, 173, 176–179, 181
shadowing, 248
Shannon–Hartley capacity equation, 29
short message service center (SMSC):
  general packet radio system (GPRS) and, 168
  global system for mobile communication (GSM) and, 145
short messaging service (SMS), 21, 80, 182
shortened burst, IS-136 and, 91, **92**
shot noise, receivers and, 44–45
signal quality estimate (SQE), 272, 371
signal strength:
  global system for mobile communication (GSM) and, insufficient (IS), 339–340
  integrated dispatch enhanced network (iDEN) and, 185
  IS-136 and, insufficient (IS), 309
signal to noise ratio (SNR), 42, 44, 47, 65
signal transfer points (STPs), auxiliary node performance in, 402
signaling connection control part (SCCP), GSM and, 143
signaling transfer point (STP), 24
simple IP, CDMA and CDMA2000 in, 127–129, **128–130**
simple network management protocol (SNMP), 538
single gateway control protocol (SGCP), voice over IP (VoIP) and, 20
site acceptance (SA), 208–209
site acceptance form (SAF), 201, 209, **210**
site activation performance, 364–374, **368, 369**
site investigations, 374–382
site qualification test (SQT), 201, 205–208, **206–208**, 358–364, **358**
site rejection (SR), 201, 209–211, **211**
611 numbers, 433
slant separation, 264, **264**
slow associated control channel (SACCH):
  global system for mobile communication (GSM) and, 148–150, **150**
  IS-136 and, 86
smart outsourcing, 555–556
soft handoffs, CDMA and CDMA2000 in, 103, 133, 135–137, **137**
software performance, system software report in, 410, 480, **481**
source antenna correction (SAC), 263
space transmit diversity (STD), 237, 238
specialized mobile radio (SMR), 3, 69, 83, 246, 495–497
spectrum (*see* frequency spectrum)
speech coding, IS-136 and, 89–90, **90**
speed of transmission/throughput, 2–3, 29, 161, 163
  cellular communication, 3
  integrated dispatch enhanced network (iDEN) and, 183–184
  wireless LAN (WLAN) and, 19
spikes in traffic load, 394–398, **395**
spread spectrum technology, 6, 101
spreading rate (SR), CDMA and CDMA2000 in, 119–120, **120**, 122–123, **123**
spurious free dynamic range (SFDR), receivers and, 42
SS7, **422**
  general packet radio system (GPRS) and, 167
  global system for mobile communication (GSM) and, 143, 152
  link traffic loading in, 403–404
  routing efficiency rating in, 406–409
  troubleshooting, 421–422, **422**
staff (*see* personnel)
stand-alone dedicated control channel (SDCCH), GSM, 148–153, 155–157, 160, 180, 506–507
standards, personal communication services (PCS), 6–7, 81
standing wave ratio (SWR), 34, 377
static, IS-136 and, 309–310
statistics reports for performance, 275, **276**

subnetwork dependent convergence protocol (SNDCP), GPRS and, 177
subscriber data, 411
subscriber identity module (SIM), 141–142, 342
subscribers:
  head count requirements in, 539, **540**
  numbers of, 438–440
  switch capacity and, 528–530
supplemental channel (SCH), CDMA and CDMA2000 in, 110
supplementary digital color code (SDCC), IS-136 and, 100, **101**
switch port capacity study in, 526–528, **527, 529**
switches, 24
  alarm setting, for switches and nodes, 401
  call processing efficiency in, 398–399, **399**
  CDR call data comparison in, 443–444
  circuit switch growth and, 524–530
  CPU loading and performance optimization in, 393–398, **396**
  dual tone multifrequency (DTMF) channels in, 400
  erlangs (total) and call volume in, 400–401
  global system for mobile communication (GSM) and, 141
  memory settings and utilization, for switches, 401
  multifrequency (MF) channels in, 400
  service circuit loading in, 400
  switch port capacity study in, 526–528, **527, 529**
  switch subscriber capacity study in, 528–530
  system circuit switch traffic forecast report in, 469, **470**
  timing source accuracy in, 401–402, **402**
symbol rates, 31
SYNC/synchronization, IS-136 and, 88
synchronization channel (SCH):
  CDMA and CDMA2000 in, 111
  global system for mobile communication (GSM) and, 147
system circuit switch traffic forecast report in, 469, **470**
system documentation and reports, 273–281, 389, 449–486
  accuracy of, checking for, 450
  bouncing congestion hour traffic report (node and service) in, 454–457, **455–459**
  circuit switch/node performance report in, 462–463, **464**
  company meetings and, 481–484
  customer care report in, 476, **477**
  descriptions for, 449–450
  engineering library for, 450–451
  exception report in, 471–476, **474, 476**
  facilities interconnect report (data) in, 467–468, **469**
  facility usage report in, 465–467, **468**
  frequency of reports in, 485–486, **486**
  hierarchical approach to, 449
  identifying documents in, 450
  IP number inventory report in, 465, **466, 467**
  network briefings and, 485
  network configuration report in, 470–471, **472**
  objectives of reports in, 452–454
  packet switch performance report in, 460–462, **462, 463**
  project status report (current and pending) in, **478**, 479–480
  report categories in, 451–452
  RF network performance report in, 457–460, **461**
  system circuit switch traffic forecast report in, 469, **470**
  system growth status report in, 471, **473**
  system software report in, 480, **481**
  system status bulletin board report in, 476–479, **477**
  telephone number inventory report in, 464–465
  upper management report in, 480, **482–483**
system expansion, 494–497
system growth status report, 471, **473**
system performance engineer's role, 390
system software report in, 480, **481**
system status bulletin board report, 476–479, **477**

588 Index

system usage and performance data, comparison, analysis, verification of, 443–447

T1/E1:
  advanced mobile phone system (AMPS) and, 71
  local multipoint distribution system (LMDS) and, 12
telco room, 23
telegraph, 2
telephone, 2
telephone number inventory report, 464–465
television, 2
temporary block flow (TBF), GPRS and, 169
temporary control channel (TCCH), iDEN and, 185
temporary flow identity (TFI), GPRS and, 169
temporary logical link identity (TLLI), GPRS and, 169
temporary mobile station identity (TMSI), GPRS and, 169
temporary mobile subscriber identity (TMSI), GSM and, 152
testing, call testing procedures, 415–418, **416, 417**, 431–432
thermal noise, receivers and, 44–45
third generation (3G) technology, 1, 67, 496–497
third order intercept (IP3), receivers and, 45–46
3XRTT, CDMA and CDMA2000 in, 106, 119, 121
throughput (see speed of transmission/throughput)
time alignment:
  global system for mobile communication (GSM) and, in handoffs in, 157–158
  IS-136 and, 90–91
time division duplexing (TDD), 6, 13
time division multiple access (TDMA), 81, 83–101, 274
  CDMA and CDMA2000 in vs., 103
  digital AMPS (D-AMPS) and, 84–101
  global system for mobile communication (GSM) and, 146
  integrated dispatch enhanced network (iDEN) and, 183
  IS-136 and, 84–10
  performance in, 315
  wireless application protocol (WAP), 17
  wireless LAN (WLAN) and, 20
time division multiplexing (TDM), 494
  CDMA and CDMA2000 in, 110
  local multipoint distribution system (LMDS) and, 13
  voice over IP (VoIP) and, 21
time slots:
  general packet radio system (GPRS) and, 163, 165–166
  global system for mobile communication (GSM) and, 147, 148–149, **149**, 157–158
time vs. money in troubleshooting, 271
timeouts, CDMA and, 328
timing source accuracy, for switches, 401–402, **402**
toll free numbers, 433
toll room, 23, 24
top-end numbers, blocking, 299
total access communication system (TACS), 3, 5, 67, 81
tower-top amplifiers (TTA), 63–64, **63**
traffic calculation:
  general packet radio system (GPRS) and, 169, 180
  global system for mobile communication (GSM) and, 160–161
traffic channel (TCH):
  general packet radio service (GPRS), 342
  global system for mobile communication (GSM) and, 148–150, **150**, 153–155, 157
  integrated dispatch enhanced network (iDEN) and, 185, 314
  voice traffic projections in, 499–500, 506
traffic engineering/traffic management:
  bouncing congestion hour traffic report (node and service) in, 454–457, **455–459**
  CDMA and, 317–318, 325, 327
  CDMA and code division multiple access (CDMA)

traffic engineering/traffic management (Cont.):
  CPU switching loading and performance optimization in, 393–398, **396**
  data traffic projections in, 511–515
  global positioning system (GPS) and, 321
  global system for mobile communication (GSM) and, 334–335
  integrated dispatch enhanced network (iDEN) and, 312
  IS-136 and, 306
  link traffic loading in, 403–404
  off-loading and, 521–522, **522, 523**
  radio voice traffic projections in, 498–511
  spikes in traffic load and, 394–398, **395**
  system and switch traffic data in, 411
  system circuit switch traffic forecast report in, 469, **470**
  traffic projections and, 497–498
  traffic tables in, 492–494
traffic off-loading, 521–522, **522, 523**
traffic projections and, 497–498
traffic tables, 492–494
training, 556–559
transatlantic cable, 2
transcoder (XCDR), iDEN and, 182
transcoding and rate adaptation unit (TRAU), GSM and, 143–144
transmitters, 26, 27, **28**
  AM, 27, **28**
  effective radiated power (ERP) and, 56–58, 205
  electromagnetic field (EMF) compliance and, 212–216, **215**
  FM, 27, **28**
  intermodulation in, 385–387
  isolation in, 254–264
  phase modulation (PM), 27, **28**
  power budget in, 212–216, **215**
  power density chart for, 213–216, **213**
  reradiators for, 246–247
  site qualification test (SQT) in, 205–208, **206–208**, 358–364
trimode subscriber units, 494–495
trouble reports, 272, 371
troubleshooting (see call delivery troubleshooting; performance issues)
trunks, 531
TRXs, voice traffic projections and, 508–511, **509, 511**
tuning resolution, receiver, 39
tuning speed, receiver, 39
tunnel system design, 247–254, **252, 253**
tunneling:
  CDMA and CDMA2000 in, 128, 129
  general packet radio system (GPRS) and, 168
turn on (see cell sites; site activation performance)
two-and-a-half generation (2.5G) technology, 67
two-way radio, 2, 63–64, **63**
type approval code (TAC), GSM and, 145

uniform record locator (URL), GPRS and, 176
universal mobile telecommunication systems (UMTS), WAP and, 17
unlicensed national information infrastructure (UNII) band, 19
uplink:
  global system for mobile communication (GSM) and, 146–147
  link budgets in, 221–224, **223, 225, 226**
uplink state flag (USF), GPRS and, 169
upper management report, 480, **482–483**
usage, 272, 278, **278**, 371

vector sum excited linear prediction (VSELP), IS-136 and, 89
verification of billing, 431–433, 440–443
vertical separation, 260–261, **261**

videoconferencing, LMDS and, 12
virtual private network (VPN):
  blocking and, 299
  CDMA and CDMA2000 in, 127, 128–129, **130, 131**, 132, **134**
  local multipoint distribution system (LMDS) and, 12
visitor location register (VLR):
  CDMA and CDMA2000 in, 126–127, 323
  general packet radio system (GPRS) and, 167, 168, 173, 179
  global system for mobile communication (GSM) and, 143, 145, 152, 155
vocoders, IS-136 and, 85–86
voice mail, 24, 402
voice over IP (VoIP), 12, 20–21, **22, 23**
voice processing:
  CDMA and, 325, 328
  general packet radio service (GPRS), 341, 346
  global system for mobile communication (GSM) and, 341, 346
voice telephony, LMDS and, 12
voice traffic, radio voice traffic projections in, 498–511
  advanced mobile phone system (AMPS), 499
  broadcast control channel (BCCH), 499, 506, 507
  code division multiple access (CDMA) and, 499, 501–505
  common control channel (CCCH), 506, 507
  general packet radio service (GPRS) and, 499, 505–511, **509**
  global system for mobile communication (GSM) and, 499, 505–511, **509**
  integrated dispatch enhanced network (iDEN), 499
  IS-136 and, 49, 500–501, **501**
  1XRTT and, 499, 503
  packet associated control channel (PACCH), 508
  packet data traffic channel (PDTCH), 107–108
  stand-alone dedicated control channel (SDCCH), 506, 507
  traffic channels (TCH), 499
  TRXs in, 508–511, **509, 511**
voltage standing wave ratio (VSWR), 318
volume benchmarks using CDR, 440–442

Walfisch propagation model (*see* Cost231 propagation model)
wall chart to display performance metrics, 279–280, **280**
wall mounted antenna, 241
Walsh codes, CDMA, 109, 110, 122–123, 322, 325, 326, 328, 502, 505, **505**
Web services, LMDS and, 12

wide area networks (WAN), LMDS and, 12
wideband PCS systems, 6
wideband CDMA (WCDM), 17, 67, 105
WiFi, 19
wireless application protocol (WAP), 14, 17
wireless cable, 13
wireless fidelity (*see* WiFi)
wireless LAN (WLAN), 18–20, **18**
  Bluetooth and, 19, **20**
  collision sense multiple access/collision avoidance (CSMA/CA) in, 19
  effective radiated power (ERP) in, 19
  802.11 protocol for, 19
  802.11a and, 19
  802.11b (WiFi) and, 19
  frequency spectrum for, 19
  HiperLan/2, 20
  media access control (MAC) and, 20
  orthogonal frequency division multiplexing (OFDM) in, 20
  private branch exchange (PBX) and, 19
  protocols for, 19
  speed of transmission/throughput in, 19
  time division multiple access (TDMA) and, 20
  (*see also* wireless LANs/in-building systems)
wireless LANs (WLAN)/in-building systems, 247–254, **247**
  antenna systems for, 250, **250**
  bidirectional antenna system for, 251
  cell size in (macro-, micro-, pico-), 249
  coverage areas for, 248
  distributed antenna system (DAS) for, 250
  diversity in, 249
  interference and distortion in, 248
  line of sight (LOS) issues in, 248–249
  loss, fade, shadowing in, 248
  planning in, 253–254
  power budget for, 248–249
  reradiators and, 249
  tunnel applications, 251–253, **252, 253**
wireless local loop (WLL), 8, **9**
wireless mobile radio technologies, 67–194
World Wide Web, 19

zoning board approval, 216–217

ABOUT THE AUTHORS

CLINT SMITH, P.E., is Vice President of Network Engineering for ASI, and is currently designing 3G migration solutions for some of the world's largest telecom companies. Previously Director of Engineering for Verizon Mobile, he is the author or co-author of six successful books on wireless communications, all published by McGraw-Hill: the classic *Cellular System Design and Optimization*, *Wireless Telecom FAQs*, *LMDS*, *Practical Cellular and PCS Design*, and *3G Wireless Networks*.

CURT GERVELIS is Senior Operations Manager for Nextel International. He was previously Director of Network Engineering for TeleCorp PCS, where he designed, built, and operated five PCS systems throughout the United States, including Puerto Rico. He has also worked with many landline-based carriers, including Verizon, BellSouth, Southwestern Bell, and AT&T. Mr. Gervelis is the co-author of *Cellular System Design and Optimization*.